Magnetic Oxides

Gerald F. Dionne

Magnetic Oxides

Gerald F. Dionne
Massachusetts Institute of Technology
244 Wood Street
Lexington, MA 02420
dionne@ll.mit.edu

ISBN 978-1-4419-0053-1 e-ISBN 978-1-4419-0054-8
DOI 10.1007/978-1-4419-0054-8
Springer New York Dordrecht Heidelberg London

Library of Congress Control Number: 2009935694

Printed on acid-free paper

Springer is part of Springer Science+Business Media (www.springer.com)

The author dedicates this book to

Rev. Hugh McPhee, S.J.,
former Dean of Science at Loyola College
in Montreal,

who once advised a liberal arts student
that science could offer a clearer window
on the world.

Preface

The inspirations for this book probably began in 1961 when I left a promising career as a semiconductor device engineer in the Rte 128 cauldron of the Boston area to pursue a new challenge at the McGill University Eaton Electronics Research Laboratory. Three years later I wrote a Ph.D. thesis on paramagnetic resonance and 3 years after I was adding to that experience as a Staff Member with MIT Lincoln Laboratory, where the scope of my obligations gradually broadened from microwave magnetic resonance to the physics and chemistry of ferrites and related magnetic oxide systems. At the time of this writing, I continue there as a resident consultant and also as a research affiliate with the MIT Department of Materials Science and Engineering.

Magnetic resonance has played a vital role in the study of magnetism in oxides and other insulating compounds that began during World War II and flourished globally for a quarter century. During this halcyon period, texts on magnetism became abundant as many of the pioneers took pen in hand to leave a treasure of elegantly presented reference literature as the 1960s drew to a close. By the mid-1970s, the once-fledgling field of semiconductor electronics that I had abandoned was overwhelming almost all competing technologies, including those with a magnetic component. For the better part of the two decades between the end of the Vietnam war and the discoveries of high-temperature superconductivity and giant magnetoresistance in transition-metal oxides in the early 1990s, fundamental investigations of magnetic compounds were nearly dormant.

The content and organization of this volume are intended to serve two purposes: (1) bridging of the intellectual gap left by the 20 years of reduced inquiry into magnetic phenomena, and (2) restoration of the molecular approach to the study of magnetic insulators that function more by local rather than the collective electron interactions that are more characteristic of metals and semiconductors.

The level of discussion presumes the reader to have some familiarity with atomic physics and basic quantum mechanics. Chapter 1 is an abbreviated introduction to magnetism. The discussion begins with a reminder of some fundamental definitions and selected subjects that can be found in most standard textbooks. However, two topics are treated in greater depth. A generic description of the quantum origins of magnetic exchange introduces the antisymmetry requirements of the hybrid eigenstates that determine the stabilization of parallel (metal) or antiparallel (insulator) spin alignment. This general theory is intended to support the later discussion of superexchange that is approached with a model that is more specific to magnetic oxides. An introduction to magnetic resonance and relaxation derived from classical Larmor precession serves a similar purpose for the examination of the broad subject of electromagnetism in ferrites. The physics and chemistry of magnetism in oxide compounds is covered in Chaps. 2, 3, and 4 in terms of localized ion and molecular-orbital models of molecular bonding. Chapter 5 addresses the secondary magnetic phenomena of anisotropy and magnetostriction from the standpoint of local orbital interactions with crystal fields and spin–orbit coupling that produces self-induced magnetoelastic effects. Traditional phenomenological theories are then reviewed in preparation for the examination of electromagnetic properties in Chap. 6. In Chaps. 7 and 8, magneto-optics and polarized spin transport that will be of increasing importance in the age of molecular-scale structures are described in the conceptual context of the earlier chapters.

As the preparation of this monograph draws to a close, I reflect on the journey that brought me to this point. From the frequency of their citations, the reader is certain to recognize the reliance on the seminal works of John Van Vleck, Maurice Pryce, John Goodenough, and Ernst Schlœmann, as well as the classic textbooks of Alan Morrish, Sōshin Chikazumi, Carl Ballhausen, Benjamin Lax and Kenneth Button, and many others. There were also collaborations with academia, industry, and government that are too numerous to list in any detail. However, I cannot pass up this opportunity to acknowledge the guidance of my doctoral thesis advisor Garnet Woonton and his colleague Maurice Pryce who was a most encouraging external examiner. Lincoln Laboratory's radar leaders John Allen, Carl Blake, Donald Temme, and Roger Sudbury who supported me and my vigorous colleagues Jerald Weiss, James Fitzgerald, Daniel Oates, and Russell West (of Trans-Tech, Inc.). Then there was the mentorship of John Goodenough and Benjamin Lax and the MIT campus affiliations with Mildred and Gene Dresselhaus, and Caroline Ross. Finally, I must mention the associations with Kristl Hathaway, Gary Prinz, and Stuart Wolf of the US Departments of the Navy and DARPA, and the assistance of Elaine Tham and Lauren Danahy of Springer US.

Lexington
MA

Gerald F. Dionne

Contents

1 Introductory Magnetism ... 1
- 1.1 Fundamental Concepts and Definitions ... 1
  - 1.1.1 Basic Electrostatics ... 2
  - 1.1.2 Basic Magnetostatics ... 3
  - 1.1.3 Demagnetization in Uniformly Magnetized Bodies ... 4
  - 1.1.4 Domains in Partially Magnetized Bodies ... 6
- 1.2 Induced Magnetism ... 8
  - 1.2.1 Diamagnetism and Paramagnetism ... 8
  - 1.2.2 Temperature Dependence of Susceptibility ... 11
- 1.3 Spontaneous Magnetism ... 15
  - 1.3.1 Classical Ferromagnetism and Antiferromagnetism ... 15
  - 1.3.2 Solutions of the Brillouin–Weiss Equation ... 16
  - 1.3.3 Quantum Origins of the Molecular Field ... 19
  - 1.3.4 The Ising Approximation ... 24
- 1.4 Gyromagnetism ... 25
  - 1.4.1 Larmor Precession and Resonance ... 26
  - 1.4.2 Phenomenological Relaxation Theory ... 27
  - 1.4.3 Complex Susceptibility Theory ... 29
  - 1.4.4 Resonance Line Shapes ... 33
- Appendix 1A Spin–Lattice Contribution to Linewidth ... 34
- References ... 35

2 Magnetic Ions in Oxides ... 37
- 2.1 The Transition Metals ... 37
  - 2.1.1 The Periodic Table ... 38
  - 2.1.2 Iron Group $3d^n$ Ions ... 40
  - 2.1.3 Rare Earth $4f^n$ Ions ... 42
  - 2.1.4 $4d^n$ and $5d^n$ Ions ... 42
- 2.2 Oxygen Coordinations ... 43
  - 2.2.1 Crystal Systems and Point Groups ... 44
  - 2.2.2 Cubic Symmetry ... 45
  - 2.2.3 Lower Symmetries ... 47
- 2.3 Crystal Electric Fields ... 48
  - 2.3.1 Angular Momentum States ... 49

2.3.2 Crystal Field Hamiltonian 50
2.3.3 Hierarchy of Perturbations 54
2.3.4 Weak-Field Solutions 55
2.3.5 Group Theory and Lower Symmetry 64
2.3.6 Strong Field Solutions and Term Diagrams 68
2.3.7 Rare-Earth Ion Solutions 71
2.4 Orbital Energy Stabilization 73
2.4.1 One-Electron Model 73
2.4.2 High- and Low-Spin States 75
2.4.3 Orbit–Lattice Stabilization (Jahn–Teller Effects) 79
2.4.4 Spin–Orbit–Lattice Stabilization 82
2.5 Covalent Stabilization 88
2.5.1 Molecular-Orbital Theory 89
2.5.2 Determinant Method 91
2.5.3 σ and π Bonds and the Molecular Orbital Diagram 95
2.5.4 Valence Bond Method 99
Appendix 2A Homonuclear Molecule Ion 102
Appendix 2B Valence-Bond Diatomic Molecule 103
References 105

3 Magnetic Exchange in Oxides 107
3.1 Interionic Magnetic Exchange 108
3.1.1 Molecular-Orbital Exchange Approximation 109
3.1.2 Valence-Bond Solutions 113
3.1.3 Spin Alignment in Oxides 119
3.1.4 Ferromagnetism by Spin Transfer 121
3.1.5 Goodenough–Kanamori Rules 125
3.2 Antiferromagnetism 129
3.2.1 Superexchange and Molecular Fields 129
3.2.2 Molecular Field Theory of Antiferromagnetism 131
3.2.3 Antiferromagnetic Spin Configurations 135
3.3 Antiferromagnetic Oxides 139
3.3.1 One-Metal Oxides 139
3.3.2 ABO_3 and A_2BO_4 Perovskites 140
3.3.3 The Mixed-Valence Manganite Anomaly 143
Appendix 3A Analysis of $M^{2+}O^{2-}$ Exchange Interactions 146
Appendix 3B Curie Temperature Model for (La,Ca) MnO_3 147
References 149

4 Ferrimagnetism 151
4.1 Ferrimagnetic Order 151
4.1.1 Generic Ferrimagnetic Systems 152
4.1.2 Molecular Field Theory of Ferrimagnetism 153
4.1.3 Magnetic Frustration and Spin Canting 157

4.2 Theory of Superexchange Dilution 161
4.2.1 Superexchange Energy Stabilization 161
4.2.2 Molecular Field Coefficients..................................164
4.2.3 Solution for Yttrium Iron Garnet 165
4.3 Ferrimagnetic Oxides..168
4.3.1 Spinel Ferrites $A\,[B_2]\,O_4$ 169
4.3.2 Garnet Ferrites $\{c_3\}\,[a_2]\,(d_3)\,O_{12}$ 175
4.3.3 Rare-Earth Garnet Ferrites.....................................180
4.3.4 Rare-Earth Canting Effect 184
4.3.5 Hexagonal Ferrites 190
4.3.6 Orthoferrites 193
Appendix 4A Molecular Field Analysis of LiZnTi Ferrite 193
Appendix 4B High-Magnetization Limits 195
Appendix 4C Brillouin Functions in Exchange Energy Format 196
References...197

5 Anisotropy and Magnetoelastic Properties 201
5.1 Quantum Paramagnetism of Single Ions 202
5.1.1 Theory of Anisotropic g Factors...............................202
5.1.2 Conventional Perturbation Solutions 205
5.1.3 The Spin Hamiltonian for 3d^n Ions 209
5.1.4 The Crystal-Field Hamiltonian for $4f^{\mathrm{n}}$ Ions..................210
5.2 Anisotropy of Single Ions 212
5.2.1 $3d^1$ and $3d^6$ D-State Triplet 213
5.2.2 $3d^4$ and $3d^9$ D-State Doublet (J–T Effect) 217
5.2.3 $3d^2$ and $3d^7$ F-State Triplet 219
5.2.4 $3d^3$ and $3d^8$ F-State Singlet...................................220
5.2.5 3d^5 S-State Singlet 222
5.2.6 $4f^n$ Ion Anisotropy 226
5.3 Magnetocrystalline Anisotropy and Magnetostriction................228
5.3.1 Phenomenological Anisotropy Theory 229
5.3.2 Phenomenological Magnetostriction Theory 231
5.3.3 Dipolar Pair Model of Magnetic Anisotropy..................234
5.3.4 Single-Ion Model of Ferrimagnetic Anisotropy 236
5.3.5 Cooperative Single-Ion Effects: Anisotropy 241
5.3.6 Cooperative Single-Ion Effects: Magnetostriction............246
5.4 Magnetization Process and Hysteresis.................................250
5.4.1 Initial Permeability and Coercivity 251
5.4.2 Anisotropy Field and Remanence Ratio.......................254
5.4.3 Approach to Saturation 256
5.4.4 Demagnetization and Permanent Magnets 258
Appendix 5A Four-Level Degenerate Perturbation Solution
for d^1 261

Appendix 5B T_{2g} Solution for d^1 in an Exchange Field....................263
Appendix 5C Orbital States of d^5 in a Cubic Field..........................265
Appendix 5D Angular Dependence of Cubic Anisotropy Fields...........267
References...269

6 Electromagnetic Properties..273
6.1 Magnetic Relaxation 274
6.1.1 Nonresonant Longitudinal Relaxation 274
6.1.2 Quantum Mechanisms of Spin–Lattice Relaxation...........278
6.1.3 Perturbation Theories of Spin–Phonon Interaction...........286
6.2 Gyromagnetic Resonance and Relaxation.............................287
6.2.1 Paramagnetic Resonance......................................288
6.2.2 Ferromagnetic Resonance 292
6.2.3 Uniform Precession Damping 295
6.2.4 Inhomogeneous Resonance Line Broadening.................297
6.2.5 Fast-Relaxing Ion Effects 300
6.2.6 The Exchange Isolation Effect.................................306
6.3 Exchange-Coupled Modes (Spin Waves)..............................307
6.3.1 Uniform Precession Decoherence (Degenerate Spin Waves) 307
6.3.2 Instability Threshold (Classical Approximation) 311
6.3.3 Instability Threshold (Nonlinear Spin Waves)................315
6.3.4 Magnetostatic Modes 317
6.4 Permeability and Propagation...318
6.4.1 Low-Frequency Longitudinal Permeability 318
6.4.2 High-Frequency Transverse Limits............................322
6.4.3 Snoek's Law Considerations..................................324
6.4.4 Circular Polarization and Nonreciprocal Properties 327
6.4.5 Linear Polarization and Faraday Rotation.....................332
Appendix 6A Transverse Permeability Tensor 333
Appendix 6B Classical Instability Threshold 336
Appendix 6C Domain Wall Susceptibility Equation........................338
References...340

7 Magneto-Optical Properties...343
7.1 Infrared Exchange Resonance 344
7.1.1 Classical Precession Model....................................344
7.1.2 Quantum Spin Transition Model 346
7.1.3 Experimental Exchange Spectra...............................351
7.2 Combined Permeability and Permittivity..............................352
7.2.1 The $[\varepsilon]\cdot[\mu]$ Tensor Solutions.................................352
7.2.2 Propagation Parameters and Faraday Rotation 353
7.3 Magneto-Optical Spectra...355
7.3.1 Electric-Dipole Transitions 355
7.3.2 Yttrium Iron Garnet Spectra (Paramagnetic) 360

7.3.3 Iron Garnets with Bismuth Ions (Diamagnetic) 366
7.3.4 Fe^{3+}–Bi^{3+} Hybrid Excited States 371
7.3.5 Intersublattice Transitions and the $\Delta S = 0$ Rule 376
Appendix 7A Magnetic Circular Birefringence and Dichroism 381
References 382

8 Spin Transport Properties 385
8.1 Polarons and Charge Transfer 386
8.1.1 Transfer Among Equivalent Energy Sites (Small Polarons) 388
8.1.2 Transfer to Higher Energy Sites (Large Polarons) 389
8.1.3 Transfer by Covalent Tunneling 392
8.1.4 The Holstein Polaron Theory 394
8.2 Metallic Oxides with Polarized Spins 396
8.2.1 Simple Oxides 397
8.2.2 Complex Oxides 397
8.2.3 Classical Resistivity–Temperature Model 400
8.3 Magnetoresistance in Oxides (CMR) 401
8.3.1 Manganese-Ion Exchange Interactions 402
8.3.2 Magnetoresistivity-Temperature Model 405
8.3.3 Dilute Magnetic Oxides 410
8.4 Superconductivity in Oxides 413
8.4.1 Classical Foundations 413
8.4.2 Zero-Spin Polarons and Magnetic Frustration 419
8.4.3 Large-Polaron Superconductivity 423
8.4.4 Normal Resistivity and Critical Temperature 426
8.4.5 Layered Cuprate Superconductors 430
8.5 Supercurrents and Magnetic Fields 439
8.5.1 Supercurrent Formation 439
8.5.2 Condensation Energy 442
8.5.3 London Penetration Depth 443
8.5.4 Critical Magnetic Field 445
8.5.5 Critical Current Density 447
8.5.6 Coherence Length 450
8.5.7 Type-II Superconductors 452
Appendix 8A Magnetic Levitation 455
References 456

Index 461

Chapter 1
Introductory Magnetism

The use of magnetic oxides in electronics technology has become so commonplace that few systems can operate effectively without some form of them making a vital contribution. Wherever magnetic materials with dielectric properties (or vice versa) are required, there is likely to be an application for a ferrite or other transition-metal oxide, from cores for inductors and transformers, to discs or tapes and read/write heads for information storage, to thin films for high-density computer memories, to nonreciprocal microwave control devices, to antennas for home electronics, to microwave antireflection coatings, to permanent magnets for automobile ignitions, to isolator devices for fiber-optical laser sources, to rubberized refrigerator magnets. In later years, exotic phenomena that jointly involve the magnetic and electrical conductivity properties have been discovered in oxides that contain magnetic ions. High-temperature superconductivity for low power loss and giant magnetoresistance effects for magnetic field sensors and magnetic random access memories (MRAMs) have been found in perovskite-based compounds. Hybrid combinations of piezoelectric and magnetic compounds have spawned a growing interest in *piezomagnetics* and *multiferroics*. Magnetic oxides have also provided a molecular-scale vehicle for fundamental investigations of the electronic and magnetic properties of the important transition-metal and rare-earth elements of the Periodic table.

1.1 Fundamental Concepts and Definitions

Although the main focus of this book will be the molecular origins of magnetism and its various manifestations in metal oxides, dielectric properties influence the magnetic behavior of these materials and in many instances can determine the limits of their applicability. To this end, we first review some of the basic relations of electrostatics and extend them to magnetostatics.

G.F. Dionne, *Magnetic Oxides*, DOI 10.1007/978-1-4419-0054-8_1,

1.1.1 Basic Electrostatics

In unrationalized electrostatic (Gaussian) units the electric field vector of a charge q at a radial distance r is derived from Coulomb's law

$$\boldsymbol{E} = \frac{q}{r^3}\boldsymbol{r}, \tag{1.1}$$

where q would be labeled e for an electron and its value in these units is 4.8×10^{-10} esu. In the mks (SI) system of units, $e = 1.6 \times 10^{-19}$ C. For an electric dipole of charge separation d, the dipole moment vector $\boldsymbol{p} = q\boldsymbol{d}$ and

$$\boldsymbol{E}_{\text{dip}} = -\nabla \Omega_{\text{e}}, \tag{1.2}$$

where the electric scalar potential $\Omega_{\text{e}} = \left(\frac{\mathbf{p} \cdot \mathbf{r}}{r^3}\right)$ and r is the distance from the center of the dipole.

In a dielectric material, the polarization vector $\boldsymbol{P}$ is proportional to $\boldsymbol{E}$, according to

$$\boldsymbol{P} = \chi_{\text{e}}\boldsymbol{E}, \tag{1.3}$$

where χ_{e} is the electric susceptibility (also defined as the electric polarizability α when referred to an individual molecule). The associated charge displacement vector $\boldsymbol{D}$ of the Maxwell equation $\nabla \cdot \boldsymbol{D} = 4\pi\rho$ for a charge density ρ is defined as

$$\boldsymbol{D} = \boldsymbol{E} + 4\pi \boldsymbol{P} = (1 + 4\pi\chi_{\text{e}})\,\boldsymbol{E} = \varepsilon \boldsymbol{E}, \tag{1.4}$$

where ε is the electric permittivity or dielectric constant, which is scaled to the permittivity of free space that is set to unity. The 4π factor is required for the convention of Gaussian units to be used in this text. In SI or mks units the dielectric constant becomes $K_{\text{e}} = \varepsilon/\varepsilon_0$, where $\varepsilon_0 = 8.85 \times 10^{-12}$ F/m. This basic definition of the permittivity follows from the assumption that the $\boldsymbol{E}$ and $\boldsymbol{P}$ vectors are parallel. For other cases, as with magneto-optical coupling discussed in Chap. 7, ε is expressed as a ε_{ij} tensor according to

$$\boldsymbol{D} = [\varepsilon]\,\boldsymbol{E} = \begin{bmatrix} \varepsilon_{xx} & \varepsilon_{xy} & \varepsilon_{xz} \\ \varepsilon_{yx} & \varepsilon_{yy} & \varepsilon_{yz} \\ \varepsilon_{zx} & \varepsilon_{zy} & \varepsilon_{zz} \end{bmatrix} \begin{pmatrix} E_x \\ E_y \\ E_z \end{pmatrix}. \tag{1.5}$$

From (1.2), the dipolar interaction energy between two neighboring dipoles $\boldsymbol{p}_1$ and $\boldsymbol{p}_2$ separated by r_{12} is expressed as

$$E_{\text{dip}} = \frac{\boldsymbol{p}_1 \cdot \boldsymbol{p}_2}{r_{12}^3} - \frac{3(\boldsymbol{r}_{12} \cdot \boldsymbol{p}_1)(\boldsymbol{r}_{12} \cdot \boldsymbol{p}_2)}{r_{12}^5}, \tag{1.6}$$

1.1.2 Basic Magnetostatics

To introduce the magnetic properties of matter, consider two principal areas where basic magnetism entities occur: magnetic dipoles and their interactions with the magnetic fields and each other, and the application of Maxwell's equations to matter containing magnetic dipoles.

Like electric fields, magnetic fields originate from electrical charges. Unlike electric fields, the electrical charges must be in motion relative to the frame of reference of the observer, as in a wire carrying electric current, an electron orbiting a nucleus, or a charged particle simply spinning. The basic unit of magnetism is not an independent charge or pole, but rather a pair of poles, termed positive and negative – a magnetic dipole with a moment defined as the vector quantity $\boldsymbol{m}$. All magnetic fields can be traced to some effective dipole moment as its source, from an individual or a collection of moments. Our attention, therefore, is focused on the magnetic fields and the energy associated with the interaction between fields and moments.

Simply stated, the field of a magnetic dipole is the gradient of the scalar potential

$$\boldsymbol{H}_{\text{dip}} = -\nabla \Omega_{\text{m}}, \tag{1.7}$$

where the magnetic scalar potential $\Omega_{\text{m}} = (\mathbf{m} \cdot \mathbf{r}/r^3)$. It can be determined by inspection that the concept of magnetic scalar potential is analogous to those of the electrostatic potential (charge/distance) of (1.1) and the gravitational potential (mass/distance).

Isolated dipoles that occur in materials can couple to an applied magnetic field and produce a collective magnetic moment. When expressed as a volume density it is termed magnetic strength or magnetization $\boldsymbol{M}$ that is proportional to the magnetic field that cause them to be polarized in the field direction, according to

$$\boldsymbol{M} = \chi_{\text{m}} \boldsymbol{H}, \tag{1.8}$$

where χ_{m} is the magnetic susceptibility. Analogous to the electrostatic displacement flux, the magnetic flux density is related through

$$\boldsymbol{B} = \boldsymbol{H} + 4\pi \boldsymbol{M} = (1 + 4\pi\chi_{\text{m}})\,\boldsymbol{H} = \mu \boldsymbol{H}, \tag{1.9}$$

where μ is the magnetic permeability that is scaled to the permeability of free space, which is set to unity in the Gaussian system. In SI or mks units, the permeability of free space is $\mu_0 = 4\pi \times 10^7$ Wb/A m and the relation in (1.9) is expressed in terms of the relative permeability $\mu_{\text{r}} = \mu/\mu_0$. Analogous to (1.4), this basic definition of the permeability follows from the assumption that $\boldsymbol{H}$ and $\boldsymbol{M}$ are parallel. When this is not true, as with gyromagnetic coupling discussed in Chap. 6, μ is expressed as a $[\mu]$ tensor according to

$$\boldsymbol{B} = [\mu]\,\boldsymbol{H} = \begin{bmatrix} \mu_{xx} & \mu_{xy} & \mu_{xz} \\ \mu_{yx} & \mu_{yy} & \mu_{yz} \\ \mu_{zx} & \mu_{zy} & \mu_{zz} \end{bmatrix} \begin{pmatrix} H_x \\ H_y \\ H_z \end{pmatrix}, \tag{1.10}$$

where $\mu_{ii} = 1 + 4\pi\chi_{ii}$ and $\mu_{ij} = 4\pi\chi_{ij}$.

The magnitude of χ_m and χ_e are constant for small densities of dipoles. As the separation between dipoles decreases and the dipoles begin to interact with each other through their fields, $\boldsymbol{M}$ and $\boldsymbol{P}$ approach saturation values. In materials with a high density of magnetic moments that couple spontaneously through short-range interactions related to the chemical bonding, a magnetization can exist without the presence of an applied $\boldsymbol{H}$ in the form of ferromagnetism, antiferromagnetism, and ferrimagnetism.

1.1.3 Demagnetization in Uniformly Magnetized Bodies

The significance of the aforementioned theory can be appreciated immediately in the examination of magnetized bodies. Application of (1.9) produces insight into properties of uniformly magnetized structures of specific geometric shapes. As illustrated in Fig. 1.1, a magnetized body in a magnetic field $\boldsymbol{H}$ will develop magnetic poles on the surfaces through which magnetic flux lines pass. In effect, the body becomes a dipole that creates an external magnetic field, which is a property of any permanent magnet. To satisfy the Maxwell equation $\nabla \cdot \boldsymbol{B} = 0$ and satisfy

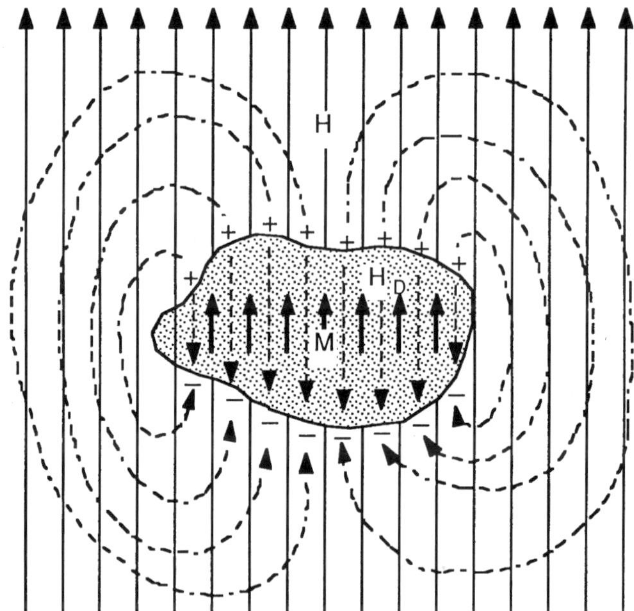

Fig. 1.1 Two-dimensional sketch of an irregularly shaped body of magnetization M in a uniform external magnetic field H. Internal demagnetization results from the presence of magnetic poles on the surface that induce an opposing field of magnitude H_D, which depends on the particular geometric details

the boundary condition that the normal component of $\boldsymbol{B}$ be continuous across any interface, the surface poles must also create an internal demagnetizing magnetic field $\boldsymbol{H}_\mathrm{D}$ that opposes the magnetization, so that the flux density becomes

$$\boldsymbol{B} = \boldsymbol{H} + 4\pi\boldsymbol{M} - \boldsymbol{H}_\mathrm{D}, \tag{1.11}$$

where $H_\mathrm{D} = N_\mathrm{D}(4\pi M)$ with $0 \leq N_\mathrm{D} \leq 1$ represents the effective demagnetizing factor that depends on the geometric details. [A frequently-used alternative has 4π absorbed into the demagnetizing factor so that $0 \leq N_\mathrm{D} \leq 4\pi$.] Therefore, (1.11) is commonly written as

$$\boldsymbol{B} = \boldsymbol{H} + (1 - N_\mathrm{D})\, 4\pi\boldsymbol{M}. \tag{1.12}$$

The flux patterns of this common situation are indicated by the sketch in Fig. 1.1.

A comprehensive discussion of demagnetizing factors, including a table of N_D values accurate for *uniformly* magnetized bodies of ellipsoidal shape, may be found in Bozorth [1]. To calculate appropriate values of N_D for use in the variety of geometrical shapes that are encountered in magnetic applications, numerical methods must be applied wherever nonellipsoidal forms are involved. Although the magnetization is uniform throughout the body in these cases, the demagnetizing fields are nonuniform and finite element or other methods that require computer aided design must be employed. Where the shapes can be approximated by an ellipsoid, however, analytical relations are available in the literature, for example, by Osborn [2]. For ellipsoids of revolution, that is, prolate (elongated) or oblate (compressed) spheroids with circular cross section, the following formulas apply:

Along the axis of symmetry for which the axis is greater than the diameter (prolate), that is, $m_l = l/d > 1$,

$$N_\mathrm{D}{}^l = \frac{1}{2(m_l^2 - 1)}\left[\frac{m_l}{\sqrt{2}} \ln\left(\frac{m_l + (m_l^2 - 1)^{1/2}}{m_l - (m_l^2 - 1)^{1/2}}\right) - 1\right]. \tag{1.13}$$

Because the sum of the factors along the three major axes of the ellipse equals unity, it follows that along any diameter,

$$N_\mathrm{D}{}^d = \frac{1}{2}\left(1 - N_\mathrm{D}{}^l\right). \tag{1.14}$$

Along any diameter for which the diameter is greater than the length of the symmetry axis (oblate), $m_d = d/l > 1$,

$$N_\mathrm{D}{}^d = \frac{1}{2(m_d{}^2 - 1)}\left\{m_d{}^2(m_d{}^2 - 1)^{-1/2}\arcsin\left[\frac{(m_d{}^2 - 1)^{1/2}}{m_d{}^2}\right] - 1\right\} \tag{1.15}$$

and

$$N_\mathrm{D}{}^l = \left(1 - 2N_\mathrm{D}{}^d\right). \tag{1.16}$$

Table 1.1 Demagnetizing factors N_D for ellipsoids of revolution with length l and diameter d

Axis	Sphere	Needle	Disc
	$l = d$	$l \gg d$	$l \ll d$
$N_D{}^l$	1/3	≈ 0	≈ 1
$N_D{}^d$	1/3	$\approx 1/2$	≈ 0

In most practical cases, there are three limiting situations listed in Table 1.1: a sphere, for which $N_D = 1/3$ isotropically, a cylindrical needle (acicular), for which $N_D = 0$ along its axis and 1/2 along any diameter, and a thin circular disc, for which $N_D = 1$ normal to its surface and 0 in its plane. For these and other less common geometries, calculated design curves can be found in the literature [2].

Another concept that is important in the magnetism of materials is that of magnetic energy and some useful relations that may be found in any standard textbook will be stated here for later reference. From the definition of $\boldsymbol{H}_{\text{dip}}$ in (1.1) and (1.2), the energy of interaction between two neighboring dipoles $\boldsymbol{m}_1$ and $\boldsymbol{m}_2$ in each other's respective magnetic field is

$$E_{\text{dip}} = \frac{\boldsymbol{m}_1 \cdot \boldsymbol{m}_2}{r_{12}^3} - \frac{3(\boldsymbol{r}_{12} \cdot \boldsymbol{m}_1)(\boldsymbol{r}_{12} \cdot \boldsymbol{m}_2)}{r_{12}^5}, \tag{1.17}$$

where r_{12} is the distance between dipole centers. The energy of interaction between a dipole moment with a magnetic field is simply

$$E_{\text{m}} = -\boldsymbol{m} \cdot \boldsymbol{H}. \tag{1.18}$$

The energy of a magnetized body in a magnetic field may be shown to be

$$E_{\text{m}} = -\frac{1}{2} \int M \cdot H_{\text{D}}\, \text{d}v - \int M \cdot H\, \text{d}v, \tag{1.19}$$

where the energy density is integrated over the volume of the magnet, first to account for the self-energy of the magnet in its own demagnetizing field $\boldsymbol{H}_{\text{D}}$, and second to add the energy of interaction with the external field $\boldsymbol{H}$.

1.1.4 Domains in Partially Magnetized Bodies

When a body resides in an external field of strength insufficient to maintain the magnetically saturated state, that is, when $H < N_{\text{D}}(4\pi M_{\text{s}})$, demagnetization begins to take place through the formation of magnetic domains. It is a property of spontaneously magnetic (ferromagnetic) materials. Theoretical analyses of domain structures may be found in most standard texts on magnetism. The basic physical reason for their occurrence is to reduce the magnetostatic energy that results from poles that are induced on the specimen surface in proportion to the strength of the magnetization. As H is reduced, $4\pi M$ decreases and the domains change accordingly.

The domain pattern is usually initiated by the appearance of 180° domains of reverse magnetization separated by narrow regions (Bloch walls) in which the magnetic moments rotate to accomplish the reversal in direction. As the magnetizing field is reduced further the reverse domains grow at the expense of the existing magnetized volume until equilibrium is reached. To remove the surface poles and suppress the attendant magnetostatic energy, 90° closure domains are created at the specimen edges.

In Fig. 1.2 the two-dimensional model of rectangular geometry illustrates the concept of how the 90° closure domain walls are set up at 45° angles to confine the flux within the specimen, with a simple and then more complex pattern. The actual arrangement is determined by balancing the magnetostatic energy with the anisotropy energy contained in the domain walls.[1] Within the domains themselves, the magnetization vectors can rotate further in their anisotropy fields to complete the demagnetization process. The phenomena of magnetic anisotropy and hysteresis are explored in Chap. 5.

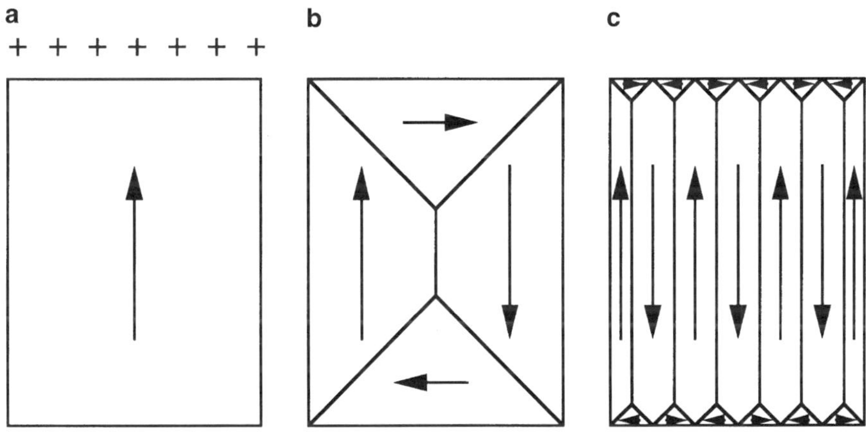

Fig. 1.2 The elimination of surface poles by 90° closure domains: (**a**) magnetized specimen with surface poles, (**b**) simplest example of two 180° domains with 90° closure domains, and (**c**) domain pattern more typical of an actual situation

[1] Where hysteresis is significant, as in permanent magnets, there are internal bias fields arising from magnetocrystalline anisotropy that must be overcome, for example, the coercive field H_c, before the domain walls can move. When the coercive field (and its related anisotropy field H_K) is taken into account, the onset of demagnetization would not begin until $H + H_c < N_D(4\pi M_s)$.

1.2 Induced Magnetism

Magnetic moments occur in all states of matter, but will be limited to solids for the discussion at hand. The fundamental types have been traditionally characterized as follows: (1) diamagnetism, from magnetic dipoles that must be induced and aligned by an applied magnetic field, (2) paramagnetism, with permanent magnetic moments that may be aligned by an applied magnetic field and are randomized by increasing temperature, (3) spontaneous magnetism in which the moments align collinearly (ferromagnetism), antiparallel (antiferromagnetism), and the important case of combined ferromagnetism and antiferromagnetism termed ferrimagnetism, all of which exist without the presence of an applied field. A fourth class of magnetic materials with unusual magnetic and electrical properties are superconductors, which now includes metal oxides.

1.2.1 Diamagnetism and Paramagnetism

All materials have the potential for diamagnetism. According to Lenz's law, the magnetic dipole moment of a current loop induced by a magnetic field will align so as to produce a magnetic field that "opposes the field producing it." Unbound electrons of any material can be arranged to form magnetic dipoles that give a magnetization defined by (1.4) with a temperature-dependent diamagnetic susceptibility $\chi_{\mathrm{dia}} < 0$. A sampling of diamagnetic susceptibilities expressed in emu is given in Table 1.2. In the discussions of sold-state magnetism presented in later chapters, ions with closed-shell electronic structures are diamagnetic. In the general literature, they are commonly referred to as nonmagnetic.

In certain of these materials, the atoms or ions already have magnetic moments that generally result from unfilled quantum states with unpaired electron spins to provide a net moment, sometimes coupled to moments associated with the electron orbital motion. The magnitude of the susceptibility of typical paramagnets is on the order of 10^{-4}, which would dominate the susceptibilities of diamagnets by more than an order of magnitude. The origin of paramagnetism in solids begins with electrons of charge e and mass m_e, which possess magnetic moments that arise from both spin and the moment generated by the simulated current orbital motion about

Table 1.2 Diamagnetic susceptibilities

Element	$\chi_{\mathrm{dia}}(\times 10^{-6})$	Element	$\chi_{\mathrm{dia}}(\times 10^{-6})$
Ag	−31	Cl	−15.6
B	−7	F	−20.1
Bi	−192	I	−44.6
C	−6	N	−5.51
P	−26.3	O	−4.61

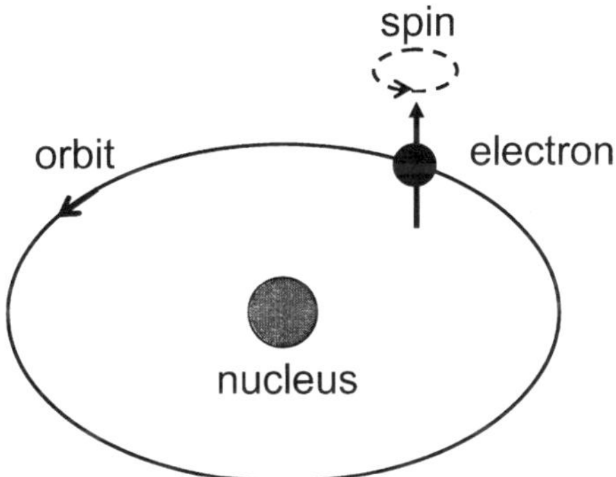

Fig. 1.3 Atomic origins of the orbital and spin magnetic moments

a nucleus, as pictured in Fig. 1.3. The most basic quantity in electronic magnetism is the magnetic moment of the electron spin m_B, called the Bohr magneton, which is defined from the principles of electrodynamics as

$$m_\mathrm{B} = \frac{e\hbar}{2m_\mathrm{e}c} = 9.27 \times 10^{-21}\ (\mathrm{emu}). \tag{1.20}$$

[For a proton, a much smaller nuclear magneton is similarly defined as m_N, with mass $= 1,836\,m_\mathrm{e}$.] Although electric charges can be distinguished by a sign difference, in neglect of rigor we shall adopt the commonly used convention that e is positive unless otherwise required.

A vital concept that also arises from quantum mechanics is the relation between magnetic moment and angular momentum. For the electron spin, the angular momentum along the z-axis of quantization is defined from the uncertainty principle as $\sigma_z = \pm\hbar/2$ and it follows from (1.20) that

$$m_\mathrm{B} = g\left(\frac{e}{2m_\mathrm{e}c}\right)\sigma_z, \tag{1.21}$$

where g (=2 for an electron spin) is the Landé or spectroscopic splitting factor (of historical significance in connection with the discovery of the Zeeman effect). Note that Planck's constant is contained in σ_z. This line of reasoning can extend to the orbital angular momentum through a simple classical analogy of a current loop formed by the electron in an orbit of radius r and angular frequency ω that constitutes a current $i = e\omega/2\pi c$ encompassing a circular area $A = \pi r^2$ with orbital angular momentum $l = m_\mathrm{e}r^2\omega$. Here, the individual orbital magnetic moment becomes

$$m_l = iA = \left(\frac{e\omega}{2\pi c}\right)\pi r^2 = g\left(\frac{e}{2m_\mathrm{e}c}\right)l = g\left(\frac{e}{2m_\mathrm{e}c}\right)l\sqrt{(l+1)}\hbar, \tag{1.22}$$

where $g = 1$ for orbital magnetism and Planck's constant is contained in l. It is therefore appropriate to adopt the notion that the magnetic moment of a free atom or ion is directly proportional to its angular momentum and the effective g factor, which varies between 1 and 2 depending on the relative weighting of the orbital and spin contributions.

In the many-electron systems of atoms and ions, the angular momenta of spins and orbits are added as scalars, provided the rules of the Russell-Saunders coupling are observed. From the standard theory of the addition of angular momentum, the squares of the resultant angular momenta expectation values of $\boldsymbol{S}$ for spins and $\boldsymbol{L}$ for orbits summed over all of the electrons in a particular orbital shell are related by

$$\begin{aligned} \boldsymbol{S}^2 &= S\,(S+1)\,\hbar^2, \\ \boldsymbol{L}^2 &= L\,(L+1)\,\hbar^2, \end{aligned} \tag{1.23}$$

where $S = \sum s$, with $s = 1/2$ and $L = \sum l$, with $l = 0, 1, 2, 3$, etc. for the s, p, d, and f shells, respectively. The values of S range from a maximum S and descend in steps of 1 to a minimum of 0 or 1/2 depending on whether the number of electrons is even or odd, that is, S, $S-1$, $S-2, \ldots$, 1/2 or 0. The values of L follow the same rule except that the minimum can only be 0 because of the integer values of l.

Of particular importance for the rare earth elements is the total angular momentum $\boldsymbol{J}$, which is formed by the coupling of $\boldsymbol{S}$ and $\boldsymbol{L}$ vectors and is given by

$$\boldsymbol{J}^2 = J\,(J+1)\,\hbar^2, \tag{1.24}$$

where $\boldsymbol{J}^2 = \boldsymbol{L}^2 + \boldsymbol{S}^2 + 2\boldsymbol{L}\cdot\boldsymbol{S}$, and J may assume values from a maximum of $|L+S|$ to $|L-S|$, again in steps of 1, thereby creating a multiplicity of values $(2S+1)$. Note that the spin–orbit coupling operator can be expressed as

$$\boldsymbol{L}\cdot\boldsymbol{S} = \frac{1}{2}\left[J\,(J+1) - L\,(L+1) - S\,(S+1)\right]\hbar^2. \tag{1.25}$$

Apart from the quantum mechanical operators $\boldsymbol{S}^2$, $\boldsymbol{L}^2$, and $\boldsymbol{J}^2$, the other group of diagonal operators are the z-components of these momenta $\mathcal{M}_j\hbar$, and their values can be observed by the spatial quantization model of Fig. 1.4 for $S = 5/2$, where $\mathcal{M}_j$ can take values from $+J$ to $-J$, with a multiplicity of $2J+1$. In magnetism, the z component represents the measurable or observable quantity and is therefore of critical importance. From these basic relations, it is now possible to construct general expressions for the total magnetic moments from a set of $\boldsymbol{S}$, $\boldsymbol{L}$, and $\boldsymbol{J}$ vectors with Planck's constant assumed by the Bohr magneton (1.20):

$$m_j = g\left(\frac{e}{2m_{\mathrm{e}}c}\right)\sqrt{J(J+1)}\hbar = g m_{\mathrm{B}}\sqrt{J(J+1)}, \tag{1.26}$$

and

$$m_{J_z} = g\left(\frac{e}{2m_{\mathrm{e}}c}\right)\mathcal{M}_j\hbar = g m_{\mathrm{B}}\mathcal{M}_j, \tag{1.27}$$

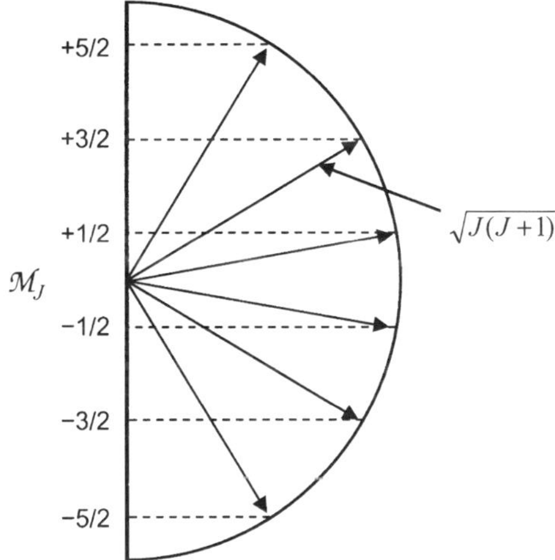

Fig. 1.4 Spatial quantization diagram of the total angular momentum for $J = 5/2$. Note that Planck's constant becomes absorbed in the Bohr magneton m_B when the magnetic moment is defined

Total Angular Momentum, $\boldsymbol{J} = \sqrt{J(J+1)}\hbar$

Magnetic Moment, $\boldsymbol{m}_J = \mathrm{g m_B}\sqrt{J(J+1)}$

z – Component, $m_{Jz} = \mathrm{g m_B}\,\mathcal{M}_J$

where

$$g = 1 + \frac{J(J+1) + S(S+1) - L(L+1)}{2J(J+1)}. \tag{1.28}$$

1.2.2 Temperature Dependence of Susceptibility

Except for the paramagnetism of a free-electron gas in metals (Pauli paramagnetism), which is temperature insensitive to first order, the most characteristic feature of paramagnetism is the inverse temperature dependence of the susceptibility. The derivation of the functions describing this dependence is born out of classical Boltzmann statistics, which are applicable provided that interactions between the individual moments are negligible. Note that the small diamagnetic component remains part of the total induced susceptibility.

Consider a collection of n independent identical magnets of moment m per unit volume subjected to a magnetic field H. At $T = 0\ K$, the moments may be expected to align with the field without perturbation. At higher temperatures, randomization due to thermal agitation from lattice vibrations would work against the magnetic alignment. The resultant effective magnetic moment would be less than nm by a factor that varies directly with H and inversely with T.

Langevin first solved this problem from an entirely classical standpoint by assuming that (1) m is the only value of magnetic moment possible, (2) all directions of m are allowed, and (3) the only aligning agent is $\boldsymbol{H}$. The reduction in energy from the interaction between a moment m forming an angle θ with z-axis-directed $\boldsymbol{H}$ vector is

$$E_m = -mH\ \cos\theta. \tag{1.29}$$

The ratio of the effective z component of the total magnetic moment to the actual moment is determined by integrating over the continuum of energy states weighted by the Boltzmann occupation probability $\exp(-E_m/kT)$ covering the complete range of θ values. The weighted average of the moment in the direction of $\boldsymbol{H}$ is

$$\bar{m}_z = m\frac{\int_{+1}^{-1} \mathrm{e}^{y\cos\theta}\ \cos\theta\,\mathrm{d}(\cos\theta)}{\int_{+1}^{-1} \mathrm{e}^{y\cos\theta}\,\mathrm{d}(\cos\theta)}. \tag{1.30}$$

The details of the integration over the energy states may be found in most standard texts on magnetism. For the purposes of this introductory exercise, a summary of the results will be sufficient:

$$\frac{\bar{m}_z}{m} = L(y) = \coth y - \frac{1}{y}, \tag{1.31}$$

where $\mathcal{L}(y)$ is the Langevin function and $y = mH/kT$. For typical fields and temperatures in the paramagnetic region, $y \ll 1$. In this limit, $\mathcal{L}(v) \approx y/3$ and the volume susceptibility may be expressed from (1.8) as

$$\chi_{\mathrm{par}} = \frac{4\pi n\bar{m}_z}{H} = \frac{4\pi nm}{H}L(y) \approx \frac{4\pi nm^2}{3kT}, \tag{1.32}$$

where n is the volume density of magnetic moments, so that $4\pi nm$ corresponds to the magnetization $4\pi M$.

Refinement to this theory was contributed by Brillouin, who introduced the $2J + 1$s multiplicity to Langevin's assumption (1). In this case, the decrease in energy of (1.29) becomes

$$E_m = -gm_{\mathrm{B}}\sqrt{J(J+1)}H = -gm_{\mathrm{B}}\mathcal{M}_j H, \tag{1.33}$$

and the total moment per unit volume in the direction of $\boldsymbol{H}$ is

$$n\bar{m}_z = \frac{ngm_{\mathrm{B}}\sum_{+J}^{-J} M_{J_j}\mathrm{e}^{M_{J_j}y}}{\sum_{+J}^{-J}{}_j\mathrm{e}^{M_{J_j}y}}, \tag{1.34}$$

where $y = gm_B H/kT$ and the sums are over the allowed $\mathcal{M}_j$ values from $+J$ to $-J$. Through a combination of integration and series summation, (1.34) is transformed to

$$n\bar{m}_z = ngm_B J \left[\frac{2J+1}{2J} \coth\left(\frac{2J+1}{2}\right) y - \frac{1}{2J} \coth\frac{y}{2} \right]. \tag{1.35}$$

To expose the influence of different J values on the maximum z-component of $ngm_B J$,[2] it is convenient to make a further substitution of $y = a/J$, which modifies (1.35) to

$$\frac{n\bar{m}_z}{ngm_B J} = \mathcal{B}(a) = \frac{2J+1}{2J} \coth\left(\frac{2J+1}{2J}\right) a - \frac{1}{2J} \coth\frac{a}{2J}, \tag{1.36}$$

where $\mathcal{B}(a)$ is the celebrated Brillouin function. Since the assumption that $y \ll 1$ also applies to a in the paramagnetic limit, (1.36) reduces to

$$\mathcal{B}(a) \approx \frac{J+1}{3J} a - \left(\frac{J+1}{3J}\right)\left(\frac{2J^2+2J+1}{30J^2}\right) a^3. \tag{1.37}$$

If the cubic term is ignored in this approximation, we can substitute $m = gm_B J$ into (1.32) and then apply (1.37) to express the susceptibility as

$$\chi_{\text{par}} = \frac{4\pi ngm_B J}{H} \mathcal{B}(a) \cong \frac{4\pi ng^2 J(J+1) m_B^2}{3kT}. \tag{1.38}$$

Equation (1.38) is also a phenomenological explanation for the empirical Curie law $\chi_{\text{par}} = C/T$, with the Curie constant

$$C = ng^2 J\,(J+1)\, m_B{}^2/3k. \tag{1.39}$$

Because $m_j = gm_B \sqrt{J\,(J+1)}$ from (1.26), (1.32) and (1.38) indicate that the Langevin and Brillouin functions are equivalent in the linear paramagnetic region at low H or high T. Where the two theories differ is in the saturation values of ngm_{Jz}, which are proportional to J in the Brillouin function instead of $\sqrt{J(J+1)}$ because of the spatial quantization of the magnetic moment. Only as $J \to \infty$ do the two curves merge, as seen in Fig. 1.5. The accuracy of the Brillouin function was verified experimentally under saturation conditions as shown in Fig. 1.6 for ions of different J values [3].

[2] Where h is absorbed into m_B in magnetic analyses, the practice of replacing $\mathcal{M}_j$ with J_z as the z-component of $\boldsymbol{J}$, or simply using J as the angular momentum counting parameter has become accepted. This is particularly true in the widespread application of the Ising approximation to the Heisenberg model of magnetic exchange. However, where angular momentum is not directly involved in magnetic concerns, for example, electric-dipole transitions discussed in Chap. 7, h must not be forgotten.

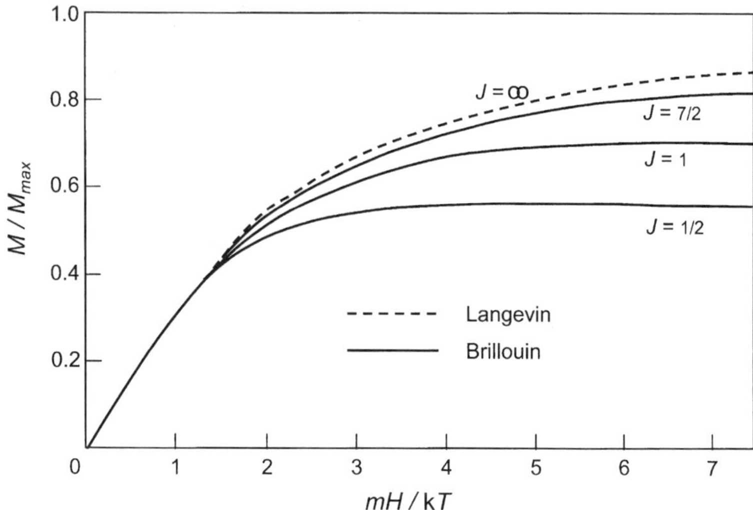

Fig. 1.5 Standard graphs of the Langevin and Brillouin models as a function mH/kT for $J = 1/2$, 1, 7/2, and ∞. Note that M remains less than $M_{\max}$ (proportional to $J\sqrt{J+1}$) because of the quantum correction, and that the two models merge only as $J \to \infty$

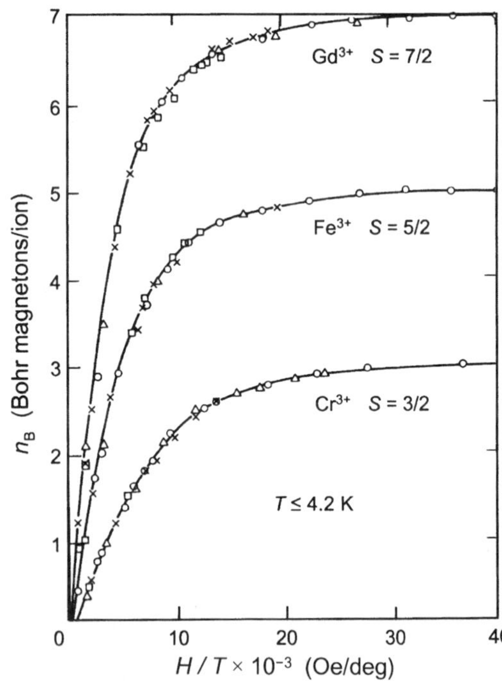

Fig. 1.6 Paramagnetic saturation in potassium chromium alum ($J = 3/2$), ferric ammonium alum ($J = 5/2$), and gadolinium sulphate octahydrate ($J = 7/2$). Data from W.E. Henry [3]

1.3 Spontaneous Magnetism

The interactions of magnetic moments in the preceding discussion have been limited to mutual coupling through dipolar fields and with external fields. The most intense manifestation of magnetic phenomena in solids occurs when the moments are brought close together where the short-range forces of chemical bonding influence strong parallel or antiparallel alignments. What distinguishes these phenomena from diamagnetism and paramagnetism is that the materials may remain magnetized in the absence of a magnetic field. For that reason, magnetism associated with the chemical bonding is referred to as spontaneous.

1.3.1 Classical Ferromagnetism and Antiferromagnetism

Although the Langevin and Brillouin functions provide a phenomenological basis for the empirical Curie law, the behavior of the susceptibility at the onset of spontaneous magnetic ordering is not accounted for. To introduce these effects, Weiss added a term to H that represents the collective influence of the neighboring magnetic moments. This modification defined an effective Weiss "molecular" field

$$H_W = H + N_W M, \tag{1.40}$$

where N_W is the resultant Weiss molecular-field coefficient, which is dimensionless when the magnetic moment is expressed as a volume density, that is, as magnetization M. By manipulation of (1.32) and (1.38), the magnetic susceptibility relation that includes the exchange field H_W becomes

$$\chi_W = \frac{C}{T} = \frac{M}{H + N_W M}, \tag{1.41}$$

which can be converted to

$$\chi_m = \frac{M}{H} = \frac{C}{T - CN_W} = \frac{C}{T - \theta_C}, \tag{1.42}$$

where θ_C is the Curie temperature parameter for ferromagnetism. An analogous relation describes the antiferromagnetic case, according to

$$\chi_m = \frac{C}{T + \theta_N}. \tag{1.43}$$

From (1.39), the paramagnetic Curie and Néel parameters are given by

$$\theta_C = CN_W = \frac{ng^2 J(J+1) m_B{}^2}{3k} N_W,$$

$$\theta_N = -\frac{C}{2} N_W = -\frac{1}{2}\frac{ng^2 J(J+1) m_B{}^2}{3k} N_W. \tag{1.44}$$

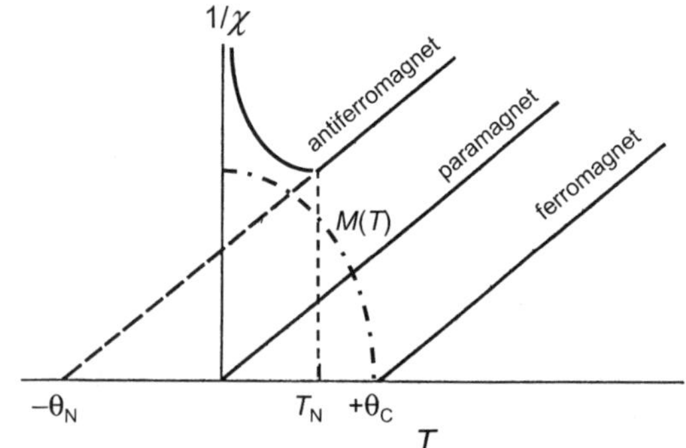

Fig. 1.7 Schematic of inverse susceptibility vs. temperature comparing the Curie–Weiss law for paramagnetism and two spontaneous cases, ferromagnetism and antiferromagnetism

In the antiferromagnetic case, N_W represents the resultant coefficient of the opposing sublattices, with the factor 2 in the denominator because the population of like spins in each sublattice is $n/2$. It must be noted that θ_C and θ_N are smaller in magnitude than the actual T_C and T_N threshold temperatures for long-range magnetic order. In most cases, $T_C \approx \theta_C$, but the relation between T_N and θ_N is more complex because of the competing inter- and intrasublattice molecular fields. Only where the intersublattice molecular field is dominant can the aforementioned relation for θ_N be equated to T_N. This subject is examined further in Chap. 3.

For $N_W > 0$, parallel alignment is the result; for $N_W < 0$, the alignment is antiparallel. By plotting measurement data of $1/\chi$ as a function of T as illustrated in Fig. 1.7, values of C and θ_C or θ_N may be determined from the slope and intercept of the straight-line graph formed from the linear function. From these results values of N_W can be deduced.

The ability to obtain basic information about the strength of spontaneous magnetism from these plots represented an important milestone in the investigation of magnetism. With the introduction of the molecular or mean field concept, the variation in magnetization as a function of temperature in the ferromagnetic state can be calculated with good accuracy.

1.3.2 Solutions of the Brillouin–Weiss Equation

The variation of magnetization with temperature and magnetic field in the ferromagnetic state may be computed by substituting the relation for the effective magnetic field of (1.40) into the previously defined expressions for E_m, y, and a. Accordingly,

$$\begin{aligned} a_{\text{eff}} &= \frac{g m_B J}{kT}\left[H + N_W M(\text{T})\right] \\ &= \frac{g m_B J}{kT}\left[H + N_W M(0)\,\mathcal{B}(a_{\text{eff}})\right]. \end{aligned} \tag{1.45}$$

Because a_{eff} is now a function of $\mathcal{B}(a_{\text{eff}})$, (1.36) can no longer be solved in closed form. The most expeditious way to solve for $\mathcal{B}$ as a function of T and H is through the use of a numerical iteration code that is managed easily by a digital computer, which is reviewed in Sects.3.2 and 4.1. In many textbooks [4, 5], however, graphical solutions are indicated, and there is merit in reviewing the results of these procedures. To investigate spontaneous magnetization, we set $H = 0$ and rearrange (1.45) to obtain a second relation between $M(T)/M(0)$ and a_{eff}, so that

$$\frac{M(T)}{M(0)} = \frac{kT}{ng^2{m_{\text{B}}}^2 J^2 N_{\text{W}}} a_{\text{eff}} \tag{1.46}$$

and from (1.36)

$$\frac{M(T)}{M(0)} = \mathcal{B}(a_{\text{eff}}), \tag{1.47}$$

where T is emphasized as the principal independent variable, and $M(0) = ngm_{\text{B}}J$. Solutions for $M(T)/M(0)$ as a function of a_{eff} are found by plotting (1.46) and (1.47) as shown in the standard graph of Fig. 1.8 and by noting the nonzero intersection of the linear curve with the Brillouin function curve. The Curie temperature, which is the highest temperature for which spontaneous magnetic order can exist, is that obtained from the slope of the linear curve that matches the asymptote of the $\mathcal{B}$ vs . a_{eff} curve at the origin. As contained in the linear term in (1.37), the slope of $\mathcal{B}$ at the limit of $a_{\text{eff}} \ll 1$ is given by $[(J+1)/3J]\, a_{\text{eff}}$, which leads directly to the limiting value of temperature for spontaneous magnetism

$$T_{\text{C}} = \theta_{\text{C}} = \frac{ng^2{m_{\text{B}}}^2 J(J+1)}{3k} N_{\text{W}}. \tag{1.48}$$

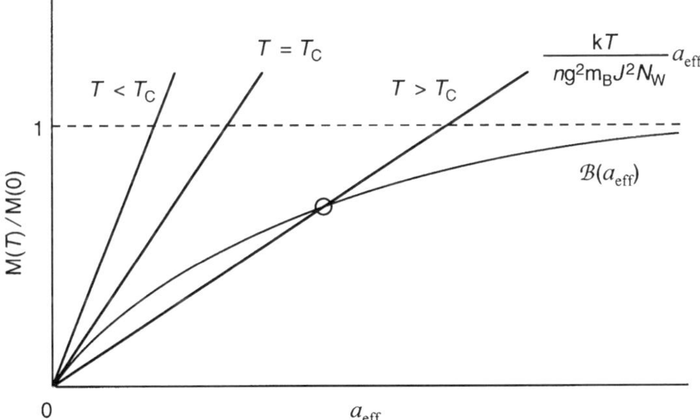

Fig. 1.8 Graphical solution for the Curie temperature, as described in text. Spontaneous magnetism breaks down at the temperature where the linear curve becomes tangential at the origin of the Brillouin curve

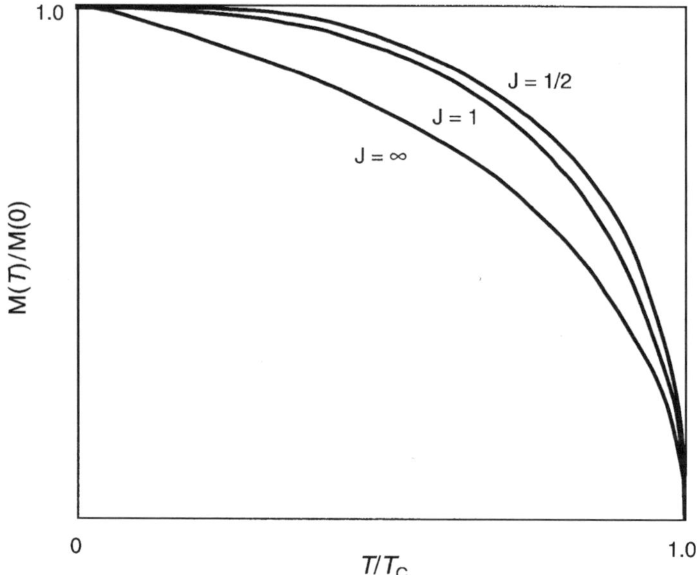

Fig. 1.9 Universal Brillouin–Weiss curves for $J = 1/2$, 1, and ∞, with $H_{ex} \gg H$

From (1.46) and (1.48) a universal relation can be constructed according to

$$\frac{M(T)}{M(0)} = \left(\frac{J+1}{3J}\right)\left(\frac{T}{T_C}\right) a_{eff}. \tag{1.49}$$

Universal curves of the Brillouin–Weiss theory are shown in Fig. 1.9, where $\mathcal{B}$ is presented for different values of J. The curve for $j \to \infty$ is the Langevin limit where $\sqrt{J(J+1)} \to J$.

Another useful relation can be obtained by combining (1.37) and (1.49) to eliminate a_{eff}:

$$\left[\frac{M(T)}{M(0)}\right]^2 = \frac{10}{3}\left(\frac{(J+1)^2}{J^2+(J+1)^2}\right)\left(1-\frac{T}{T_C}\right)\left(\frac{T}{T_C}\right)^2$$
$$\approx \frac{10}{3}\left(\frac{(J+1)^2}{J^2+(J+1)^2}\right)\left(1-\frac{T}{T_C}\right) \quad \text{for } \frac{T}{T_C} \to 1. \tag{1.50}$$

This relation indicates that $M(\mathrm{T})$ is a continuous function of temperature up to the Curie temperature. It also reveals that the $M(T)/M(0)$ follows a $(1 - T/T_C)^{1/2}$ near the Curie temperature, and that its slope tends to infinity at $T = T_C$. The exponent 1/2 was subsequently defined by the parameter β, sometimes called the scaling constant, which usually assumes values lower than 1/2 in other models of the approach to the Curie temperature.

At temperatures near 0 K, a relation that has its origins in spin-wave theory has proven to be the more effective at fitting data than the Brillouin function. It is commonly referred to as the Bloch $T^{3/2}$ law and is given by

$$\frac{M\,(T)}{M(0)} = 1 - AT^{3/2}, \tag{1.51}$$

where the exchange constant J enters the relation through the parameter

$$A = \frac{0.1174}{f}\left(\frac{k}{J}\right)^{3/2}$$

and $f = 1$, 2, or 4 for the simple, body-centered, or face-centered cubic lattice, respectively.

The values of $\mathcal{B}$ as a function of T can also be computed [6] by means of a convergent iterative procedure that is described in Chap. 4 in relation to multiple sublattice ferrites. *The importance of the* Brillouin function will revisited in the discussions of antiferromagnetism. In Chap. 7, an external field H is included as an active variable in combination with the molecular field to illustrate the origin of magnetoresistance properties of magnetic oxides.

1.3.3 Quantum Origins of the Molecular Field

The origin of the Weiss molecular field was first proposed by Heisenberg [7], who postulated that spontaneous spin alignments are determined by short-range interactions between adjacent spins that are made possible by the coupling of orbital wave functions as part of the chemical bonding. The effect is electrostatic and arises from electron exchange between bonding atoms. In a quantum mechanical format, the Heisenberg exchange energy is expressed as a Hamiltonian function with adjacent spins at sites i and j related by

$$\mathcal{H}_{\text{ex}} = -2\sum_{i>j} J_{ij}\,\boldsymbol{S}_i \cdot \boldsymbol{S}_j, \tag{1.52}$$

where J_{ij} is the exchange constant. Phenomenologically, the sign of J_{ij} determines the type of spin alignment; intuitively, it is seen that $J_{ij} > 0$ will cause ferromagnetism and $J_{ij} < 0$ will create antiferromagnetism depending on the cosine of the angle between $\boldsymbol{S}_i$ and $\boldsymbol{S}_j$.

The orbital interactions the determine J_{ij} are analyzed through the use of the one-electron Hartree–Fock approximation to the wave functions of a many electron system, with modifications required to account for the indistinguishability of fermions and satisfy the Pauli exclusion principle [8]. An example that can illustrate the quantum mechanical origin of the effect is the case of two one-electron atoms a and b with corresponding electrons labeled (1) and (2). This is the hydrogen molecule (H_2) employing the *valence-bond* approach of Heitler and London [9]. The Hamiltonian (here representing a local Madelung energy) for this pair before taking into account the electron–electron repulsion is therefore

$$\mathcal{H}_0 = -\frac{\hbar^2}{2m_e}\left[\nabla_a^2 + \nabla_b^2\right] - \frac{Z_a e^2}{r_{a1}} - \frac{Z_b e^2}{r_{b2}}, \tag{1.53}$$

where Z_a and Z_b are the respective nuclear charges. There are two possible solutions to this equation, $\varphi_0 = \varphi_a(1)\,\varphi_b(2)$ where the two electrons remain on their "home" atoms, and $\varphi_{ex} = \varphi_a(2)\,\varphi_b(1)$ where the electrons are exchanged between the two atoms.

To this point in the analysis, the electrons have been treated as "distinguishable." However, orthogonality requires that their eigenfunctions satisfy the relation

$$|\langle\varphi_0 \mid \varphi_0\rangle|^2 = |\langle\varphi_{ex} \mid \varphi_{ex}\rangle|^2, \tag{1.54}$$

which means that two solutions exist:

$$\varphi_0 = \pm\varphi_{ex}. \tag{1.55}$$

Because neither of the basic wavefunctions φ_0 nor φ_{ex} can fulfill the indistinguishability requirement of (1.55), the usual linear combinations are constructed,

$$\begin{aligned}\varphi_{sym} &= \frac{1}{\sqrt{2}}(\varphi_0 + \varphi_{ex}),\\ \varphi_{anti} &= \frac{1}{\sqrt{2}}(\varphi_0 - \varphi_{ex}).\end{aligned} \tag{1.56}$$

To satisfy the indistinguishability, wavefunctions must always be antisymmetric, meaning that exchange of two fermions must always involve a sign reversal [8]. However, only φ_{anti} meets this criterion. To render both eigenfunctions antisymmetric, the orbital functions must be completed by attaching the spin matrix functions χ_{sym} (for parallel spins) and χ_{anti} (for antiparallel spins) to create the final set

$$\begin{aligned}\psi_{\uparrow\uparrow} &= \varphi_{anti}\chi_{sym},\\ \psi_{\uparrow\downarrow} &= \varphi_{sym}\chi_{anti}.\end{aligned} \tag{1.57}$$

For a two-electron molecule, the total spin for χ_{sym} is $S = 1$ from parallel spin vectors, thereby forming a spin angular momentum triplet ($S_z = 1, 0,$ and -1 in units of $\hbar$). Conversely, χ_{anti} is the singlet $S = 0$ from antiparallel spins.

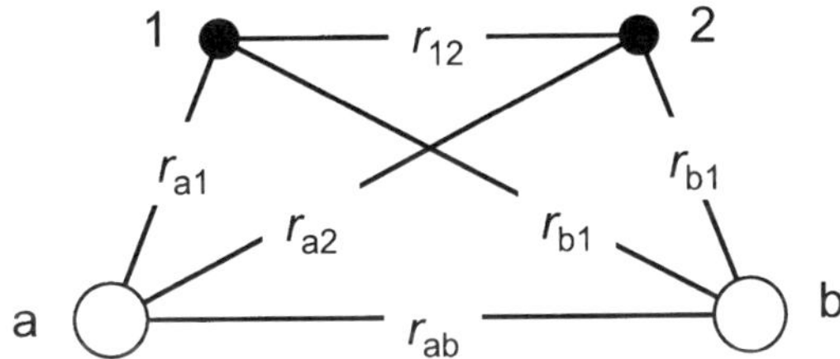

Fig. 1.10 Basic diagram of magnetic exchange interaction identifying the various linkage distances among positively charged nuclei a and b and their bonding electrons 1 and 2

The problem can now be solved by considering a perturbed Hamiltonian in the form of $\mathcal{H} = \mathcal{H}_0 + \mathcal{H}_1$ and by calculating the eigenvalues of the two states for the various electrostatic interactions among the nuclei and electrons, according to

$$\mathcal{H}_1 = \underset{\text{repulsion}}{\left[\frac{e^2}{r_{12}} + \frac{Z_\mathrm{a} Z_\mathrm{b} e^2}{r_\mathrm{ab}}\right]} - \underset{\text{attraction}}{\left[\frac{Z_\mathrm{a} e^2}{r_\mathrm{a2}} + \frac{Z_\mathrm{b} e^2}{r_\mathrm{b1}}\right]}, \tag{1.58}$$

as depicted in Fig. 1.10. Computing the expectation values of the integrals $\langle\psi_{\uparrow\downarrow}|\,\mathcal{H}_1\,|\psi_{\uparrow\downarrow}\rangle, \langle\psi_{\uparrow\downarrow}|\,\mathcal{H}_1\,|\psi_{\uparrow\uparrow}\rangle$, and $\langle\psi_{\uparrow\uparrow}|\,\mathcal{H}_1\,|\psi_{\uparrow\uparrow}\rangle$ produces the energy eigenvalues given in a form that exposes the spin exchange operator as determined by the procedure of second quantization using Fermi operators [10, 11],

$$E = K - J\left(\frac{1}{2} + 2\boldsymbol{s}_1 \cdot \boldsymbol{s}_2\right), \tag{1.59}$$

where

$$\begin{aligned} K &= \langle\varphi_0|\,\mathcal{H}_1\,|\varphi_0\rangle \quad \text{(Coulomb integral)}\,, \\ J &= \langle\varphi_0|\,\mathcal{H}_1\,|\varphi_\mathrm{ex}\rangle \quad \text{(exchange integral)}\,. \end{aligned} \tag{1.60}$$

To calculate the exact quantum mechanical values of the $\boldsymbol{s}_1 \cdot \boldsymbol{s}_2$ scalar product (in units of $\hbar^2$), the standard relation for the angular momentum vector addition $\boldsymbol{s}_1 + \boldsymbol{s}_2 = \boldsymbol{S}$ can be used:

$$2\boldsymbol{s}_1 \cdot \boldsymbol{s}_2 = S\,(S+1) - s_1\,(s_1+1) - s_2\,(s_2+1)\,. \tag{1.61}$$

For the parallel and antiparallel values of $S = 1$ and 0, respectively, and the values of $s_1 = s_2 = 1/2$, (1.59) reduces to

$$\begin{aligned} E &= K - J \quad \text{for } S = 1 \quad \text{(ferromagnetism)}\,, \\ E &= K + J \quad \text{for } S = 0 \quad \text{(antiferromagnetism)}\,. \end{aligned} \tag{1.62}$$

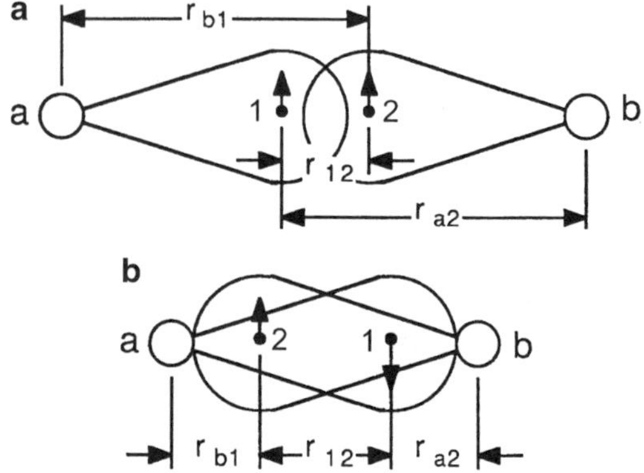

Fig. 1.11 Orbital interaction diagram contrasting the competing roles of mutual electron repulsion and electron-nuclear attraction as nuclear separation decreases: (**a**) ferromagnetism in an antibonding state, and (**b**) antiferromagnetism in a bonding state. Note the reversed locations of electrons 1 and 2 relative to the two nuclei

The choice of ground state in (1.62) is therefore determined by the sign of J, which follows the same rules that are stated below (1.52), that is, ferromagnetism for $J > 0$, antiferromagnetism for $J < 0$.[3]

Computation of J is not a trivial task, but some insight may be gained by inspecting the relative sizes of the terms in (1.58) and comparing them with the diagrams in Fig. 1.11. For overlapping orbital wave function lobes of the type shown in Fig. 1.11a, the close proximity of the electron charge concentrations can make the repulsive e^2/r_{12} term dominant and render $J > 0$, thereby establishing the ferromagnetic $\psi_{\uparrow\uparrow}$ as the ground state. If the overlap is more extreme because of a smaller r_{ab} distance, the situation of Fig. 1.11b would arise and the attractive $-Z_{\mathrm{a}}e^2/r_{\mathrm{a}2}$ and $-Z_{\mathrm{b}}e^2/r_{b2}$ terms could be large enough to make $J < 0$ and produce the antiferromagnetic $\psi_{\uparrow\downarrow}$ ground state.

Overlapping wavefunction lobes are characteristic of d and f electron shells. Ferromagnetism is more likely to occur in the upper half of the iron group elements because of their larger populations of unpaired and itinerant electrons. The graph in Fig. 1.12 illustrates the qualitative support for this model, where J values were estimated from Curie and Néel temperature measurements. Positive J values occur for larger r_{ab} distances in the familiar Fe, Co, and Ni of the $3d^n$ transition group and members of the $4f^n$ rare-earth series.

An early phenomenological description of magnetic exchange was introduced by Stoner [12]. Because of its success in interpreting magnetism in metals, it is known as the theory of *collective electron ferromagnetism*. The result of this approach is the band theory model that has helped to explain some features of the $3d^n$ transition series of elements. The theory predicts that ferromagnetic metals have moments that are lower than anticipated based on the number of unpaired Bohr magnetons of the

[3] Since the spins of an antiferromagnet are antisymmetric, it follows that the orbital functions must be symmetric, and vice-versa for the ferromagnet. The former case is referred to as a "bonding" state because of the orbital wavefunction overlaps, which is why most ionic compounds are intrinsically antiferromagnetic. The antibonding state then becomes a ferromagnet, as explained in Chaps. 2 and 3.

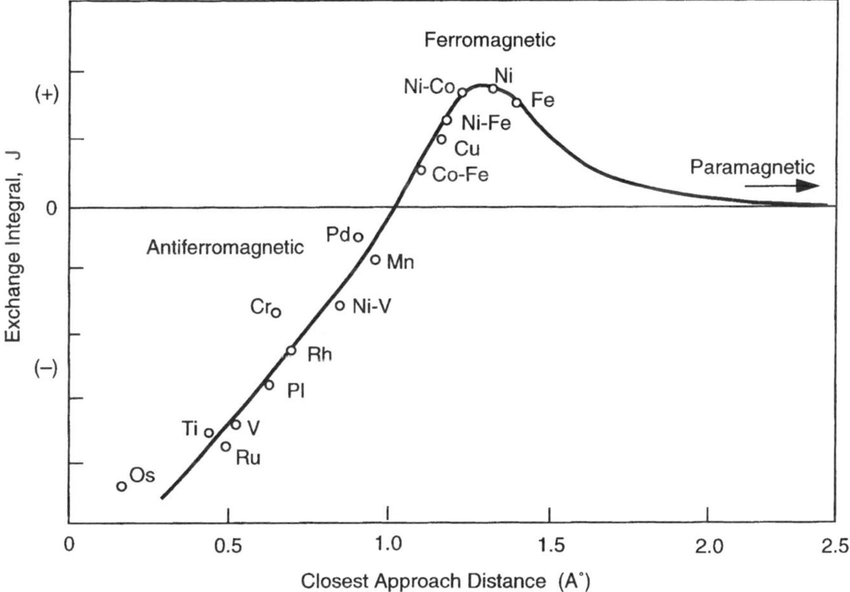

Fig. 1.12 Chart of exchange constant data compiled as a function of the distance of closest approach of electron of neighboring atoms. Adapted from Lax and Button [14]

Table 1.3 Magnetic moments of ferromagnetic atoms

Transition element	Free atom configuration	Solid state distribution	Net $3d$ or $4f$ Bohr magnetons
Fe	$3d^6 4s^2$	$3d^{7.4} 4s^{0.6}$	$+4.8 - 2.6 = 2.2$
Co	$3d^7 4s^2$	$3d^{8.3} 4s^{0.7}$	$+5 - 3.3 = 1.7$
Ni	$3d^8 4s^2$	$3d^{9.4} 4s^{0.6}$	$+5 - 4.4 = 0.6$
Gd	$4f^7 5d^1 6s^2$	$\sim 4f^{7.1} 5d^2 6s^{0.9}$	7.10

free atom, as shown in Table 1.3. The reasons lie in the details of the overlapping of $4s$ and $3d$ electronic bands in metals [13], as sketched in the density of states of the split magnetic bands in Fig. 1.13. Further support for the band concept is given by the result for Gd^{3+}, which has seven unpaired electrons that occupy the shielded $4f$ inner shell and are thus exempted from the distractions of the $6s$ bonding states.

An effective approximation for insulator compounds is based on single-electron molecular-orbital theory of covalent bonding. In contrast to the valence bond method described earlier, molecular orbital theory focuses specifically on the electron–nuclear interaction terms of (1.58) and ignores the mutual repulsion e^2/r_{12}, thereby favoring antiparallel spin alignment in the ground state. The spin stabilization state that emerges is called indirect or superexchange because the interactions between the positive magnetic ions (spin cations) occur through the mediation of negative ligands (anions). It is particularly well suited for describing transition-metal insulating compounds where the basic exchange is

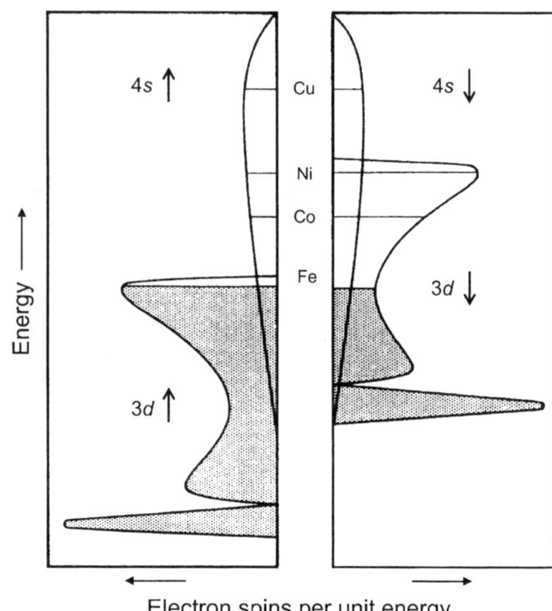

Fig. 1.13 Band model of electron spin occupancy as a function of energy for the overlapping 3*d* and 4s bands axis [12]. The bands are filled to the levels indicated for iron metal. The concept of dividing the bands into collective up and down spins separated in energy by the exchange stabilization [11] resembles the application of Hund's rule in formulating the ground-state spin configuration of transition-metal ions, as diagrammed in Figs. 2.3 and 2.19. Adapted from Lax and Button [14]

antiferromagnetic. However, there are varying conditions of superexchange (including charge transfer or itinerance) that apply to different cation–anion–cation situations.

This subject is placed into a broader context in Chap. 3, where collective and localized spontaneous magnetism are contrasted. In a normally antiferromagnetic structure, ferromagnetism requires electron delocalization in the form of Fermi gas conductivity or large-polaron charge transfer between magnetic cations in mixed-valence compounds.

1.3.4 The Ising Approximation

The exact solutions of the total spin scalar product $\boldsymbol{S}_j \cdot \boldsymbol{S}_j$ eigenvalues require the application of Fermi operators that lead to the mathematical and conceptual complexities indicated by the stated result for the exchange energy E_{ex} taken from (1.59). An effective simplification called the Ising approximation has proven to be accurate for most practical situations. If the exchange constant is the same for all spin interactions, J_{ij} of (1.52) may be replaced by J and the scalar product can be expanded and summed over the nearest neighbors S_j of spin S_i to give

$$H_{\mathrm{ex}} = -2J \sum_{i>j} \left(S_{xi} S_{xj} + S_{yi} S_{yj} + S_{zi} S_{zj}\right). \tag{1.63}$$

If the spin components are replaced by their time averages, and a total of z identical nearest neighbors are included in the sum, (1.63) may be expressed as

$$H_{\mathrm{ex}} = -2zJ\left(\bar{S}_{xi}\bar{S}_{xj} + \bar{S}_{yi}\bar{S}_{yj} + \bar{S}_{zi}\bar{S}_{zj}\right). \tag{1.64}$$

If the classical Larmor theory is used (see Sect. 1.5.1), the spin vectors will precess about an effective molecular (or exchange) field directed along the z-axis of quantization. As a consequence, the time averages of the x and y spin components are assumed to be zero, and

$$H_{\mathrm{ex}} \approx -2zJ\,\bar{S}_{zi}\bar{S}_{zj}. \tag{1.65}$$

From this approximation a relation between the exchange integral J and the Weiss molecular field coefficient N_{W} may be derived. Sincc we consider only the z component of the magnetization $M_z\left(=ngm_{\mathrm{B}}S_{zj}\right)$ and (1.65) becomes

$$H_{\mathrm{ex}} \approx -2\left(\frac{z}{n}\right)\frac{JS_{zi}M_z}{g_i m_{\mathrm{B}}}. \tag{1.66}$$

For $H = 0$, H_{ex} can also be expressed in terms of the Weiss molecular field by equating to E_m from (1.33), according to

$$H_{\mathrm{ex}} \approx -g_i S_{zi} m_{\mathrm{B}} N_{\mathrm{w}} M_z, \tag{1.67}$$

where M_j becomes S_{zj} and the applied field H is replaced by the exchange field $N_{\mathrm{W}}M_z$. From (1.66) and (1.67)

$$N_{\mathrm{W}} = \frac{2zJ}{n{g_i}^2{m_{\mathrm{B}}}^2}, \tag{1.68}$$

and we have for the Curie temperature from (1.48), after (1.68) is substituted,

$$T_{\mathrm{C}} = \frac{2zJS\left(S+1\right)}{3k}, \tag{1.69}$$

with S replacing the total angular momentum (J) for our discussion of spin exchange. Equations (1.68) and (1.69) are important in the models for thermomagnetism and magnetoresistance.

1.4 Gyromagnetism

To this point in the discussion, there has been little mention of the spectroscopic splitting factor g that was introduced as part of the quantum mechanical definition of the magnetic moment. Before leaving this introductory chapter, it is appropriate to explain the classical origins of the magnetic resonance phenomenon that is not only an important topic for later parts of this text, but also provides a direct means of measuring g.

1.4.1 *Larmor Precession and Resonance*

Magnetic resonance is based on a theorem that the motion of an electron under a central force, for example, Coulomb attraction to a nucleus, and a magnetic field $\boldsymbol{H}$ in a fixed coordinate system is identical to that of an electron under the same central force with $\boldsymbol{H} = 0$ in a coordinate system that is rotating about the $\boldsymbol{H}$ axis with angular frequency

$$\omega_0 = \gamma H = \frac{ge}{2m_e c} H, \tag{1.70}$$

where ω_0 is the Larmor precession frequency and g equals 1 for an orbiting electron and 2 for an electron spin; $\left(\gamma = 1.76 \times 10^7 \text{ rad/s/Oe}\right)$ is usually referred to as the gyromagnetic constant.[4] The derivation of this relation may be found in many standard textbooks [14, 15]. For the diamagnetic case with $g = 1$, the sense of rotation in the precession orbit is to establish a magnetic moment that is directed against $\boldsymbol{H}$, thereby providing the basic mechanism for diamagnetism.

From the general case of a fixed magnetic moment $\boldsymbol{m}$ in a field $\boldsymbol{H}$, the rotation direction would be consistent with the moment and field in parallel. The basic theory of magnetic resonance may be derived from the equation of motion

$$\frac{\mathrm{d}\boldsymbol{m}}{\mathrm{d}t} = \frac{ge}{2m_e c} \left(\boldsymbol{m} \times \boldsymbol{H}\right), \tag{1.71}$$

where the vector $\boldsymbol{m} \cdot \boldsymbol{H}$ is normal to the plane containing $\boldsymbol{m}$ and $\boldsymbol{H}$. Consequently, the resulting torque causes the clockwise precession of $\boldsymbol{m}$ about $\boldsymbol{H}$ viewed along the direction chosen as the z-axis of the magnetic field. The solutions of (1.71) for the individual components of $\boldsymbol{m}$ may be expressed as

$$\begin{aligned} m_x &= m \sin\theta \cos\omega_L t, \\ m_y &= m \sin\theta \sin\omega_L t, \\ m_z &= m \cos\theta = \text{constant}. \end{aligned} \tag{1.72}$$

where $\boldsymbol{m}$ forms a constant angle θ with the z axis in establishing a cone of precession at the Larmor frequency. Recalling the earlier discussion of the quantum mechanical constraints on the values of the projection of $\boldsymbol{m}$ on the axis of quantization, we express $m_{Jz} = g m_B J_z$ since Planck's constant $\hbar$ is absorbed in the Bohr magneton. Note also that from a quantum energy standpoint, the Zeeman splitting

$$\hbar\omega_0 = g m_B H \Delta J_z = g m_B H \quad \text{for } \Delta J_z = 1. \tag{1.73}$$

The equivalence of (1.70) and (1.73) can be determined by inspection.

[4] Where the frequency is designated by the symbols ν or f expressed in cycles/s (Hz), an alternative constant γ' (=2.78 GHz/kOe) is defined. However, confusion can arise when the symbol ω is used for ν, usually in microwave engineering literature. In such instances, γ' must be employed with H. Note also that the negative sign of the electron charge e has been absorbed in the definition of ω_0 in (1.70), thereby allowing γ to be treated as positive wherever frequency and field or magnetization are related.

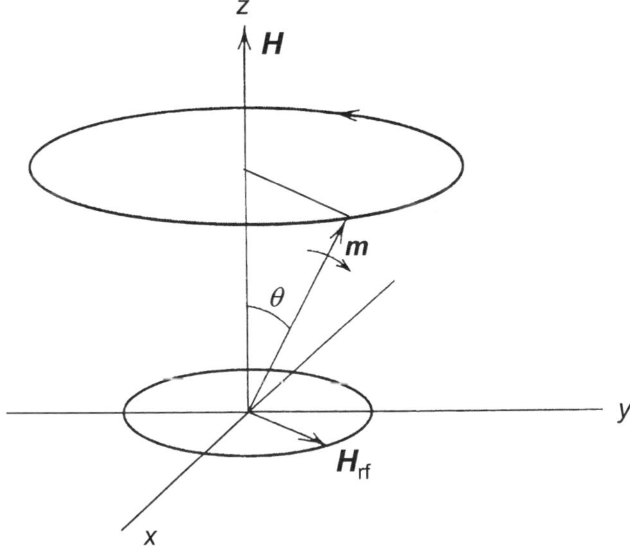

Fig. 1.14 Standard classical diagram of magnetic resonance showing the precession of a magnetic moment $\boldsymbol{m}$ about a field $\boldsymbol{H}$ at the Larmor frequency. When a circularly polarized oscillating field $\boldsymbol{H}_{\mathrm{rf}}$ matches the frequency and sense of the precession, $\boldsymbol{m}$ is subjected to a torque that rotates it away from the $\boldsymbol{H}$ axis, thereby producing a complex susceptibility

Magnetic resonance occurs where an alternating (usually radio frequency) magnetic field $\boldsymbol{H}_{\mathrm{rf}}$ is applied in a direction perpendicular to H. Because a linearly polarized signal can be decomposed into two counterrotating circularly polarized signals, the physical situation resembles that depicted in Fig. 1.14, where only the component that rotates in the direction of the precessing moment is capable of continuously influencing the angle of $\boldsymbol{m}$ relative to the z axis by creating a second torque $\boldsymbol{m} \times \boldsymbol{H}_{\mathrm{rf}}$ normal to the $\boldsymbol{m} \times \boldsymbol{H}$ direction. By setting the frequency of the alternating field at $\omega = \omega_0$, $\boldsymbol{H}_{\mathrm{rf}}$ will synchronize with the precession and apply a constant torque that will cause the cone half-angle θ to oscillate from full alignment with $\boldsymbol{H}$ ($\theta = 0$) to its opposite limit ($\theta = \pi$). The rotation of $\boldsymbol{m}$ away from $\boldsymbol{H}$ represents the absorption of energy. Where a quantum mechanical model can be applied, the two extreme values of θ represent two energy levels of a degeneracy that is split by $\boldsymbol{H}$ (Zeeman effect). This view of magnetic resonance is explored in a discussion of electron paramagnetic resonance in Chap. 6.

1.4.2 Phenomenological Relaxation Theory

An important feature of all resonating systems is the effect of damping. It occurs most visibly in vibrating mechanical systems that reduce their amplitudes exponentially because of air resistance or friction. The decay is a manifestation of the loss of energy as the system relaxes back to equilibrium. For systems under constant excitation, it is the energy that must be supplied to maintain the resonance condition at a given intensity. In magnetic resonance, the relaxation arises principally from two sources: (1) the transfer of magnetic energy imparted by the alternating magnetic fields to the system of magnetic moments, commonly referred to as the "spin system," back to the environment, which is the lattice in the case of solids and (2) the tendency of the precessing moments to lose phase coherence due to perturbations of the external magnetic field by local dipolar fields. The first is loss of signal energy directly to the lattice; the second is a loss of signal by decoherence of the precessing spins. Eventually this energy will also be transformed into heat or radiation.

In mathematical terms, following the approach developed by Bloch and Bloembergen [16, 17], the relaxation of magnetization $\boldsymbol{M}$ $(=\Sigma \boldsymbol{m})$ back toward the z axis (longitudinal relaxation) following the removal of the $\boldsymbol{H}_{\rm rf}$ field is determined by the relaxation rate τ_z^{-1} of its M_z component

$$\frac{\mathrm{d}M_z}{\mathrm{d}t} = -\frac{M_z - M}{\tau_z} = -\frac{M_z - M}{\tau_1}, \tag{1.74}$$

where τ_1 $(=\tau_z)$ is the spin–lattice relaxation time and M $(=M_{\rm s})$ is the equilibrium value of M_z. In the stationary frame of reference of Fig. 1.15, the decay process of the individual $\boldsymbol{m}$ vector can be visualized as following an inwardly precessing spiral. Accordingly,

$$M - M_z(t) = [M - M_z(t)] \exp(-t/\tau_1) . \tag{1.75}$$

The second relaxation phenomenon of importance concerns the phase decoherence of M_x and M_y among individual spin vectors as they precess about $\boldsymbol{H}$ (transverse relaxation), which occurs in paramagnetic systems through dipole–dipole interactions. In ferromagnetic systems, the decoherence represented by τ_2 is seen in the form of *spin waves*, which occur under special conditions. In this case the vector components of the spins that are perpendicular to $\boldsymbol{H}$, that is, M_x and M_y lose their collective coherence at a rate defined as τ_2^{-1}, once the alternating drive field is re-

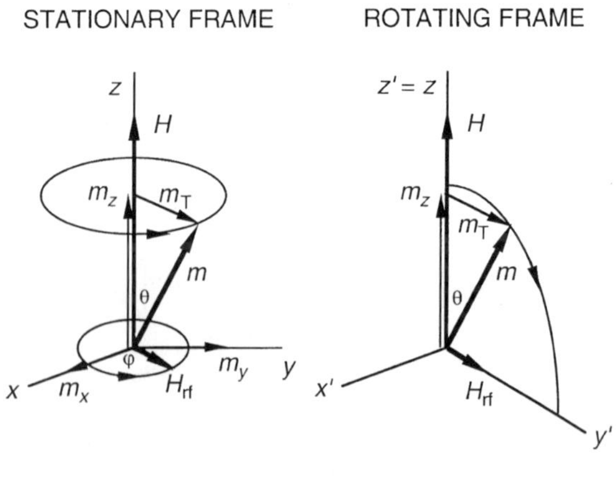

Fig. 1.15 Two frames of reference for magnetic moment precession: stationary and rotating at the precession angular frequency

moved. Because the amplitudes of M_x and M_y must also decay at a rate proportional to τ_1^{-1}, as analyzed in Appendix 1A, the phenomenological damping rate for these components from (1.93) can be expressed as

$$\tau_{x,y}{}^{-1} \cong \tau_2{}^{-1} + f(\theta)\,\tau_1{}^{-1}, \tag{1.76}$$

where $f(\theta) = \frac{|\cos\theta|}{1+|\cos\theta|}$,

which varies between 0 and 1/2. The relation for the limiting case of $f(0) = 1/2$ was stated without proof in [18].

We can therefore express

$$\frac{\mathrm{d}M_{x,y}}{\mathrm{d}t} = -\frac{M_{x,y}}{\tau_{x,y}} = -M_{x,y}\left(\frac{1}{\tau_2} + f(\theta)\frac{1}{\tau_1}\right) \cong -\frac{M_{x,y}}{\tau_2}, \tag{1.77}$$

under the usual paramagnetic condition that $\tau_2 \ll \tau_1$ and the influence of spin–lattice relaxation is moot. In terms of the Heisenberg uncertainty principle, the decoherence rate $1/\tau_2 \sim \Delta\omega \approx \gamma\Delta H$ of the resonance line, that is, the half-linewidth at half maximum. Thus, greater dipolar interactions cause broader linewidths. For paramagnetic systems $\tau_2 \sim 10^{-10}$ s and produces line broadening of a few kHz at microwave frequencies of several GHz.

By contrast, τ_1 is usually not short enough to affect the linewidth unless fast-relaxing impurities are present. However, the spin–lattice interaction is a sensitive function of temperature, with τ_1 being shorter $\left(\sim 10^{-6}\text{ s}\right)$ at room temperature, and lengthening into the millisecond range to produce line narrowing for easy measurement only at liquid helium temperatures of 4 K and below. In the paramagnetic limit, line broadening due to spin precession decoherence from dipole–dipole interactions can therefore be observed in systems where broadening effects are determined by $\tau_2{}^{-1}$. Where concentrations of spins are large enough, exchange coupling produces an effective field that can align the spins into a coherent collective magnetic moment that renders $\tau_2{}^{-1} \to 0$ under ideal gyromagnetic resonance conditions.

Spatial decoherence of the $M_{x,y}$ component can also occur in ferromagnetic systems through spin waves where conditions permit degeneracy of rf signal and spin wave frequencies. Line broadening by a $\tau_2{}^{-1}$ effect can reappear and even deteriorate into irreversible nonlinear transfer of energy from signal to spin wave system.

1.4.3 Complex Susceptibility Theory

There are facets to the mysteries of magnetic resonance that have challenged and fascinated researchers from the time of the phenomenon's discovery. One topic that has received attention involves the rapid rotation of the $\boldsymbol{m}$ vector when the amplitude H_{rf} is very large, often administered in the form of a high-power pulse. From the perspective of the rotating frame of $\boldsymbol{H}_{\mathrm{rf}}$ (x', y', z', where $z' = z$ if the now-static $\boldsymbol{H}_{\mathrm{rf}}$ is along the x axis), we can visualize the polar rotation of $\boldsymbol{m}$ as confined to the fixed z'-y' plane, for which the corresponding angular precession frequency becomes γH_{rf}, as sketched in Fig. 1.15. If $\gamma H_{\text{rf}} \gg \tau_1{}^{-1}$, $\tau_2{}^{-1}$, the two relaxation rates, the reversal of $\boldsymbol{m}$ through an angle π is termed "adiabatic fast passage" because the damping effects associated with the relaxation do not have time to take effect. In the sense of the previous description of dispersion with $\boldsymbol{H}_{\text{rf}}$ parallel to $\boldsymbol{H}$, it would mean that $\omega\tau \gg 1$. This regime of rf drive fields leads to the classical studies of resonance-line "hole burning" and "spin-echo" effects from which estimates of precession-phase decoherence influenced by dipolar spin interactions and spin-flip damping by spin–lattice relaxation can be obtained directly from experiment [19]. In a medium with an inhomogeneously broadened resonance line, spin diffusion can be investigated by spin-echo methods with $\gamma H_{\text{rf}} \ll \tau_1{}^{-1}$, $\tau_2{}^{-1}$ [20]. However, these examples are esoteric enough to fall outside of the scope of our immediate interests. For an insightful treatment of this general topic, the reader is referred to standard texts, for example, by Pake [21] and Abragam [22]. In Chap. 6, the subject of spin–lattice relaxation is examined further in relation to the general microwave properties of magnetic oxides.

To explain high frequency and microwave signal propagation as functions of frequency, magnetic field, and temperature, the behavior of the complex rf susceptibility must be understood. In very high dc magnetic fields, transmission properties at optical wavelengths can also be influenced by magnetic resonance. If the rf field is small enough to be treated as a perturbation, so that fast-passage effects are not present, that is, $\gamma H_{\text{rf}} \ll \tau_1{}^{-1}$, $\tau_2{}^{-1}$, it is convenient to work within the stationary frame x, y, z. Here, we can call the regime "slow passage" and include all of the damping effects, and then derive the complex susceptibility relations for a circularly polarized rf signal confined to the x-y plane, as indicated in Fig. 1.15.

Consider the a wave of amplitude H_{rf} linearly polarized along the x axis of the laboratory frame of reference given by

$$\begin{aligned} H_x &= H_{\text{rf}} \cos \omega t = \frac{1}{2} H_{\text{rf}} \left[\exp(i\omega t) + \exp(-i\omega t)\right], \\ H_y &= i H_{\text{rf}} \sin \omega t = \frac{1}{2} H_{\text{rf}} \left[\exp(i\omega t) - \exp(-i\omega t)\right], \end{aligned} \tag{1.78}$$

where the two exponential functions represent the counter-rotating modes of circular polarization. For the resonance condition only the positive exponent has influence when $\omega \to \omega_L$. If (1.78) is combined with (1.71), (1.74), and (1.77), the following general relations can be constructed:

$$\begin{aligned} \frac{\mathrm{d}M_z}{\mathrm{d}t} &= \gamma \left(M_y H_{\text{rf}} \cos \omega t - M_x H_{\text{rf}} \sin \omega t\right) + \frac{M - M_z}{\tau_1}, \\ \frac{\mathrm{d}M_x}{\mathrm{d}t} &= \gamma \left(M_z H_{rf} \sin \omega t - H M_y\right) - \frac{M_x}{\tau_2}, \\ \frac{\mathrm{d}M_y}{\mathrm{d}t} &= \gamma \left(H M_x - M_z H_{\text{rf}} \cos \omega t\right) - \frac{M_y}{\tau_2}. \end{aligned} \tag{1.79}$$

A complex rf susceptibility can then be expressed in the standard form of

$$\chi_{\mathrm{rf}} = \chi_{\mathrm{rf}}' - i\chi_{\mathrm{rf}}'' \tag{1.80}$$

and the components of $\boldsymbol{M}$ are expressed in terms of χ_{rf} by the relations

$$\begin{aligned} M_x &= \chi_{\mathrm{rf}}' H_{\mathrm{rf}} \cos\omega t + \chi_{\mathrm{rf}}'' H_{\mathrm{rf}} \sin\omega t \\ M_y &= i\left(\chi_{\mathrm{rf}}' H_{\mathrm{rf}} \sin\omega t + \chi_{\mathrm{rf}}'' H_{\mathrm{rf}} \cos\omega t\right). \end{aligned} \tag{1.81}$$

For the remainder of this derivation, we leave the manipulations of the relations as an exercise or to be found in standard texts on magnetic resonance [21–23]. Solutions for the z component of $\boldsymbol{M}$ under the resonance conditions reduce to

$$M_z = \chi_m H \frac{1 + \tau_2{}^2 (\gamma H - \omega)^2}{1 + \tau_2{}^2 (\gamma H - \omega)^2 + \gamma^2 H_{\mathrm{rf}}{}^2 \tau_1 \tau_2}, \tag{1.82}$$

where χ_m is the dc susceptibility that would apply far from resonance. For the x and y components we obtain

$$M_{x,y} = (\gamma M)\,\tau_2 \frac{\tau_2 (\gamma H - \omega) H_{\mathrm{rf}} \cos\omega t + H_{\mathrm{rf}} \sin\omega t}{1 + \tau_2{}^2 (\gamma H - \omega)^2 + \gamma^2 H_{\mathrm{rf}}{}^2 \tau_1 \tau_2}, \tag{1.83}$$

recalling that $M = \chi_m H$. Combination with (1.80) and (1.81) enables the identification of χ_{rf}' and χ_{rf}'' according to

$$\begin{aligned} \chi_{\mathrm{rf}}' &= \frac{1}{2} (\gamma M)\,\tau_2 \frac{\tau_2 (\gamma H - \omega)}{1 + \tau_2{}^2 (\gamma H - \omega)^2 + \gamma^2 H_{\mathrm{rf}}{}^2 \tau_1 \tau_2}, \\ \chi_{\mathrm{rf}}'' &= \frac{1}{2} (\gamma M)\,\tau_2 \frac{1}{1 + \tau_2{}^2 (\gamma H - \omega)^2 + \gamma^2 H_{\mathrm{rf}}{}^2 \tau_1 \tau_2}. \end{aligned} \tag{1.84}$$

Note that the factor 1/2 from (1.78) reappears because only the right-hand circular polarization component (RHCP) of the linearly polarized signal H_{rf} enters the susceptibility resonance expressions of (1.84). In Fig. 1.16, rough sketches of χ_{rf}' and χ_{rf}'' about the resonance frequency illustrate two important features: (1) the Lorentzian tails of the derivative (dispersion) curve representing phase shift extend further than those of the absorption curve that represents loss, and (2) the corresponding linewidth is defined at the *half-power* point of the absorption curve, which is approximately equal to the separation of the two peaks of the dispersion curve. The design parameter in the selection of a gyromagnetic property for a microwave application is critically dependent on the effective intrinsic linewidth. For the discussions of ferromagnetic resonance in Chap. 6 where both circular polarization modes are important, the factor of 1/2 is not included.

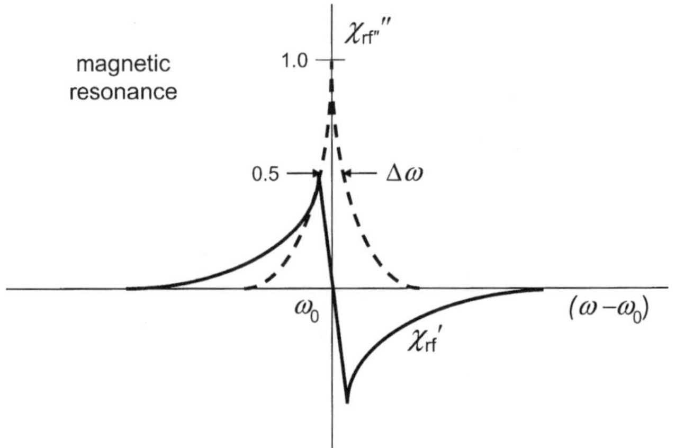

Fig. 1.16 Conceptual illustration of the real χ_{rf}' and imaginary χ_{rf}'' components of the complex rf susceptibility. Note the definition of linewidth at the 0.5 point and the longer extension of the dispersion tails, which are important for high efficiency in signal transmission

If the $\boldsymbol{H} \cdot (\mathrm{d}\boldsymbol{M}/\mathrm{d}t)$ is integrated over one cycle an expression for the power absorbed form the out-of-phase component is obtained:

$$A = (\omega/2\pi) \int_0^{2\pi/\omega} \boldsymbol{H} \cdot (\mathrm{d}\boldsymbol{M}/\mathrm{d}t)\,\mathrm{d}t = 2\omega\chi_{rf}'' H_{rf}^2$$

$$= \chi_m H_{rf}^2 \frac{(\gamma H)^2 \Delta\omega}{(\gamma H - \omega)^2 + (\Delta\omega)^2 + (\gamma H_{rf})^2 \tau_1 \Delta\omega}, \qquad (1.85)$$

which simplifies to

$$A \cong \frac{\gamma M \Delta\omega}{(\gamma H - \omega)^2 + (\Delta\omega)^2} H_{rf}{}^2 \qquad (1.86)$$

for small H_{rf}. For large values of H_{rf} at resonance with $\omega = \gamma H$, we obtain a limiting relation that exposes the dependence on the spin–lattice relaxation time and provides the "saturation method" for its measurement [24]:

$$A = \chi_m H^2 \tau_1{}^{-1} = M H \tau_1{}^{-1}. \qquad (1.87)$$

In these relations, $\Delta\omega$ or $\gamma\Delta H \left(= \tau_2{}^{-1}\right)$ is the half-linewidth at half-maximum. Note that the previous condition for "fast passage" can now be expressed as $H_{rf} \gg \Delta H$. When $\tau_x \leq \tau_y$ due to fast spin–lattice relaxation, τ_2 should be replaced by $\tau_{x,y}$ from (1.76) to account for τ_1 influence on ΔH broadening.

1.4.4 Resonance Line Shapes

Although the limitations of the scope of this text will not permit an extensive description of the intricacies of magnetic resonance, some general statements about line shapes and their causes will prove useful in later topics. Line broadening mechanisms may be divided into homogeneous (from broadening processes inherent in the line itself) and inhomogeneous (from overlapping resonances offset in frequency by an irregular arrangement of anisotropic resonance centers). For the basic resonance phenomenon under discussion, only the homogeneous case applies. The effect of inhomogeneities will be introduced in a review of the resonance behavior of porous ferrimagnetic ceramics.

In general, there are two basic lineshape functions that are used as reference extremes: Lorentzian and Gaussian. In the sketches of Fig. 1.17, the Lorentzian shape is identified by a sharper center and sweeping tails; the Gaussian has the opposite characteristics. For analytical convenience, the Lorentzian curve is characteristic of homogeneous broadening and is generally applicable for widely separated or uniformly spaced moments. Paramagnetic resonance line broadening due to dipolar interactions was examined by Van Vleck [25], and the dependence of the shape function on the concentration of individual magnetic moments was studied by Kittel and Abrahams [26], who reported that increased concentrations of like paramagnetic ions tend to drive the lineshape toward a Gaussian function, suggesting an inhomogeneous distribution. When the concentration is sufficiently high to allow exchange forces to narrow the line as mentioned earlier, the lineshape tends to be Lorentzian. This latter observation is important in the analysis of microwave and optical effects in ferrimagnets to be examined in later chapters.

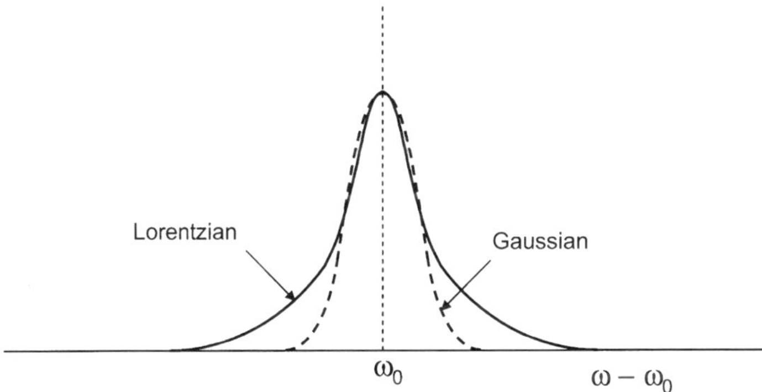

Fig. 1.17 Normalized Lorentzian and Gaussian resonance lines emphasizing their contrasting shapes. Homogeneous systems feature Lorentzian contours with sharper peaks and longer tails of a single resonance frequency; inhomogeneous systems have the opposite characteristics due to multiple resonance frequencies and are often described by a Gaussian function

Appendix 1A Spin–Lattice Contribution to Linewidth

The basic phenomenological analysis of magnetic relaxation considered the phase coherence of M_x and M_y, but neglected the decrease in M_z amplitude, which arises from the canting angle θ between the vector $\boldsymbol{M}$ and the magnetic field direction (z axis). The standard relations in spherical polar coordinates are

$$\begin{aligned} M_z &= M \cos\theta, \\ M_x &= M \sin\theta \cos\varphi = \left(M^2 - M_z{}^2\right)^{1/2} \cos\varphi, \\ M_y &= M \sin\theta \sin\varphi = \left(M^2 - M_z^2\right)^{1/2} \sin\varphi. \end{aligned} \tag{1.88}$$

By definition, the spin–lattice damping time constant τ_1 of longitudinal relaxation follows from

$$\frac{\mathrm{d}M_z}{\mathrm{d}t} = \frac{M - M_z}{\tau_1}. \tag{1.89}$$

If the x component is used as the example of transverse relaxation

$$\frac{\mathrm{d}M_x}{\mathrm{d}t} = \left(M^2 - M_z{}^2\right)^{1/2} \frac{\mathrm{d}(\cos\varphi)}{\mathrm{d}t} + \cos\varphi \frac{\mathrm{d}\left(M^2 - M_z{}^2\right)^{1/2}}{\mathrm{d}t}. \tag{1.90}$$

The first term represents the phase decoherence of the precessing spins and the second is the effect of the amplitude relaxation. Therefore, (1.90) can be expressed as

$$\frac{\mathrm{d}M_x}{\mathrm{d}t} = -\frac{M_x}{\tau_2} - \frac{M_x}{(M^2 - M_z{}^2)^{1/2}} \times \frac{M_z}{(M^2 - M_z{}^2)^{1/2}} \frac{\mathrm{d}M_z}{\mathrm{d}t}. \tag{1.91}$$

After substituting from (1.89) for $\mathrm{d}M_z/\mathrm{d}t$ and simplifying,

$$\begin{aligned} \frac{\mathrm{d}M_x}{\mathrm{d}t} &= -\frac{M_x}{\tau_2} - \frac{M_x}{\tau_1}\left(\frac{M_z}{M - M_z}\right) \\ &= -M_x\left(\tau_2^{-1} + f(\theta)\,\tau_1^{-1}\right), \end{aligned} \tag{1.92}$$

where

$$f(\theta) = \frac{|\cos\theta|}{1 + |\cos\theta|}.$$

Note that $0 \leq f(\theta) \leq 1/2$, for which the extremes represent the case of a wide-open spin cone, sometimes used to measure τ_2 in spin-echo experiments conducted with $\theta = \pi/2$, and the $\theta = 0$ case for the complete spin alignment of ferro- or ferrimagnetism. In general, where $\tau_1 \leq \tau_2$, the linewidth would then be related by

$$\Delta H \sim \tau_2^{-1} + f(\theta)\,\tau_1^{-1}, \tag{1.93}$$

so that spin–lattice relaxation broadening could become dominant, as in the cases where rare-earth or other fast-relaxing ions are present in large enough concentrations.

References

1. R.M. Bozorth, *Ferromagnetism*, (D. Van Nostrand, New York, 1951)
2. J.A. Osborn, *Phys. Rev.* **67**, 351 (1945)
3. W.E. Henry, *Phys. Rev.* **88**, 559 (1952)
4. N. Cusack, *The Electrical and Magnetic Properties of Solids*, (Longmans, Green and Co., New York, 1958)
5. A.H. Morrish, *The Physical Principles of Magnetism*, (Wiley, New York, 1965), Chapter 2
6. G.F. Dionne, *J. Appl. Phys.* **41**, 4874 (1970)
7. W. Heisenberg, *Z. Phys.* **49**, 619 (1928)
8. L.D. Landau and E.D. Lifshitz, *Quantum Mechanics*, (Addison-Wesley, Reading, MA, 1958), Chapter IX
9. W. Heitler and F. London, *Z. Physik* **44**, 455 (1927)
10. K. Yosida, *Theory of Magnetism*, (Springer-Verlag, New York, 1996)
11. R.M. White, *Quantum Theory of Magnetism*, (Springer-Verlag, New York, 1985)
12. E.C. Stoner, *Proc. Leeds Phil. Soc.* **2**, 391 (1933)
13. J.C. Slater, *Phys. Rev.* **49**, 537 (1936)
14. A.H. Morrish, *The Physical Principles of Magnetism*, (Wiley, New York, 1965)
15. B. Lax and K.J. Button, *Microwave Ferrites and Ferrimagnetics*, (McGraw Hill, New York, 1962)
16. F. Bloch, *Phys. Rev.* **70**, 460 (1946)
17. N. Bloembergen, *Phys. Rev.* **78**, 572 (1950); also N. Bloembergen, *Proc. IRE* **44**, 1259 (1956)
18. A.H. Morrish, *The Physical Principles of Magnetism*, (Wiley, New York, 1965), p. 198
19. E.L. Hahn, *Phys. Rev.* **80**, 580 (1950)
20. J.C. Dyment, *Can J. Phys.* **44**, 637 (1966)
21. G.E. Pake, *Paramagnetic Resonance*, (W.A. Benjamin, New York, 1962)
22. A. Abragam, *The Principles of Nuclear Magnetism*, (Clarendon Press, Oxford, 1962)
23. C.P. Slichter, *Principles of Magnetic Resonance*, (Springer, New York, 1996)
24. A.L. Kipling, P.W. Smith, J. Vanier, and G.A. Woonton, *Can. J. Phys.* **39**, 1859 (1961)
25. J.H. Van Vleck, *Phys. Rev.* **74**, 1168 (1948)
26. C. Kittel and E. Abrahams, *Phys. Rev.* **90**, 238 (1953)

Chapter 2
Magnetic Ions in Oxides

To establish the pattern of this book, a logical first step is to review the Periodic table of chemical elements, to identify the transition groups within it, and to explain the local influences of the chemical bonding environment in which the various ions reside in a crystal lattice of an oxide. By the terminology of transition groups is meant those elements for which the inner shells remain unfilled while electrons occupying outer shells participate in chemical bonding. Consequently, the electrons of the unfilled inner shells are responsible for a variety of magnetic properties because of the magnetic moments carried by their unpaired spins.

From the theory of atomic spectra, the angular momentum of the electron spin is coupled to the angular momentum that is derived from the motion of the electron in its orbit about the nucleus, that is, the orbital angular momentum. The strength of spin–orbit coupling is a key factor in determining the extent to which the orbital moment contributes to the magnetic properties and conversely, to what extent the spins interact with the lattice. When placed in a crystal lattice, the magnetic ion is subjected to two effective fields that separately influence the spin and orbital momenta – the crystal electric field of the lattice site that captures or "quenches" the orbital moment by a Stark effect, and the exchange interaction that orders the spin into a collective ferromagnetic or antiferromagnetic state. The origin of the crystal and exchange fields is reviewed first. The role of spin–orbit coupling is examined later in relation to magnetocrystalline anisotropy and magnetostriction.

2.1 The Transition Metals

In the introduction to orbital angular momentum in Sect. 1.2, each electron has an orbital quantum state encoded by n and L. The n label identifies the basic Coulomb state energies of the Bohr atomic model and L specifies the particular orbital angular momentum state. For an atom without perturbation, including that from spin–orbit coupling, the array of orbital angular momentum states are initially quantum mechanically degenerate (of same energy), but are identified according to the scheme described in Sect. 1.2.1, with $2L + 1$ states for the sequence labeled $s\ (L = 0)$, $p\ (L = 1)$, $d\ (L = 2)$, $f\ (L = 3)$, $g\ (L = 4)$, etc. Because there are two spin states

G.F. Dionne, *Magnetic Oxides*, DOI 10.1007/978-1-4419-0054-8_2,

for each orbital state, in filled orbital shells there are correspondingly 2 s electrons, 6 p electrons, 10 d electrons, 14 f electrons, and 17 g electrons. As a general rule, all inner orbital shells are complete, that is, the only unfilled orbital states are in the outermost shell. Certain important exceptions exist, however, and it is from these metal elements that most magnetic effects originate. They are called the *transition-metal* groups.

2.1.1 The Periodic Table

Figure 2.1 is modeled from the Mendelief Periodic table that has been tailored to highlight the transition element groups. The definition of a transition group is one in which inner electron shells are unfilled while outer shells are occupied by electrons. Characteristically, in the building of the table there is an orderly progression of state occupations from hydrogen with a single 1s electron all the way out to the first transition series that begins with a single electron in the ten-state 3d shell of scandium Sc (atomic number 21) outside a completed argon (Ar) core, but inside a completed two-state 4s shell, as sketched in Fig. 2.2a. The series ends at number 29 with copper (Cu) when the ten electrons fill the 3d shell (designated 3d^{10}). This series is commonly referred to as the "iron group" and will be the central feature of this book.

In a similar manner, the second series is built on a krypton core (Kr) with a filled outer 4p^6. It begins at number 39 with yttrium (Y) and ends with number 47 silver (Ag), but notably features a completely unfilled 4f shell, shown in Fig. 2.2b. Next in importance to the 3d shell, the "rare-earth" or lanthanide transition series

$d^1 \rightarrow d^{10}$

$f^2 \rightarrow f^{14}$

1 H																	2 He
3 Li	4 Be											5 B	6 C	7 N	8 O	9 F	10 Ne
11 Na	12 Mg	1	2	3	4	5	6	7	8	9	10	13 Al	14 Si	15 P	16 S	17 Cl	18 Ar
19 K	20 Ca	21 Sc	22 Ti	23 V	24 Cr	25 Mn	26 Fe	27 Co	28 Ni	29 Cu	30 Zn	31 Ga	32 Ge	33 As	34 Se	35 Br	36 Kr
37 Rb	38 Sr	39 Y	40 Zr	41 Nb	42 Mo	43 Tc	44 Ru	45 Rh	46 Pd	47 Ag	48 Cd	49 In	50 Sn	51 Sb	52 Te	53 I	54 Xe
55 Cs	56 Ba	57 La	58 Ce	59 Pr	60 Nd	61 Pm	62 Sm	63 Eu	64 Gd	65 Tb	66 Dy	67 Ho	68 Er	69 Tm	70 Yb		
		71 Lu	72 Hf	73 Ta	74 W	75 Re	76 Os	77 Ir	78 Pt	79 Au	80 Hg	81 Tl	82 Pb	83 Bi	84 Po	85 At	86 Rn
87 Fr	88 Ra	89 Ac	90 Th	91 Pa	92 U	93 Np	94 Pu	95 Am	96 Cm	97 Bk	98 Cf	99 Es	100 Fm	101 Md	102 No		
		103 Lr	104 Rf	105 Db	106 Sg	107 Bh	108 Hs	109 Mt	110	111	112	113	114	115	116	117	118

Fig. 2.1 Periodic table configured to highlight the transition element groups

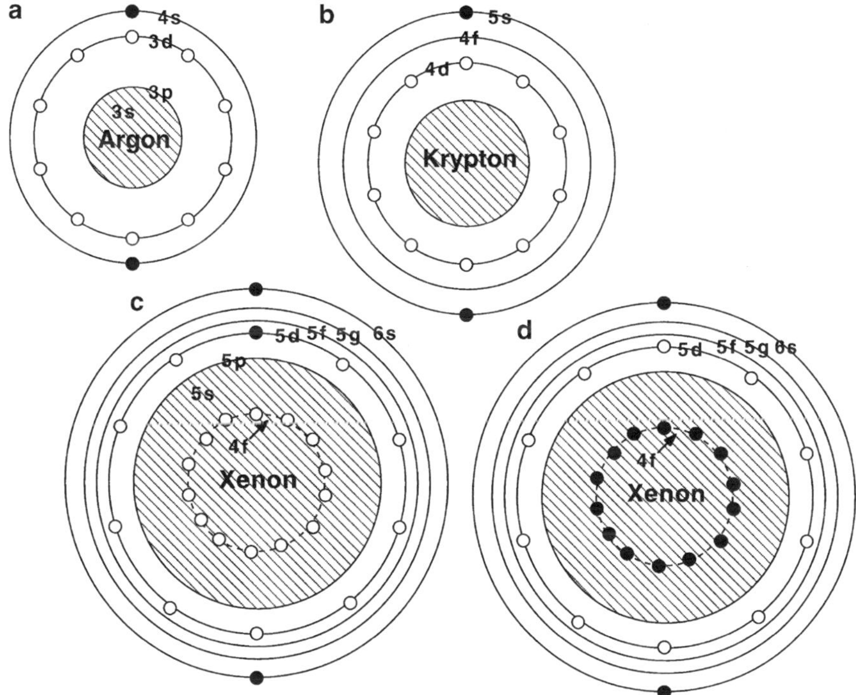

Fig. 2.2 Electron shell diagrams of the transition elements: (**a**) $3d^n$, (**b**) $4d^n$, (**c**) $4f^n$, (**d**) $5d^n$

defined by the filling of the $4f$ shell begins with number 57 lanthanum (La) and continues to number 70 ytterbium (Yb) with a complete $4f^{14}$. In Fig.2.2c, it is shown schematically that the rare-earth series is built on an xenon (Xe) core, even though the $4f$ shell is shown pictorially inside the Xe $5s^2$ and $5p^6$ filled shells (not shown). Beyond the Xe core there are also empty $5d$, $5f$, and $5g$ shells before the outermost $6s^2$ shell completes the electronic configuration of these elements. The transition series shown in Fig. 2.2d begins with number 71 lutetium (Lu) and ends with number 79 gold (Au) and is fashioned from the completion of the $5d$ shell outside of the Xe core (now with the $4f$ shell filled), still inside the $6s^2$ shell. A final transition series may be considered as beginning with number 89 actinium (Ac) and ending with number 101 mendelivium (Md), formed on a radon (Rn), but with an incomplete $5f$ shell inside the radon core, analogous to the configurations of the rare earth series. This group comprises many synthetic elements and is not of any practical importance in magnetism.

Among the four groups of interest, only certain members become interesting for their magnetic properties in the ionic states, and only the iron group and rare-earth group have so far made a significant impact on the practical properties of oxide compounds.

2.1.2 Iron Group $3d^n$ Ions

In the periodic table of Fig. 2.1, the elements with ions that produce significant magnetic effects in combination with oxygen are distinguished by shading. Their common feature is an incomplete inner d^n of f^n shell. In general, the atoms surrender their outer s electrons to form ionic bonds with atoms that accept them to complete their own unfilled s and p shells. In an analogy to electron tubes with a cathode (emitter) and anode (plate), the metal atom that donates electrons is called a cation, while the "anode" atom becomes an anion. Because oxygen normally has an anion valence of 2−, the cations formed from the transition metals usually have valence charges of at least 2+, particularly in the complex oxide compounds that produce the magnetic properties of practical interest.

One immediate observation is that once the outer $4s$ electrons are stripped from the iron group elements, the partially filled $3d^n$ shell is exposed to the molecular environment, which is a crystal lattice of specific symmetry comprising electric and magnetic fields of its own. Figure 2.3 is a diagram of the electron occupancy of the ten d states (two for each of the five orbitals in compliance with the Pauli exclusion principle), indicating the formation of the multi-electron L, S, and J quantum states for each ground term based on Hund's rules. The relevant parameter data for the iron group are summarized in Table 2.1.

The ions of the lower third of the group $3d^1$, $3d^2$, and $3d^3$ are generally paramagnetic and exhibit little collective properties, usually only causing perturbing

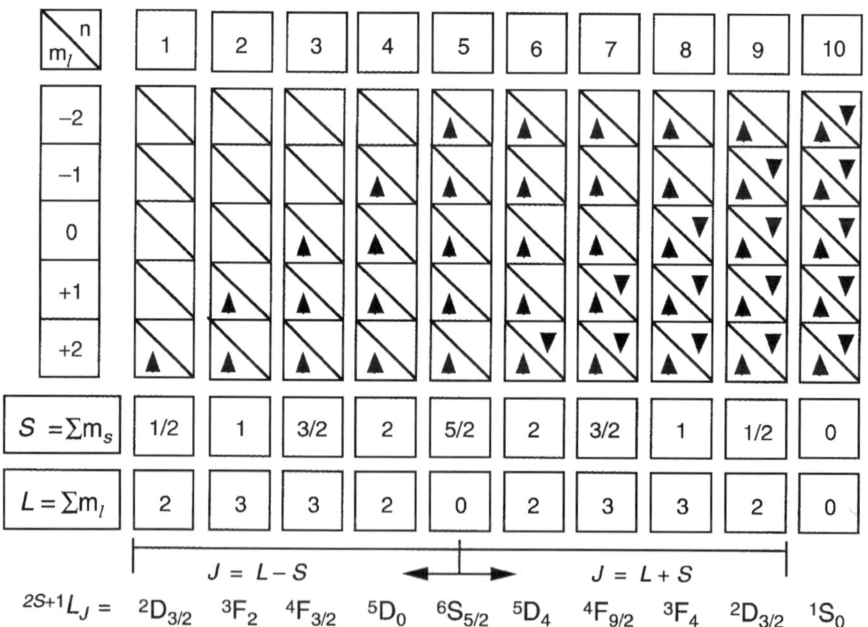

Fig. 2.3 Application of Hund's rule in the formation of angular momentum ground terms in the $3d^n$ shell. Terms are either D ($L = 2$), F ($L = 3$), or in the $3d^5$ case, S ($L = 0$)

Table 2.1 Parameters of the iron group $3d^n$ ion series in oxygen sites

		Ground state				Radius[a]	O_n^{2-}	
Electrons	Ion	Term	L	S	J	(Å)	n	Remarks
$3d^0$	Sc^{3+}	$^1S_0$	0	0	0	0.73	6	Diamagnetic ion
$3d^1$	Ti^{3+}	$^2D_{3/2}$	2	1/2	3/2	0.67	6	Metallic with Ti^{4+}, forms blue sapphire in Al_2O_3
$3d^2$	V^{3+} Ti^{2+}	$^3F_2$	3	1	2	0.64 0.86	6 6	Large ion
$3d^3$	Cr^{3+} Mn^{4+}	$^4F_{3/2}$	3	3/2	3/2	0.61 0.54	6 6	Forms red ruby in Al_2O_3
$3d^4$	Cr^{2+} Mn^{3+}	$^5D_0$	2	2	0	0.82 0.65	6 6	J–T ion, magnetostrictive. metallic with Mn^{4+}
$3d^5$	Mn^{2+} Fe^{3+}	$^6S_{5/2}$	0	5/2	5/2	0.82 0.64, 49	6 6, 4	S-state ion, low-spin $S = 1/2$
$3d^6$	Fe^{2+} Co^{3+}	$^5D_4$	2	2	4	0.77 0.61	6 6	Magnetostrictive, low-spin $S = 0$
$3d^7$	Co^{2+} Ni^{3+}	$^4F_{9/2}$	3	3/2	9/2	0.73 0.60	6 6	Spin–orbit stabilized, highly anisotropic, fast-relaxing, low-spin $S = 1/2$
$3d^8$	Ni^{2+} Cu^{3+}	$^3F_4$	3	1	4	0.70 -	6 6	Magnetostrictive low-spin $S = 0$
$3d^9$	Cu^{2+}	$^2D_{5/2}$	2	1/2	5/2	0.73	6	J–T ion, magnetostrictive, metallic conductor with $S = 0\ (Cu^{3+})$[b]
$3d^{10}$	Cu^{1+}	$^1S_0$	0	0	0	0.96	6	large diamagnetic ion, metallic with Cu^{2+}

[a] Based on radius of divalent oxygen of 1.40 Å

[b] J.B. Goodenough, G. Demazeau, M. Pouchard, and P. Hagenmüller, *Solid State Chem.* **8**, 325 (1973)

effects on the remainder of the series. The magnetic moments of ions from $3d^4$ through $3d^9$ are capable of producing strong spontaneous magnetism when in sufficient densities to allow exchange coupling to the order magnetically as ferro-, ferri-, or antiferromagnets. In all cases where the ion is chemically bonded in an anion lattice, the combined orbital angular momentum $\boldsymbol{L}$ is uncoupled from the spin $\boldsymbol{S}$ by the electrostatic fields of the lattice, and the spin moments dominate the magnetic properties. Where this occurs J is no longer a meaningful quantum number. Consequently, the spectroscopic g factor is approximately equal to 2 for the spin angular momentum of a free electron in all but a couple of special situations.

2.1.3 Rare Earth $4f^n$ Ions

The second most important transition group from a magnetic standpoint has an unfilled $4f^n$ shell. These elements are commonly referred to as the rare earths or lanthanides because La is the first member of the series. The higher group with unfilled $5f^n$ shell, called the actinides also has magnetic properties. As mentioned earlier an important distinction between the $4f^n$ ions and those of the $3d^n$ iron group is the shielding of the $4f$ shell inside the $5s^2$ and $5p^6$ outer shells of the Xe core. In other words, the magnetically active electrons are buried inside the Xe core and are therefore shielded from electrostatic fields of the molecular environment. As a result, the ions act largely independent of one another, even in highly concentrated compounds, and are generally paramagnetic because the multiple lobes of the $4f$ orbital wavefunctions do not extend far enough for covalent bonding and magnetic exchange to be significant.

Inspection of Table 2.2 reveals that the J values of the ions are divided into a lower and upper group, based on whether L is larger or smaller than S. Following Hund's rule, for the lower half from Ce^{3+} to Eu^{3+}, $J = |L - S|$; the upper half from Gd^{3+} to Yb^{3+} features larger spin values, i.e., $J = |L + S|$. It will become evident later that the J values of the rare earths are the important angular momentum parameters because of strong spin–orbit coupling energies. Consequently, the g factors also are heavily dependent on J through the L contribution. As computed from (1.18), the g factors listed in Table 2.2 are exclusively less than 2. Rare-earth ions contribute a number of important effects that include the tailoring of magnetization vs. temperature behavior in magnetic garnets, the control of high-power properties of microwave ferrites, and the Faraday rotation of magnetic garnets and other compounds for optical applications.

2.1.4 $4d^n$ and $5d^n$ Ions

The unfilled $4d$ and $5d$ shells of the other two transition series have ions that resemble the $3d$ series in magnetic properties and can be used as alternatives for them in certain cases. Tetravalent ruthenium (Ru^{4+}) with a $4d^4$ configuration, for example,

Table 2.2 Parameters of the rare earth $4f^n$ ion series

		Ground state					Radius[a]	
Electrons	Ion	Term	L	S	J	g	(Å)	Remarks
$4f^0$	La^{3+}	$^1S_0$	0	0	0	–	1.18	Diamagnetic ions
	Ce^{4+}						0.97	
$4f^1$	Ce^{3+}	$^2F_{5/2}$	3	1/2	5/2	6/7	1.14	Strong magneto-optical properties
	Pr^{4+}						0.99	
$4f^2$	Pr^{3+}	$^3H_4$	5	1	4	4/5	1.14	Strong magneto-optical properties
$4f^3$	Nd^{3+}	$^4I_{9/2}$	6	2	4	8/11	1.12	Strong magneto-optical properties
$4f^4$	Pm^{3+}	$^5I_4$	6	2	4	3/5	0.98	Synthetic element
$4f^5$	Sm^{3+}	$^6H_{5/2}$	5	5/2	5/2	2/7	1.09	–
$4f^6$	Eu^{3+}	$^7F_0$	3	3	0	–	1.07	Diamagnetic ion
$4f^7$	Eu^{2+}	$^8S_{7/2}$	0	7/2	2	2	1.25	S-state ions
	Gd^{3+}						1.06	
$4f^8$	Tb^{3+}	$^7F_6$	3	3	6	3/2	1.04	Fast-relaxing ion
$4f^9$	Dy^{3+}	$^6H_{15/2}$	5	5/2	15/2	4/3	1.03	Fast-relaxing ion
$4f^{10}$	Ho^{3+}	$^5I_8$	6	2	8	5/4	1.02	Fast-relaxing ion
$4f^{11}$	Er^{3+}	$^4I_{15/2}$	6	3/2	15/2	6/5	1.00	Fast-relaxing ion
$4f^{12}$	Tm^{3+}	$^3H_6$	5	1	6	7/6	0.99	Fast-relaxing ion
$4f^{13}$	Yb^{3+}	$^2F_{7/2}$	3	1/2	7/2	8/7	0.98	Fast-relaxing ion
$4f^{14}$	Lu^{3+}[b]	$^1S_0$	0	0	0	–	0.97	Fast-relaxing ion

[a] Based on an oxygen coordination of 8

[b] Lutiteum is included here to complete the $4f$ shell. It also represents the beginning of the $5d^n$ series

has magnetoelastic properties similar to those of trivalent manganese $\left(Mn^{3+}\right)$ with a $3d^4$ occupancy. In general, these ions have not attracted much interest for their magnetic properties because the compounds formed from them do not exhibit strong spontaneous magnetic properties. Moreover, several of them, such as rhodium (Rh), palladium (Pd), osmium (Os), iridium (Ir), and platinum (Pt) are not available in sufficient abundance to be considered for low-cost applications. Intermetallic compounds of niobium (Nb), however, have found important uses as superconductors.

2.2 Oxygen Coordinations

Transition metal ions are chemically reactive and occur naturally bonded to anions of the seventh or eighth columns in compounds that are made up of distinct crystal structures. Such compounds can be viewed as comprising separate metal (cation) and anion lattices. In the discussions to follow, emphasis is placed on the immediate surroundings of the metal ions, specifically the disposition of the oxygen or ligand coordinations relative to the transition ion.

2.2.1 Crystal Systems and Point Groups

The subject of crystallographic symmetry is important to the study of magnetic oxides, and a brief review is essential for the understanding of the concepts that determine the properties of the transition ions in oxygen coordinations. For a thorough treatment of the subject the reader is referred to the more general literature and text books [1, 2]. In this text, the focus is on the crystal field and molecular orbital theories by which most of the magnetic-related properties are examined.

Table 2.3 lists the seven systems that comprise all the naturally occurring types of crystalline structures. Corresponding cell sketches are presented in Fig. 2.4. These systems are in turn composed of 32 crystallographic "point groups" or crystal classes, which describe the basic symmetry of crystallographic building blocks or unit cells. A point group, therefore, consists of a collection of symmetry operators that serve to define particular crystal structures. There are three such basic operations: rotation, whereby the structure repeats itself upon rotation around a particular direction or axis, for example, fourfold meaning that it repeats the image of its projection along the axis every 90°, reflection about a plane, and inversion, meaning that the crystal retains its appearance after undergoing reversal of each of the x,

Table 2.3 Crystal systems and their symmetry elements

System	Generic point group	Unit cell	Symmetry
Triclinic	–	$a \neq b \neq c$ $\alpha \neq \beta \neq \gamma \neq 90$	No axes, no planes
Monoclinic	–	$a \neq b \neq c$ $\alpha = \beta = 90 \neq \gamma$	One twofold axis or one plane
Orthorhombic	D_{2h}	$a \neq b \neq c$ $\alpha = \beta = \gamma = 90$	Three orthogonal 2-fold axes two planes intersecting a 2-fold axis
Tetragonal	D_{4h}	$a = b \neq c$ $\alpha = \beta = \gamma = 90$	One 4-fold axis or a 4-fold inversion axis
Rhombohedral (trigonal)	D_{3d}	$a = b = c$ $\alpha = \beta = \gamma \neq 90$	One 3-fold axis
Hexagonal	D_{6h}	Three axes a in $x - y$ plane at $\alpha = 120$; $c \neq a$	One 6-fold axis
Cubic (isometric)	O_h	$a = b = c$ $\alpha = \beta = \gamma = 90$	Four 3-fold axes

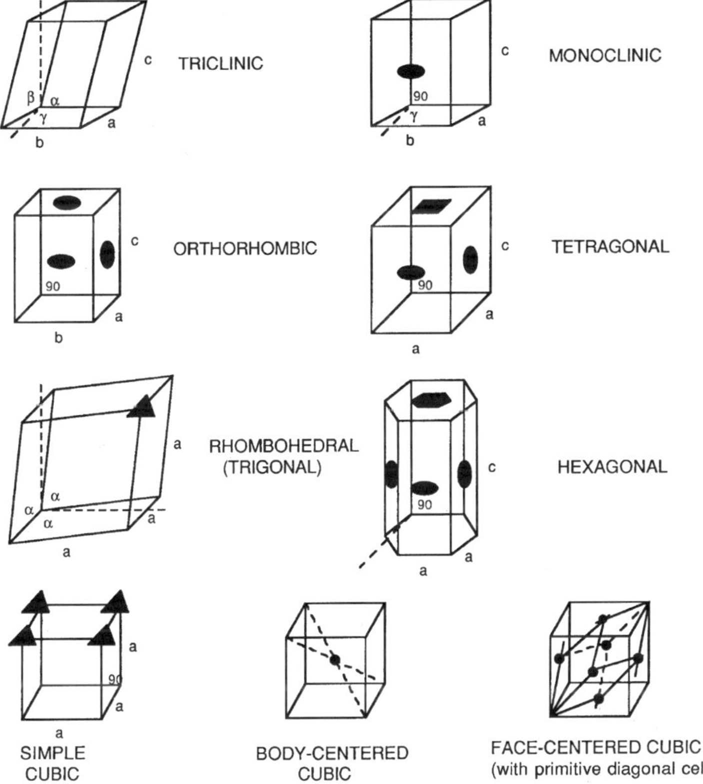

Fig. 2.4 Crystal system diagrams

y, and z coordinates. An even further refinement categorizes the point groups into 240 "space groups," but this level of detail will not be necessary for the purposes of this text.

There will be no attempt to explain in detail the nomenclature of the point groups beyond the occasional labeling of particular structures for identification purposes. For the scope of this text, the rotation operators will be sufficient to designate the symmetries that will be encountered.

2.2.2 *Cubic Symmetry*

In the magnetic oxides of interest in this text, the cations usually reside in lattice sites comprising oxygen arrangements referred to as coordinations. The basic unit of building block is generally of cubic (isometric) symmetry or one that is derived from it. In Fig. 2.5 the most common situations are sketched in relation to the orthogonal Cartesian axes and are of four types: tetrahedral, with four sides and four anions on alternate corners as shown; octahedral with eight sides and six anions on the

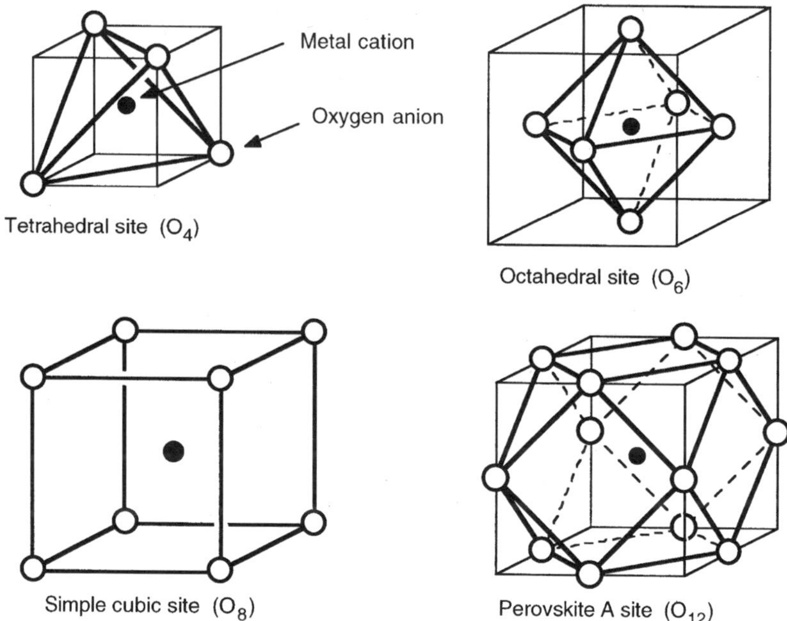

Fig. 2.5 Cation sites with ligand coordinations of cubic symmetry

corners of an octahedron formed with anions at the cube face centers; simple cubic with six sides and eight anions at the cube corners; and dodecahedral, with 12 sides and 14 anions located at the eight corners of a half-sized cube and six more at the face centers of the full cube. This latter structure does not actually occur in the magnetic oxides of interest here. In reality the "dodecahedral" site of the garnets is a 12-sided cell formed from a twisted cube with only eight corner anions. Each of these coordinations can have the required four threefold symmetry axes directed along body diagonals, referred to as the <111> family according to the convention of the Miller indices.[1] The corresponding indices for the three fourfold cubic axes (along x, y, and z directions) are the <100> family and for the face diagonals, it is the <110> family.

The most common oxygen coordination is octahedral, here labeled as O_6 for the six anions. The octahedral site is of paramount importance in spinels, garnets, and the various perovskite-related compounds. The tetrahedral coordination, designated O_4, is the alternate cation site in the spinels and garnets and is generally occupied by smaller metal ions. Because the smaller separation between anions leads to higher

[1] The Miller indices were developed to identify the various planes in a crystallographic lattice. The system is based on the values of the three intercepts of the plane with the x, y, and z axes expressed as the lowest integer values. The labeling convention for family of planes is {hkl} and an individual plane is (hkl). Alternatively, the normal axes to the planes are labeled < hkl > for the family and [hkl] for an individual axis.

mutual repulsive forces, the higher coordination numbers result in larger site volumes to minimize these bonding energies. For this reason the larger ions, such as the lanthanide rare earths, usually occupy O_{12} as in the cubic perovskites, or the dodecahedrally distorted O_8 sites in the garnets.

2.2.3 Lower Symmetries

In all but the simplest oxides, the cubic symmetry is usually reduced by lattice distortions that are often large enough to be considered as part of a phase transition to another point group even though the effect may be local and involving only an isolated cation complex. Even when the distortions are subtle, their effects on the magnetic properties can be significant. Since the issue of lattice distortions and symmetry changes will recur regularly throughout the balance of the text, it is appropriate that it be introduced in an orderly manner.

Departures from cubic symmetry vary from slight to catastrophic. However, there is no immediate need to discuss more than the few that are depicted in Fig. 2.6. The most convenient vehicle to examine these distortions is the octahedral site, which may undergo extensions or compressions along any of the principal axes of symmetry. In most cases, the <111> and <100> groups shown in Fig. 2.6a are the ones of concern. If the distortion is along a <111> axis as pictured in Fig. 2.6b, the symmetry is reduced form cubic O_h to trigonal or rhombohedral D_{3d} or C_{3v}, and the immediate environment of the cation has one threefold symmetry axis. Figure 2.6c, d indicates the effects of a tetragonal D_{4h} compression along the [001] axis and then the addition of a second distortion, this time an extension along the

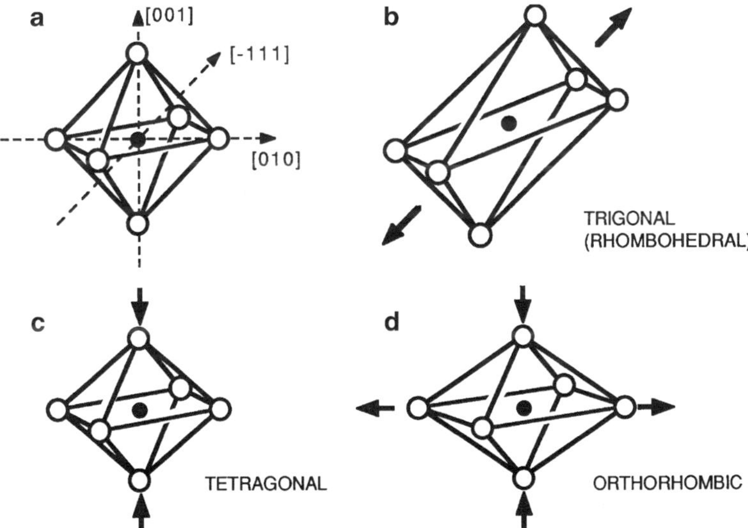

Fig. 2.6 Cubic cation sites with simple distortions

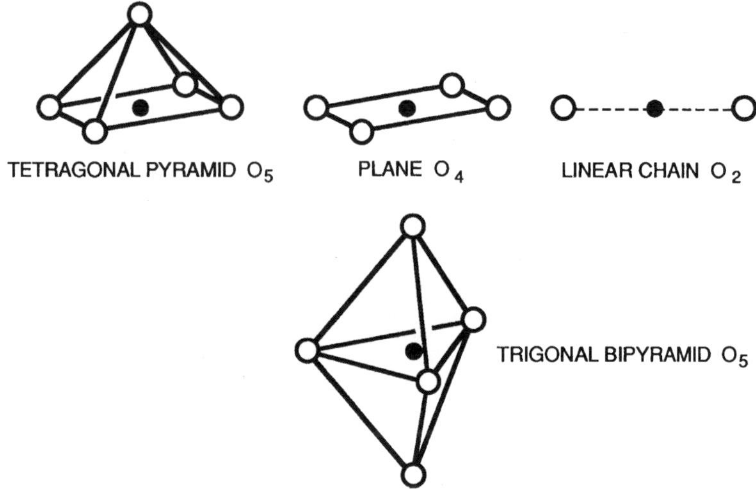

Fig. 2.7 Other ligand coordinations in oxides

[010] axis to create a lower orthorhombic D_{2h} symmetry. Combinations of trigonal and tetragonal distortions that can occur through spontaneous local distortions called Jahn–Teller effects can reduce the symmetry even further, as will be described in Chap. 5.

There are four other situations that are commonly encountered in magnetic oxides, shown in Fig. 2.7. A tetragonal pyramid, which occurs when one of the z-axis oxygen ions is missing, provides an O_5 coordination of C_4 symmetry (no reflection plane in this case); a square or rectangular planar configuration is formed by the removal of both dihedral anions and provides the ultimate tetragonal or orthorhombic D_{4h} or D_{2h} symmetry, and a one-dimensional chain that remains after two opposite planar oxygens are removed gives pure cylindrical symmetry. In recent years, these abbreviated octahedra have been detected in the superconducting layered perovskite-type compounds. A fourth cation site takes the form of a trigonal bipyramid and occurs in the hexagonal structure originally called "magnetoplumbite." This site occurs as only one of 13 in the type-M hexagonal ferrite compounds and is a strong contributor to the highly anisotropic properties of these important compounds.

2.3 Crystal Electric Fields

For an ionic lattice, the site of the cation can be approximated by an electrostatic trap formed by a "cage" of anion neighbors. If a "point charge" approximation is used for both the cation and the coordination of anions, the stabilization energy of the cation relative to free space is determined by the strength of the resulting ionic bond. When the electrons orbiting the cation nucleus are considered, the separate states of a degenerate orbital term are split by the perturbation in the manner of a Stark effect, which typically amount to about 10% of the lattice bonding energy, depending on the relative proximities of the various wavefunction lobes to the anion charges. This perturbation field from the anion coordination is called the *crystal field*. When the orbital wavefunctions of the anions are taken into account, the point charges are elevated to the status of ligands, and the eigenfunctions of the orbital states are hybridized to include both cation and anion contributions. The crystal-field model then forms the basis of *ligand field theory* that in turn serves as the foundation for the molecular-orbital concepts described in later sections.

Although the bonding is principally ionic in oxide compounds, the smaller covalent component is critically important for the electronic and magnetic properties of compounds with cations of a transition series. To analyze states of an ion for which the orbiting d electrons interact with the negative charges of the anion coordination, the effects of point-charge crystal fields on orbital angular momentum of the cation are examined first.

2.3.1 Angular Momentum States

To introduce the quantum mechanical effects of Stark splittings to the free ion orbital angular momentum states, it is necessary to review the formation of the multielectron orbital terms that are usually governed by Hund's rules, which state that the lowest energy multiplet term has the following:

1. The maximum possible combined spin value S, and
2. Within the maximum S manifold, the maximum combined L.

These rules originate from the Pauli exclusion principle and the quantum mechanical necessity for spins to align parallel when dispersed among the set of orthogonal orbital wavefunctions by mutual electrostatic repulsion. To visualize the "laddering" exercise, Fig. 2.3 illustrates schematically the situation among the states of a d^n series. The rows of stacked boxes represent an orbital angular momentum value m_l of operator l_z. Each box can hold two electrons, one for each up or down spin orientation as required by the Pauli exclusion principle. Beginning with the lower half of the series from d^1 to d^5, the electrons are added sequentially, obeying the spin polarization requirement to fill the first five up spin compartments and produce a half-filled set of orbitals with the maximum spin value of $S = 5/2$ when the d^5 limit is reached.

From an energy standpoint, the half-filled shell is most stable because each d electron occupies a separate orthogonal orbital state, and the destabilizing effect of the mutual repulsion is a minimum. This correlated spatial dispersal of the polarized spins beyond a random distribution reduces the screening of the nucleus and stabilizes the spins in proportion to their numbers (or their combined S). As the upper half of the shell begins to fill, the sixth electron must now share an orbit with its spin antiparallel to the net spin of the lower half in order to satisfy the Pauli principle. The natural consequence is an abrupt increase in energy for d^6 that is the direct result of the e^2/r_{ij} correlated repulsion. From d^6 to d^{10}, the process is repeated with a positive energy increment until the d shell is filled and the net spin returns to $S = 0$, at which point the d states can be considered part of the closed-shell ion core. The effect on energy from the filling of the d-orbital shell obeying Hund's rule can be seen in the ionization potentials (plotted negative relative to free space) in Fig. 2.8 as a function of n for different ionic valences across the transition series. Note that the departures from the baseline for random dispersal are consistent with the concept that the internal alignment (intraexchange) energy U_{ex} is proportional to the net spin value of the ion, and that the maximum destabilization between up and down spins is on the order of 2–3 eV.

More important for the immediate discussion are the combined values of L, which can be calculated by straightforward additions of m_l in each column. The resultant designations for the ground terms of each free ion ${}^{2S+1}L_j$ indicate that only three orbital degeneracies occur in the d-electron transition series: S, D, and F, but not P. Since the S represents an $L = 0$ state, d^5 automatically becomes a spin-only magnetic entity, which greatly simplifies analysis of magnetic properties, at least to first-order approximation.

When a positive magnetic ion is subjected to the electric field of the negative anion charges, the lobes of the electron orbital wavefunctions react to repulsive forces that either stabilize or destabilize the different orbitals depending on their relative proximity to the orbital lobes of the ligands, for example, $2p_{x,y,z}$ orbital functions of oxygen. In the broadest of contexts, the orbital angular momentum is captured by the crystal field, and in the process it is decoupled from the magnetic moments of the electron spins. A sketch of this quenching effect and its relation to spin–orbit coupling is presented in Fig. 2.9 for a uniaxial crystal field that separates the $\boldsymbol{L}$ and $\boldsymbol{S}$ vectors when it is not collinear with a magnetic field vector $\boldsymbol{H}$. Crystal-field theory is applied through quantum mechanical methods to determine the energy level structures of the resultant orbital states and their associated eigenfunctions.

2.3.2 Crystal Field Hamiltonian

Before the effects of the crystalline environment on the cation energy states are considered, the Hamiltonian of the free or unperturbed ion must be reviewed. A more complete discussion can be found in other texts [3–6] that have emanated from the treatise by Condon and Shortley [7]. If the terms involving neighboring nuclear charges are omitted, the Hamiltonian for a free ion of angular and spin momentum quantum numbers L and S

$$\mathcal{H} = [\mathcal{H}_{\text{Coul}} + \mathcal{H}_{\text{Hund}}] + \mathcal{H}_{LS}, \tag{2.1}$$

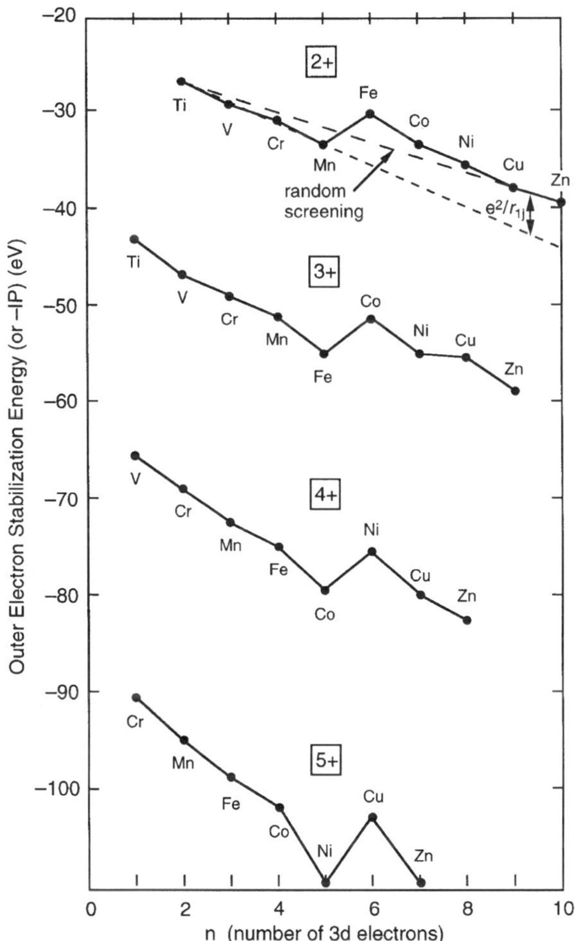

Fig. 2.8 Electronic energies of d-shell ions plotted as reverse ionization potentials. Note stabilizing effects of Hund's rule spin polarization and the destabilization by $U_{\text{ex}} = e^2/r_{ij}$ needed to establish Pauli spin pairing, as the upper half of shell fills. The electrons of the half-filled shell d^5 (Fe^{3+}) are the most stable, and the d^6 configuration is the least stable, which explains why Fe^{2+} ions frequently act as electron "donors" in charge transfer phenomena in mixed-valence situations with Fe^{3+} ions. Conversely, the unfilled upper half shell can be the source of holes, as in the case of d^4 (e.g., Mn^{2+}), which then act as "acceptors." Data are from C.E. Moore, NSRDS-NBS 34, Office of Standard Reference Data, National Bureau of Standards, Washington, DC

where the bracketed terms are the free-ion energies comprising

$$\mathcal{H}_{\text{Coul}} = -\frac{\hbar^2}{2m_{\text{e}}}\sum_i \nabla_i^2 - \sum_i \frac{Ze^2}{r_i}\,\left(\sim 10^5\ \text{cm}^{-1}\right),$$

the basic relation containing the Coulomb attractive energy between the Z electrons and the nuclear charge separated by r_i, and $\mathcal{H}_{\text{Hund}} = \sum_{i>j} e^2 / r_{ij}$, the energy of

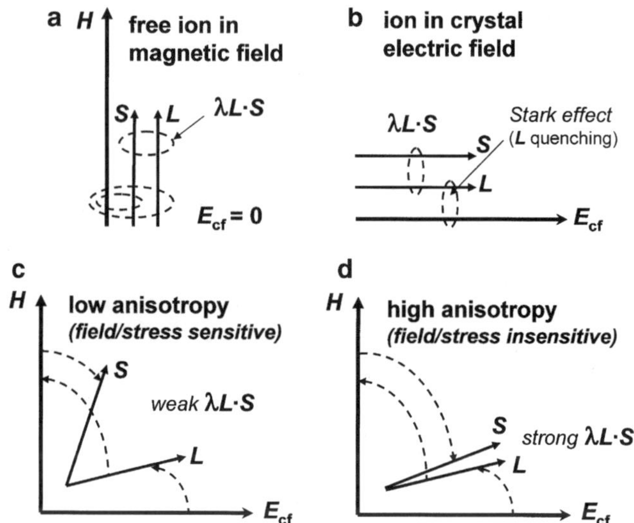

Fig. 2.9 Orbital angular momentum and spin–orbit coupling in a uniaxial crystal field: (**a**) magnetic field $\boldsymbol{H}$ acting on a free magnetic ion aligns the orbital $\boldsymbol{L}$ and spin $\boldsymbol{S}$ angular momentum vectors already made collinear through spin–orbit coupling $\lambda\ \boldsymbol{L}\cdot\boldsymbol{S}$, (**b**) independent of $\boldsymbol{H}$, lattice crystal-field $\boldsymbol{E}_{\mathrm{cf}}$ (presented as orthogonal to $\boldsymbol{H}$ axis) couples with $\boldsymbol{L}$, creating a Stark effect that partially "quenches" the orbital magnetic moment and causes an elastic distortion of the ion site, (**c**) magnetic polarization by $\boldsymbol{H}$ modifies and converts the lattice distortion from Stark effect into a magnetostrictive effect that creates high magnetic sensitivity to both field and external stress if spin–orbit coupling is weak. This is the case of "soft" magnetization, and (**d**) similar to (**c**) but with spin–orbit interaction strong enough to produce substantial magnetocrystalline anisotropy that reduces magnetic sensitivity and creates a condition for "hard" magnetization. The Stark distortion effect can also be the result of spontaneous local orbit–lattice (Jahn–Teller) stabilizations in which spin–orbit coupling mediates interactions between orbit–lattice stabilizations and a magnetically polarized spin system

mutual repulsion between the Z electrons orbiting the same nucleus and separated by r_{ij}. The latter quantity is a concern for ions with multiple unpaired electrons in an unfilled shell, particularly in the presence of strong crystal fields. In combination with the crystal field, the $\mathcal{H}_{\mathrm{Hund}}$ operator is responsible for the distribution of electrons among the various orbital states of the unfilled shell and therefore the ordering and separation of the orbital energy terms, which have been computed and thoroughly documented in the literature of atomic physics [7]. The eigenfunctions of this free-ion Hamiltonian are the familiar solutions of the Schrödinger equation in the form of exponentially decaying radially symmetric functions $R(r)$ combined with spherical harmonics $Y_{\ell}^{m_{\ell}}$ that shape the various wavefunction lobes, according to

$$\phi_{\ell}^{\mathrm{m}\ell} = R(r)\, Y_{\ell}^{\mathrm{m}\ell}. \tag{2.2}$$

The $R(r)$ function is in part a decaying exponential that is common to all orbitals within a main Bohr "n" shell. For a given transition series, it is treated as a scale constant. The various terms formed from the spherical harmonics tend to be ordered energetically according to Hund's ^{2S+1}L rule, which states that the lower energies favor first the highest multiplicities $2S+1$ and then the highest L within each $2S+1$ group. For the $3d^n$ series, the $3d^1$ case with $L = 2$ and $S = 1/2$, ^2D is the only orbital term because the influence of the $\mathcal{H}_{\text{Hund}}$ mutual repulsion energy is moot.

The solutions for the multiple electron cases, which are sorted out by the influence of $\mathcal{H}_{\text{Hund}}$, are listed in Table 2.4. An example of the important five-electron case $3d^5$ corresponding to Fe^{3+} is shown in Fig. 2.10, with the ^{6}S ground term and

Table 2.4 $3d^n$ (iron-group) free-ion energy terms (lowest 5)

$d^1\,(d^9)$	$d^2\,(d^8)$	$d^3\,(d^7)$	$d^4\,(d^6)$	$(d^5)^b$
–	^{1}S	^{2}F	^{3}D	^{4}F
–	^{1}G	^{2}G	^{3}F[a]	^{4}D
–	^{3}P	^{2}H	^{3}G	^{4}P
–	^{1}D	^{4}P	^{3}H	^{4}G
^{2}D	^{3}F	^{4}F	^{5}D	^{6}S

[a] There are two values for this term

[b] For this case in particular the order of the term energies does not follow the approximation of Hund's rule. This is characteristic of the higher energy terms in configurations with greater numbers of d electrons

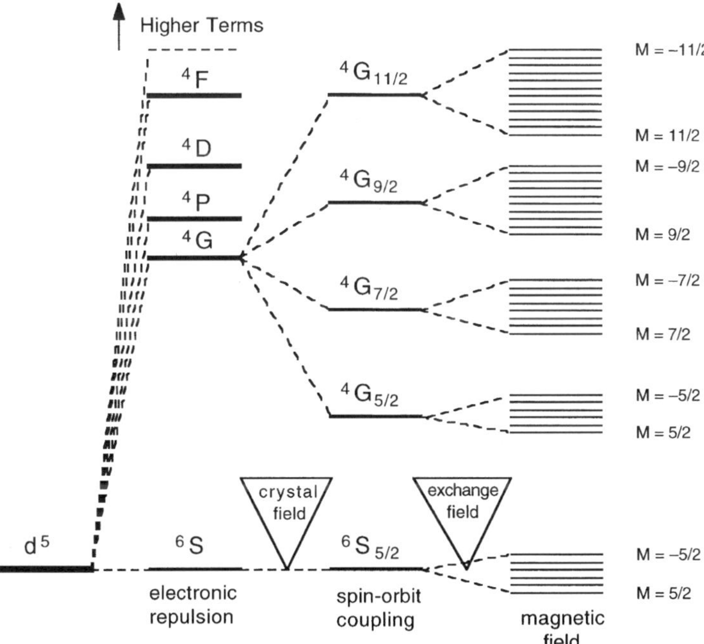

Fig. 2.10 Generic model of energy-level structure of five-d electron (d^5) configuration, typical of the Fe^{3+} ^{6}S-state ion

the first excited term ^{4}G with its subsequent multiplet splittings and eventual Zeeman splittings in a magnetic field. A physical picture of the ^{4}G state would have the $m_l = -2$ state electron shown in the occupancy diagram of Fig. 2.3 reversing its spin sense to form a pair in the $m_l = 2$ orbit and provide a resultant $L = 4$ with an $S = 3/2$. This configuration will be shown in Chap. 5 to provide the basis for magnetic anisotropy of iron in cubic crystal fields. The spin–orbit coupling energy is the third term in (2.1); $\mathcal{H}_{LS} = \sum_i \xi_i(r)\,\boldsymbol{l}_i \cdot \boldsymbol{s}_i$ ($\sim 10^2\,\text{cm}^{-1}$ for the iron group and $10^3\,\text{cm}^{-1}$ for the rare-earth group) is the perturbation that produces the multiplet structure observed in atomic spectra. Where the coupling functions $\xi_i(r)$ are sufficiently invariant among the states, they are usually combined into a semiempirical constant λ_{so}.

When the ion is situated in a crystal lattice, a crystal field term $\mathcal{H}_{\text{cf}}$ must be added to (2.1) to account for the interactions between the electron charges and the electric field of the crystal lattice environment:

$$\mathcal{H} = [\mathcal{H}_{\text{Coul}} + \mathcal{H}_{\text{Hund}}] + \mathcal{H}_{\text{cf}} + \mathcal{H}_{LS}. \tag{2.3}$$

In the simplest approximation, the source of $\mathcal{H}_{\text{cf}}$ is represented as point charges fixed at the locations of the particular ligands (anions) surrounding the cation. The purpose is to simulate a Stark effect coupling between the orbital angular momentum L and the crystal field that competes with the spin–orbit coupling between S and L as depicted in Fig. 2.9. The immediate effects are to make S as the principal source of the magnetic moment and remove J as a "good" quantum number. This action by the crystal field is called "quenching" of the orbital magnetism and results in $g \approx 2$ when it is dominant.

2.3.3 Hierarchy of Perturbations

As suggested by the order of terms in (2.3), $\mathcal{H}_{LS}$ is usually smaller than the lattice-related perturbation terms. At this point it becomes both convenient and instructive to define three crystal field regimes, defined loosely as weak and strong for the d^n series, and the shielded case of the $4f^n$ rare-earths, according to

$$\begin{aligned} &\mathcal{H}_{\text{Hund}} > \mathcal{H}_{\text{cf}} > \mathcal{H}_{LS} && (3d^{\text{n}}\ \text{series}),\\ &\mathcal{H}_{\text{cf}} \geq \mathcal{H}_{\text{Hund}} && (4d^{\text{n}}\ \text{and}\ 5d^{\text{n}}\ \text{series}),\\ &\mathcal{H}_{\text{cf}} < \mathcal{H}_{LS} && (4f^{\text{n}}\ \text{series}). \end{aligned} \tag{2.4}$$

The first of these is the one of principal interest because it applies to the most commonly encountered iron group $3d^n$ series. In this "weak field" case, the crystal field is smaller than the energy term separations due the $\mathcal{H}_{\text{Hund}}$ repulsive energy of (2.1) listed in Table 2.4. Consequently, the starting free-ion terms in a perturbation calculation are not mixed, only their degeneracies are split into fine structures by the crystal fields. Moreover, only the ground terms need to be considered for interpreting most magnetic effects. These operations and their implications on the magnetic properties will be the main topic of this text.

The "strong field" second case is also important, perhaps more for the high-energy transitions to be examined in a later discussion of magneto-optical properties. It is analytically more challenging than the "weak field" case because the $\mathcal{H}_{\rm cf}$ magnitudes are equal or greater than the free-ion term splittings set by $\mathcal{H}_{\rm Hund}$ and are therefore strong enough to mix the starting orbital terms prior to the removal of their degeneracies. As a result, the various possible electron distributions among the individual d orbital states, that is, the excited states, must be included as separate energy levels prior to application of the symmetry constraints imposed by the $\mathcal{H}_{\rm cf}$ operator. The strong field situation is sometimes referred to as the covalent limit because the strong $\mathcal{H}_{\rm cf}$ potential energy is produced by the overlap of the cation and anion orbital lobes. It is more common among the $4d^n$ and $5d^n$ transition series ions with larger ionic radii, but can also apply in the $3d^n$ series when the anion complex provides a locally stronger crystal field than that of the standard O^{2-} coordinations. In certain cases the crystal-field splitting can be large enough to cause a breakdown in Hund's maximum S rule by producing what is called a "low-spin" state that then leads to a change in the orbital ground term.

The third is the rare-earth $4f^n$ case, in which the Stark effect of the crystal field is not great enough to decouple L from S because of the shielding by the filled $5s^2$ and $5p^6$ shells. Here, $\lambda \boldsymbol{L} \cdot \boldsymbol{S}$ remains a constant of the motion and the $\mathcal{H}_{LS}$ operation creates the various multiplet terms now identified by ${}^{2S+1}L_j$, where L represents the orbital angular momentum of the orbital term designated by S, P, D, F, G, etc, with respective values of L being 0, 1, 2, 3, 4. For the rare earths, the total angular momentum $\boldsymbol{J}$ and its specific g value as defined by (1.29), rather than simply $\boldsymbol{S}$ with its fixed $g = 2$, determine the individual ion contributions to the magnetic properties.

2.3.4 *Weak-Field Solutions*

The subject of crystal field theory has been presented in many excellent texts [3–5]. Historically, the seminal work was carried out by Kramers [8], Van Vleck [9], and Schlapp and Penney [10], who treated the combined effects of the various lattice charges at a given cation site as the result of repulsive electrostatic fields from negative point charges that represent the effects of the anions or ligands. Because the potential $\mathcal{V}_{\rm cryst}$ at the cation site from the assembly of neighboring charges satisfies Laplace's equation $\nabla^2 \mathcal{V}_{\rm cf} = 0$, $\mathcal{H}_{\rm cf}$ $(=e\mathcal{V}_{\rm cf})$ may be expressed as an expansion of generalized Legendre polynomials, which take the same familiar form of spherical harmonics comprising (2.2). The problem of applying quantum perturbation theory to determine the electronic states of the cation in a particular crystal field is then reduced to the solving of a secular equation,

$$\mathcal{H}_{\mathrm{cf}}^{k} = \left|\mathcal{H}_{\mathrm{cf}ij} - E_{\mathrm{cf}}^{k}\delta_{ij}\right| = 0, \tag{2.5}$$

where $\mathcal{H}_{\mathrm{cf}ij} = \langle\varphi_i|\,\mathcal{H}_{\mathrm{cf}}\,|\varphi_j\rangle$, i and j are integers that run from 1 to k. E_{cf}^{k} are the k eigenvalue solutions of the matrix, each representing new energy states depending on the extent of the degeneracy removal. In this case it is the orbital angular momentum degeneracies of the spherical harmonic parts of the free ion wavefunctions of (2.2) that determine the order of the splittings.

To illustrate the method, the example of a single$d$ electron will be reviewed. The spherical harmonic functions (also expressed in the d_{m_l} abbreviations) for the ^{2}D term are given in Cartesian coordinates by

$$\begin{aligned} Y_2^{-2} &= d_{-2} = \sqrt{\frac{5}{4\pi}}\sqrt{\frac{3}{8}}\frac{(x-iy)^2}{r^2},\\ Y_2^{-1} &= d_{-1} = \sqrt{\frac{5}{4\pi}}\sqrt{\frac{3}{2}}\frac{z(x-iy)}{r^2},\\ Y_2^{0} &= d_{0} = \sqrt{\frac{5}{4\pi}}\sqrt{\frac{1}{4}}\frac{3z^2-r^2}{r^2},\\ Y_2^{1} &= d_{1} = -\sqrt{\frac{5}{4\pi}}\sqrt{\frac{3}{2}}\frac{z(x+iy)}{r^2},\\ Y_2^{2} &= d_{2} = \sqrt{\frac{5}{4\pi}}\sqrt{\frac{3}{8}}\frac{(x+iy)^2}{r^2}. \end{aligned} \tag{2.6}$$

For an octahedral (O_6) site, the crystal field potential energy is given by [11]

$$\mathcal{V}_{\mathrm{cf}}^{\mathrm{oct}} = D_4\left(x^4+y^4+z^4-\frac{3}{5}r^4\right) + \text{ higher order terms.} \tag{2.7}$$

Expressed in spherical harmonics, (2.7) becomes

$$\mathcal{V}_{\mathrm{cf}}^{\mathrm{oct}} = \left(\frac{7}{2}\right) D_4\left[Y_4^0+\sqrt{5/14}\,(Y_4^4+Y_4^{-4})\right]+\left(\frac{3}{4}\right) D_6\left[Y_6^0-\sqrt{7/2}\,(Y_6^6+Y_6^{-6})\right] \tag{2.8}$$

where $D_4 = (35/4)\,Ze^2/a^6$ and $D_6 = (21/2)\,Ze^2/a^7$ and a is the cation to anion distance. The additional spherical harmonics are expressed as [12]

$$\begin{aligned} Y_2^0 &= \sqrt{\frac{5}{4\pi}}\sqrt{\frac{1}{4}}\frac{3z^2-r^2}{r^2},\\ Y_4^0 &= \sqrt{\frac{9}{4\pi}}\sqrt{\frac{1}{64}}\frac{35z^4-30z^2r^2+3r^4}{r^4},\\ Y_4^4+Y_4^{-4} &= \sqrt{\frac{9}{4\pi}}\sqrt{\frac{70}{64}}\frac{x^4-6x^2y^2+y^4}{r^4}, \end{aligned} \tag{2.9}$$

$$Y_6^0 = \sqrt{\frac{1}{4\pi}}\sqrt{\frac{13}{256}}\frac{231z^6 - 315z^4r^2 + 105z^2r^4 - 5r^6}{r^6},$$

$$Y_6^6 + Y_6^{-6} = \sqrt{\frac{231}{4\pi}}\sqrt{\frac{27}{512}}\frac{x^6 - 15x^4y^2 + 15x^2y^4 - y^6}{r^6}.$$

For the tetrahedral (O_4) and cubic (O_8) coordinations only the Y_4 terms of (2.8) enter the calculation. Their crystal field energies scale according to

$$\begin{aligned} \mathcal{V}_{\text{cf}}^{\text{tet}} &= -(4/9)\,\mathcal{V}_{\text{cf}}^{\text{oct}}, \\ \mathcal{V}_{\text{cf}}^{\text{cub}} &= -(8/9)\,\mathcal{V}_{\text{cf}}^{\text{oct}}. \end{aligned} \tag{2.10}$$

If the radial part of (2.2) is folded into the scale factor of the matrix elements within the $n = 3$ shell, we may work with only the Y_l^m functions of (2.6) to set up a 5×5 matrix based on (2.5). Following this step, diagonalization with the aid of rotational symmetry considerations (group theory) or by solution of the secular equation will separate the 5×5 matrix into a 2×2 and a 3×3 matrix, according to

$$\begin{aligned} e_{\text{g}}^{\text{b}} &= \frac{1}{\sqrt{2}}(d_2 + d_{-2}), \\ e_{\text{g}}^{\text{a}} &= d_0. \\ t_{2\text{g}}^{+} &= d_1, \\ t_{2\text{g}}^{-} &= d_{-1}, \\ t_{2\text{g}}^{0} &= \frac{1}{\sqrt{2}}(d_2 - d_{-2}). \end{aligned} \tag{2.11}$$

The degree of crystal-field quenching of the orbital angular momentum about the [001] axis may be checked by the expectation values of the l_z operator from the appropriate inner products to show that, in addition to d_0, the e_{g}^{b} and $t_{2\text{g}}^{0}$ states have $m_l = 0$, while the remaining two $t_{2\text{g}}$ states retain $m_l = \pm 1$. If the function set of (2.11) are formed into the set of linear combinations in real form sketched in Fig. 2.11, they are expressed as

$$\begin{aligned} &e_{\text{g}}\left\{\begin{array}{l} d_{x^2-y^2} = \frac{\sqrt{3}}{\sqrt{2}}(d_2 + d_{-2}) = \frac{\sqrt{3}}{\sqrt{2}}\left(Y_2^2 + Y_2^{-2}\right) = \frac{\sqrt{3}}{2}\left(x^2 - y^2\right) \\ d_{z^2} = d_0 = Y_2^0 = \frac{1}{2}\left(3z^2 - r^2\right) \end{array}\right\}, \\ &t_{2\text{g}}\left\{\begin{array}{l} d_{xy} = \frac{\sqrt{3}}{i\sqrt{2}}(d_2 - d_{-2}) = \frac{\sqrt{3}}{i\sqrt{2}}\left(Y_2^2 - Y_2^{-2}\right) = \sqrt{3}xy \\ d_{xz} = -\frac{\sqrt{3}}{\sqrt{2}}(d_1 - d_{-1}) = -\frac{\sqrt{3}}{\sqrt{2}}\left(Y_2^1 - Y_2^{-1}\right) = \sqrt{3}xz \\ d_{yz} = -\frac{\sqrt{3}}{i\sqrt{2}}(d_1 + d_{-1}) = -\frac{\sqrt{3}}{i\sqrt{2}}\left(Y_2^1 + Y_2^{-1}\right) = \sqrt{3}yz \end{array}\right\}, \end{aligned} \tag{2.12}$$

where the radial factor $R(r)$ and other common factors have been dropped for convenience. The designations e_{g} and $t_{2\text{g}}$ are from group theory conventions for individual electron orbitals. (A more general nomenclature for these states that is

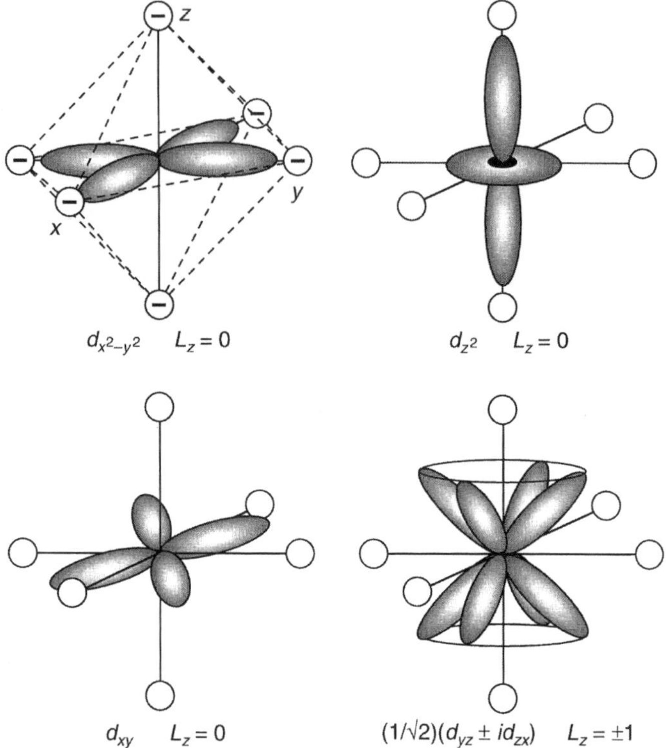

Fig. 2.11 Eigenfunctions lobes of the d-electron shell with orbital angular momentum quenched by a crystal field of tetragonal (D_{4h}) symmetry. Only the d_{xz}, d_{yz} states remain degenerate with nonzero $L_z = \pm 1$

used where multiple electrons are involved is A_{1g}, A_{2g} for singlets (also B_{1g}, B_{2g} in lower symmetry refinements), E_g for doublets, and T_{1g}, T_{2g} for triplets). With this set of wavefunctions, the matrix is diagonal, so that the eigenvalues of (2.5) become

$$E\left(e_g\right) = \left\{ \begin{array}{l} \varepsilon_0 + \left\langle d_{x^2-y^2}\right| \mathcal{V}_0^{\mathrm{T}} \left| d_{x^2-y^2}\right\rangle = \varepsilon_0 + \varepsilon_1 \\ \varepsilon_0 + \left\langle d_{z^2}\right| \mathcal{V}_0^{\mathrm{T}} \left| d_{z^2}\right\rangle = \varepsilon_0 + \varepsilon_1 \end{array} \right\} \tag{2.13a}$$

and

$$E\left(t_{2g}\right) = \left\{ \begin{array}{l} \varepsilon_0 + \left\langle d_{xy}\right| \mathcal{V}_0^{\mathrm{T}} \left| d_{xy}\right\rangle = \varepsilon_0 + \varepsilon_2 \\ \varepsilon_0 + \left\langle d_{xz}\right| \mathcal{V}_0^{\mathrm{T}} \left| d_{xz}\right\rangle = \varepsilon_0 + \varepsilon_2 \\ \varepsilon_0 + \left\langle d_{yz}\right| \mathcal{V}_0^{\mathrm{T}} \left| d_{yz}\right\rangle = \varepsilon_0 + \varepsilon_2 \end{array} \right\}, \tag{2.13b}$$

$$\text{where } \mathcal{V}_0^T = Y_4^0 + \sqrt{5/14}\left(Y_4^4 + Y_4^{-4}\right). \tag{2.14}$$

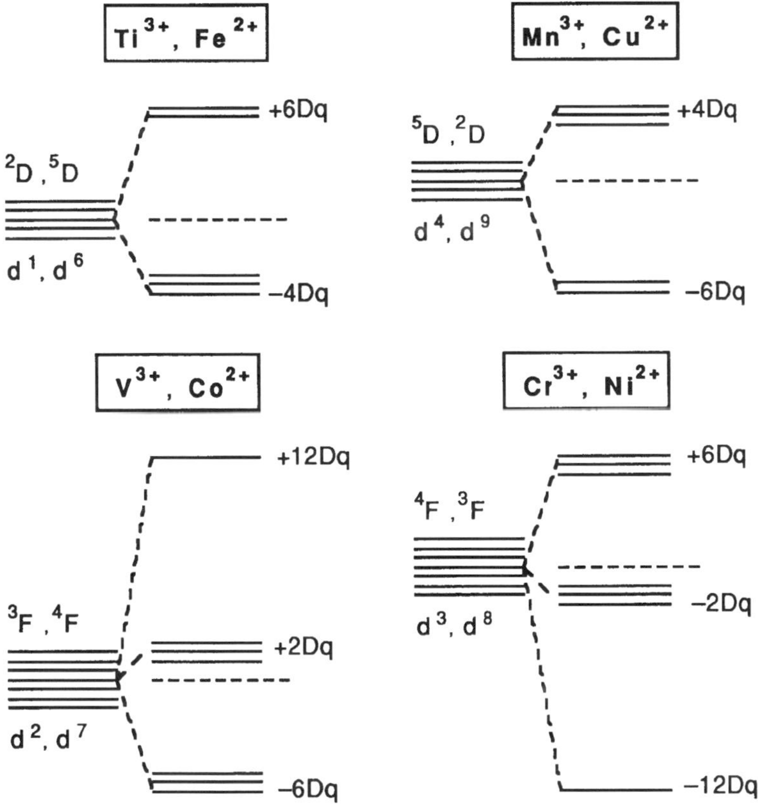

Fig. 2.12 Multiple electron crystal-field energy levels for D and F terms indicating correspondence between members of the lower and upper halves of the $3d^n$ transition series

which is derived from the octahedral potential energy of (2.8), without the normalizing factors. Equation (2.13) indicates that the fivefold degeneracy of the ^{2}D or ^{5}D term ($L = 2$) is split into a doublet and triplet. If the free ion ground term is ^{3}F or ^{4}F ($L = 3$), there are seven orbital states that are split into a singlet and two triplets as sketched in the four basic octahedral crystal-field diagrams of Fig. 2.12. The energy ordering of the levels, that is, upright or inverted, is a matter determined by the spin occupancy and Hund's rule and is examined in Sect. 2.4. The correspondence between d^n and d^{10-n} configurations can be reasoned by recognizing that d^n features electrons and d^{10-n} holes, which become distinguishable under the influence of the ligand charges. For this reason, level inversion occurs between the d^n and d^{10-n} ions, that is, equal numbers of unpaired electron spins vs. "hole" spins. A further convention is to label the overall splitting equal to $10Dq$, where $Dq > 0$. Then

$$\varepsilon_1 - \varepsilon_2 = 10Dq. \tag{2.15}$$

Since diagonal elements remains unchanged before and after the perturbation is applied, for the five ${}^2D$ orbital states

$$2(\varepsilon_0 + \varepsilon_1) + 3(\varepsilon_0 + \varepsilon_2) = 5\varepsilon_0, \tag{2.16}$$

and it follows that, if ε_0 is arbitrarily set to 0,

$$\begin{aligned} E\left(e_g\right) &= {}^2E_g = \varepsilon_1 = 6Dq, \\ E\left(t_{2g}\right) &= {}^2T_{2g} = \varepsilon_2 = -4Dq. \end{aligned} \tag{2.17}$$

The value of Dq is determined semiempirically, that is, by experiment, but an expression for it can be arrived at analytically. As explained in [13],

$$\begin{aligned} Dq &= \pm\left(\frac{2}{63}\right) \mathrm{D}_4 \left\langle r^4 \right\rangle \quad \text{for } D \text{ states} \\ &= \mp\left(\frac{2}{315}\right) \mathrm{D}_4 \left\langle r^4 \right\rangle \quad \text{for } F \text{ states,} \end{aligned} \tag{2.18}$$

where the second signs apply to ions of the lower half of the d^n series.

From the eigenfunctions of (2.12) that are plotted in Fig. 2.11, the ordering of the energy levels can be determined by inspection of the relative positions of the negatively charged lobes in relation to the negative ligand point charges. Note that the e_g orbitals are directed toward the ligands and therefore will assume the higher energy states.

From these orbital sketches, the existence of an unquenched l_z angular momentum component may be also discerned. A test for deciding whether an orbital momentum about an axis can still be present is whether the eigenstate can be transformed into another eigenstate within its degenerate manifold by a rotation about that axis. In this case of the [001] as axis of quantization, it can be seen by visual (or analytical) inspection that $d_{x^2-y^2}$ rotates into d_{xy} by a 45° rotation about z and that the same applies to d_{xz} and d_{yz}. Only the latter pairs are eigenstates in a cubic field, however, which means that the remaining three states have their l_z fully quenched, including the degenerate e_g orbitals. This latter condition will be shown to be significant in the discussion of the Jahn–Teller effect.

If the z axis of quantization is taken as the [111] direction, where threefold symmetry is prevalent, the appropriate basis vectors may be constructed from the pure set of (2.6) as

$$\begin{aligned} e_g &= \begin{cases} \frac{1}{\sqrt{3}} d_{-2} - \frac{\sqrt{2}}{\sqrt{3}} d_1 \\ \frac{1}{\sqrt{3}} d_2 + \frac{\sqrt{2}}{\sqrt{3}} d_{-1}, \end{cases} \\ t_{2g} &= \begin{cases} \frac{\sqrt{2}}{\sqrt{3}} d_{-2} + \frac{1}{\sqrt{3}} d_1 \\ \frac{\sqrt{2}}{\sqrt{3}} d_2 - \frac{1}{\sqrt{3}} d_{-1}. \\ d_0 \end{cases} \end{aligned} \tag{2.19}$$

As with the earlier eigenfunction set with the z axis along the [001] direction, a straightforward application of the l_z operator along the [111] direction will verify that d_0 and the two e_g states have zero angular momentum, while the remaining two t_{2g} states retain $m_l = \pm 1$.

By taking linear combinations of the basis vectors in real form from (2.12), we obtain for D_{3d} or C_{3v} with the z axis along a <111> direction

$$e = \begin{cases} \frac{1}{\sqrt{3}} d_{x^2-y^2} + \frac{\sqrt{2}}{\sqrt{3}} d_{xz} = e^+ \\ \frac{1}{\sqrt{3}} d_{xy} - \frac{\sqrt{2}}{\sqrt{3}} d_{yz} = e^-, \end{cases}$$

$$t_2 = \begin{cases} \frac{\sqrt{2}}{\sqrt{3}} d_{x^2-y^2} - \frac{1}{\sqrt{3}} d_{xz} = t_2^+ \\ \frac{\sqrt{2}}{\sqrt{3}} d_{xy} + \frac{1}{\sqrt{3}} d_{yz} = t_2^-. \\ d_{z^2} \qquad\qquad = t_2{}^0. \end{cases} \tag{2.20}$$

The functions of (2.20) are expressed in a coordinate system with z directed along the [111] direction of the cube body diagonal. If an analytical problem that involved a trigonal or rhombohedral perturbation along the [111] axis was to be solved with this combination of basis vectors expressed in the regular cubic coordinate system with x, y, and z transformed back into the x', y', z' coordinates set up coincident with the (001) family of axes, eigenfunctions for this purpose have been reported by Pryce and Runciman [14] and Dionne and Palm [15].

For this set the appropriate crystal field potential energy is given by a relation [16] analogous to (2.14).

$$\mathcal{V}_0^{\tau} = Y_4^0 + \sqrt{10/7}\left(Y_4^3 - Y_4^{-3}\right), \tag{2.21}$$

where
$$Y_4^3 - Y_4^{-3} = -\sqrt{\frac{9}{4\pi}}\sqrt{\frac{35}{4}}\frac{z\left(x^3 - 3xy^2\right)}{r^4}. \tag{2.22}$$

The second case to be discussed is the three electron ^4F term, which is of greater historical importance than the one electron case because it was the basis for the invention of the maser (microwave amplification by stimulated electron radiation). For this situation the orbital spherical harmonics are the Y_3 group [12],

$$Y_3^{-3} = \sqrt{\frac{7}{4\pi}}\sqrt{\frac{3}{8}}\frac{(x-iy)^3}{r^3},$$

$$Y_3^{-2} = \sqrt{\frac{7}{4\pi}}\sqrt{\frac{15}{8}}\frac{z(x-iy)^2}{r^3},$$

$$Y_3^{-1} = \sqrt{\frac{7}{4\pi}}\sqrt{\frac{3}{16}}\frac{(x-iy)\left(5z^2-r^2\right)}{r^3},$$

$$Y_3^{0} = \sqrt{\frac{7}{4\pi}}\sqrt{\frac{1}{4}}\frac{z\left(5z^2-r^2\right)}{r^3}, \tag{2.23}$$

$$Y_3^1 = -\sqrt{\frac{7}{4\pi}}\sqrt{\frac{3}{16}}\frac{(x+iy)\left(5z^2-r^2\right)}{r^3},$$
$$Y_3^2 = \sqrt{\frac{7}{4\pi}}\sqrt{\frac{15}{8}}\frac{z\,(x+iy)^2}{r^3},$$
$$Y_3^3 = -\sqrt{\frac{7}{4\pi}}\sqrt{\frac{5}{16}}\frac{(x+iy)^3}{r^3}.$$

For the F states, the Y_6 terms of (2.8) must be included in the calculation. Solutions of the resulting secular equation are a singlet ground state A_{2g} and two higher triplets T_{1g} and T_{2g} and those of the inverted case shown in Fig. 2.12, with corresponding eigenfunctions

$$\begin{aligned}
&T_{1g}\left\{\begin{array}{l}\sqrt{\frac{3}{8}}Y_3^1+\sqrt{\frac{5}{8}}Y_3^{-3}\\ \sqrt{\frac{3}{8}}Y_3^{-1}+\sqrt{\frac{5}{8}}Y_3^{3}\\ Y_3^0\end{array}\right\},\\
&T_{2g}\left\{\begin{array}{l}\sqrt{\frac{3}{8}}Y_3^1-\sqrt{\frac{5}{8}}Y_3^{-3}\\ \sqrt{\frac{3}{8}}Y_3^{-1}-\sqrt{\frac{5}{8}}Y_3^{3}\\ \frac{1}{\sqrt{2}}\left(Y_3^2+Y_3^{-2}\right)\end{array}\right\},\\
&A_{2g}\left\{\frac{1}{\sqrt{2}}\left(Y_3^2-Y_3^{-2}\right)\right\}.
\end{aligned}\tag{2.24}$$

Following the reasoning leading up to (2.14), for ${}^4F$ of d^3 and ${}^4F$ of d^7, these term energies are

$$\begin{aligned}
T_{1g} &= 6Dq,\\
T_{2g} &= -2Dq,\\
A_{2g} &= -12Dq.
\end{aligned}\tag{2.25}$$

Equation (2.24) is a convenient example of the meaning of orbital angular momentum quenching. The ground state is a linear combination of two spherical harmonics of the $l = 2$ manifold that yield the singlet A_{2g} term under the influence of the cubic crystal field. As such, it has the symmetry properties of a singlet s orbital and therefore would carry the properties of $l = 0$; its only contribution to the magnetic moment of the ion must come from the ion spin. Conversely, if the orbital levels are inverted in energy, the triplet T_{2g} becomes the ground state and the orbital angular momentum would have the characteristics of a degenerate p state with $l = 1$. In this case, l is reduced from the d state value of 2 down to 1 and the result is only partial quenching because not all of the ground state degeneracy has been lifted. This residual degeneracy is an important factor in the properties of certain transition $3d^n$, for example, Co^{2+} and a number of $4f^n$ rare-earth ions. Furthermore, where the crystal field splitting parameter Dq is small enough to allow the influence of the upper terms in a subsequent perturbation calculation, appropriate

Table 2.5 $3d^n$ states and energies in weak octahedral fields

d Electrons	Orbital ground state	States and energies in Dq[a]
d^1	^{2}D	$^2E_g\,(+6)$
		$^2T_{2g}\,(-4)$
d^2	^{3}F	$^3A_{2g}\,(+12)$
		$^3T_{2g}\,(+2)$
		$^3T_{1g}\,(-6)$
d^3	^{4}F	$^4T_{1g}\,(+6)$
		$^4T_{2g}\,(-2)$
		$^4A_{2g}\,(-12)$
d^4	^{5}D	$^5T_{2g}\,(+4)$
		$^5E_g\,(-6)$
d^5	^{6}S	$^6A_{1g}\,(0)$
d^6	^{5}D	$^5E_g\,(+6)$
		$^5T_{2g}\,(-4)$
d^7	^{4}F	$^4A_{2g}\,(+12)$
		$^4T_{2g}\,(+2)$
		$^4T_{1g}\,(-6)$
d^8	^{3}F	$^3T_{1g}\,(+6)$
		$^3T_{2g}\,(-2)$
		$^3A_{2g}\,(-12)$
d^9	^{2}D	$^2T_{2g}\,(+4)$
		$^2E_g\,(-6)$

[a] For tetrahedral (O_4) coordinations, multiply Dq by $-4/9$; for cubic (O_8), multiply by $-8/9$

orbital contributions to the magnetic moment will enter into the eigenfunctions after spin–orbit and magnetic field perturbations are applied.

Recalling the ground state terms of the $3d^n$ series listed in Table 2.4, we can now point out that the two example solutions for ^2D and ^4F will apply equally to the ^5D and ^3F cases. Table 2.5 lists these crystal field terms for the d^n series in units of Dq, with their signs adjusted to take into account the sign reversal for the upper half of the series. An insightful commentary on the correspondence and contrast among these ions was given by Van Vleck [17].

The point charge calculation may also be approached by another powerful technique called "operator equivalents" developed by Stevens [18]. This method is based on the replacement of the Cartesian operator functions of the $\mathcal{V}_{\text{cf}}$ potential energy with the equivalent L or J (whichever is applicable) angular momentum operators. Expressed in operator equivalents, the first part of the D_4 term of the octahedral field given by (2.7)

$$
\begin{aligned}
\mathcal{V}_{\text{cf}}^{\text{oct}} &= D_4\left(x^4+y^4+z^4-\frac{3}{5}r^4\right) \\
&= D_4\left\{L_x^4+L_y^4+L_z^4-\frac{1}{5}L\,(L+1)\,[3L\,(L+1)-1]\right\}. \qquad (2.26)
\end{aligned}
$$

Since the eigenfunctions of these angular momentum operators are linear combinations of the spherical harmonics, the calculation of matrix elements is straightforward. Operator equivalents can be very useful for quantitative calculations of more complex symmetries that involve higher order terms in E_{cf} and also for cases of higher L (or J) values. An introduction to these techniques is given in Ballhausen [3] and Low [13] and a more comprehensive discussion including many tables of matrix elements may be found in Hutchings [16]. To continue with this discussion, the theory of symmetry groups will be introduced as a powerful tool for finding crystal field solutions.

2.3.5 Group Theory and Lower Symmetry

Conventional perturbation calculations to determine crystal field states can become arduous for more complicated systems. The solutions for the simple cases outlined in the previous section will prove almost sufficient for our discussion of the various magnetic properties. To cope with the frequently encountered trigonal, tetragonal, and orthorhombic distortions of the cubic coordinations, however, the solutions are found by a shortcut that is derived from symmetry considerations. In the point charge calculations, diagonalization of matrices by solutions of higher order secular determinants may be accomplished by applying group theory to determine not only the best linear combinations of wavefunctions but also the degeneracies of the different eigenstates, for example, the A_{2g}, E_g, T_{1g}, and T_{2g} terms of the type defined by (2.21).

Unfortunately, the scope of this text will not permit a detailed exposition of group theory. The interested reader is directed toward any number of excellent treatments of this subject, including those cited earlier [3–5, 19]. For the purposes at hand, we need to recognize that the diagonalization process involves the construction of wavefunction combinations that conform to the symmetry of the perturbation operator. Group theory provides a method for predetermining the correct eigenfunction combinations for a particular perturbation problem and is useful in solving for the eigenfunctions of lower symmetry fields.

There are some terminologies that should be mentioned because they will recur throughout this volume. Energy levels or eigenvalues are often referred to as irreducible representations or energy terms. Their corresponding eigenfunctions are called basis vectors. Early development of this discipline was conducted by Bethe [20] and Mulliken [21], and two nomenclatures of the representations have survived, although the Mulliken version seems to have gained some preference. It has already been introduced in the designations of the spherical harmonic combinations in (2.24). Table 2.6 lists the notations for these two systems with the corresponding degeneracies. The results of group theory analysis of the crystal field problems have been well documented, and the cubic field representations for the various orbital terms are summarized in Table 2.7.

For the simple case of a descent in symmetry from cubic O_h to tetragonal D_{4h} to orthorhombic D_{2h}, the relation of the basis vector lobes to the changing locations

Table 2.6 Comparison of Mulliken and Bethe representation notations

Mulliken	Bethe	Degeneracy
A_1	Γ_1	1
A_2	Γ_2	1
E	Γ_3	2
T_1	Γ_4	3
T_2	Γ_5	3
$E_{1/2}$	Γ_6	2
$E_{5/2}$	Γ_7	2
G	Γ_8	4

Table 2.7 Irreducible representations for cubic symmetry

Ground Term	l	Mulliken	Bethe
S	0	A_1	Γ_1
P	1	T_1	Γ_4
D	2	$E + T_2$	$\Gamma_3 + \Gamma_5$
F	3	$A_2 + T_1 + T_2$	$\Gamma_2 + \Gamma_3 + \Gamma_5$
G	4	$A_1 + E + T_1 + T_2$	$\Gamma_1 + \Gamma_3 + \Gamma_4 + \Gamma_5$
H	5	$E + 2T_1 + T_2$	$\Gamma_3 + 2\Gamma_4 + \Gamma_5$
I	6	$A_1 + A_2 + E + T_1 + 2T_2$	$\Gamma_1 + \Gamma_2 + \Gamma_3 + \Gamma_4 + 2\Gamma_5$

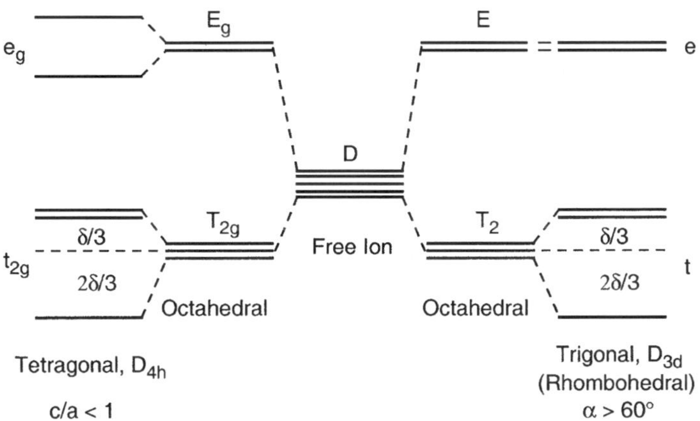

Fig. 2.13 Comparison of d-electron energy levels in crystal fields of tetragonal ($c/a < 1$) and trigonal ($\alpha > 60°$) symmetries [15]

of the negatively charged ligands can be visualized in Fig. 2.11. For a tetragonal distortion shown in Fig. 2.6c, T_{2g} (d_{xy}, d_{xz}, and d_{yz}) splits in the same way as the trigonal case, but the upper E_g doublet is now also split because of the relation of the $d_{x^2-y^2}$ and $d_z{}^2$ lobes to the octahedral ligands. The orthorhombic distortion D_{2h} of Fig. 2.6d will remove the final degeneracy and split the d_{xz} and d_{yz} states. The splittings of the orbital D term in these axially distorted cubic fields are compared in Fig. 2.13.

The energy level structure for the special cases of the tetragonal or orthorhombic distortions that occur with pyramidal O_5 and planar O_4 coordinations shown in Fig. 2.7 may be inferred by extrapolating the results for the weak field solutions. The descent in symmetry from cubic to the planar structure is particularly important in the cuprate superconductors to be examined in Chap. 8. To obtain a quantitative sense of the influence of a strong tetragonal field, we return briefly to the point charge calculation and examine the effects of a tetragonal component $\mathcal{V}_{\mathrm{T}}$ to the octahedral crystal field energy $\mathcal{H}_{\mathrm{cf}} \approx \mathcal{V}_{\mathrm{cf}}^{\mathrm{oct+T}} = \mathcal{V}_{\mathrm{cf}}^{\mathrm{oct}} + \mathcal{V}_{\mathrm{T}}$, where [22]

$$\mathcal{V}_{\mathrm{T}} = f_2(r)\, R(r)\, Y_2^0 + f_4'(r)\, R(r)\, Y_4^0, \tag{2.27}$$

where $f_2(r)$ and $f_4'(r)$ are radially-dependent coefficients. With the eigenfunction set of (2.11), diagonal matrix elements may be obtained by straightforward integral computations. Part of this process involves the defining of two additional splitting parameters representing the integrals of the radial components of the respective matrix elements over space, according to

$$\begin{aligned} Ds &= \int [R(r)]^2 \frac{3}{2} f_2(r)\, d\tau, \\ Dt &= \int [R(r)]^2 \frac{3}{2} f_4'(r)\, d\tau. \end{aligned} \tag{2.28}$$

From this definition, it may be shown that the energy states of the tetragonal perturbation follow directly from the diagonal matrix elements. The splitting of the upper doublet E_g of the $O_h + D_{4h}$ group becomes

$$\begin{aligned} \left\langle d^*_{x^2-y^2} \right| \mathcal{V}_{\mathrm{cf}}^{\mathrm{oct}} + \mathcal{V}_{\mathrm{T}} \left| d_{x^2-y^2} \right\rangle &= 6Dq + 2Ds - Dt, \\ \left\langle d^*_{z^2} \right| \mathcal{V}_{\mathrm{cf}}^{\mathrm{oct}} + \mathcal{V}_{\mathrm{T}} \left| d_{z^2} \right\rangle &= 6Dq - 2Ds - 6Dt. \end{aligned} \tag{2.29a}$$

and that of the lower T_{2g} triplet is

$$\begin{aligned} \left\langle d^*_{xy} \right| \mathcal{V}_{\mathrm{cf}}^{\mathrm{oct}} + \mathcal{V}_{\mathrm{T}} \left| d_{xy} \right\rangle &= -4Dq + 2Ds - Dt, \\ \left\langle d^*_{xz,yz} \right| \mathcal{V}_{\mathrm{cf}}^{\mathrm{oct}} + \mathcal{V}_{\mathrm{T}} \left| d_{xz,yz} \right\rangle &= -4Dq - Ds + 4Dt. \end{aligned} \tag{2.29b}$$

where the lowest d_{xz} and d_{yz} orbitals retain their degeneracy. For Ds and $Dt > 0$, the order of energy levels is shown in Fig. 2.14. As drawn, the structure is shown with the doublet as the ground state, but this may not necessarily be the case since the relative individual values of Ds and Dt would determine the correct order. Note that the uppermost state is still d_{z^2} and that it reaches a maximum separation of $10Dq$ from the next highest state, which is now d_{xy} instead of d_{z^2}. The crossover point where this upper state splitting becomes equal to $10Dq$ can be attained with a large tetragonal distortion, but may not necessarily require a complete removal of the two apical ligands along the z axis that would leave only an O_4 planar coordination. This point is discussed further in relation to the superconductivity of cuprates in Chap. 8.

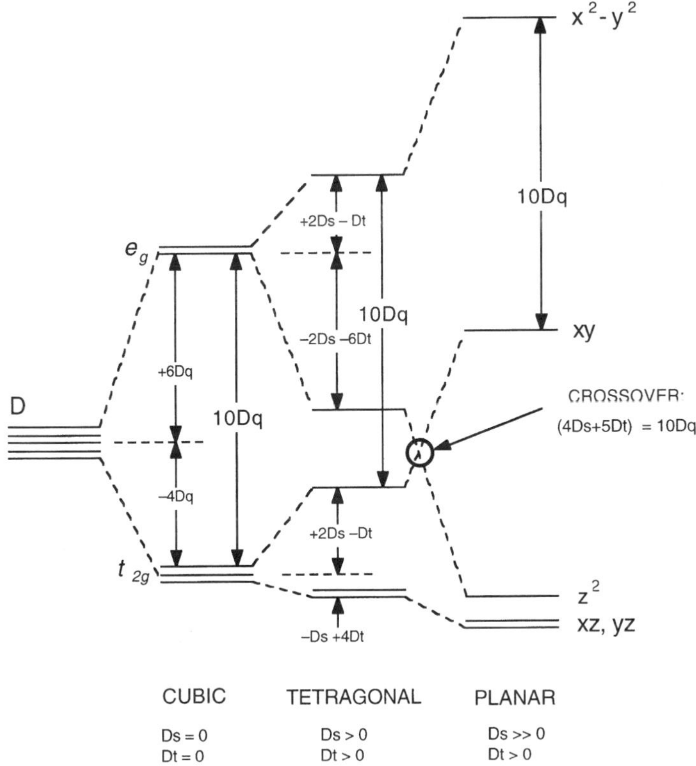

Fig. 2.14 Details of orbital energy level splittings as a tetragonal crystal evolves from cubic to planar, showing the $10Dq$ destabilization of the highest e_g level. Diagram is based on Fig. A.47 of [4]

Another demonstration of energy level determinations by group theory is realized by the descent in symmetry from $O_h \rightarrow D_{3d}$ that is commonly encountered in magnetic oxides. From the basis vectors for a trigonal distortion of an octahedral site reveal that the lower triplet T_{2g} is split into a doublet and a singlet, while the degeneracy of the upper doublet E_g is unchanged, as shown in Fig. 2.13. The fact that the upper doublet remains degenerate will be shown to be important in a later discussion of the Jahn–Teller effect.

For a trigonal distortion, the crystal field potential energy is$\mathcal{V}_{\mathrm{cf}}^{\mathrm{oct}+\tau} = \mathcal{V}_{\mathrm{cf}}^{\mathrm{oct}} + \mathcal{V}_\tau$, where $\mathcal{V}_\tau$ is applied in a similar fashion to that of the tetragonal field component $\mathcal{V}_{\mathrm{T}}$ given by (2.27), except that the set of orbital functions that it operates on are the group of (2.20). Upon application of this perturbation, the matrix elements are

$$\left\langle e^{\pm} \right| \mathcal{V}_{\mathrm{cf}}^{\mathrm{oct}} + \mathcal{V}_\tau \left| e^{\pm} \right\rangle = 6Dq + \frac{7}{3} D\tau,$$

$$\left\langle t_2^{\pm} \right| \mathcal{V}_{\mathrm{cf}}^{\mathrm{oct}} + \mathcal{V}_\tau \left| t_2^{\pm} \right\rangle = -4Dq + D\sigma + \frac{2}{3} D\tau,$$

$$\langle t_2^{\mathrm{o}} | \mathcal{V}_{\mathrm{cf}}^{\mathrm{oct}} + \mathcal{V}_\tau | t_2^{\mathrm{o}} \rangle = -4Dq - 2D\sigma - 6D\tau, \tag{2.30}$$

$$\left\langle t_2^{\pm} \right| \mathcal{V}_{\mathrm{cf}}^{\mathrm{oct}} + \mathcal{V}_\tau \left| e^{\pm} \right\rangle = \sqrt{2} D\sigma - \frac{5\sqrt{2}}{3} D\tau.$$

where $D\sigma$ and $D\tau$ are defined analogously to Ds and Dt of (2.28).

At this point it is instructive to compare (2.29) and (2.30). The tetragonal and trigonal cases are similar in that the T_{2g} (and T_2) group is split into a singlet and doublet, but as illustrated in Fig. 2.13, the E_g (and E) term remains degenerate in the trigonal field. Moreover, we now see that the $t_2^{\pm}$ and $e^{\pm}$ states mix under the V_τ perturbation. Pryce and Runciman [14] have studied this question in detail, but for our purposes, we assume that $D\sigma \sim (5/3)\ D\tau$ and that the off-diagonal elements are negligible, so that the matrix may be approximated as diagonal in later discussions.

2.3.6 Strong Field Solutions and Term Diagrams

If the crystal field is strong enough to compete with the mutually repulsive interactions among orbiting electrons in ions with multiple d electrons, or if the crystal field influence on the excited terms is important, all of the terms listed in Table 2.4 must be included as part of any thorough perturbation calculation. This situation is encountered in cases where the crystal-field splitting is larger, for example, for certain ligands such as the cyanide radical CN_6, or with the larger radii $4d^n$ and $5d^n$ ions. The most common need for the full energy term diagrams occurs in the interpretation of optical transitions from the ground state, as discussed in Chap. 7. Although the ground term is usually the only part of the free-ion energy level structure that is needed to explain the properties of magnetic oxides, the reader should appreciate the meaning of the term diagrams and the basis of their theoretical origins.

If the procedure outlined for the weak field is extended to include the upper terms in the conventional way, large matrices result and solutions to the complete term picture must be worked out by solution of the corresponding equations, simplified wherever possible by the use of group theory and any other methods for reducing the complexity of the matrices. To this end, Orgel [23] reported matrices and computations for the d^n series expressed in terms of the single parameter Dq. His results for the important d^5 case of Fe^{3+} or Mn^{2+} (symmetric in sign for any of the cubic coordinations in this particular instance) are shown in Fig. 2.15.

An alternative approach to the strong field problem is to consider the effects of the ligands on the orbital electrons prior to the energy of their mutual interaction E_{term} that determines the free-ion term splittings, that is, the intra-atomic e^2/r_{12} repulsive energies. It is then assumed that the distributions of electrons among the d orbital states are determined first by the repulsive forces of the ligands, with the mutual interactions among the electrons treated as the perturbation. In this situation, the starting energy states are no longer influenced by Hund's rule of orbital ordering, but rather by the various electron distributions among the t_{2g} and e_g orbital states as dictated by the crystal field, in this case anticipated as octahedral.

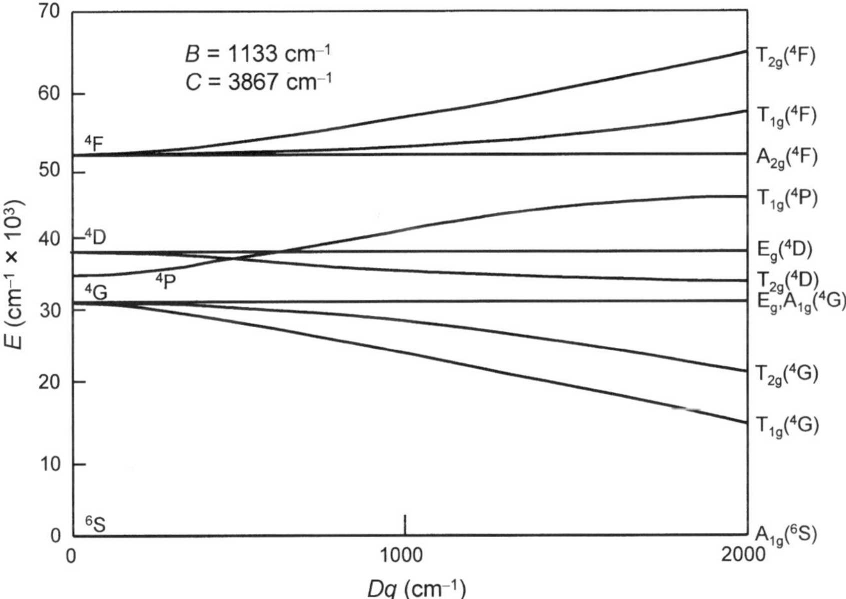

Fig. 2.15 Multielectron energy level term diagram for a $3d^5$ Mn^{2+} ion in an octahedral crystal field of O^{2-} ligands. Author's computations were based on the model of Orgel [24]

In the strong field limit, it is assumed that the mutual repulsion among the electrons in the ligands dominates the distribution of electrons among the d orbital states. As a consequence, the system of free-ion terms is broken down, and the energy states are selected according to occupation numbers of electrons in the t_{2g} and e_g shells, first by filling the lower t_{2g} levels and then the less stable e_g states. For the simplest multiple electron case of two d electrons, there are three possible orbital configurations within the octahedral Dq separation, $(t_{2g})^2$, $(t_{2g})^1(e_g)^1$, and $(e_g)^2$. Within each of these distributions are many possible combinations that are set by number of electrons and the number of individual states [24]. The respective energies of these groups of states is given to a first approximation by assigning $-4\,Dq$ to each electron in a t_{2g} level and $+6Dq$ to the remaining e_g electrons. To complete the calculation, the mutual repulsion energies, which separate the free-ion term energies through the E_{rep} perturbation, are recalculated based on mixtures of the specific octahedral t_{2g} and e_g wave functions of (2.12) and then added as combinations of K (Coulomb) and J (exchange) determinant element integrals in the manner of those introduced previously in relation to interatomic electron repulsion in Sect. 1.3.3. As an example calculation, consider the interaction between the d_{xz} and d_{yz} orbitals of (2.12), for which the elements (of the Slater determinant) are simply stated from the definition in Ballhausen [25]:

$$\begin{aligned} K(d_{xz}, d_{yz}) &= \langle d_{xz}(1)^* d_{yz}(2)^* | \frac{1}{r_{12}} | d_{xz}(1) d_{yz}(2) \rangle \\ &= F_0 - 2F_2 - 4F_4 \quad \text{(Slater integrals)} \\ &= A - 2B + C \qquad \text{(Racah parameters)}, \end{aligned} \tag{2.31a}$$

$$\begin{aligned} J(d_{xz}, d_{yz}) &= \langle d_{xz}(1)^* d_{yz}(2)^* | \frac{1}{r_{12}} | d_{yz}(1) d_{xz}(2) \rangle \\ &= 3F_2 + 20F_4 \quad \text{(Slater integrals)} \\ &= 3B + C \qquad \text{(Racah parameters)}. \end{aligned} \tag{2.31b}$$

From combinations of these integrals together with the help of group theory, the "promotional" energies are computed and added to the octahedral energies already determined from the respective electron distribution of the particular state to complete the energy term diagram. The F_n parameters are the Slater integrals of the various d orbital functions expressed in their basic spherical harmonics. Their identities have been documented in the original work [26]. Alternatively, Racah [27] defined parameters A, B, and C that can be reduced to the Slater integrals. In tabular form, these interaction energies and related term information of the d^2 case are listed in Table 2.8 for the strong-field method [7]. The evolution from the term diagram to the weak field result is sketched in Fig. 2.16. For additional comparison with the listings in Table 2.8, the term energies from the weak field approach are presented in Table 2.9.

The strong field approach to crystal field theory will be revisited in our discussion of S-state ion magnetoelastic properties in Chap. 5 and in the general discussion of electric dipole optical transitions for Faraday rotation in the magnetic garnets in Chap. 7. For future reference, the Racah parameter relations for the lowest five d^n terms listed in Table 2.4 are recorded in Table 2.10. In most situations, however, a simpler approximation based on the ground state stabilization energies is all that is necessary to sort out the causes of the various magnetic properties.

Table 2.8 d^2 Term splittings for strong field octahedral coordination

Free-ion term	E_{rep}	Crystal-field term	E_{cf}
^{1}S	$A + 14B + 7C$	$^1A_{1g}$	$2\varepsilon_0$
^{1}G	$A + 4B + 2C$	$^1A_{1g}$	$2\varepsilon_0 + 4Dq$
		^1E_g	$2\varepsilon_0 + 4/7Dq$
		$^1T_{1g}$	$2\varepsilon_0 + 2Dq$
		$^1T_{2g}$	$2\varepsilon_0 - 26/7Dq$
^{3}P	$A + 7B$	$^3T_{1g}$	$2\varepsilon_0$
^{1}D	$A - 3B + 2C$	^1E_g	$2\varepsilon_0 + 24/7Dq$
		$^1T_{2g}$	$2\varepsilon_0 - 16/7Dq$
^{3}F	$A - 8B$	$^3A_{2g}$	$2\varepsilon_0 + 12Dq$
		$^3T_{2g}$	$2\varepsilon_0 + 2Dq$
		$^3T_{1g}$	$2\varepsilon_0 - 6Dq$

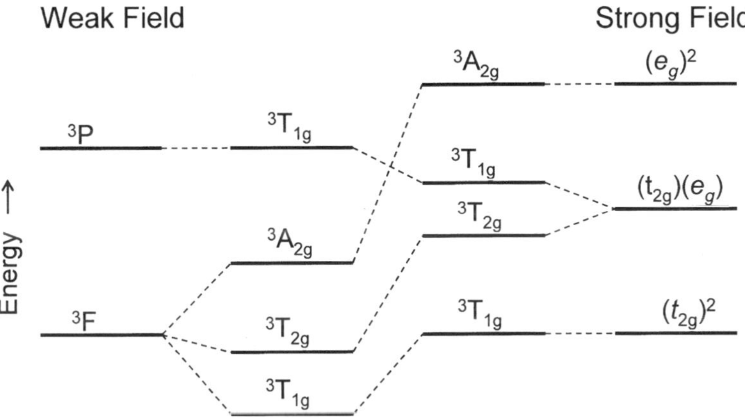

Fig. 2.16 Schematic of energy levels of a d^2 ion in an octahedral crystal field of increasing strength

Table 2.9 d^2 Term energies for weak field octahedral coordination

Crystal-field configuration	E_{cryst}	Crystal-field term	E_{term}
$(e_g)^2$	$2\varepsilon_0 + 12Dq$	${}^1A_{1g}$	$A + 8B + 4C$
		${}^1E_g$	$A + 2C$
		${}^3A_{2g}$	$A - 8B$
$(t_{2g})^1 (e_g)^1$	$2\varepsilon_0 + 2Dq$	${}^1T_{1g}$	$A + 4B + 2C$
		${}^1T_{2g}$	$A + 2C$
		${}^3T_{1g}$	$A + 4B$
		${}^3T_{2g}$	$A - 8B$
$(t_{2g})^2$	$2\varepsilon_0 - 8Dq$	${}^3T_{1g}$	$2\varepsilon_0$

2.3.7 *Rare-Earth Ion Solutions*

Crystal-field stabilizations of the orbital angular momentum must be treated differently in the rare-earth $4f^n$ series because $\mathcal{H}_{LS} >> V_{cf}$, which means that J is the quantum number that defines the angular momentum. Because of the stronger $\lambda \boldsymbol{L} \cdot \boldsymbol{S}$ energy, the terms are first split into a multiplet of states with J values running from $|L - S|$ to $|L + S|$ in the standard notation of atomic spectra. As a consequence the term splittings caused by the crystal field are labeled according to the J instead of the L degeneracies (although in the physical reality only L is quenched). Nonetheless, effects of the crystal field are significant in magnetic properties of oxides, and the various term splittings established with the aid of group theory and operator equivalents as determined by Lea et al. [28] are summarized in Table 2.11. These eigenstates of the crystal field represent raising of the J degeneracy and therefore can cause a decrease in the effective J value as

Table 2.10 Racah parameter energy relations for d^n and d^{10-n} with n = 2,3,4, and 5

d^2	d^3
$^1S = A + 14B + 7C$	$^2F = 3A + 9B + 3C$
$^1D = A - B + 2C$	$^2G = 3A - 11B + 3C$
$^1G = A + 4B + 2C$	$^2H = {}^2P = 3A - 6B + 3C$
$^3P = A + 7B$	$^4P = 3A$
$^3F = A - 8B$	$^4F = 3A - 15B$
d^4	$(d^5)^a$
$^3D = 6A - 5B + 4C$	$^4F = 10A - 13B + 7C$
$^3F = 6A - 5B + (11/2)C \pm (3/2)\left(68B^2 + 4BC + C^2\right)^{1/2}$	$^4D = 10A - 18B + 5C$
$^3G = 6A - 12B + 4C$	$^4P = 10A - 28B + 7C$
$^3H = 6A - 17B + 4C$	$^4G = 10A - 25B + 5C$
$^5D = 6A21B$	$^6S = 10A - 35B$

The derivation of these relations can be found in [5], Sect. 4.6

[a] These terms are listed differently from that anticipated by Hund's rule in that the ^{4}P and ^{4}F terms are exchanged in position on the energy ladder. This is an artifact of this configuration and results for the relative values of B and C, which are found to occur in the ratio of $C/B \sim 4.5$. There are other departures from the Hund's rule norm that will be fall beyond the scope of this discussion

Table 2.11 $4f^n$ (Rare-earth group) term splittings in an oxygen coordination of O_h symmetry

J	Term representations
1/2	$E_{1/2g}$
1	T_{1g}
3/2	G_g
2	$E + T_{2g}$
5/2	$E_{5/2g} + G_g$
3	$A_{2g} + T_{1g} + T_{2g}$
7/2	$E_{1/2g} + E_{5/2g} + G_g$
4	$A_{1g} + E_g + T_{1g} + T_{2g}$
9/2	$E_{1/2g} + 2G_g$
5	$E_g + 2T_{1g} + T_{2g}$
11/2	$E_{1/2g} + E_{5/2g} + 2G_g$
6	$A_{1g} + A_{2g} + E_g + T_{1g} + 2T_{2g}$
13/2	$E_{1/2g} + 2E_{5/2g} + 2G_g$
7	$A_{2g} + E_g + 2T_{1g} + 2T_{2g}$
15/2	$E_{1/2g} + E_{5/2g} + 3G_g$
8	$A_{1g} + 2E_g + 2T_{1g} + 2T_{2g}$

The contents of this table are included here because they cover the full scope of the $4f^n$ series that is important in the context of this book. In addition, these entries could also the first complete listing available in the public domain measured magnetically, just as in the case of orbital angular momentum quenching of L in the d^n series.

The situation regarding the rare-earth ions becomes more complex not only because the high J multiplicities render analytical solutions more laborious, but also because magnetic exchange fields interact only with the spin component of J. Moreover, since the crystal field is about two orders of magnitude smaller than that of the d^n series, the degree of J quenching can been influenced by externally applied magnetic fields of magnitudes (>10 T) that are now attainable with modern magnet technology, for example, superconducting magnets. These effects of competing crystal fields, exchange fields, and applied magnetic fields are examined in the context of the magnetic properties of the rare-earth iron garnets in Sect. 4.3.3, and later in relation to electron spin resonance in Chap. 6.

2.4 Orbital Energy Stabilization

When cations are placed in a lattice, the crystal field from the anion charges is derived from the ionic part of the chemical bonding. As a consequence, the spin occupancies pictured in Fig. 2.3 for free ions will undergo modifications. Within the point charge approximation, the resultant electrostatic potential between neighboring ions (often approximated by a lattice energy calculation) and the ionization potentials of the various ionic species determine the electronic energies of the outermost electrons. For the $3d^n$-electron series, the local crystal field of the immediate anion neighbors provides additional binding energy by lowering the electronic ground state energy as part of the splitting of the orbital degeneracy that was reviewed in the previous section. The additional electronic stabilization is therefore the result of a perturbation by the point-charge field, which may be further enhanced by a covalent interaction.

2.4.1 One-Electron Model

In the weak-field regime, a one-electron model can be constructed for the $3d^n$ series as a general qualitative approximation to provide an occupancy map of the ground state electronic structures sketched. Instead of computing the electron interaction energies in terms of the Slater integrals or Racah parameters, the ground state is first pictured as a distribution of electrons among the cubic field terms [29]. The energy levels of the T_{2g} and E_g terms from the ^{2}D and ^{5}D cases of Fig. 2.12 for single-d electron are used as a "floor plan" to keep track of the likely d-electron spin distributions in the ground states, as diagrammed in Figs. 2.17 and 2.18 for octahedral and tetrahedral coordinations, respectively, including the low-spin states that can

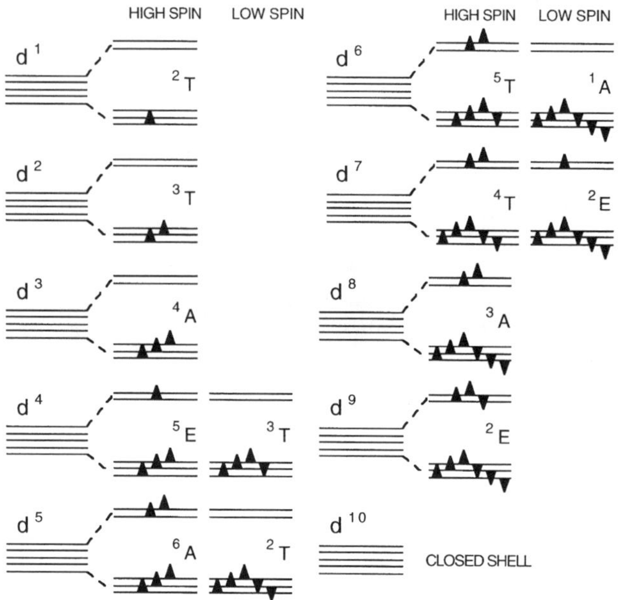

Fig. 2.17 One-electron d-orbital occupancy diagrams: octahedral site

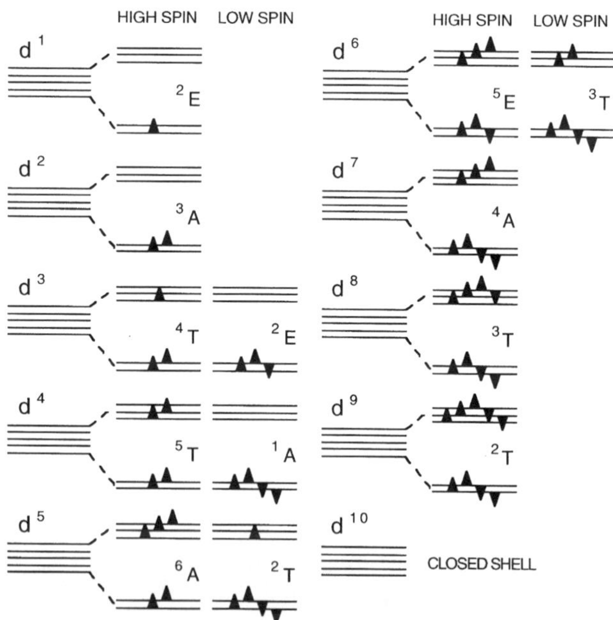

Fig. 2.18 One-electron d-orbital occupancy diagrams: tetrahedral site

occur in strong crystal fields (violation of Hund's rule).[2] In this abbreviation, the "Aufbau method" is applied by adding d electrons sequentially beginning with the orbital state of lowest energy, while observing Hund's rule of spin polarization for the first (α) half of the d shell and the Pauli principle of spin pairing as spins are added to the second (β) half.[3]

Although not shown explicitly, the crystal-field stabilization energy of each configuration is offset by intraorbital electron repulsion energy e^2/r_{ij}, which is manifested by the particular ionization potential (IP) in Fig. 2.19a (replotted from Fig. 2.8 for 2+ and 3+) as a function of n across the transition series.[4] Note that the destabilization energy (characterized by the parameter U_{ex}) of β spins relative to α spins in the same orbital is on the order of 2–3 eV, which exceeds the $10Dq$ values of many oxide sites.

It is appropriate here to point out that the Aufbau approximation can apply with rigor to only the d^1 or d^9 cases. For that reason, the model is often called the "one-electron approximation." Where the effects of intraelectron repulsion must be taken into account, multielectron solutions with their attendant complexities in dealing with other important perturbations such as spin–orbit coupling, magnetic exchange, and Zeeman effects in an external magnetic field can be considered.

2.4.2 *High- and Low-Spin States*

Where $\mathcal{H}_{Hund}$ exceeds $\mathcal{H}_{cf}$, this procedure establishes a ground state with the maximum available spin number. The electrons are distributed among the orbital states with Pauli spin pairing permitted only after each of the five orbitals are half-filled (Hund's rule). Since the energy distribution also changes with the assignment of electrons, one of the main insights gained from this approximation is the relative magnitudes of the cation site stabilization energies. In Figs. 2.17 and 2.18, the one-electron ground-state configurations for the octahedral and tetrahedral sites include

[2] The Aufbau concept can be used here directly because all of the electrons occupy orthogonal crystal-field states of the same orbital term under the influence of the same nuclear charge. When applied to molecular bonding that involves Coulomb fields of multiple nuclei, the applicability is limited by the covalent sharing of orbital states that are not fully orthogonal.

[3] The reader is cautioned that these diagrams are used to sort out the electron occupancies of the orbital ground state in order to anticipate the quantum designation of the ground state. The virtue of the one-electron models is the ready insight that they can provide without the necessity of complex mathematical analysis and computation.

[4] In collective-electron band theory that was introduced by Stoner [30], the Fermi level is used as the reference energy for electrical properties, and it has been found phenomenologically convenient to separate the spin populations into up (majority α) and down (minority β) spin bands based on the difference in energy between the upper and lower parts of the d-shell spin ladders depicted in Fig. 2.3. This model is then used to explain the net collective moment in the manner of a ferrimagnetic spin system.

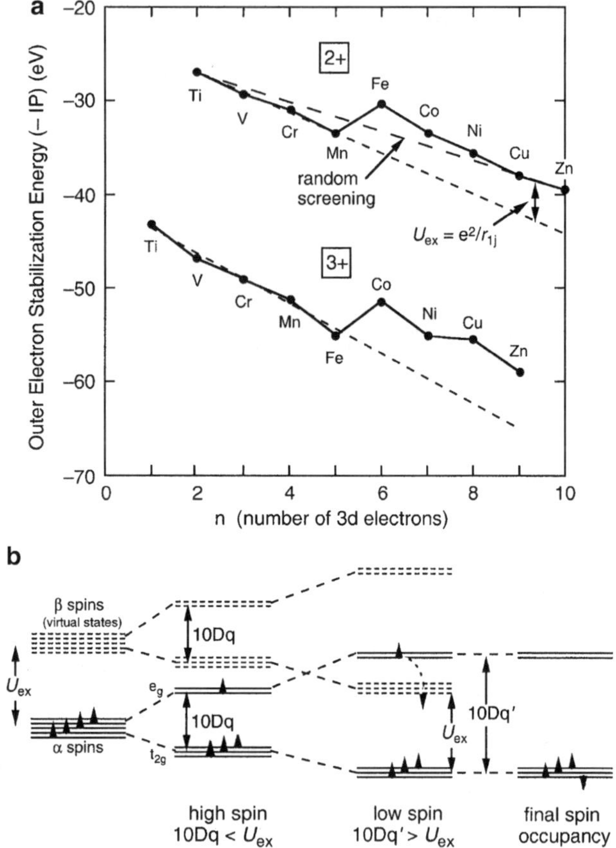

Fig. 2.19 Origin of the low-spin state of d^4 shown in (**b**), where the upper β half is depicted as a virtual shell of energy U_{ex}. (**a**) is extracted from Fig. 2.8 to serve as an energy reference. For $10Dq < U_{ex}$, all four spins are aligned parallel, producing $S = 2$ in the α states. When the crystal-field splitting increases to $10Dq' > U_{ex}$, the single $e_g\alpha$ spin will flip as it can now stabilize in the t_{2g} states of the β shell, both satisfying the Pauli principle and creating a low-spin value of $S = 1$. Note that in the one-electron approximation, the IP energy information from (**a**) associated with the $U_{ex} = e^2/r_{ij}$ correlation energy has been discarded

the "low-spin" states that occur in strong crystal fields (in violation of Hund's rule when the destabilization from the $10Dq$ splitting energy overrides the e^2/r_{ij} spin polarization energy that we designate U_{ex}).

In the weak-field limit, the spins are polarized to the maximum extent and the result is logically called the "high-spin" state. Although these spin alignments are the usual situations in the $3d^n$ series, low-spin occupancies may also occur in selected instances where $10Dq > U_{ex}$. When this occurs, the lower half of the $3d^n$ shell begins to fill before the upper one is half-filled, as illustrated by the example of $3d^4$ in Fig. 2.19b. In most situations, stable low-spin condensations are not expected unless cation valences exceed 3+ in octahedral sites. The example chosen here ($3d^4$), the Jahn–Teller case to be discussed in the next section, is not usually expected to condense into a low-spin configuration unless the valence increases to 4+, for example, the infrequently encountered Fe^{4+} ion. A more likely situation where U_{ex} can exceed $10Dq$ would be found with $n = 6$ in an octahedral site. Co^{3+} $(3d^6)$ in a perovskite lattice is analyzed for high, low, and intermediate spin configurations by Ibarra et al [31]. A detailed diagram in the manner of Fig. 2.19b is included to illustrate the application of these concepts to a challenging problem. The magnetoelastic implications of this family of ions are discussed further in Chap. 5 (see Fig. 5.7).

In general, low-spin states do not occur in tetrahedral sites because of the 4/9 reduction factor in the value of Dq, but the hypothetical configurations are included here for completeness. The respective d-electron diagrams are included in Figs. 2.17 and 2.18, and the corresponding stabilization energies in units of Dq are listed in Tables 2.12 and 2.13. Comparison of these estimates of site stabilization energy can be used to decide the likely preference for ions in lattices such as spinels and garnets

Table 2.12 High- and low-spin d-electron stabilization energies in an octahedral coordination

d Electrons	High-spin configuration	Stabilization energy	Low-spin configuration	Stabilization energy
1	$(t_{2g})^1$	$4Dq$	$(t_{2g})^1$	$4Dq$
2	$(t_{2g})^2$	$8Dq$	$(t_{2g})^2$	$8Dq$
3	$(t_{2g})^3$	$12Dq$	$(t_{2g})^3$	$12Dq$
4	$(t_{2g})^3(e_g)^1$	$6Dq$	$(t_{2g})^4$	$16Dq$
5	$(t_{2g})^3(e_g)^2$	0	$(_{2g})^5$	$20Dq$
6	$(t_{2g})^4(e_g)^2$	$4Dq$	$(t_{2g})^6$	$24Dq$
7	$(t_{2g})^5(e_g)^2$	$8Dq$	$(t_{2g})^6(e_g)^1$	$18Dq$
8	$(t_{2g})^6(e_g)^2$	$12Dq$	$(t_{2g})^6(e_g)^2$	$12Dq$
9	$(t_{2g})^6(e_g)^3$	$6Dq$	$(t_{2g})^6(e_g)^3$	$6Dq$
10	$(t_{2g})^6(e_g)^4$	0	$(t_{2g})^6(e_g)^4$	0

Table 2.13 High- and low-spin d-electron stabilization energies in a tetrahedral coordination

d Electrons	High-spin configuration	Stabilization energy	Low-spin configuration	Stabilization energy
1	$(e_g)^1$	$6Dq$	$(e_g)^1$	$4Dq$
2	$(e_g)^2$	$12Dq$	$(e_g)^2$	$8Dq$
3	$(e_g)^2(t_{2g})^1$	$8Dq$	$(e_g)^3$	$18Dq$
4	$(e_g)^2(t_{2g})^2$	$4Dq$	$(e_g)^4$	$24Dq$
5	$(e_g)^2(t_{2g})^3$	0	$(e_g)^4(t_{2g})^1$	$20Dq$
6	$(e_g)^3(t_{2g})^3$	$6Dq$	$(e_g)^4(t_{2g})^2$	$16Dq$
7	$(e_g)^4(t_{2g})^3$	$12Dq$	$(e_g)^4(t_{2g})^3$	$12Dq$
8	$(e_g)^4(t_{2g})^4$	$8Dq$	$(e_g)^4(t_{2g})^4$	$8Dq$
9	$(e_g)^4(t_{2g})^5$	$4Dq$	$(e_g)^4(t_{2g})^5$	$4Dq$
10	$(e_g)^4(t_{2g})^6$	0	$(e_g)^4(t_{2g})^6$	0

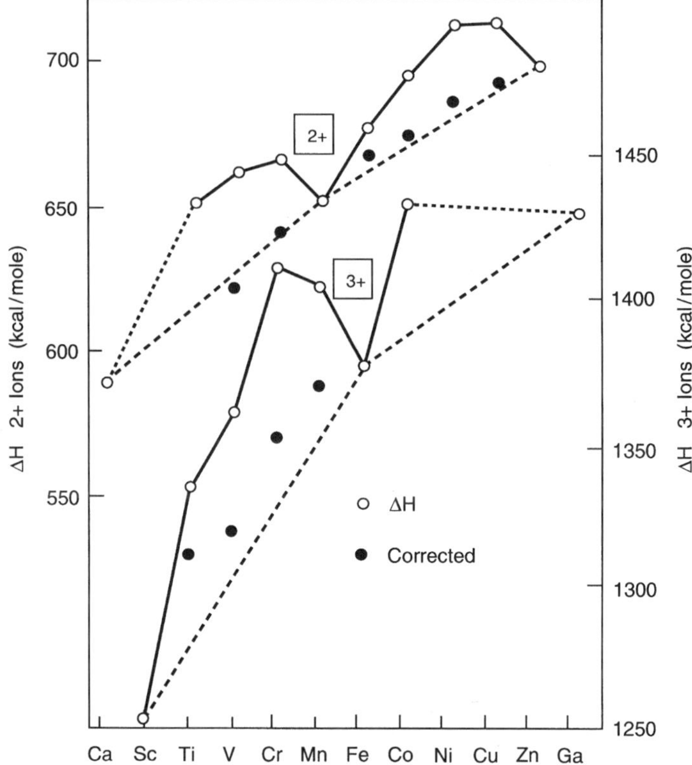

Fig. 2.20 Heats of hydration of divalent and trivalent transition metal ions as a function of atomic number. Ligand field energies are subtracted from total values to reveal the expected smooth monotonic increase in binding energy with increasing atomic number. Data are from Holmes and McClure [32]

where both octahedral and tetrahedral sites are present. For the specific example of Cr^{3+}, a site preference energy of $4Dq$ can be deduced from the stabilization energies listed in the tables.

The success of this simple model may be observed from the heat of hydration data of Holmes and McClure [32] plotted Fig. 2.20, where the excess binding energy attributed to the octahedral ligand field stabilization is seen to follow the values predicted in Table 2.12 for the high-spin case. Of the various $3d^n$ ions, the most dramatic example of a transition from high spin to low spin is $3d^5$, i.e., Fe^{3+} or Mn^{2+} whereby the spin value can decrease from $S = 5/2$ to 1/2 when $Dq/B>3$ in Fig. 2.21, adapted from the calculations of Tanabe and Sugano [33]. Such a low-spin state occurs for isolated Fe^{3+} ions in K_3Co^{3+} (CN_6) as observed in electron spin resonance (EPR) measurements [34]. With $(CN_6)^{6-}$ ligands, this intriguing situation is examined in relation to spin–lattice relaxation in Sect. 6.2. $Co^{3+}(3d^6)$ also has a low-spin configuration in an octahedral site [35], where all six electrons fill the t_{2g} shell producing $S = 0$ instead of 2.

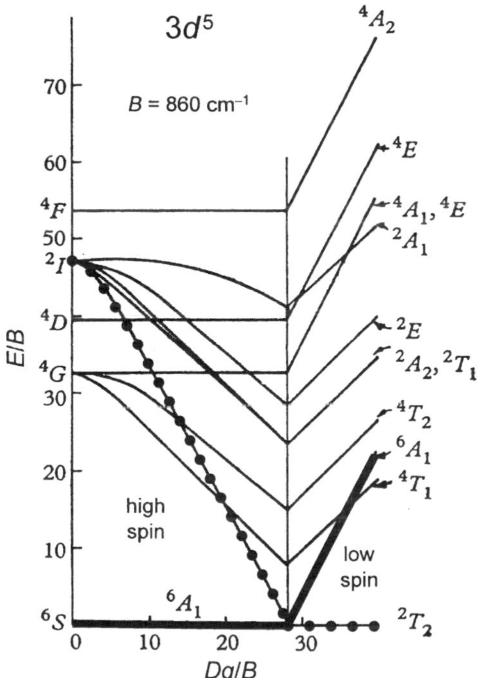

Fig. 2.21 Splitting of states of the d^5 configuration by an octahedral field (abbreviated and modified from Tanabe and Sugano calculations [33]). The Racah parameter $B\ (=860\,\mathrm{cm}^{-1})$ represents the effect of the internal spin polarization energy U_{ex} of Fig. 2.19b. Note the transition from a high-spin orbital singlet $^6A_1$ ground term to a low-spin orbital triplet $^2T_2$ when $Dq/B > 3$

From the tables and figures presented so far, it is evident that the cubic fields alone fail to remove all of the ground-state degeneracy. Further stabilization can occur, and there are two mechanisms by which this can take place locally: (1) orbit–lattice coupling manifested in the now-celebrated Jahn–Teller (J–T) spontaneous distortion effects, and (2) the more conventional spin–orbit (S–O) coupling that can override and reverse the J–T distortion in select cases. Both of these phenomena are responsible for important magnetic and electronic behavior.

2.4.3 Orbit–Lattice Stabilization (Jahn–Teller Effects)

In Figs. 2.17 and 2.18 that were discussed in relation to the single d electron approximations in octahedral and tetrahedral sites, it is evident that in many cases not all of the degeneracy is removed by quenching of the orbital angular momentum in cubic fields. Group theory dictates that lower symmetry arrangements of the oxygen ligands can lift the remaining degeneracies and eliminate all vestiges of unquenched orbital angular momentum in the ground states.[5] Except for situations

[5] It should be pointed out that this lifting of degeneracies does not include those of Kramers doublets, which are spin degeneracies that can be split only by magnetic fields. Kramers doublets occur

where the lattice itself furnishes the lower symmetry through its own point group, the sources for such lower symmetry field components are not always apparent. In other cases there may be crystal field components of lower symmetry that exist locally because of lattice vacancies or variations in neighboring cation charges situated along the relevant crystal axes. In two particular situations of great importance in determining many magnetic and magnetoelastic properties, the distortion may occur spontaneously at the cation site because of orbit–lattice or spin–orbit–lattice interactions peculiar to the electron configuration of the transition ion.

In studies of the paramagnetic resonance behavior of Cu^{2+} $(3d^9)$ ions in hydrated salts [36], Jahn and Teller [37, 38] noted the presence of a tetragonal component to the crystal field in a normally cubic octahedral site. Their explanation for the phenomenon was as elegant as it was important. In situations where a molecule has a degenerate orbital angular momentum state, the immediate environment of the site will be found to have lower symmetry if such a distortion will lift the degeneracy to provide a state of lower energy for the electrons that occupy the degenerate state, that is, an increased site stabilization energy. In its elementary definition, the Jahn–Teller (J–T) effect produces a singlet orbital ground state, which may still retain is spin degeneracy, for example, a Kramers doublet. Such Jahn–Teller distortions can be locally static and can spontaneously deform the lattice from cubic to tetragonal (or lower) through cooperative involvements of many sites if the density of local distortions is high enough. If the J–T interaction is weaker, it can manifest a dynamic behavior in which vibronic normal modes of the ligand complex are active. This topic is discussed in detail in references already cited, and this text will not dwell on esoteric material that might serve only to distract the reader. There are, however, some important points that should be made clear about variations of the distortion effects that can take place for specific members of the transition ion series. Their influence on magnetoelastic effects and resonance relaxation processes is reviewed in Chaps. 5 and 6.

The Jahn–Teller effect in its essential form may be seen from the simple schematic picture of Fig. 2.22. The example shows a single electron in the e_g orbital doublet of an octahedral crystal field of energy E_{cf}. A tetragonal crystal field component from a <100> z-axis distortion splits the doublet to create an electronic stabilization energy ΔE_{el} that is equal to the product of E_{cf} and the z-axis strain, that is, $\Delta E_{el} = E_{cf}\,(\Delta z/z_0)$, where z_0 is the initial length of the octahedron. To establish equilibrium in the localized system, ΔE_{el} is offset by an increase in lattice elastic energy ΔE_{lat} for the site of volume v_{site}, which is approximated as quadratic in the form of $\Delta E_{lat} = (1/2)\,c_{lat}\,(\Delta z/z_0)^2\,v_{site}$. The reduction in energy of the electron that occupies the lower half of the split doublet is determined by

with a noninteger spin quantum number, that is, $S = 1/2$, 3/2, 5/2, etc., resulting from ions with odd numbers of unpaired electron spins. Care must be exercised in the use of the exchange field concept. It is not a true magnetic field in the Maxwell sense. It is born out of covalent bonding and the Pauli exclusion principle of indistinguishability and is therefore of electrostatic origin. It can be only a scalar and has neither the ability to polarize the spins that it gathers along a chosen direction nor to split Kramers doublets.

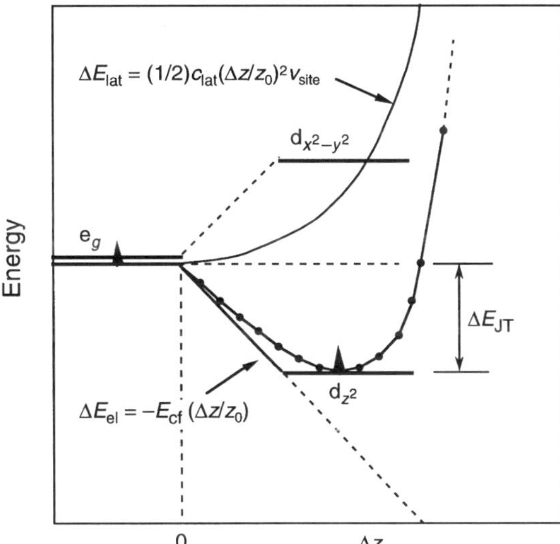

Fig. 2.22 Generic model of the Jahn–Teller effect

minimizing the net J–T stabilization defined as $\Delta E_{\mathrm{JT}} = \Delta E_{\mathrm{lat}} - \Delta E_{\mathrm{el}}$, plotted as a function of the strain $\Delta z / z_0$. From this simple analysis, the equilibrium value of $z/z_0 = E_{\mathrm{cf}}/c_{\mathrm{lat}} v_{\mathrm{site}}$, and the static Jahn–Teller stabilization energy becomes

$$\Delta E_{\mathrm{JT}} = (1/2)\, E_{\mathrm{cf}}^2 / c_{\mathrm{lat}} v_{\mathrm{site}}. \tag{2.32}$$

For a typical oxide compound $E_{\mathrm{cf}} \sim 10Dq \sim 2\,\mathrm{eV}$, the lattice elastic constant $c_{\mathrm{lat}} \sim 2 \times 10^{12}$ dyne cm^{t2}, and the site volume $v_{\mathrm{site}} \sim 10^{-23}\,\mathrm{cm}^3$. These values then predict a J–T strain ~ 0.15 and a $\Delta E_{\mathrm{JT}} \sim 0.2\,\mathrm{eV}$ or $1,500\,\mathrm{cm}^{-1}$.

The double-valued solution for Δz confirms that either expansion or contraction can occur along the z axis, causing the energies of the $d_{x^2-y^2}$ and d_{z^2} levels to reverse accordingly. For this reason, the J–T effect can be temperature dependent, diminished thermally by the growth of the random lattice phonon population, and specifically by the vibronic modes of the local cation–anion configuration [39]. When the latter conditions dominate, the stabilization of the electronic ground state is compromised by the blurring of the e_g orbital levels if the occupying electrons cannot keep up with the vibronic fluctuations, and doublet degeneracy can be restored. Otherwise, the electronic stabilization is preserved and the resulting condition is termed "quasistatic" in a manner that is difficult to distinguish from the simple static effect prevalent at low temperatures [40]. What distinguishes the pure J–T effect from other spontaneous lattice distortions is that the $d_{x^2-y^2}$ and d_{z^2} states involved in the splitting each have zero orbital angular momentum, that is, $m_l = 0$. For J–T splitting of the e_g doublet, there are four situations among the high-spin d^n electron transition groups where this can occur in an octahedral site. As diagrammed

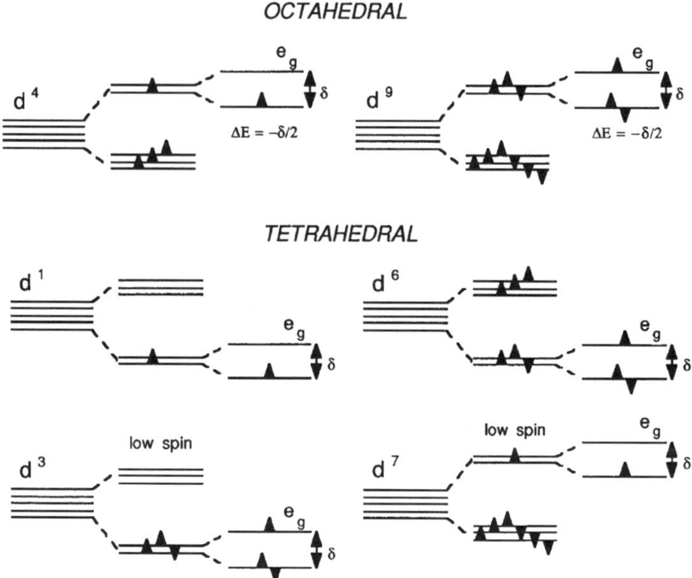

Fig. 2.23 One-electron model of Jahn–Teller stabilizations of the e_g states

in Fig. 2.23, they are d^4 and d^9 in octahedral and d^1 and d^6 in tetrahedral sites.[6] If the low-spin configurations are considered, d^3 and d^7 ions in tetrahedral sites are also subject to Jahn–Teller e_g splittings. The different J–T effects reveal themselves in various situations that involve magnetoelastic properties as well as electrical conductivity related to polaron activity that is discussed in Chap. 8.

2.4.4 Spin–Orbit–Lattice Stabilization

In the context of local magnetoelastic effects, it is appropriate here to point out an important implication of the spin–orbit coupling perturbation energy. In reviewing the crystal field orbital occupancy diagrams for the ground state in Figs. 2.17 and 2.18, one observes that certain t_{2g} configurations remain degenerate because of partially filled states. Unlike the e_g doublet with $m_l = 0$ that undergoes a pure J–T

[6] A point worthy of note concerns the infrequent situation of d^6 in a tetrahedral (O_4) site. Such a case is Fe^{2+} substituting for Zn^{2+} in a ZnO lattice. According to Fig. 2.23, a pure Jahn–Teller effect is expected in the e_g orbital ground states. The energy of this stabilization, however, would be significantly less than that of Mn^{3+} or Fe^{4+} in an O_6 site because of the 4/9 reduction in the crystal-field strength combined with the lower valence charge of the Fe^{2+} cation. This occurrence of the J–T effect is analyzed by Goodenough [41].

splitting independent to first order of spin–orbit interactions, the t_{2g} triplet contains both a singlet d_{xy} with $m_l = 0$ and doublet $d_{xz,yz}$ orbitals with nonzero $m_l = \pm 1$ because they are hybrids of $t_{2g}^{+}(d_1)$ and $t_{2g}^{-}(d_{-1})$ from (2.11), as explained in Sect. 2.3.3. Therefore, by definition, the t_{2g} states also can be split by a J–T effect if the d_{xy} singlet emerges as the ground state. However, if the spin–orbit coupling $\lambda \boldsymbol{L} \cdot \boldsymbol{S}$ exceeds the J–T δ splitting, an orbitally degenerate ground state can also occur if the hybridized $d_{xz,yz}$ orbitals are stabilized. Furthermore, unlike the e_g splitting that requires a tetragonal or orthorhombic distortion along a <100> axis, the t_{2g} J–T splitting can occur by either <100> or trigonal axis <111> distortions, as indicated by Fig. 2.13. This latter condition has implications for the design of anisotropy and magnetostriction in ferrites.

Sketches of the doublet orbital lobes are shown in Fig. 2.24. The preference for one sign of distortion over the other lies in the relative magnitudes of δ and the spin–orbit coupling energy λ. Where an exchange field H_{ex} imposes ordering of the spins along the axis of the unquenched orbital angular momentum, spin–orbit coupling can stabilize the doublet as an alternative ground state that is manifested as a collective distortion of the opposite sign, as depicted in the tutorial sketch of the perturbed orbital states in Fig. 2.25. This possibility is represented throughout the various t_{2g} spin occupancy situations shown in Figs. 2.26 and 2.27. The sign of the lattice deformations can be a guide to which effect is occurring in a particular situation.

The basic requirement for S–O stabilization is the retention of a nonzero L in the ground state, which would mean a degenerate state in the ideal case. As indicated in Fig. 2.28 for the example of a single d electron in an octahedral site, both an axial crystal field that produces a $\delta < 0$ and a spin–orbit splitting, each acting

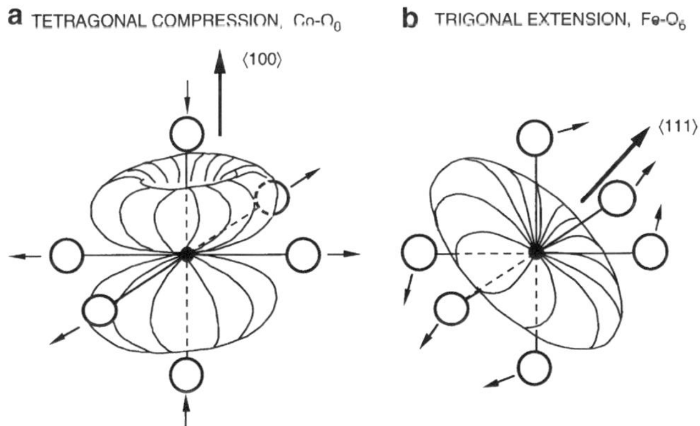

Fig. 2.24 Octahedral-site deformations from stabilization of spin–orbit doublet create ground states in the partially quenched t_{2g} shell: (**a**) a single electron in degenerate hybrid $d_{yz} \pm i d_{xz}$ causes a ⟨100⟩ tetragonal compression (characteristic of CoO), and (**b**) a single electron in degenerate hybrid of a trigonal field produces a ⟨111⟩ extension (characteristic of FeO) (Based on Figs. 44 and 53 of [40])

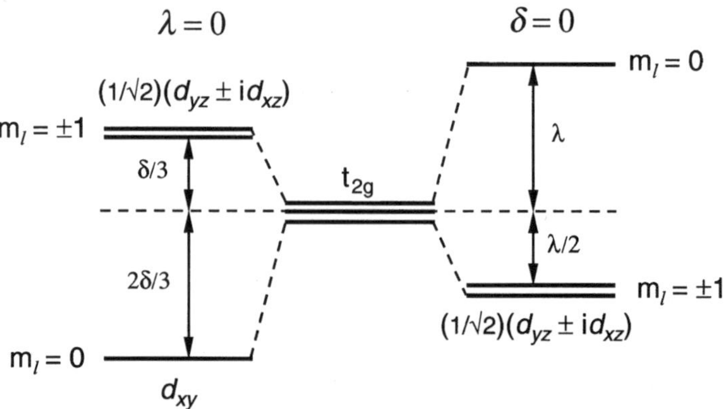

Fig. 2.25 Spin–orbit and Jahn–Teller stabilizations of the t_{2g} states

independently, can stabilize parts of the same doublet hybrid $(1/\sqrt{2})\left(d_{yz} \pm i d_{xz}\right)$, while $\delta > 0$ will produce the opposite effect. (The signs of δ reverse for a single d "hole" in the t_{2g} shell because of the inverted level structure). In the presence of a significant H_{ex}, the threshold condition for S–O dominance based on this rudimentary analysis is [40]:

$$(\lambda \boldsymbol{L} \cdot \boldsymbol{S} + \delta/3) \geq \Delta E_{\text{JT}} = 2\delta/3. \tag{2.33}$$

For a single d electron, $S_z = 1/2$, and since $L_z = \text{m}_l = \pm 1$, (2.33) reduces to

$$\begin{aligned} &\lambda/2 + \delta/3 \geq 2\delta/3 \\ &\text{or} \\ &\lambda/\delta \geq 2/3. \end{aligned} \tag{2.34}$$

This subject is examined in greater detail for the specific $3d^n$ ions in relation to the magnetoelastic properties in Chap. 5, supported by a quantum perturbation analysis of the T_{2g} term under the influence of the parameters δ, λ, and H_{ex}.

In Fig. 2.29, two common examples of the J–T and S–O stabilizations (d^4 and d^7 in octahedral sites) are compared in relation to observed ligand displacements. Note that the z-axis of a compressive distortion of the S–O case has the sign needed to create a degenerate ground state stabilized with the $m_l = \pm 1$ orbital angular momenta. There are many instances where potential spin–orbit–lattice stabilizations could take place. A summary of these possibilities is given in Table 2.14 with known occurrences highlighted.

The distinction between these two spontaneous stabilization effects should be restated because it can have important influence on the properties of magnetic oxides. Because the J–T effect overrides spin–orbit coupling and stabilizes the $m_l = 0$ singlet, cooperative manifestations of it might appear less dependent on a spin ordering condition (Curie or Néel). However, the S–O effect requires collinearity of

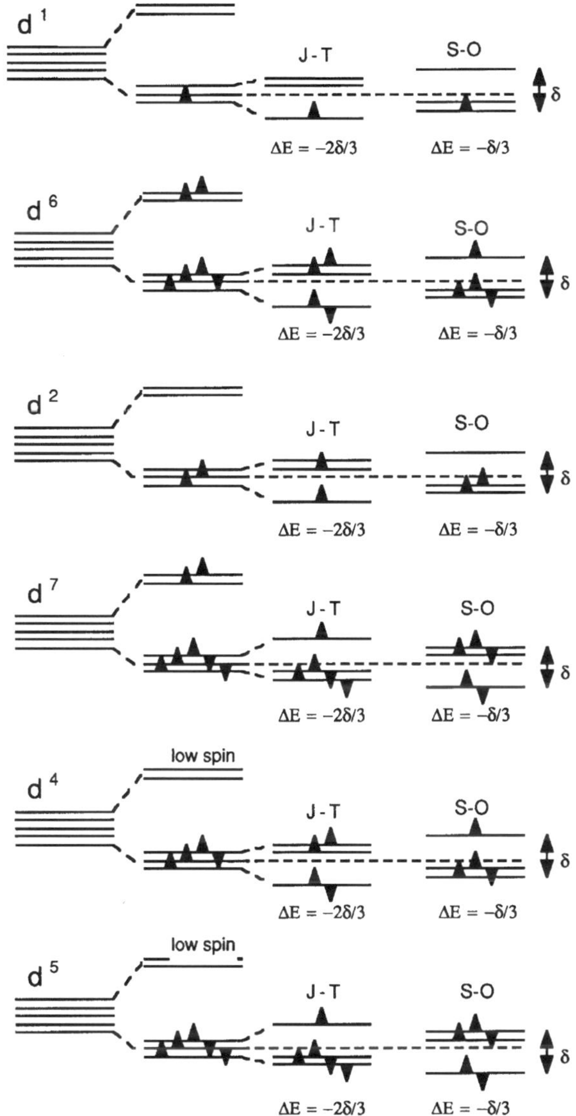

Fig. 2.26 One-electron model of J–T/S–O stabilizations of the t_{2g} shell: octahedral site

spins for cooperative distortions to occur [40] and is therefore expected only in systems where the energy gain from H_{ex} through $\lambda \boldsymbol{L}\cdot\boldsymbol{S}$ is large enough to offset the J–T singlet stabilization. The attendant local distortions of either sign can produce spin–lattice magnetostriction effects in spin-ordered systems, one depending on a dominant orbit–lattice interaction with less spin–orbit coupling, and the other depending on a strong spin–orbit coupling with a smaller orbit–lattice (crystal-field)

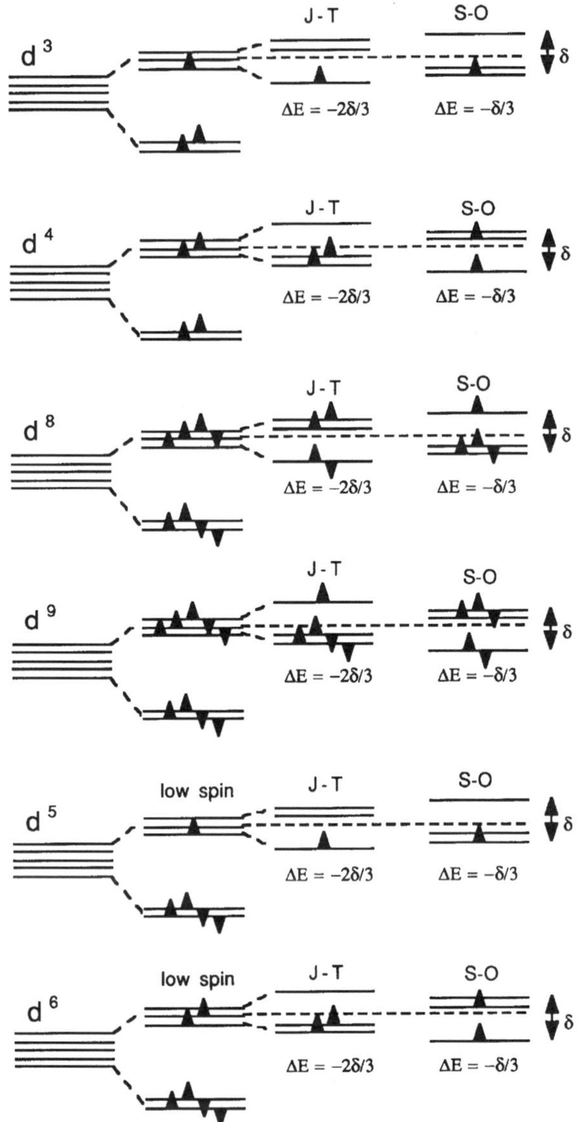

Fig. 2.27 One-electron model of J–T/S–O stabilizations of the t_{2g} shell: tetrahedral site

stabilization (see Fig. 2.9). If the concentration of J–T or S–O stabilized ions is high enough, a crystallographic phase transition can take place when the temperature decreases below a condensation threshold.[7]

[7] For a comprehensive discussion of this effect with Mn^{3+} and Co^{2+} in octahedral sites, the reader is directed to Chap. III, Sect. 1E2 of Goodenough [40].

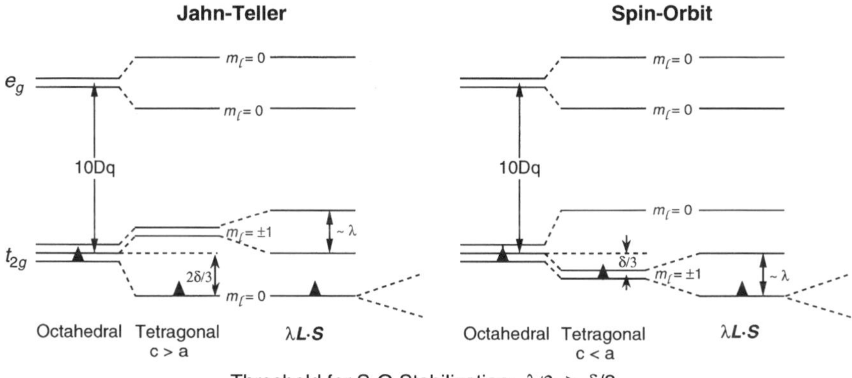

Fig. 2.28 c-Axis octahedral site stabilizations of a single d spin in the context of **Fig. 2.25**

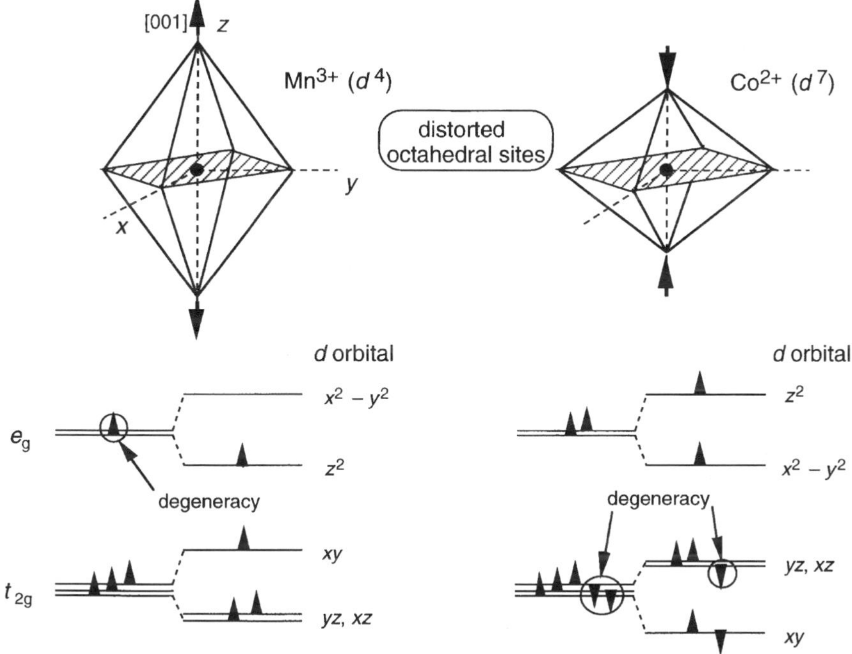

Fig. 2.29 Examples of J–T (d^4, Mn^{3+}) and S–O (d^7, Co^{2+}) stabilizations

In addition to the lattice distortions, spin–orbit stabilization also brings strong spin–lattice effects that manifest themselves in magnetoelastic properties of ferrites, particularly with Co^{2+} (d^7) ions' octahedral sites. It should also be remarked that any lower symmetry crystal field, regardless of origin, that leaves an unquenched orbital ground state can have great influence on spin–lattice relaxation properties. These subjects are revisited in relation to the magnetoelastic and microwave behavior of specific members of the $3d^n$ ion series in Chaps. 5 and 6.

Table 2.14 Jahn–Teller and spin–orbit Stabilized d^n Ions

Pure J–T effect e_g shell		S–O (or J–T) effect t_{2g} shell	
Octahedral	Tetrahedral	Octahedral	Tetrahedral
–	d^1	d^1	–
–	–	d^2	–
–	d^3 (low spin)	–	d^3
d^4	–	d^4 (low spin)	d^4
–	–	d^5 (low spin)	d^5 (Low spin)
–	d^6	d^6	d^6 (Low spin)
d^7 (Low spin)	–	d^7	–
d^8 (Low spin)	–	–	d^8, d^8 (Low spin)
d^9	–	–	d^9

These phenomena can be even more intriguing with ions of the $4f^n$ rare-earth series because of the combination of a weak crystal field and a stronger spin–orbit coupling that leaves the total angular momentum J largely unperturbed, retaining most of their orbital and spin degeneracies in a crystal lattice. In some instances, for example, paramagnetic $TmPO_4$ with a degenerate E_g doublet ground term from the partial quenching by the cubic crystal field, the distinction between J–T and S–O stabilization is somewhat blurred. Many observed spin–orbit–lattice effects such as structural phase transitions induced by a high magnetic field are viewed as part of a broad generic J–T category [42] although spin–orbit effects remain dominant in other properties such as spin–lattice relaxation.

2.5 Covalent Stabilization

The general topic of chemical bonding is expansive, and its evolution as a prime vehicle for the application of quantum theory to electronic structure and properties has grown with the availability and capability of high-speed digital computers. In a primarily ionic crystal lattice, for example, Na^+Cl^-, the positive cation is bonded electrostatically to its negative anion neighbors through direct Coulomb forces of attraction after the valence electrons of the cation are transferred to the anion. Where the electronic wavefunctions extend far enough to overlap in the interatomic spaces, bonding can then be stabilized by attractive forces between electrons and opposing nuclei, as in the case of the H_2 molecule. This situation would represent the purest example of covalent bonding in which the orbiting electrons remain partially delocalized in their shared orbital states. In an extreme case where the s and p electrons correlate into itinerant charge clouds that provide the electrostatic "glue" between the nuclei, the bonding is defined as metallic because the electrons are then conducting.

When the metal is from a transition series with unpaired electron spins in orbital states that do not participate directly in the bonding, their action is treated as a perturbation already introduced by the point-charge crystal field effects. To probe the influence of these unfilled shells, hybrid orbital states formed between adjacent neighbors must be established. The process therefore involves two approaches to examining magnetic properties of crystal lattices. To determine the geometrical structure and bonding energy of the lattice based on ionic and covalent stabilization energies, the valence-bond method is used. For the purposes of describing the electronic origins of the spontaneous effects involving d- and f-shell electrons, an extension of the crystal field or point-charge model called molecular orbital theory is the most useful method for analyzing local magnetic properties. In this approximation the point charges are replaced by the actual anions (or ligands) and their orbital states (usually $2p$) interact with the d or f states of the transition metal cations. Analogous to the one-electron model of an individual ion that sorts out high- and low-spin states by determining the internal spin ordering, molecular orbitals can provide insight into the stabilization of interionic magnetic ordering prior to attempting more elaborate solutions.

2.5.1 Molecular-Orbital Theory

To examine the role of interactions among electrons occupying individual orbital states of adjacent ions, hybrid orbital states belonging to the overall nuclear skeleton of the molecule are constructed from the individual orbitals. In this concept, the combined orbitals are assigned to the molecule rather than the individual atoms or ions, and the available bonding electrons among the ions can then be distributed among these "molecular orbitals" with electron spin directions following the dictates of the Pauli principle and Hund's rule, analogous to the formation of electronic configurations in atomic structure. Once the molecular orbital scheme is established, the Aufbau principle that was applied to map the electron distributions that make up the crystal-field ground states of the transition-metal ions in Figs. 2.17 and 2.18 can be adopted for the molecule [43]. The particular molecule structure can be determined a priori from the solution of a valence-bond analysis, with the initial electronic orbital state energies determined, for example, by a Madelung energy computation in the case of a well-defined ionic molecule.

In contrast to the valence-bond method that provides information about the chemical and macroscopic physical properties, including mechanical and thermal behavior, that are controlled by the strength and geometry of the bonding, a molecular orbital approximation can add a window to the specific orbital (and electron spin) interactions that determine the electronic and magnetic properties. This approach culminates in the formation of hybrid wave functions that describe the resulting density distributions of the actual charge clouds – hence the term molecular orbitals. It begins with the introduction of covalence in the form of linear combinations of the atomic orbitals (LCAO) of the individual atoms comprising the molecule. In transition-metal cations, these orbital states are the eigenfunctions of the particular crystal field in which the ion resides. As a result, the focus of the molecular orbital analysis is specifically the interaction between individual electron orbitals rather than the combined result of multielectron orbital wave functions, in the manner of the weak rather than strong crystal field approach. Attractions between nuclei and electrons are therefore not treated as competing with the electron–electron repulsive perturbation, but rather grouped as part of the overall crystal lattice energy.

For a generic heteronuclear diatomic molecule with metal ions a and b, the individual orbital wavefunctions are hybridized into linear combinations (LCAO) from the corresponding one-electron (nonorthogonal) orbital functions φ_a and φ_b according to

$$\varphi_- = N_- (\lambda_- \varphi_a - \varphi_b) \text{ (antibonding state)} \tag{2.35a}$$

$$\text{and } \varphi_+ = N_+ (\varphi_a + \lambda_+ \varphi_b) \text{ (bonding state)}, \tag{2.35b}$$

where $N_+ = \left(1 + \lambda_+^2 + \lambda_+ \sigma_{ab}\right)^{-1/2}$ and $N_- = \left(1 + \lambda_-^2 - \lambda_- \sigma_{ab}\right)^{-1/2}$ are the normalization coefficients, if desired, and $\sigma_{ab} = \langle \phi_a \mid \phi_b \rangle$ is the density of the orbital wavefunction overlap or simply the overlap integral. It represents a probability that can vary from 0 to 1. Thus, the covalent sharing of orbital states by localized electrons within the overall ionic bonding scheme is represented by the terms $\lambda_{+,-}\sigma_{ab} \leq 1$. (If the molecule is homonuclear, such as H_2, $\lambda_{+,-} = 1$). The spatial probability densities φ_+^2 and φ_-^2 are depicted by the sketches in Fig. 2.30. Where the electrons can be delocalized in the overlap volume between the nuclei, spins are

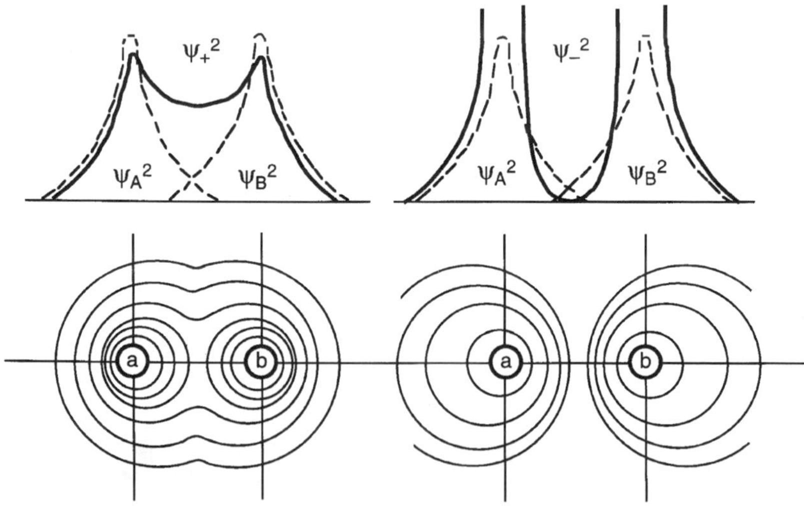

Fig. 2.30 Schematic of bonding and antibonding states for spherically symmetric s-electron wavefunctions of two similar atoms A and B, showing the probability distributions of electron charge between them. Except in ferromagnetic cases with partially filled d shells, the more stable is called the bonding state with greater charge density between the nuclei

aligned antiparallel in observance of the Pauli principle, and the energy is reduced because the electrons screen the repulsive forces between the two positively charged nuclei. For this reason, φ_+ is called the "bonding" state; the opposite condition occurs in the "antibonding" state φ_- because wavefunction charge clouds repel each other, thereby restoring orthogonality.[8] In this case, spin directions can return to the more natural parallel alignment in observance of Hund's rule of maximum spin polarization. Note also that the unnormalized density function of the bonding state $|\varphi_+|^2$ is greater than that of the antibonding state $|\varphi_-|^2$, consistent with a net stabilization energy gained by the wavefunction overlap (see Appendix 2A). From this elementary approximation it can be concluded that antiparallel spin alignment requires the nonorthogonality of φ_a and φ_b.

2.5.2 Determinant Method

Estimates of the molecular orbital eigenfunctions can be obtained by solving two-level perturbation problems in the standard way, guided by the abbreviated Hamiltonian for the interaction of two empty orbital states

$$\mathcal{H} = \mathcal{H}_a + \mathcal{H}_b = -\left(\hbar^2/2m_e\nabla_a^2 + \frac{Z_a e^2}{r_a} + V_a\right) - \left(\hbar^2/2m_e\nabla_b^2 + \frac{Z_b e^2}{r_b} + V_b\right), \tag{2.36}$$

where the V_a and V_b represent latent contributions from the respective cross-nuclear terms $Z_a e^2/r_b$ and $Z_b e^2/r_a$, here treated as constant along the bond axis. The interionic energies are treated in this manner because a one-electron solution, for example, for H_2^+, is sought as a molecular-orbital base analogous to the $3d^1$ crystal field model. These interactions are also diminished by screening from the electron (1)–electron (2) repulsive energy (normally stated as e^2/r_{12}) that is taken into account by a semiempirical factor that modifies the off-diagonal matrix term defined later. For the example of overlapping orbital functions of ions with effective nuclear (valence) charge Z_a and Z_b and respective ionic radius r_a and r_b, the stabilization energy and coefficients of hybridization can be approximated from a matrix perturbation method applied to a diatomic molecule, as described in Ballhausen [44], with a detailed derivation in Ballhausen and Gray [45]. In this method, a self-consistent term appears in the secular equation $\left|\mathcal{H}_{ij} - E(\delta_{ij} - (1-\delta_{ij})\,\sigma_{ij}\right| = 0$ for the determinant

$$\begin{vmatrix} \mathcal{H}_{aa} - E & \mathcal{H}_{ab} - E\sigma_{ab} \\ \mathcal{H}_{ba} - E\sigma_{ba} & \mathcal{H}_{bb} - E \end{vmatrix} = \begin{vmatrix} E_a - E & b_{ab} - E\sigma_{ab} \\ b_{ba} - E\sigma_{ba} & E_b - E \end{vmatrix} = 0, \tag{2.37}$$

[8] φ_+ is called the "bonding" state and is often designated by a subscript g (for gerade or even). The opposite effect occurs with the "antibonding" state φ_-, which would be indicated by a u (for ungerade or odd) subscript.

where

$$\begin{aligned}\mathcal{H}_{aa} &= \langle\varphi_a|\mathcal{H}|\varphi_a\rangle = E_a\langle\varphi_a\mid\varphi_a\rangle = E_a,\\ \mathcal{H}_{bb} &= \langle\varphi_b|\mathcal{H}|\varphi_b\rangle = E_b\langle\varphi_a\mid\varphi_a\rangle = E_b,\\ \mathcal{H}_{ab} &= \mathcal{H}_{ba} = b_{ab} \approx \langle\varphi_a|h_{ab}(E_a+E_b)|\varphi_b\rangle = h_{ab}(E_a+E_b)\langle\varphi_a\mid\varphi_b\rangle\\ &\approx h_{ab}(E_a+E_b)\sigma_{ab}.\end{aligned} \tag{2.38}$$

where b_{ab} is the electron exchange or transfer integral between states of energy E_a and E_b,[9] $\sigma_{ab} = \langle\varphi_a\mid\varphi_b\rangle \leq 1$ is the orbital overlap integral, and h_{ab} is an interionic screening factor that typically has a value of unity for ionic bonds. This relation for $\mathcal{H}_{ab}$ was recommended by Wolfsberg and Helmholtz [44,46].

To implement the solution of this equation when applied to the molecular orbital problem, it must first be recognized that φ_a and φ_b are not part of an orthonormal set. The respective energies E_a and E_b of the φ_a and φ_b states appear on the diagonal of the determinant in (2.37) and correspond to those of the outermost electrons under the influence of the charges from their respective nuclear skeletons and valence electrons. In an ionic molecule, the electron energy E_a of the cation at site a can be estimated from the ionization potential of the cation outer electron destabilized by the repulsive field of the negative anion (the source of the anisotropic crystal-field perturbation), and E_b from the electron affinity of the anion outer electron stabilized by the attractive field of the positive cation.[10] To be Hermitian, however, $\mathcal{H}_{ab}$ must equal $\mathcal{H}_{ba}$. Because these determinant elements represent the transfer or exchange integral b_{ab} of the tight-binding approximation, it is reasonable to adopt a singular value of $b_{ab} = b$.

This expression for $\mathcal{H}_{ab}$ in (2.38) is based on the assumption that E_a and E_b are nearly uniform within the overlap region as suggested by Fig. 2.30. The values of E_a and E_b are defined graphically in the simple molecular orbital diagram of Fig. 2.31 as the stabilization energies of the outer electrons on the respective ions. Note that the values of E_a and E_b are negative, with E_a chosen to be of lower energy. Regardless of what the exact expression for $\mathcal{H}_{ab}$ is, the earlier exercise points out that the magnitude of the transfer integral b is jointly dependent on the atomic stabilization energy of the electrons involved in covalent bond and the volume fraction of overlap σ.

General solutions of (2.37) for the bonding (+) and antibonding (−) states are as follows:

[9] In this model, all electronic energies are referenced to the zero energy of the free ion. E_a or E_b is the algebraic sum of its ionic stabilization energy (the cation ionization potential or anion electron affinity) and the electrostatic potential from the charge on its neighboring ion.

[10] When applied to ionic bonds, the covalent electrons are treated as initially localized on their nuclei, as in the case of the O^{2-} anion with its filled $2p$ shell. As a result, the Hund's rule repulsion arising from a dominant e^2/r_{ij} internal exchange term is absent, which then precludes the possibility of itinerant ferromagnetism from an antibonding band.

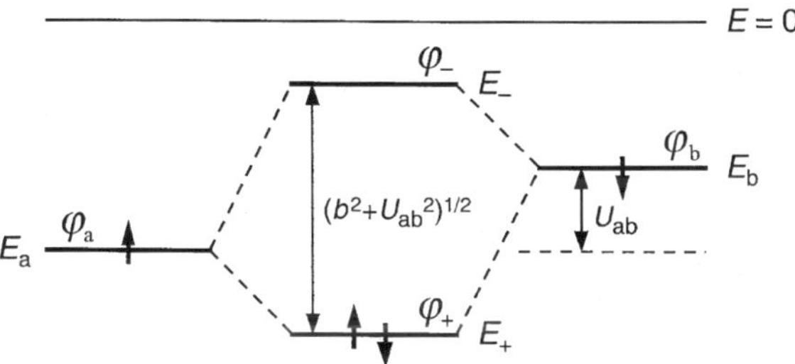

Fig. 2.31 Basic two-ion molecular orbital diagram, with one electron per ion. According to the Aufbau method, both would be expected to favor the lower energy bonding state φ_+

$$E_\pm = \frac{E_a + E_b - 2b\sigma \pm \sqrt{(E_a + E_b - 2b\sigma)^2 - 4\,(E_a E_b - b^2)\,(1-\sigma^2)}}{2\,(1-\sigma^2)}. \tag{2.39}$$

If we now substitute $b = h\,(E_\mathrm{a} + E_\mathrm{b})\,\sigma$, (2.39) will be expressed as

$$E_\pm = \frac{E_\mathrm{F}\left(1 - 2h\sigma^2\right) \pm \frac{1}{2}\sqrt{U^2\,(1-\sigma^2) + b^2\,(2h-1)^2\,/\,h^2}}{1-\sigma^2}, \tag{2.40}$$

where $E_\mathrm{F} = (E_\mathrm{a} + E_\mathrm{b})\,/2$ is the unscreened average electron energy, and the unscreened excitation energy $U = E_\mathrm{a} - E_\mathrm{b}$, so chosen to set its value negative and therefore consistent with the other energy parameters. The influence of h is illustrated in Appendix 2A for a homonuclear molecule, for example, H_2, for which $E_\mathrm{F} = E_\mathrm{a} = E_\mathrm{b}$. If $h < 0$, the antibonding state can be stabilized, thereby implying the implicit screening effect of the e^2/r_{12} repulsive energy. Based on the findings of Wolfsberg and Helmholz [46], the parameter $h \approx 1$ for ionic compounds, which selects the positive solution to be the ground orbital state. An antiferromagnetic spin pairing with the symmetrical hybridization is required for Pauli antiparallel spin ordering of a bonding state. For oxides, (2.40) simplifies to

$$E_\pm = \frac{E_\mathrm{F}\left(1 - 2h\sigma^2\right) \pm \frac{1}{2}\sqrt{U^2\,(1-\sigma^2) + b^2}}{1-\sigma^2}. \tag{2.41}$$

A further approximation can be made for $\sigma^2 << 1$,

$$E_\pm = E_\mathrm{F} \pm \frac{1}{2}\sqrt{U^2 + b^2}. \tag{2.42}$$

There is a procedure to determine the value of σ by computation [46], but in practice $\sigma < 1$ can be treated as a semiempirical parameter to be deduced from experiment [47, 48].

Eigenfunctions of the E_+ and E_- levels can now be determined by forming the hybrid linear combinations

$$\begin{aligned} \varphi_- &= N_-\left(c_{\mathrm{ba}}\varphi_{\mathrm{a}} - c_{\mathrm{bb}}\varphi_{\mathrm{b}}\right) \quad \text{(antibonding)} \\ \varphi_+ &= N_+\left(c_{\mathrm{aa}}\varphi_{\mathrm{a}} + c_{\mathrm{ab}}\varphi_{\mathrm{b}}\right) \quad \text{(bonding)}, \end{aligned} \tag{2.43}$$

and $N_- = \left(c_{\mathrm{ba}}^2 + c_{\mathrm{bb}}^2 - 2c_{\mathrm{ba}}c_{\mathrm{bb}}\sigma\right)^{-1/2}$ and $N_+ = \left(c_{\mathrm{aa}}^2 + c_{\mathrm{ab}}^2 + 2c_{\mathrm{aa}}c_{\mathrm{ab}}\sigma\right)^{-1/2}$. Expressed in terms of the corresponding LCAO coefficients, $c_{\mathrm{ab}}=c_{\mathrm{ba}} = \lambda_{+,-}$. The c_{ij} coefficients are then determined from the relation $\sum_{ij}\{H_{\mathrm{ij}}-E[\delta_{ij}-\left(1-\delta_{ij}\right)\sigma_{ij}]\}c_{ij}=0$ for each solution value of E. Accordingly, we can write

$$\begin{aligned} &(b - E_-\sigma)\,c_{\mathrm{ba}} + (E_{\mathrm{b}} - E_-)\,c_{\mathrm{bb}} = 0, \\ &(E_{\mathrm{a}} - E_+)\,c_{\mathrm{aa}} + (b - E_+\sigma)\,c_{\mathrm{ab}} = 0. \end{aligned} \tag{2.44}$$

After recognizing the normalizing conditions $N_-{}^2\left(c_{\mathrm{ba}}{}^2 + c_{\mathrm{bb}}{}^2\right) = 1$ and $N_+{}^2\left(c_{\mathrm{aa}}^2 + c_{\mathrm{ab}}^2\right) = 1$, we obtain the general relations

$$\begin{aligned} N_-{}^2c_{\mathrm{ba}}{}^2 &= \left[\frac{(E_{\mathrm{b}} - E_-)^2}{(E_{\mathrm{b}} - E_-)^2 + (b - E_-\sigma)^2}\right], \\ N_-{}^2c_{\mathrm{bb}}{}^2 &= \left[\frac{(b - E_-\sigma)^2}{(E_{\mathrm{b}} - E_-)^2 + (b - E_-\sigma)^2}\right], \\ N_+{}^2c_{\mathrm{aa}}{}^2 &= \left[\frac{(b - E_+\sigma)^2}{(E_{\mathrm{a}} - E_+)^2 + (b - E_+\sigma)^2}\right], \\ N_+{}^2c_{\mathrm{ab}}{}^2 &= \left[\frac{(E_{\mathrm{a}} - E_+)^2}{(E_{\mathrm{a}} - E_+)^2 + (b - E_+\sigma)^2}\right]. \end{aligned} \tag{2.45}$$

The eigenvalues and hybrid eigenfunctions for the two extreme approximations can now be determined for use in (2.43), recalling the definitions of b, E_{F}, and U.

For strongly covalent molecules with $b^2/U^2 >> 1$, (2.42) becomes mainly a first-order perturbation, with a small off-diagonal correction

$$E_\pm \approx E_{\mathrm{F}} \pm \frac{1}{2}\sqrt{U^2 + b^2} \approx E_{\mathrm{F}} \pm b/2 \pm U^2/4\mathrm{b}. \tag{2.46}$$

In this limit, $U^2/4b$ is the additional stabilization of the bonding state and can therefore represent the trap barrier for an electron transfer between φ_{a} and φ_{b}. It is the bonding state energy that must be gained to escape the trap. As a consequence, $U^2/4b$ is also an effective activation energy for polaronic charge transport to be discussed in Chap. 8.

For metal-oxide compounds, the bonds are primarily ionic, with covalent electron sharing (delocalization) more of a perturbation, giving a ratio $b^2/U^2 << 1$. In this extreme, the covalent interaction is a second-order perturbation,

$$E_\pm \approx E_{\mathrm{F}} \pm \frac{1}{2}U\left(1 + b^2/2U^2\right) = E_{\mathrm{F}} \pm \frac{1}{2}U \pm b^2/4U, \tag{2.47}$$

where

$$
\begin{aligned}
E_- &\approx E_b - b^2/4U \quad \text{antibonding}, \\
E_+ &\approx E_a + b^2/4U \quad \text{bonding}, \\
E_+ - E_- &\approx U + b^2/2U.
\end{aligned}
\tag{2.48}
$$

Because the transfer integral b has also been designated as a negative quantity consistent with E_a and E_b, the $b^2/4U$ term represents the amount by which the hybrid energy is stabilized by the transfer energy. For the oxide case, (2.45) simplifies to

$$
\begin{aligned}
N_-{}^2 c_{ba}{}^2 &= \left(\frac{E_F}{E_a}\frac{b}{2U}\right)^2 \left[1 + \left(\frac{E_F}{E_a}\frac{b}{2U}\right)^2\right]^{-1} \approx \left(\frac{E_F}{E_a}\frac{b}{2U}\right)^2, \\
N_-{}^2 c_{bb}{}^2 &= \left[1 + \left(\frac{E_F}{E_a}\frac{b}{2U}\right)^2\right]^{-1} \approx 1, \\
N_+{}^2 c_{aa}{}^2 &= \left[1 + \left(\frac{E_F}{E_b}\frac{b}{2U}\right)^2\right]^{-1} \approx 1, \\
N_+{}^2 c_{ab}{}^2 &= \left(\frac{E_F}{E_b}\frac{b}{2U}\right)^2 \left[1 + \left(\frac{E_F}{E_b}\frac{b}{2U}\right)^2\right]^{-1} \approx \left(\frac{E_F}{E_b}\frac{b}{2U}\right)^2.
\end{aligned}
\tag{2.49}
$$

for direct application to (2.43), which can then be approximated by

$$
\begin{aligned}
\varphi_- &\approx \left(\frac{E_F}{E_a}\frac{b}{2U}\right)\varphi_a - \varphi_b \approx \left(\frac{b}{2U}\right)\varphi_a - \varphi_b, \\
\varphi_+ &\approx \varphi_a + \left(\frac{E_F}{E_b}\frac{b}{2U}\right)\varphi_b \approx \varphi_a + \left(\frac{b}{2U}\right)\varphi_b.
\end{aligned}
\tag{2.50}
$$

Note the difference between the N_+c_{ab} and N_-c_{ba} coefficients of (2.49), which is similar to (2.45). This arises from the nonzero overlap integral ($\sigma \neq 0$), as described in Appendix 2A together with the implications of normalization. For preliminary estimates, E_F/E_b and E_F/sE_a can be set to unity. The square of coefficient $b/2U$ then represents the lesser fractional shares of the starting φ_a and φ_b functions in their respective φ_+ and φ_- hybrid eigenfunctions.

2.5.3 *σ and π Bonds and the Molecular Orbital Diagram*

A thorough discussion of the symmetry contributions to covalent bonding will not be attempted in this text, but an appreciation of its importance can be introduced through the definition of the primary types of overlap interactions. For the transition-metal oxides, we need only to consider the influence of an unfilled d electron shell and its interaction with the $2p$ shell of the O^{2-} ligands. Although the electrons in the outer p and s shells of the cation are principally responsible for the chemical bond, the magnetic properties enter through the d shell. In Fig. 2.32, four examples are sketched for the $d-p$ orbital lobe arrangements characteristic of transition-metal ions in octahedral oxygen coordinations.

The two principal factors that determine the magnitude of σ are as follows: (1) the amount of spatial overlap of the cation d and ligand p orbital lobes, designated according to whether the ligand lobe is directed along the axis joining the two nuclei (σ bonds) or at a significant angle to it (π bonds), and (2) the relative signs of the lobes. A rule often followed is that a *nonzero overlap integral σ will occur if both orbitals have the same symmetry about the axis joining the two nuclei*. From inspection of the examples in Fig. 2.32 we observe that $e_g\left(d_{x^2-y^2}\right)-p_{xy}\sigma$ provides the largest σ and strongest bond while $t_{2g}\left(d_{xy}\right)-\ p_{x,y}\pi$ can also contribute, but through a significantly weaker interaction. Similar constructions can also be made for the $e_g\left(d_{z^2}\right)-p_{xy}\sigma,\ \pi$ couplings. The possible influence from $e_g-p\pi$ and $t_{2g}-p\pi$ is deemed negligible because the orthogonality of the lobe polarities would produce cancellation. For a first-order estimate of covalent energy in octahedral complexes, only $e_g-p\sigma$ is generally taken into account in determining the bonding and anti-

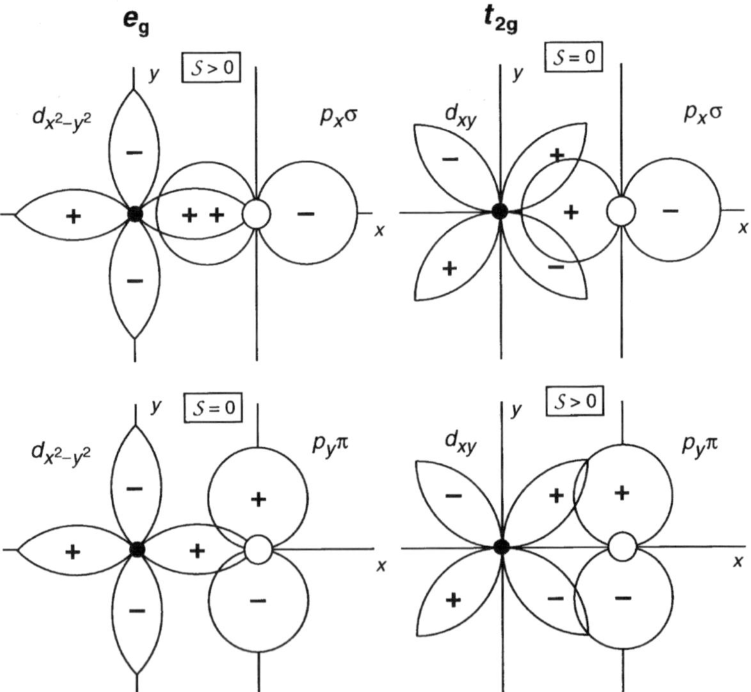

Fig. 2.32 Illustrations of the difference between σ (direct overlap with oxygen 2p orbital lobes directed at the cation nucleus) and π (oblique overlap with oxygen p orbital lobes directed orthogonal to the cation nucleus). Example is given for the $x-y$ plane

bonding states, while the t_{2g} orbitals in octahedral sites are treated as inactive or "nonbonding." In Fig. 2.33, schematic energy level diagrams illustrate the relation between the point-charge crystal-field approximation described in Sects. 2.3 and 2.4 and the molecular orbital model for a generic $3d$–$2p$ octahedral-site compound, typically a transition-metal oxide, for example, Mn^{3+}–O_6. In this case, the anion is shown with the lower energy. The bonding (b) and antibonding (ab) hybrid states for the separate e_g and t_{2g} shells are sketched in conformance with Fig. 2.31. To emphasize the Aufbau occupancy expectation, the bonding states are presented as occupied mainly by the $2p$ paired spins from the fully populated anion states. The $3d$ spins from Mn^{3+} $(3d^4)$ would then occupy the e_g–$2p\sigma$ antibonding states, beginning with the t_{2g}–$2p\pi$ nonbonding levels separated by $\sim b_\sigma^2/4U$ from the e_g–$2p\sigma$ antibonding levels, which can be shown to be consistent with measured values of $10Dq \sim 2\,\text{eV}$. To extend these concepts further, note that large Dq stabilizes low-spin (spin-paired) configurations. In cases of two ions with unpaired spins, large $b^2/4U$ will tend to stabilize an antiferromagnetic ground state. In addition, large U means high ionicity and large b strong covalence.

In tetrahedral sites, the bond angle between cation and anion differs substantially from those of octahedral sites, and the t_{2g} orbitals become stronger contributors because the ligands reside along the cube body diagonals, as evident in Fig. 2.5. It is also for this reason that the actual value of Dq is negative, which is accounted for by showing with the doublet stabilized relative to the triplet, as illustrated in the one-electron models of Figs. 2.17 and 2.18. In later sections dealing with ferrimagnetic oxides, bond angles will be seen to play an important part in the magnetic exchange properties.

In Fig. 2.34 a complete molecular orbital energy level diagram including $4s$ and $4p$ states is shown for a $3d^4$ transition metal ion and oxygen ligands in an octahedral coordination. To a first approximation, the diagram could be constructed by computing bonding and antibonding states for each of the metal levels linked covalently to the oxygen $2p$ states in the manner of Fig. 2.33. Once this diagram is formed, the

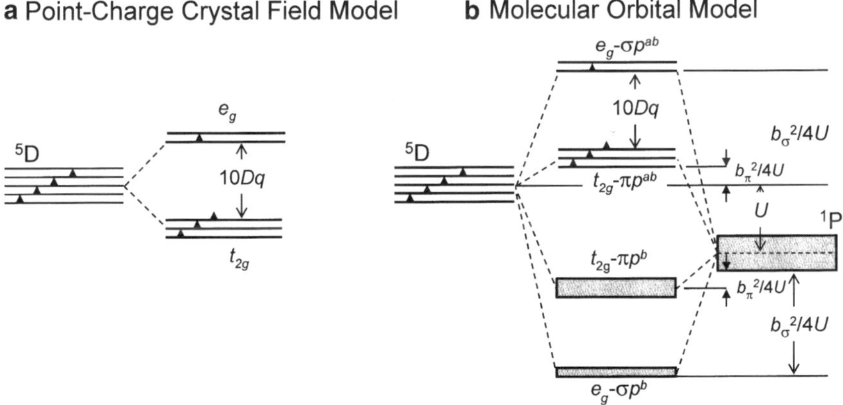

Fig. 2.33 One electron point-charge crystal-field energy-level structure contrasted with molecular-orbital model for a d^4 configuration

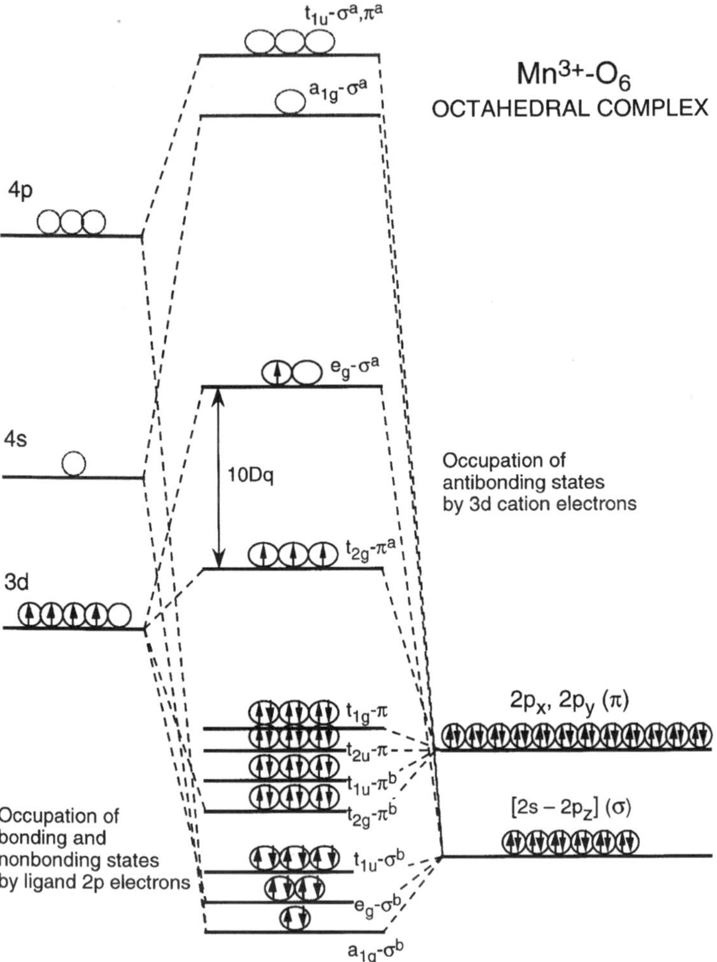

Fig. 2.34 Basic molecular orbital energy diagram for an $\left(Mn^{3+}\text{–}O_6\right)$ complex. Note that the Mn^{3+} four d electrons occupy mainly the nonbonding $\left(t_{2g}\text{–}\pi\, p^{ab}\right)$ and antibonding $\left(e_g\text{–}\sigma\, p^{ab}\right)$ states

individual molecular orbital states are then assigned electrons from the two participating ions according to Hund's rule and the Aufbau principle that was employed previously in the one-electron ground state models of the crystal-field theory. As indicated, the low-energy bonding states are mainly $2p$ states that fill with spin-paired electrons from the oxygen ligands in the fractional amount $N_+{}^2c_{aa}{}^2$ for each hybrid orbital combination, while the antibonding distributions would favor the σ-bonding e_g states in the corresponding fraction $N_-{}^2c_{bb}{}^2$. Note also that the weaker π-bonding t_{2g} electrons are designated as nonbonding and are considered to remain in their initial cation orbital states. To place the crystal-field orbital angular momentum quenching in the context of the full molecular orbital scheme, the $3d$ and $2p$ spin occupancies of an Mn^{3+} $\left(d^4\right)$ ion in an oxygen octahedral complex are included in Fig. 2.33b. Note how the $10Dq$ splitting between the t_{2g} and e_g states is retained, although relabeled as antibonding t_{2g}–π^a and e_g–σ^a to conform to the expanded nomenclature. Because of the weak $t_{2g}-p\pi^a$ bonding, the $e_g-p\sigma^a-t_{2g}-p\pi^a$ energy splitting $10Dq$ can be roughly equated to antibonding destabilization energy $b^2/4U$ from (2.48).

Although situations in which the earlier two-level model can be applied with some degree of rigor are relatively uncommon, the approximation can be very helpful in providing insight on a qualitative basis. In later chapters, such applications will be demonstrated in problems that involve magnetic exchange, magnetoelasticity, magneto-optical effects, polarized spin transport, and superconductivity.

2.5.4 Valence Bond Method

Although the one-electron molecular orbital method is sufficient for most of the discussion of spin ordering and polarized spin transport on a qualitative level, the valence bond concept is used as the basis for quantitative analyses that involve multielectron interactions [43]. This is analogous to the difference between the one-electron crystal field approximation and the multielectron solutions required for interpretation of paramagnetic resonance spectra. The valence bond method was introduced by Heitler and London [7] in their analysis of the H_2 molecule and later generalized by Pauling [49] for application to a wide variety of molecular structures comprising dissimilar atoms. For bonding energy calculations, each atom is assigned a valence state after a specific lattice structure is defined. Depending on the location of the atomic components in the Periodic table, the lattice bonding is described in terms of varying degrees of ionic and covalent.

In a diatomic molecule made up of radically different elements such as an alkali halide, the bond is almost entirely ionic. The basic interaction of ionic bonding is the Coulomb attraction between positive and negative charges of the cations (+) and anions (−). The simplest example is sodium chloride NaCl with Na^+ and Cl^- ions occupying alternating sites at the corners of a cubic unit cell shown in Fig. 2.35,

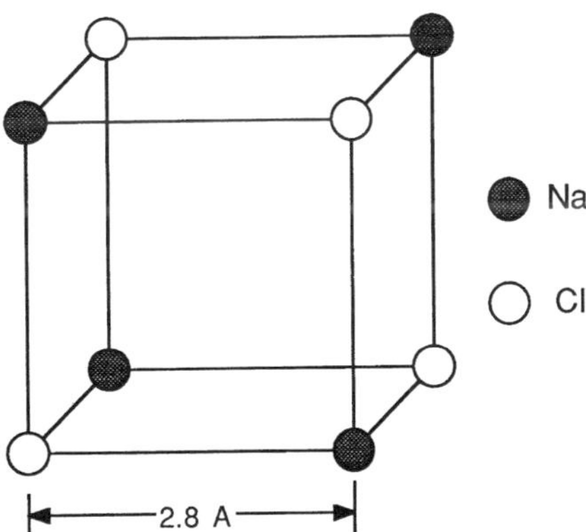

Fig. 2.35 Basic unit cell of an Na^+Cl^- ionic lattice

in which the single $3s$ electron of Na is transferred to fill the single empty state in the $3p$ shell of Cl. In the most straightforward approach, the energy of stabilization among neutral atoms is determined by the resultant of two parts: (1) the Coulomb energy, which arises from the mutual attraction between the positive and negative ions, tempered by the shielding of filled shells, and (2) the algebraic sum of the Na $3s^1$ electron ionization potential energy (+5.14 eV) and the Cl $3p^5$ electron affinity energy (−3.71 eV) that may produce additional stabilization by the elimination of partially filled shells. From the example worked out in Kittel [50]:

$$\begin{aligned} \mathrm{Na} &\rightarrow \mathrm{Na}^+ + \mathrm{e}^- - 5.14\,\mathrm{eV}, \\ \mathrm{Cl} &\rightarrow \mathrm{Cl}^- - e - +3.71\,\mathrm{eV}, \\ \mathrm{Na}^+ + \mathrm{Cl}^- &\rightarrow \mathrm{NaCl} - 1.43\,\mathrm{eV} + \text{bond energy}. \end{aligned} \tag{2.51}$$

A first approximation to the bond energy is obtained from a simple Coulomb attraction given by $E_{\mathrm{Coul}} = \mathrm{e}^2/R_0$ (see Fig. 2.32), where $R_0(=2.81\,\text{Å})$ is the final separation between ions as determined by minimization of the total energy that encompasses the long-range ionic attraction opposed by the short-range repulsion among the electron charge clouds, which begins to dominate where the nuclear separation $R < R_0$. This calculation yields E_{Coul} of 5.1 eV, which is a factor of about 1.5 less than the experimentally determined value of 7.9 eV. The discrepancy may be accounted for by a more complete calculation that includes summations over more of the ionic lattice. Such methods are also described in Kittel where the various ways of computing the Madelung energy constant E_{Mad} are also described. For the NaCl lattice, $E_{\mathrm{Mad}} = 1.75$, as compared with the 1.55 value determined from a fit to experiment.

The valence-bond analysis of two single-electron atoms designated a and b with respective unpaired electrons (1) and (2) begins with the definition of the full Hamiltonian that includes explicitly all of the interactions embodied in the screening parameter h_{ab} of H_{ab} in (2.38)

$$\mathcal{H} = \mathcal{H}_0 + \mathcal{H}_1, \tag{2.52}$$

where

$$\mathcal{H}_0 = \mathcal{H}_{\mathrm{a}} + \mathcal{H}_{\mathrm{b}} = -\frac{\hbar^2}{2m_{\mathrm{e}}}\left[\nabla_{\mathrm{a}}^2 + \nabla_{\mathrm{b}}^2\right] - \frac{Z_{\mathrm{a}}e^2}{r_{\mathrm{a}1}} - \frac{Z_{\mathrm{b}}e^2}{r_{\mathrm{b}2}} \tag{2.53}$$

and

$$\mathcal{H}_1 = -\frac{Z_{\mathrm{a}}e^2}{r_{\mathrm{a}2}} - \frac{Z_{\mathrm{b}}e^2}{r_{\mathrm{b}1}} + \frac{e^2}{r_{12}} + \frac{Z_{\mathrm{a}}Z_{\mathrm{b}}e^2}{r_{\mathrm{ab}}}. \tag{2.54}$$

The next step is the formation of the two-electron degenerate wavefunctions $\varphi_0 = \varphi_{\mathrm{a}}(2)\,\varphi_{\mathrm{b}}(1)$ and $\varphi_{\mathrm{ex}} = \varphi_{\mathrm{a}}(2)\,\varphi_{\mathrm{b}}(1)$. The first of these is the normal state with the respective electrons on their host ions; the second is the energy equivalent state with the electrons exchanged between the two atoms. As discussed in Sect. 1.3.3, the eigenfunction solutions are linear combinations constructed according to

$$\varphi_{\mathrm{anti}} = \frac{1}{\sqrt{2}}\left[\varphi_{\mathrm{a}}(1)\,\varphi_{\mathrm{b}}(2) - \varphi_{\mathrm{a}}(2)\,\varphi_{\mathrm{b}}(1)\right], \tag{2.55}$$

$$\varphi_{sym} = \frac{1}{\sqrt{2}} \left[\varphi_a (1) \varphi_b (2) + \varphi_a (2) \varphi_b (1)\right].$$

The energy stabilization of the bond can be seen from the electron probability distributions expressed as

$$\varphi_{anti}{}^2 = \frac{1}{2} \left(\varphi_0{}^2 + \varphi_{ex}{}^2 - 2\varphi_0\varphi_{ex}\right), \tag{2.56}$$

$$\varphi_{sym}{}^2 = \frac{1}{2} \left(\varphi_0{}^2 + \varphi_{ex}{}^2 + 2\varphi_0\varphi_{ex}\right).$$

and illustrated in Fig. 2.31. Bonding therefore can be explained by the reduction of the repulsive forces between the nuclei due to the screening effect of the electrons. For that reason φ_{sym} is the lower energy hybrid orbital state. These definitions will be related to the concept of stabilization energy of the crystal fields when the molecular orbital approach is introduced.

A basic principle of the valence bond theory is that the number of bonds in the structure be equal to the number of available valence electrons that occupy the outer shell. In the broader picture of the Periodic table, the ionic alkali halides may be viewed as 1–7 compounds according to the number of occupied states in the outer *s* and *p* shells of the neutral atoms. As the combinations compress toward the center columns of the table, the distinction between cation and anion becomes less real and covalent bonding similar to that of like atoms such as the H_2 molecule becomes the preferred model. In the balanced case of the 4–4 of carbon, silicon, germanium, and to a lesser extent their 3–5 gallium arsenide cousin, the bonds result from the exchange of electrons among the *s* and *p* shells of neutral atoms. From the quartet of four-electron orbital combinations designated by $s^1 p^3$, the diamond lattice structures of pure tetrahedral bonding (see Fig. 2.5) are produced. With this concept, therefore, the practice of bond formation by hybridization of atomic orbitals was initiated.[11]

Solutions to bonding problems by the valence-bond method are carried out from first principles by constructing a Hamiltonian that includes all of the nuclear and electronic interactions necessary to determine the binding energy of the molecule and its equilibrium atomic configuration. It has therefore proven to be a powerful tool in determining molecular structures, even for complicated multielectron systems. In spite of the overall effectiveness of the valence-bond method, however, the model suffers from the shortcoming that it deemphasizes the electron transfer interactions, that is, the "ionic" exchange terms, by relying on the Madelung energy to account for these contributions. Even when hybridized, for example, $s^1 p^3$, the wave functions remain combinations of the orbitals within each atom, unperturbed by interactions with neighboring charge clouds. As a consequence, the basic method also

[11] A qualitative review of the valence-bond concept, featuring vivid illustrations of hybrid bond formations, can be found in Chang [51].

does not address directly the effects of crystal fields on the partially filled d electron states, which provide a critical part of the electrostatic stabilization that determines the magnetic properties. For this analysis, wavefunctions that represent the shared electrons of the molecule must be obtained. Appendix 2B presents examples of how solutions molecular-orbital hybrid eigenfunctions can be used to refine the results of a valence-bond analysis.

Appendix 2A Homonuclear Molecule Ion

Because the starting wavefunctions of a molecular orbital analysis of necessity lack orthogonality, it is important to consider the implications of these conditions. To this end, we review the special case of homonuclear diatomic molecule. From (2.44),

$$E_{\pm} = \frac{E_{\mathrm{F}}\left(1 - 2h\sigma^2\right) \pm \frac{1}{2}\sqrt{U^2\left(1-\sigma^2\right) + b^2\left(2h-1\right)^2/h^2}}{\left(1-\sigma^2\right)}. \tag{2.57}$$

For a homonuclear diatomic molecule, $U = 0$, $b \approx hE_{\mathrm{F}}\sigma$ and (2.57) can simplified further as

$$E_{\pm} = \frac{E_{\mathrm{F}}\left(1 - 2h\sigma^2\right) \pm E_{\mathrm{F}}\left(2h-1\right)\sigma}{\left(1-\sigma^2\right)} = \frac{E_{\mathrm{F}}\left(1 - 2h\sigma^2 \pm 2h\sigma \mp \sigma\right)}{\left(1-\sigma^2\right)}. \tag{2.58}$$

Factoring of the numerator and canceling of common factors yields the results for the antibonding and bonding energies

$$\begin{aligned} E_{-} &= \frac{E_{\mathrm{F}}\left(1 - 2h\sigma\right)}{1-\sigma}, \\ E_{+} &= \frac{E_{\mathrm{F}}\left(1 + 2h\sigma\right)}{1+\sigma}. \end{aligned} \tag{2.59}$$

Since E_{F} is chosen to be negative relative to the zero reference energy of ions in their uncombined "free" state, it is evident that E_{+} will be the ground state provided that $h > 0$, that is, the symmetric orbital bonding state that would require antiparallel spins to satisfy the Pauli principle requirement of combined asymmetry, discussed in Chap. 1. Based on the sign convention adopted, this means that the transfer energy b is negative, which is expected for attractive cross-ionic electrostatic potentials V_{a} and V_{b} in (2.36). Conversely, where the mutual electron repulsion is large, $h < 0$ and the antisymmetric orbital antibonding state would be lower, offering the possibility of ferromagnetism.

Further study of (2.59) reveals also that the destabilization of E_{-} is greater than the stabilization of E_{+}, which leaves the solution with an increase in overall energy that violates the invariance of the diagonal sum, as an apparent consequence of the lack of orthogonality of the starting wavefunctions. If one now considers the total energy when both levels are populated according to the unnormalized eigenfunctions [43],

$$\varphi_-{}^2 = (1-\sigma) \quad \text{(antibonding)}\,,$$
$$\varphi_+{}^2 = (1+\sigma) \quad \text{(bonding)}\,. \tag{2.60}$$

Then the total net energy becomes

$$\sum_{\pm} E_{\pm}\varphi_{\pm}^2 = E_+\varphi_+^2 + E_-\varphi_-{}^2 = E_{\mathrm{F}}\,(1-2h\sigma) + E_{\mathrm{F}}\,(1+2h\sigma) = 2E_{\mathrm{F}}, \tag{2.61}$$

which is unchanged. It is therefore concluded that only the unnormalized functions can restore the energy balance. This suggests that a more appropriate normalization condition is $\varphi_+^2 + \varphi_-^2 = 2$.

Appendix 2B Valence-Bond Diatomic Molecule

The usefulness of the molecular orbital approximation and how it augments the valence-bond method (including the Madelung ionic bond limit) can be appreciated by applying it to the case of a two-electron diatomic molecule. The exercise begins with the construction of two-electron products of the single-electron hybrid molecular orbitals of (2.35) that now become $\varphi_+ = \varphi_+\,(1)\,\varphi_+\,(2)$ and $\varphi_- = \varphi_-\,(1)\,\varphi_-\,(2)$. In expanded form, the bonding and antibonding states are expressed as

$$\varphi_- = N_-{}^2\,[\varphi_{\mathrm{a}}\,(1) - \lambda_-\varphi_{\mathrm{b}}\,(1)]\,[\varphi_{\mathrm{a}}\,(2) - \lambda_-\varphi_{\mathrm{b}}\,(2)]\,,$$
$$\varphi_+ = N_+{}^2\,[\varphi_{\mathrm{a}}\,(1) + \lambda_+\varphi_{\mathrm{b}}\,(1)]\,[\varphi_{\mathrm{a}}\,(2) + \lambda_+\varphi_{\mathrm{b}}\,(2)]\,. \tag{2.62}$$

If the bonding state is expanded by multiplication, ionic and covalent contributions can be separated according to

$$\varphi_+ = \varphi_+\,(\text{ionic}) + \varphi_+\,(\text{covalent})\,, \tag{2.63}$$

where

$$\varphi_+\,(\text{ionic}) = \mathrm{N}_+{}^2\left[\varphi_{\mathrm{a}}\,(1)\,\varphi_{\mathrm{a}}\,(2) + \lambda_+{}^2\varphi_{\mathrm{b}}\,(1)\,\varphi_{\mathrm{b}}\,(2)\right],$$
$$\varphi_+\,(\text{covalent}) = \mathrm{N}_+{}^2\lambda_+\,[\varphi_{\mathrm{a}}\,(1)\,\varphi_{\mathrm{b}}\,(2) + \varphi_{\mathrm{a}}\,(2)\,\varphi_{\mathrm{b}}\,(1)]\,. \tag{2.64}$$

The normalizing relation for (2.64) is $N_+ = \left(1+\lambda_+{}^2 + 2\lambda_+\sigma\right)^{-1/2}$, where $\sigma = \langle\varphi_{\mathrm{a}}|\varphi_{\mathrm{b}}\rangle$.

From (2.64) the different components of the chemical bond can now be examined. Ψ_+(ionic) is the part of the wavefunction that accounts for the two electrons occupying the bonding orbital of the same atom, thereby converting either the a or b atom into a negatively charged ion. This part is overlooked in the valence bond method as described in the text of Sect. 2.5.1 (although later versions have attempted to introduce it through a variational perturbation [52]). For ${N_+}^2 = 1/2$ and $\lambda_+ = 1, \varphi_+$ (covalent) will be recognized as the Heitler–London H_2 valence bond function φ_{sym} that combines the φ_0 *and* φ_{ex} terms in (2.55).

There are two special cases to examine:

(1) *Homonuclear molecule*, $\lambda_+ = 1$: This is the H_2 molecule case, for which the bonding state becomes

$$\phi_+ = N_+^2 \left\{ \left[\phi_a(1)\,\phi_a(2) + \phi_b(1)\,\phi_b(2) \right] + \left[\phi_a(1)\,\phi_b(2) + \phi_a(2)\,\phi_b(1) \right] \right\}, \tag{2.65}$$

where the normalizing relation reduces to $N_+ = (2 + 2\sigma)^{-1/2}$. The second term is precisely the result of the valence bond analysis for φ_{sym} given by (2.55). Most notable in this molecular orbital function is the inclusion of the "ionic" term, which would represent the ${H_a}^+{H_b}^-$ and ${H_a}^-H_b^+$ possibilities that are given equal weight to the covalent terms. It has been pointed out that although the valence bond solution errs in its exclusion of these ionic terms, the molecular orbital theory exaggerates their importance. In either case, however, the magnitude of the orbital overlap integral σ maintains its influence.

(2) *Ionic molecule*, $\lambda_+ >> 1$ (for a nearly pure ligand orbital): This situation could represent the NaCl molecule with $\varphi_{Na} = \varphi_a$ and $\varphi_{Cl} = \varphi_b$. The bonding state is then expressed as

$$\varphi_+ \cong {N_+}^2 \left\{ \begin{array}{l} \left[\varphi_{Na}(1)\,\varphi_{Na}(2) + \lambda_+^2 \varphi_{Cl}(1)\,\varphi_{Cl}(2) \right] \\ \quad + {N_+}^2 \lambda_+ \left[\varphi_{Na}(1)\,\varphi_{Cl}(2) + \varphi_{Na}(2)\,\varphi_{Cl}(1) \right] \end{array} \right\}, \tag{2.66}$$

where $N_+ \approx \left(\lambda_+^2 + 2\lambda_+ \sigma \right)^{-1/2}$. Because $\sigma \leq 1$, $N_+ \approx {\lambda_+}^{-1}$ in the particular case of NaCl, (2.66) can be further simplified to

$$\varphi_+ \cong \varphi_{Cl}(1)\,\varphi_{Cl}(2) + \lambda_+^{-1} \left[\varphi_{Na}(1)\,\varphi_{Cl}(2) + \varphi_{Na}(2)\,\varphi_{Cl}(1) \right], \tag{2.67}$$

where the covalent term can now be treated as a perturbation that is inversely proportional to λ_+. Note that the dominant term in the ionic part of (2.64) is the one that represents both electrons occupying the $3p$ orbital shell of Cl, that is, the Na^+Cl^- component. To a first approximation then,

$$\varphi_+ \cong \varphi_{Cl}(1)\,\varphi_{Cl}(2)\,. \tag{2.68}$$

The potential energy of the different contributions to the bonding energy of NaCl is sketched as a function of the distance between nuclei [53] in Fig. 2.36. Note in particular the difference between the covalent and ionic curves at the minimum energy and at larger separations between the two nuclei, above the crossover of the

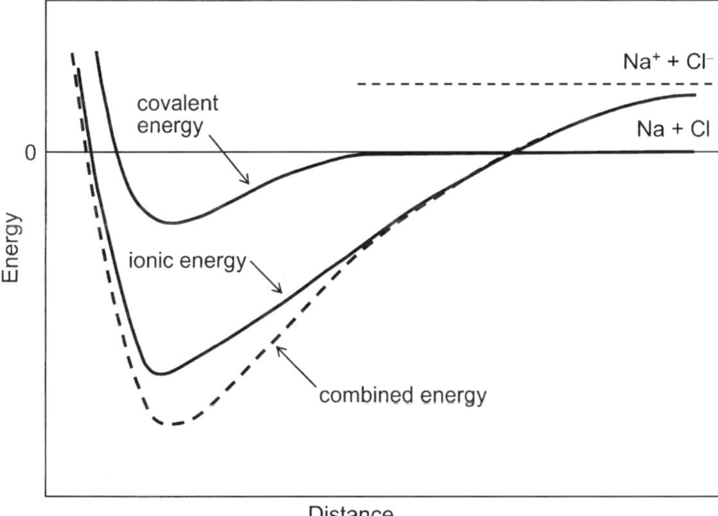

Fig. 2.36 Bonding energy contributions to the stabilization of the NaCl molecule as a function of separation

two energy curves. In this regime the combination of neutral atoms Na+Cl becomes more stable than the separated ions $Na^+ + Cl^-$ by the amount of the net ionization energy 1.43 eV. In purely covalent compounds this correction does not apply.

References

1. M.J. Buerger, *Elementary Crystallography*, (John Wiley, New York, 1956)
2. W. Borchardt-Ott, *Crystallography*, (Springer, New York, 1995)
3. C.J. Ballhausen, *Introduction to Ligand Field Theory*, (McGraw-Hill, New York, 1962)
4. H.L. Schläfer and G. Gliemann, *Basic Principles of Ligand Field Theory*, (Wiley-Interscience, New York, 1969)
5. J.S. Griffith, *The Theory of Transition-Metal Ions*, (Cambridge University Press, London, 1961)
6. J.H. Van Vleck, *The Theory of Electric and Magnetic Susceptibilities*, (Oxford University Press, London, 1932)
7. E.U. Condon and G.H. Shortley, *The Theory of Atomic Spectra*, (Cambridge University Press, London, 1963)
8. H.B. Kramers, *Proc. Amsterdam Acad. Sci.* **32**, 1176 (1929)
9. J.H. Van Vleck, *Phys. Rev.* **41**, 208 (1932)
10. W.G. Penney and R. Schlapp, *Phys. Rev.* **41**, 194 (1932); R. Schlapp and W.G. Penney, *Phys. Rev.* **42**, 666 (1932)
11. W. Low, *Paramagnetic Resonance in Solids*, (Academic Press, New York, 1960), p. 15
12. C.J. Ballhausen, *Introduction to Ligand Field Theory*, (McGraw-Hill, New York, 1962), p. 93
13. W. Low, *Paramagnetic Resonance in Solids*, (Academic Press, New York 1960) p. 22
14. M.H.L. Pryce and W.A. Runciman, *Disc. Faraday Soc.* **26**, 34 (1958)
15. G.F. Dionne and B.J. Palm, *J. Magn. Reson* **68**, 355 (1986)

16. M.T. Hutchings, *Solid State Phys.* **16**, 227 (1964)
17. J.H. Van Vleck, *Discuss. Faraday Soc.* **26** 90 (1958)
18. K.W.H. Stevens, *Proc. Phys. Soc.* **A65**, 209 (1952)
19. M.S. Dresselhaus, G. Dresselhaus, and A. Jorio, *Group Theory, Applications to the Physics of Condensed Matter*, (Springer, 2008)
20. H.A. Bethe, *Ann. Phys* **3**, 133 (1929)
21. R.S. Mulliken, *J. Chem. Phys.* **3**, 375 (1935)
22. C.J. Ballhausen, *Introduction to Ligand Field Theory*, (McGraw-Hill, New York, 1962), p. 100
23. L.E. Orgel, *J. Chem. Phys.* **23**, 1004 (1955)
24. H.L. Schläfer and G. Gliemann, *Basic Principles of Ligand Field Theory*, (Wiley-Interscience, New York, 1969), Chapter 1
25. C.J. Ballhausen, *Introduction to Ligand Field Theory*, (McGraw-Hill, New York, 1962), p. 20
26. J.C. Slater, *Phys. Rev.* **35**, 509 (1930)
27. G. Racah, *Phys. Rev.* **62**, 438 (1942); G. Racah also *Phys. Rev.* **63**, 367 (1943)
28. K.R. Lea, M.J.M. Leask, and W.P. Wolf, *J. Chem. Phys. Solids* **23**, 138 (1967)
29. L.E. Orgel, *Introduction to Transition-Metal Chemistry: Ligand-Field Theory*, (John Wiley, New York, 1959)
30. E.C. Stoner, *Proc. Leeds Phil. Soc.* **2**, 391 (1933)
31. M.R. Ibarra, R. Mahendiran, C. Marquina, B. Garcia-Landa, and J. Blasco, *Phys. Rev. B* **57**, R3217 (1998 II)
32. O.G. Holmes and D.S. McClure, *J. Chem Phys.* **26**, 1686 (1957)
33. Y. Tanabe and S. Sugano, *J. Phys. Soc. Japan* **9**, 753 (1954)
34. J.M. Baker, B. Bleaney, and K.D. Bowers, *Proc. Phys. Soc. (London)* **69**, 1205 (1956); also A.L. Kipling, P.W. Smith, J. Vanier, and G.A. Woonton, *Can. J. Phys.* **39**, 1859 (1961)
35. M.M. Schieber, *Experimental Magnetochemistry*, (John Wiley, New York, 1967), p. 250
36. A. Abragam and M.H.L. Pryce, *Proc. R. Soc.* **A205**, 135 (1951)
37. H.A. Jahn and E. Teller, *Proc. R. Soc. (London)* **A161**, 220 (1937)
38. H.A Jahn, *Proc. R. Soc. (London)* **A164**, 117 (1938)
39. J.H. Van Vleck, *Phys. Rev.* **57**, 426 (1940)
40. J.B. Goodenough, *Magnetism and the Chemical Bond*, (Wiley Interscience, New York, 1963), Chapter III, Sections IE and IF
41. J.B. Goodenough, *J. Phys. Chem. Solids* **25**, 151 (1964)
42. M. Kaplan and B. Vekhter, *Cooperative Phenomena in Jahn–Teller Crystals*, (Plenum, New York, 1995)
43. E. Cartmell and G.W.A. Fowles, *Valency and Molecular Structure*, (Butterworths, London, 1961), Chapter X
44. C.J. Ballhausen, *Introduction to Ligand Field Theory*, (McGraw-Hill, New York, 1962), p. 161
45. C.J. Ballhausen and H.B. Gray, *Molecular Electronic Structures*, an Introduction, (Benjamin/Cummings, Reading, MA, 1980)
46. M. Wolfsberg and L. Helmholz, *J. Chem Phys.* **20**, 837 (1952)
47. G.F. Dionne, *Covalent Electron Transfer Theory of Superconductivity*, (MIT Lincoln Laboratory Technical Rept. **885**, 1992), NTIS No. ADA2539757
48. G.F. Dionne, *J. Appl. Phys.* 99, 08M913 (2006)
49. L. Pauling, *The Nature of the Chemical Bond*, (Cornell University Press, New York, 1960)
50. C. Kittel, *Introduction to Solid State Physics*, (Wiley, New York, 1966), p. 90
51. R. Chang, *Chemistry in Action*, (Random House, New York, 1988), Chapter 9
52. E. Cartmell and G.W.A. Fowles, *Valency and Molecular Structure*, (Butterworths, London, 1961), pp. 92–94
53. C.J. Ballhausen and H.B. Gray, *Molecular Electronic Structures, an Introduction*, (Benjamin/Cummings, Reading, MA, 1980), p. 84

Chapter 3
Magnetic Exchange in Oxides

In the previous chapter, quenching of cation orbital angular momentum by the anion charges and the origin of the energy stabilization by covalent bonding was introduced by elementary crystal field and molecular orbital theory. In magnetically dilute compounds, the isolated $3d^n$ ions are influenced next by the weakened spin-orbit coupling perturbations and multiplet structures that determine the magnetoelastic properties to be examined in Chap. 5. These effects are initially local but can become cooperative when concentrations increase to levels where percolation can occur, e.g., a cooperative Jahn-Teller effect. However, because the spin alone is the agent of magnetic ordering in this series, the multiplet energies can be largely ignored in the discussion of spontaneous magnetism. Consequently, the next important effective field in a ligand lattice to be addressed is the magnetic exchange field that arises from the transfer integral linking magnetic cations. Magnetic exchange, therefore, is the term used to describe the energy stabilization gained from spin ordering (parallel or antiparallel) of atoms or ions covalently coupled in an ionic crystal lattice.

Although the magnetic behavior is determined by the disposition of unpaired electron spins, the underlying mechanisms are of electrostatic origin. In Chap. 1, the concepts of direct magnetic exchange were outlined for the traditional case of metallic elements of the transition groups. The basis for that discussion was the origin of spontaneous magnetism from unfilled shells within collective electron systems, and the formalism is based on the Heitler–London solution of the H_2 molecule [1] that opened the way for the valence-bond approach to chemical bonding outlined in Sect. 2.5.

For oxides and other insulating compounds, however, direct exchange is usually not an important contributor to magnetic properties because the metal ions are too far apart for the mutual electron repulsion term e^2/r_{ij} to establish an antibonding state with a Hund's rule parallel spin alignment. If the bonding is dominated by cation–anion interactions, the unpaired cation spins couple through the mediation of the anion orbital states to create indirect or "superexchange" that observes the Pauli exclusion requirement of antiparallel alignment.

This chapter is an overview of some underlying theoretical concepts of magnetic exchange between two transition-metal cations with unpaired electron spins, and the relation between the bonding stabilization and antibonding destabilization in

G.F. Dionne, *Magnetic Oxides*, DOI 10.1007/978-1-4419-0054-8_3,

determining the type of spontaneous magnetic order. More specifically, the case of $3d$ electrons linking covalently through the mediation of $2p$ ligand orbitals will be emphasized.

3.1 Interionic Magnetic Exchange

In an elementary context, the spontaneous magnetism in a crystal lattice can take two forms: parallel spins (ferromagnetism) and antiparallel or paired spins (antiferromagnetism). These two extremes are analogous to the high and low-spin states of the individual ions, and the interactions that select one in preference to the other are essentially the same. In Fig. 3.1, schematic diagrams indicate how the different spin-couplings evolve as two magnetic cations are related through a chemical bond, beginning with (a) paramagnetism when the spins fully localized to their parent

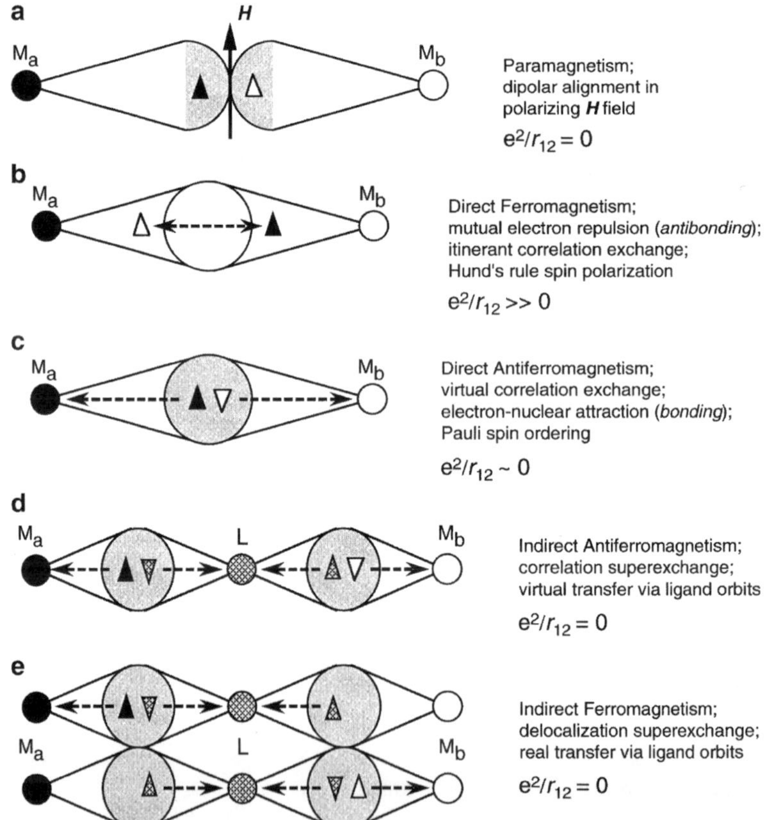

Fig. 3.1 Orbital-lobe diagrams for the different magnetic interaction and bonding situations

nuclei align in a polarizing H field with an inverse square dependence on separation. In Fig. 3.1b, direct exchange ferromagnetism arises from mutual electron repulsion that stabilizes an antibonding state. Initially, the ground state spin distributions are parallel, correlated into an $S = 1$ high-spin state as dictated by Hund's rule.[1] As the two nuclei are moved closer together, electron–nuclear cross-term attraction energies can overcome the e^2/r_{ij} energy. In Fig. 3.1c, antiferromagnetism is stabilized when the electron–nuclear attraction becomes dominant in forming a bonding ground state, and the orthogonality becomes compromised by the overlapping orbital functions of the now-delocalized spins. As the transfer integral b increases and the splitting between the bonding and antibonding states becomes larger, the orbital energy stabilization favors the low-spin $S=0$ antiparallel configuration in conformance with the Pauli's exclusion principle. In the case of a metal–oxygen–metal bond in Fig. 3.1d, the antiferromagnetic ordering of the two localized metal spins is mediated indirectly by orbital overlaps to a common anion (termed superexchange). Analogous in some respects to the one d-electron model used to display the spin occupancies of the crystal-field ground states determined by the Aufbau principle, these bonding diagrams can be used as rough guides to the spin alignments in certain cases.

3.1.1 Molecular-Orbital Exchange Approximation

For a transition-metal ion (M) of the $3d$ series in an octahedrally coordinated oxygen (O) or ligand (L) site, one-electron energy-level models are useful platforms for semiempirical estimates of the crystal-field stabilization energy [2,3]. The ground-state electron spin distributions are presumed to fill the levels sequentially across the series (Aufbau method), in accordance with Hund's rule and the Pauli's principle. If the $3d$-electron orbital basis functions are hybridized with the $2p$ states of the O^{2-} ligands, a molecular-orbital energy level model (Fig. 3.2) can represent the contribution of the covalent bonding in these primarily ionic compounds. For a single orbital state of a metal–ligand diatomic molecule (M–L), the stabilization that occurs for an orbital transfer energy $b_{\mathrm{ML}} = h \langle \phi_{\mathrm{M}} | \, H_{\mathrm{M}} + H_{\mathrm{L}} \, | \phi_{\mathrm{L}} \rangle$, as defined in terms of the Hamiltonian energies H_{M} and H_{L}. By the Wolfsberg–Helmholtz approximation with $h = 1$ for oxides from (2.38), $b_{\mathrm{ML}} = (E_{\mathrm{M}} + E_{\mathrm{L}})\,\sigma_{\mathrm{ML}}$, where E_{M} and E_{L} are the in-lattice energies of the respective electrons before covalent perturbations are applied, and the orbital overlap integral $\sigma_{\mathrm{ML}} = \langle \phi_{\mathrm{M}} \mid \phi_{\mathrm{L}} \rangle$. For $\sigma_{\mathrm{ML}}^2 << 1$, the bonding (+) and antibonding (–) state energy solutions reduced from (2.39) are

[1] This initial ferromagnetic bias arises from the electrostatic repulsion of two electrons that is not included in the one-electron molecular–orbital Hamiltonian (2.38) and is implicit in the screening parameter h. It is convenient to treat it as a high-spin state analogous to the d^1 electron models that retain parallel spin alignments until stronger crystal fields (or larger b integrals) overcome the mutual repulsion and allow antiparallel spin alignment to stabilize in the lower orbital states producing low-spin configurations.

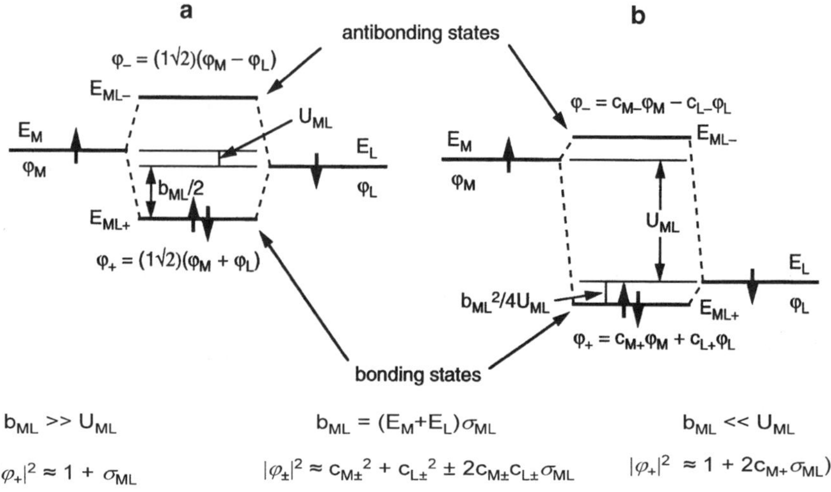

Fig. 3.2 One-electron metal–ligand molecular orbital diagram illustrating the stabilization of an electron spin pair and the formation of bonding and antibonding hybrid eigenstates for (**a**) $b_{\mathrm{ML}} >> U_{\mathrm{ML}}$, and (**b**) $b_{\mathrm{ML}} << U_{\mathrm{ML}}$

$$E_{\pm} \approx \frac{1}{2}\left[(E_{\mathrm{M}} + E_{\mathrm{L}}) \pm \sqrt{U_{\mathrm{ML}}^2 + b_{\mathrm{ML}}^2}\right]$$
$$\approx \frac{(E_{\mathrm{M}} + E_{\mathrm{L}})}{2} \pm \left(\frac{U_{\mathrm{ML}}}{2} + \frac{b_{\mathrm{ML}}^2}{4U_{\mathrm{ML}}}\right) \text{ for } b_{\mathrm{ML}}^2/U_{\mathrm{ML}}^2 << 1, \qquad (3.1)$$

where $U_{\mathrm{ML}} = E_{\mathrm{L}} - E_{\mathrm{M}}$. Hybrid eigenfunctions are $\phi_{\mathrm{ML}\pm} = c_{\mathrm{M}\pm}\phi_{\mathrm{M}} \pm c_{\mathrm{L}\pm}\phi_{\mathrm{L}}$, where

$$c_{\mathrm{M}\pm} = c_{\mathrm{L}\mp} \approx \frac{1}{\sqrt{2}}\left(1 \pm \frac{U_{\mathrm{ML}}}{\sqrt{U_{\mathrm{ML}}^2 + b_{\mathrm{ML}}^2}}\right)^{1/2}. \qquad (3.2)$$

This approximation can be used as a building block to create an artificial M–L–M′ molecule that represents the actual situations sketched in Fig. 3.3. By attaching a second cation to the opposite lobe of the same ligand orbital function, we can consider forming an illustrative model by pursuing a qualitative line of reasoning. To reduce this three-body problem diagrammed in Fig. 3.4 to a manageable two-body case, an effective overlap integral between M and M′ must be deduced from the bonding state of the M–L–M′ molecule. Once this interaction parameter is determined, the M–M′ coupling can be approximated in the manner of a simple diatomic molecule.

From the MO diagram of Fig. 3.2, the unpaired $3d$ spins in the e_g states are localized to their respective nuclei in the $\varphi_{\mathrm{ML}-}$ and $\varphi_{\mathrm{M'L}-}$ antibonding states. The smaller $b_{\mathrm{ML}}{}^2/4U_{\mathrm{ML}} \sim 10Dq$ is, the closer the levels will approach those of a "free" ion. As long as the $b_{\mathrm{ML}}, b_{\mathrm{ML}}{}' \neq 0$, however, a portion of the $|\varphi_{\mathrm{M}}|^2$ and $|\varphi_{\mathrm{M'}}|^2$ densities is hybridized with the $|\varphi_{\mathrm{L}}|^2$ density in the lower energy bonding state. As a

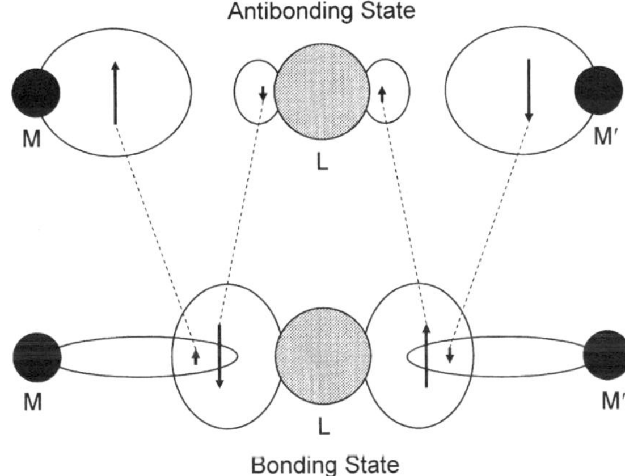

Fig. 3.3 One-dimensional sketches of metal–oxygen–metal molecular orbital hybrid states illustrating the fractional spin distributions of the respective $3d$ metal and $2p$ ligand wavefunctions

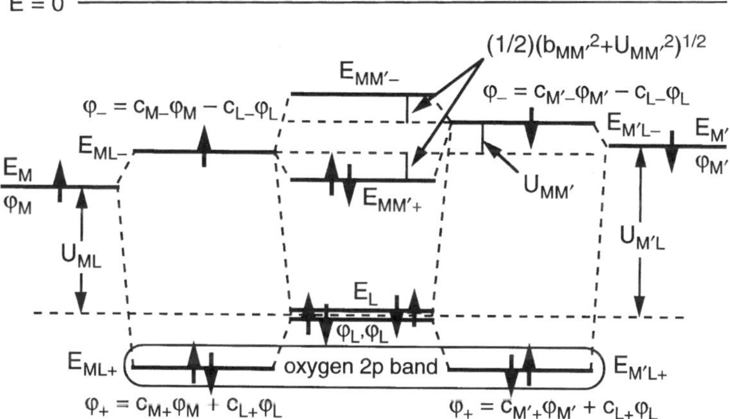

Fig. 3.4 One-electron metal–oxygen–metal molecular orbital diagram illustrating the formation of the basic bonding and antibonding hybrid eigenstates that produce superexchange in magnetic oxides. Figure reprinted from G.F. Dionne, *J. Appl. Phys.* **99**, 08M913 (2006) with permission. © 2006 by the American Institute of Physics

result, the two $3d$ spins are partially delocalized into antiferromagnetic states by an effective overlap integral $\sigma_{\mathrm{MLM'}}$ that is mediated through the ligand $2p$ orbital function containing its original paired spins. As depicted schematically in Figs. 3.1d, e, excess densities in the regions between the cations and anion, i.e., $(1 + \sigma_{\mathrm{ML}})$, will define the volume of bonding interaction, while depleted densities $(1 - \sigma_{\mathrm{ML}})$ in the corresponding antibonding states will restore the population total (see Appendix 3A of Chap. 2). From this diagram, it is evident that the orbital stabilization occurs in the bonding state, despite the fact that the $3d$ spins are mainly localized in the antibonding state. The effective overlap integral $\sigma_{\mathrm{MLM'}}$ can then be approximated from the volumes of the two $3d$ wavefunctions represented in the bonding states $|\phi_{\mathrm{ML+}}|^2$ and $|\phi_{\mathrm{M'L+}}|^2$, i.e.,

$$\sigma_{\mathrm{MLM'}} \approx \frac{c_{\mathrm{M+}}c_{\mathrm{L+}}\sigma_{\mathrm{ML}} + c_{\mathrm{M'+}}c_{\mathrm{L+}}\sigma_{\mathrm{M'L}}}{c_{ML+}^2 + c_{\mathrm{M'L+}}^2} \approx \frac{1}{2}\left(c_{\mathrm{M+}}c_{\mathrm{L+}}\sigma_{\mathrm{ML}} + c_{\mathrm{M'+}}c_{\mathrm{L+}}\sigma_{\mathrm{M'L}}\right), \tag{3.3}$$

where

$$\begin{aligned} c_{\mathrm{M+}}c_{\mathrm{L+}} &\approx \frac{1}{2}\left[\left(1 - \frac{U_{\mathrm{ML}}}{\sqrt{U_{\mathrm{ML}}^2 + b_{\mathrm{ML}}^2}}\right)\left(1 + \frac{U_{\mathrm{ML}}}{\sqrt{U_{\mathrm{ML}}^2 + b_{\mathrm{ML}}^2}}\right)\right]^{1/2} \\ &= \frac{1}{2}\left(\frac{b_{\mathrm{ML}}^2}{U_{\mathrm{ML}}^2 + b_{\mathrm{ML}}^2}\right)^{1/2}. \end{aligned} \tag{3.4}$$

For a magnetic oxide with strong ionic bonding, $b_{\mathrm{ML}}^2/U_{\mathrm{ML}}^2 << 1$ and $c_{\mathrm{M+}}c_{\mathrm{L+}} \approx b_{\mathrm{ML}}/2U_{\mathrm{ML}}$, so that

$$\begin{aligned} \sigma_{\mathrm{MLM'}} &\approx \frac{1}{2}\left(\frac{b_{\mathrm{ML}}}{2U_{\mathrm{ML}}}\sigma_{\mathrm{ML}} + \frac{b_{\mathrm{M'L'}}}{2U_{\mathrm{M'L'}}}\sigma_{\mathrm{M'L'}}\right) \\ &\approx \frac{b_{\mathrm{ML}}}{2U_{\mathrm{ML}}}\sigma_{\mathrm{ML}} \qquad \text{for homonuclear M} = \mathrm{M'}. \end{aligned} \tag{3.5}$$

Thus, the effective overlap integral between M and M′ can now be labeled as $\sigma_{\mathrm{MM'}} = \sigma_{\mathrm{MLM'}}$, and the simplified two-body problem depicted in Fig. 3.5 can be solved with $\phi_\pm = c_{\mathrm{M}\pm}\phi_{\mathrm{M}} \pm c_{\mathrm{M'}\pm}\phi_{\mathrm{M'}}$ and a bonding stabilization energy

$$\begin{aligned} \Delta E_{\mathrm{MM'}}^{\mathrm{bond}} &\approx \frac{1}{2}\sqrt{U_{\mathrm{MM'}}^2 + b_{\mathrm{MM'}}^2} \\ &\approx \frac{b_{\mathrm{MM}}}{2} \qquad \text{for homonuclear M} = \mathrm{M'}, \end{aligned} \tag{3.6}$$

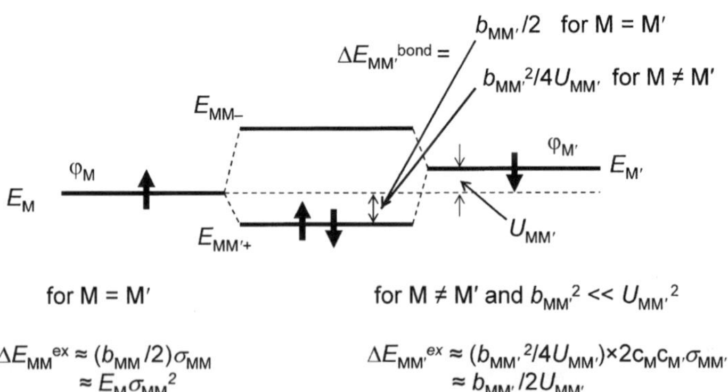

Fig. 3.5 Two-body approximation to the metal–oxygen–metal molecular orbital extracted from the upper half of Fig. 3.4

where $b_{MM'} = (E_M + E_{M'})\sigma_{MM'}$. Because only the delocalized fraction of the $3d$ spin population in $\langle\varphi_+ \mid \varphi_+\rangle$ is required to satisfy Pauli spin pairing, the portion of the covalent-bonding energy that stabilizes the antiparallel spin alignment is weighted by the delocalized fraction $2c_{M+}c_{M'+}\sigma_{MM'}$. From (3.4) and (3.6)

$$\begin{aligned}\Delta E^{ex}_{MM'} &= \Delta E^{bond}_{MM'} \times 2c_{M+}c_{M'+}\sigma_{MM'}\\ &= \frac{1}{2}\sqrt{U^2_{MM'} + b^2_{MM'}} \times \left(\frac{b^2_{MM'}}{U^2_{MM'} + b^2_{MM'}}\right)^{1/2} \sigma_{MM'} = \frac{b_{MM'}}{2}\sigma_{MM'}, \quad (3.7)\end{aligned}$$

which is a first-order function of $b_{MM'}$.

In terms of the Heisenberg exchange energy between ions M and M′ for combined electron spins $\Delta E^{ex}_{MM'} = -2J_{MM'}S_M S_{M'}$, (3.7) for a single orbital spin pair where $S_M = 1/2$ can be expressed as $J_{MM'} = b_{MM'}\sigma_{MM'}$. Because the screening parameter $h_{MM'}$ is presumed to be positive based on the ground state spin occupation in Fig. 3.5, antiferromagnetism should prevail and $J_{MM'}$ would have a negative sign in the Heisenberg model. However, in situations where direct overlap between other M and M′ orbitals can compete, $h_{MM'}$ could be negative, causing an inversion of the bonding/antibonding states and a weakening or even reversal in sign of $J_{MM'}$ leading to an antibonding ferromagnetic ground state.

To ensure the ferromagnetic alignment where $b_{MLM'}$ is small, the missing contribution from the e^2/r_{ij} repulsion would have to be introduced by a two-electron exchange calculation, which would make a positive addition to the transfer element and lead to a reversal in the cation–cation bonding/antibonding states φ_M and φ_-. It is important to recognize that although these parameters are not the same as those used in the more formal analysis of superexchange, at a qualitative level the model provides helpful insight as well as providing an estimate of the effective transfer integral $b_{MM'}$. It also allows the addition of competing direct cation–cation b integrals to arrive at a combined result. A comparison of the two approaches is offered in the next section. Recall also that since b_{ML} varies directly as the first power of the overlap integral σ_{ML}, by definition in (2.38), both $\sigma_{MM'}$ and the quasi-direct transfer integral $b_{MM'}$ are proportional to ${\sigma_{ML}}^2$ as indicated by (3.5). When the actual exchange stabilization energy $\Delta E^{ex}_{MM'}$ in (3.7) is considered, its dependence on the cation–anion orbital overlap becomes ${\sigma_{ML}}^4$, as pointed out also by Ballhausen [4]. This high sensitivity of the indirect exchange coupling confirms that the stability of long-range spin ordering in a crystal lattice is severely diminished by magnetic dilution.

3.1.2 *Valence-Bond Solutions*

Analysis of interionic electronic and magnetic properties in ionic crystal lattices by the molecular-orbital (MO) approach is largely conceptual, based on probability estimates of the ground-state electron spin occupancy. As the name implies, a molecule is first built from the strong Coulomb electric fields between cations and anions, and then as demonstrated in the previous section, a second cation is added to estimate the strength of magnetic exchange coupling that is more than an order of magnitude weaker than the electrostatic bonding. Because the stabilization from electronic hybridization is controlled by the wavefunctions overlap volumes, the orbital overlap integral σ represents a measure of the delocalization and is retained explicitly in the formalism. In its purest form, however, the method does not readily extend to collective electron situations unless the presumption of long-range homogeneity allows the adoption of Hartree-Fock formalisms that lead to band models. An important feature of the MO approach is the preservation of the internal electronic energy-level information of local cations from which the sensitivity of the orbital angular momentum to lower symmetry crystal fields affects the multiplet structure that in turn determines the local magnetoelastic, magneto-optical, and magnetic relaxation properties.

For the discussion of specific molecules, particularly as their complexities increase, the more rigorous valence-bond (VB) approach offers a more quantitative methodology [5]. Computation of bonding and exchange energies for specific structures can be carried out by a method that allows the introduction of multi-electron wavefunctions and includes the mutual electron repulsive energy e^2/r_{ij}. A two-electron Hamiltonian that retains the nuclear and electronic interactions is a convenient starting point. The effects of crystal fields on the partially filled d-electron states can also be included if the appropriate molecular orbitals are employed as the starting wavefunctions. For the analysis of direct magnetic exchange (applied originally to the H_2 molecule [1]), the valence-bond method is used to compute the orbital interaction energies that dictate the high-spin (parallel) and low-spin (antiparallel) states based on the indistinguishability requirement (Pauli principle) of the two electrons sharing hybrid orbital states of both atoms. For this purpose, the Coulomb energy terms E_{M} and E_{L} of (3.1) are replaced by the actual interionic electron–nuclear attraction and electron–electron mutual replusion terms similar to obtain a two-electron Hamiltonian $\mathcal{H} = \mathcal{H}_0 + \mathcal{H}_1$, where

$$\begin{aligned} \mathcal{H}_0 &= -\left(\frac{\hbar^2}{2m_e}\right)\left(\nabla_{\mathrm{a}}^2 + \nabla_{\mathrm{b}}^2\right) - \frac{Z_{\mathrm{a}}e^2}{r_{\mathrm{a1}}} - \frac{Z_{\mathrm{b}}e^2}{r_{\mathrm{b2}}} + \frac{Z_{\mathrm{a}}Z_{\mathrm{b}}e^2}{r_{\mathrm{ab}}}, \\ \mathcal{H}_1 &= -\frac{Z_{\mathrm{a}}e^2}{r_{\mathrm{a2}}} - \frac{Z_{\mathrm{b}}e^2}{r_{\mathrm{b1}}} + \frac{e^2}{r_{12}}. \end{aligned} \tag{3.8}$$

In the MO solutions, such as the hydrogen molecule ion sketched in Fig. 3.6a, H_1 contains only the nuclear terms. Under this restriction, the results for the MO and VB methods are usually identical. However, that quickly changes when the electron correlation energy is introduced in a formal manner. As defined in Fig. 3.6b, the distances r_{a2} and r_{b1} are the electron–nuclear separation of the exchanged electrons labeled 1 and 2 on nuclei a and b, respectively, and r_{12} is the separation between the electrons themselves. Two-electron degenerate eigenfunctions $\varphi_0 = \varphi_{\mathrm{a1}}\varphi_{\mathrm{b2}}$ (initial occupancy) and $\varphi_{\mathrm{ex}} = \varphi_{\mathrm{a2}}\varphi_{\mathrm{b1}}$ (exchanged occupancy) of this Hamiltonian are constructed from products of single-electron functions for computation of the matrix

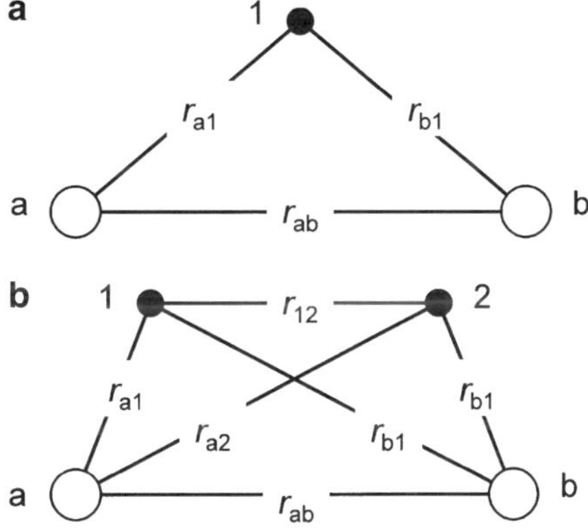

Fig. 3.6 Electrostatic linkages between nuclei a and b and electrons 1 an 2 for (**a**) the single electron hydrogen molecule ion H_2^+, and (**b**) the hydrogen molecule H_2

elements K_{12} (Coulomb) $= \langle\varphi_0|\mathcal{H}_1|\varphi_0\rangle$ and J_{12} (exchange) $= \langle\varphi_0|\mathcal{H}_1|\varphi_{ex}\rangle$ of the 2×2 Slater determinant described in Sect. 1.3. The solutions of the secular equation give bonding and antibonding states energies according to $K_{12}\pm J_{12}$. Consequently, the sign of J_{12} determines whether the spins in the ground state are parallel ($S = 1$) for $J_{12} > 0$, or antiparallel ($S = 0$) for $J_{12} < 0$.

The computation of J for a superexchange coupling between two magnetic ions (spin cations) M_a^{n+} and M_b^{n+} bonded to a common O^{2-} ligand begins with the assumption that unfilled orbital states are present in the M_a–O–M_b molecule. Since the electrostatic stabilization among ions is maximized if the oxygen ligands carry 2– charges, meaning that the oxygen $2p$ band is filled, the unpaired electron spins reside on the metal cations. Because of the ionic character of the bonding and the important influence of crystal fields that remove the degeneracies of the $3d$ shell states, as described in Chap. 2, the analysis of the interactions between M_a and M_b are treated on an individual orbital state basis.

A physical rationale is proposed to serve as a basis for superexchange second-order perturbation analysis, which requires the existence of an excited state that can be mixed with a state of lower energy by the electrostatic perturbation of the molecular bonding, i.e., a nonzero off-diagonal matrix element created by a transfer integral, to produce separate bonding and antibonding states. Maintaining observance of the Pauli principle, the model sketched in Fig. 3.7a shows the transfer of spins from the O^{2-} $2p$ states onto the unfilled orbitals of the cations to produce stabilization of an antiferromagnetic state between the spins localized on the two cations through the creation of a *virtual* excited state energy U. Initially both spins are resident on the respective cations, with the relevant $2p$ orbital state of the O^{2-} ion filled. The excited state is postulated by creating a virtual transfer of the M_a spin to the equivalent M_b orbital state, resulting in the reaction ${M_a}^{n+} + {M_b}^{n+} \rightarrow {M_a}^{(n+1)+} + {M_b}^{(n-1)+} + \mathrm{U}$. A return reaction in the will complete the transfer of spins from $\varphi_0 = \varphi_{a1}\phi_{b2}$ (initial occupancy) to $\varphi_{ex} = \varphi_{a2}\varphi_{b1}$ (exchanged occupancy), and stabilize antiferromagnetic spin alignment.

If the starting wavefunctions are molecular orbitals that reflect the effects of the cross electron–nuclear terms of (3.8), the electrostatic perturbation $\mathcal{H}_1$ can now be

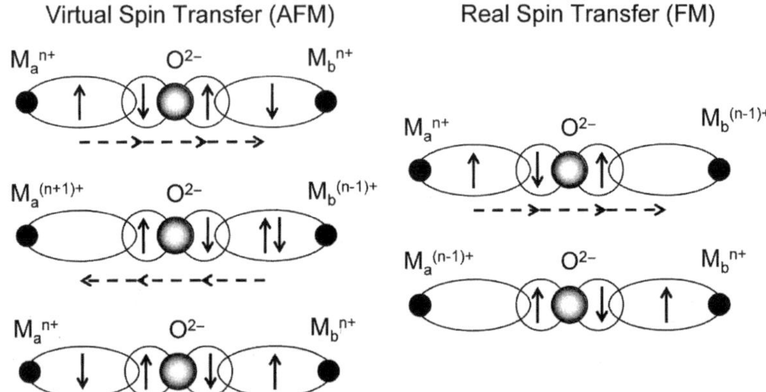

Fig. 3.7 Tutorial illustration of superexchange virtual and real spin transfer: The former produces antiferromagnetism, the latter, somewhat weaker ferromagnetism accompanied by electrical conductivity

approximated by the e^2/r_{12} electron repulsion term. This approach was used by Anderson in his original formalization of the superexchange interaction [6]. For a pair of $3d$ orbital states of cations linked by a $2p$ orbital of an oxygen ligand, second-order perturbation theory is applied to the system to compute the energy of the stabilized ground state. The standard correction to the orbital ground-state energy for the orbital energy is then

$$\Delta E = \frac{b^2}{U}, \tag{3.9}$$

where $b^2 = \langle\phi_1|\frac{e^2}{r_{12}}|\phi_2\rangle\langle\phi_2|\frac{e^2}{r_{12}}|\phi_1\rangle$, $U = \langle\phi_0|V_{12}|\phi_{ex}\rangle\langle\phi_{ex}|V_{12}|\phi_0\rangle$, and $V_{12} = \frac{e^2}{r_{12}}\exp(-\xi r_{12})$, with ξ as a screening parameter [7]. After the addition of the electron spin exchange operator $s_1 \cdot s_2$ of negative sign (for antiparallel alignment) from Anderson's formal solution by second quantization theory and the use of the four Fermi spin operators to account for the two anion spins occupying the $2p$ orbital, (3.9) becomes

$$\Delta E_{ab}^{virtual} = -2J_{ab}^{virtual} s_1 \cdot s_2 = 2\frac{2b^2}{U} s_1 \cdot s_2, \tag{3.10}$$

where $J_{ab}^{virtual} = -\frac{2b^2}{U}$. This result has the reverse sign from the first-order result for direct exchange

$$\Delta E_{ab}^{direct} = -2J_{ab}^{direct} s_1 \cdot s_2, \tag{3.11}$$

where $J_{ab}^{direct} = \langle\varphi_0|H_1|\varphi_{ex}\rangle > 0$.

In generalized format for n orbital levels linking two cations i and j, the exchange stabilization energy derived from the spin operator part of (3.9) is

$$\Delta E_{ij}^{virtual} = -2J_{ij}^{virtual} \boldsymbol{S}_i \cdot \boldsymbol{S}_j = 2\sum_n \frac{2b_{ij}^{n2}}{U_n} s_i \cdot s_j, \tag{3.12}$$

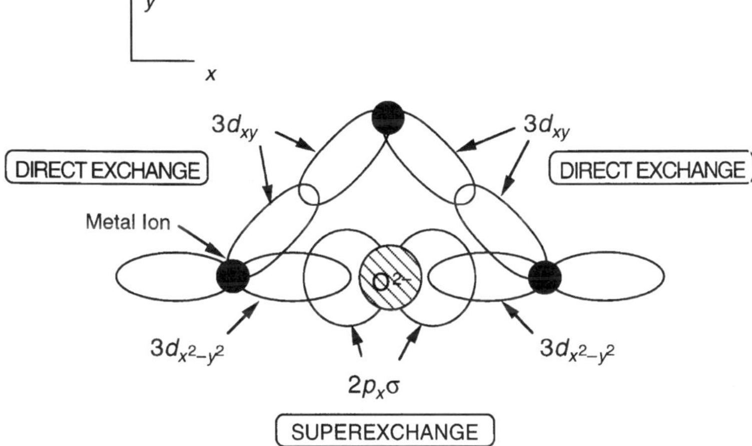

Fig. 3.8 Metal–oxygen exchange bonding showing the comparison between direct and indirect or superexchange. Superexchange stabilizes antiferromagnetism; direct exchange is commonly ferromagnetic

where J_{ij} is the direct exchange contribution suggested by Fig. 3.8. If ions have total spins $\boldsymbol{S}_i$ and $\boldsymbol{S}_j$ summed over n orbital states with $s_i = s_j = 1/2$, we can write

$$\boldsymbol{s}_i \cdot \boldsymbol{s}_j = \frac{1}{4S_iS_j}\boldsymbol{S}_i \cdot \boldsymbol{S}_j, \tag{3.13}$$

and the combined exchange constant for the molecule is determined by applying (3.11) and (3.12):

$$J_{ij} = J_{ij}^{\text{direct}} + J_{ij}^{\text{virtual}} = \frac{1}{4S_iS_j}\sum_n \left(J_{ij}^{n\text{direct}} - \frac{2b_{ij}^{n2}}{U_{ij}^n}\right),$$

or simply

$$J_{ij} = \frac{1}{4S_iS_j}\sum_n \left(J_n - \frac{2b_n^2}{U_n}\right), \tag{3.14}$$

where the subscript in (3.12) have been replaced by n for the parameters J and b.

For superexchange, we can extract from (3.12) for $S_i = S_j = S$,
or

$$J_{ij}^{\text{virtual}} = -\frac{1}{2S_iS_j}\sum_n \frac{b_{ij}^{n2}}{U_n}$$

$$J_n^{\text{virtual}} \approx -\frac{1}{n}\frac{2b_n^2}{U_n} \quad \text{for } n \text{ equivalent orbital states with equal } S = n/2, \tag{3.15}$$

Table 3.1 Bonding and exchange parameters

Molecular orbital(MO)		Valence bond(VB)	
σ_{ML} [a]	0.3		
b_{ML} [b]	$(E_M + E_L)\,\sigma_{ML} \approx -10\,\text{eV}$	b_n^2	$\left\langle \varphi_1 \left\| \frac{e^2}{r_{12}} \right\| \varphi_2 \right\rangle \left\langle \varphi_2 \left\| \frac{e^2}{r_{12}} \right\| \varphi_1 \right\rangle \approx 1.5\,\text{eV}$
U_{ML}	$E_L - E_M \approx -15\,\text{eV}$	U_n^c	$\langle \varphi_0 \| V_{12} \| \varphi_{ex} \rangle \approx 15\,\text{eV}^c$
$10Dq$	$\left\| b_{ML}^2 / 4U_{ML} \right\| = 1.7\,\text{eV}$		
σ_{MM}	$(b_{ML}/2U_{ML})\,\sigma_{ML} \approx 0.1$		
b_{MM}	$2E_M\sigma_{MM}\sigma_{ML} \approx -2\,\text{eV}$		
ΔE_{ex}^{Super}	$n\,(1/2)\,b_{MM}\sigma_{MM} \approx -0.1\,\text{eV}\ (n = 1)$	ΔE_{ex}^{Super}	$nb_n^2/U_n \approx -0.1\,\text{eV}\ (n = 1)$
J_{ex}^{Super}	$n\,b_{MM}\sigma_{MM} = -0.19\,\text{eV}\ (n = 1)$	J_{ex}^{Super}	$(1/n)\,2b_n^2/U_n = -0.2\,\text{eV}\ (n = 1)$

[a]Per e_g orbital in an octahedral site, 180° bond angles
[b]$E_M = -10eV,\ E_L = -25\,\text{eV}$
[c]These values are on the order of ionization potentials and electron affinities. See Ref. [7]

and we can also write

$$\Delta E_{ex}^{virtual} \approx -\sum_n \frac{b_n^2}{U_n} \approx -n\frac{b_n^2}{U_n}. \tag{3.16}$$

The details of the quantum mechanical solution that include the explicit representation of the Fermi spin operators that control the alignment of the respective electron spins s_1 and s_2 vectors are discussed in many standard texts [7, 11] and will not be repeated here. However, it is important to compare the initial assumptions of the MO and VB models and how they influence the results relevant to crystal field and magnetic exchange energies (Table 3.1). [2]

In typical situations, the spin directions are not precisely aligned, but form an angle θ_{ij} with each other. To introduce this angular dependence into Anderson's result, the use of the exact relation for the scalar product $\boldsymbol{S}_i \cdot \boldsymbol{S}_j$ is not convenient because of the quantum correction $S(S+1)$ in the angular momentum addition. If the Ising approximation is adopted, whereby only the z components of $\boldsymbol{S}$ are retained in the vector product, $S_z = S\cos\left(\theta_{ij}/2\right)$ and the exchange energy for two spins with z_{ij} equivalent neighbors can be expressed as

$$\begin{aligned}\Delta E_{ex}^{ferro}\,(\text{VB}) &= -2z_{ij}J_{ij}\boldsymbol{S}_i \cdot \mathbf{S}_j = -2z_{ij}J_{ij}S_{zi}S_{zj}\cos^2\left(\frac{\theta_{ij}}{2}\right)\\ &= -z_{ij}J_{ij}S_{zi}S_{zj}\left[1 + \cos\,\theta_{ij}\right],\end{aligned} \tag{3.17}$$

[2] There is a distinction between $U_{MM'}$ in the MO case and U of the VB method. In the MO case, $U_{MM'}$ represents the *in-lattice* difference in energy between the two outermost electrons that remain with their parent ions, and exchange is viewed simply as a small *first-order* perturbation without allowing for the formation of new ionic states by formal charge transfer. In the VB case, an excitation defines U as the net change of two ionic stabilization energies involved in the *virtual* spin transfer, approximated by the difference in the ionization potentials between the initial and final valence states of the ions. Here the perturbation is treated mathematically as *second order*. As a result, $U_{MM'} \approx 0$ in the localized MO treatment for *like* cations M = M′, but for the virtual spin transfer of the VB method, $U \neq 0$. See **Table 3.1**.

and for the antiparallel case,

$$\Delta E_{ex}^{\text{antiferro}}(\text{VB}) = -2z_{ij}J_{ij}\boldsymbol{S}_i \cdot \mathbf{S}_j \approx -2z_{ij}J_{ij}S_{zi}S_{zj}\sin^2\left(\frac{\theta_{ij}}{2}\right) \qquad (3.18)$$
$$= z_{ij}J_{ij}S_{zi}S_{zj}\left[1 - \cos\,\theta_{ij}\right].$$

To provide stabilization for ferromagnetism ($\theta_{ij} = 0$), $J_{ij} > 0$ and for antiferromagnetism $(\theta_{ij} = \pi)$, $J_{ij} < 0$. From this general result, the various spin ordering situations that occur in transition-metal oxides can be examined.

This result was also obtained by Goodenough [7], who reasoned that Hubbard's exchange integral b_{ij} [12] could be modified to obey the Pauli exclusion principle by constructing a transfer integral t_{12} for a particular pair of orbital states that included the canting angle θ_{ij} between ionic spins. For this exercise, the relations are $t_{ij} = b_{ij}\cos\left(\theta_{ij}/2\right)$ and $b_{ij}\sin\left(\theta_{ij}/2\right)$ for antiparallel and parallel alignment, respectively. Installation of the t_{12} expressions in place of b_{ij} in the superexchange terms of (3.12) will yield results equivalent to (3.17) and (3.18).

If the MO model is used, the transfer integral is a first-order perturbation and the respective relations of (3.17) and (3.18) simplify to

$$\Delta E_{\text{ex}}(\text{MO}) = -2z_{ij}J_{ij}S_{zi}S_{zj}\cos\,\theta_{ij}, \qquad (3.19)$$

where parallel alignment ($\theta_{ij} = 0$) requires $J_{ij} > 0$, and antiparallel alignment $(\theta_{ij} = \pi)$ sets $J_{ij} < 0$. This approach can be simply the analysis when first-order double exchange interactions are also present.

3.1.3 Spin Alignment in Oxides

The foregoing analyses for estimating spin alignment can be summarized. *Ferromagnetism (parallel spins)* occurs when the collective electron repulsion energy $\left(\sum e^2/r_{ij}\right)$ is dominant in $\mathcal{H}_1$ of (3.8). This is expected when the nuclear separation r_{ab} is large enough that the electron orbits overlap enough to disperse the spins according to Hund's rule without being unduly influenced by the attraction fields of their opposite nucleus (antibonding state is lowest). To comply with (3.12), this condition requires that $J > 0$. It should be noted in Fig. 1.12 that the light members of the 3d series provide fewer numbers of electrons to screen the attractive effects of the nuclei for ferromagnetism to occur. This observation is also consistent with the shorter bond lengths found for Ti, V, Cr, and Mn. *Antiferromagnetism (opposing spins)* occurs when the interionic electron–nuclear terms of (3.8) are dominant in $\mathcal{H}_1$. This is expected when the nuclear separation is small enough that the electron charge clouds overlap greatly ($r_{\text{ab}} << r_{\text{a2}}, r_{\text{b1}}$) and the dispositions of the spins are controlled by the attractive fields of the opposite nuclei (bonding state is lowest). For this case, the electrons can be more stable by filling states as spin pairs according to the Pauli principle, and $J < 0$. Experimental results that support these predictions are shown in Fig. 1.12 for a variety of direct exchange coupled compounds [13].

In oxides and other ionic compounds, the magnetic ions are not usually nearest neighbors. Therefore, any direct exchange effects should be weak. If the collective-electron formalism that is successful for metals is applied to oxides, r_{a2} and r_{b1} are seen as too large to create a negative J, thereby predicting that a weak ferromagnetic spin alignment might be expected. However, in most instances, this conclusion is not supported by experiment. The cation spin ordering is most often antiferromagnetic. Following the early suggestion of Kramers [14] that *indirect* exchange must occur, the general concept sketched in Fig. 3.1d was adopted. Indirect or *superexchange* is the coupling of cations spins through the medium of nearest-neighbor anion ligands. Ferromagnetism can also be produced by direct exchange ($J > 0$) between next-nearest neighbors, however, shown as $3d_{xy}$ orbitals overlapping across the diagonals of a two-dimensional model, for example, a cube face.

In addition to the mechanism where spins of like cations with half-filled orbitals correlate through common ligands by virtual charge transfer of two electrons in a periodic structure to produce antiferromagnetic order analogous to itinerant ferromagnetism in metals, real charge transfer between dissimilar ions can induce ferromagnetism by the sharing of a single electron. Because the spins are initially local to their parent ions, these short range transfers are commonly described as *delocalized* or *semicovalent*. Figure 3.9 illustrates two delocalization conditions: (a) half-filled to empty and (b) filled to half-filled orbital transfers [15]. A spe-

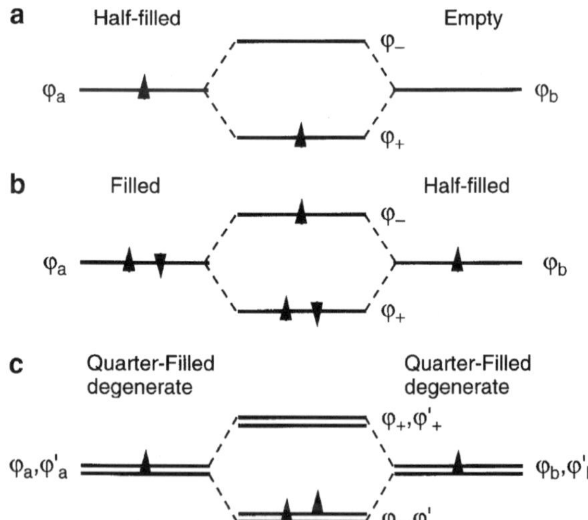

Fig. 3.9 Molecular-orbital diagrams of ferromagnetic superexchange: (**a**) delocalization from half-filled to empty, (**b**) delocalization from filled to half-filled, and (**c**) Hund's rule ferromagnetism in quarter-filled degenerate orthogonal metal–ligand states

cial case of ferromagnetic delocalization exchange[3] occurs between neighboring cations of the same atomic element occupying identical lattice sites, but with different valence charges. When this mechanism operates in an antiferromagnetic or ferrimagnetic system in which opposing magnetic sublattices are present, the transfers are intrasublattice (homonuclear, $H_2{}^+$ for the active orbital), and the resulting energy stabilization is called *double* exchange [16–19].

A somewhat related situation can occur in crystal fields (cubic, trigonal, or tetragonal) that allow orbital degeneracy to survive unquenched. In Fig. 3.9c, a single electron occupying a doublet, for example the t_{2g} spin-orbit stabilized case of d^2 or d^7 in a $c/a < 1$ tetragonal site, presents a degenerate antibonding state that allows Hund's rule to be sustained and provides metallic ferromagnetic superexchange.

A fourth type of indirect exchange occurs where the spins couple to each other through polarization of the "charge clouds" of their mutual environment rather than by a formally defined covalent bond. This mechanism, called RKKY after its collective authors Ruderman-Kittel [20], -Kasuya [21], and -Yosida [22] and described in most standard texts on magnetism, is generally associated with the rare-earth or lanthanide series where the $4f^n$ states are denied direct exchange interactions with like-orbital states of neighboring cations or with crystal fields of anions by the shielding from filled outer shells. Unpaired electron spins of the partially filled $4f^n$ inner shell interact indirectly with the immediate ligand field by polarizing the charges of their filled outer $5s^2$ and $5p^6$ shells that provide shielding from the crystal fields of the ligands, as evidenced by the weaker superexchange effects of rare-earth ions when occupying the c sublattice of magnetic garnets.

3.1.4 Ferromagnetism by Spin Transfer

Although less important from a purely magnetic standpoint, the case of *real* transfer by delocalization is of interest because of the electrical conductivity implications [3]. In the analysis of the magnetism created by a single mobile spin, it must be recognized from the outset that there can be no spin-dependent stabilization associated solely with the transfer between the two orbital states. However, if the two cations each have a net spin or reside in a cluster of lattice spins that would provide an exchange field to dictate the orientation of the transferring spin, the net spins of the cations involved in the transfer would likely be ferromagnetically aligned. Otherwise the transfer electron would undergo a spin reversal to obey Hund's first

[3] As a primer to the terminology used in the discussion of magnetic exchange in insulators, we first define two-electron exchange as *virtual* (spins do not actually switch nuclei) and therefore covalent, with correlated spins to create antiferromagnetic order by correlation exchange, or simply *superexchange*. One-electron semicovalent exchange is described as *real* (since the spin can actually switch nucleus) with delocalized spins to create ferromagnetic order exchange.

rule, violating the $\Delta S = 0$ requirement. If the spin retains its orientation in opposition to the polarization dictates of Hund's rule, its energy of the e^2/r_{ij} repulsion represents a destabilization that would offset any antiparallel alignments that are favored by unpaired spins occupying other orbital states on the same ion. The result of an antiparallel alignment between the cations would be the discouragement of the spin transfers through the creation of a spin trap or magnetic activation energy that would result in a loss of the energy that would otherwise be gained by the sharing of the spin.

As a consequence, in cases where only one of the orbital states is half-filled, the Pauli principle is no longer the main concern because only one electron can be involved in the coupling, and the two-electron exchange effects do not apply. Because the second orbital would be either empty or filled with no net spin in either case, the e^2/r_{ij} repulsive term is not involved in the charge transfer. For practical purposes, it can be convenient to view these two orbital combinations as analogous, with electron transfer into the empty orbital and "hole" transfer into the filled orbital. Except for the extremes of d^0, d^{10} or certain low-spin configurations that are diamagnetic because of low-spin $S = 0$ configurations, Hund's rule is followed for ions with $S = 0$ to prevent the loss of transfer energy that would otherwise occur through the formation of excited states within the $3d$-electron shell.

The various superexchange interactions between magnetic cations involve anion mediation and can be described in greater detail. Two basic situations can be expressed in chemical ionic notation as follows:

1. Two-electron covalent exchange (superexchange) occurs between half-filled orbitals depicted in Fig. 3.10a, where an electron from cation M_a is undergoes a virtual transfer to cation M'_b, requiring an excitation energy U according to

$$M_a{}^{n+} + M_b'{}^{n+} \rightarrow M_a^{(n+1)+} + M_b'{}^{(n-1)+} + U$$

followed by a corresponding return transfer of the electron from M'_b back to M_a to complete the exchange. Here U is for the general case of dissimilar ions of the same valence charge. The term correlation is frequently used to describe the antiparallel spin alignment that occurs cooperatively on opposite sides of the cation–anion covalent bonds through *virtual* transfer mentioned qualitatively in Sect. 3.1.1 and confirmed by the analytical formalism [5–10].

2. One-electron semicovalent exchange occurs between dissimilar ions M and M′ (with d-shell occupancy differing by only one electron spin) through *real* spin delocalization between half-filled → empty or filled → half-filled orbital combinations:

$$M_a^{n+} + M_b'^{(n+1)}+ \rightarrow M_a^{(n+1)+} + M_b'^{n+} + U \pm U_{ex},$$

where U_{ex} is the adjustment for internal exchange energy of spin stabilization (or destabilization if a spin flip is required) to comply with Hund's rule where unpaired spins occupy other states of the $3d$ shell, shown schematically in Fig. 3.10b. If the spin transfers within a ferromagnetic configuration, the M_b state

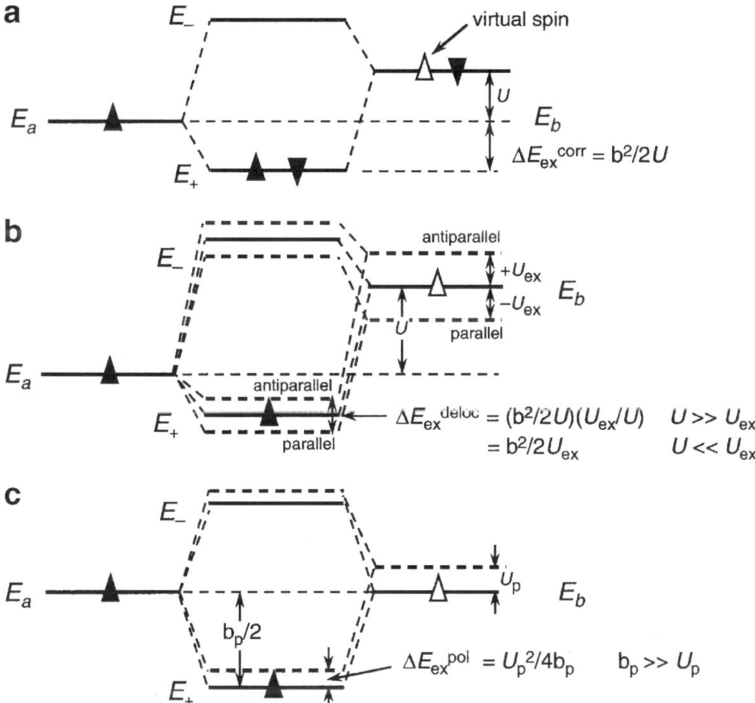

Fig. 3.10 Molecular-orbital diagrams indicating the origin of the exchange stabilization with Aufbau spin order preferences indicated for: (**a**) of two-electron correlation antiferromagnetism, (**b**) one-electron delocalization ferromagnetism (general case), and (**c**) one-electron delocalization ferromagnetism, special case of mixed-valence charge transfer

becomes more stabilized by U_{ex}, which will tend to offset U. However, if the spin transfers within an antiparallel configuration, the M_b state becomes destabilized by a similar energy.

General semicovalent exchange (Case 2) where the ions have different electron configurations results from the delocalization of a single electron as described by (2.46). Note that in its pure molecular form it is basically the H_2^+ molecule ion, i.e., a one-electron case, but with U_{ex} added. This situation can occur with ions of different atomic elements, e.g., Mn^{3+} (d^4) and Cr^{3+} (d^3) in octahedral ligand coordinations. The energy reduction from the sharing of an electron between a half-filled/empty hybrid state is also given by (3.16), but with half of the energy because of one electron instead of two involved in the exchange. Since there is no a priori requirement for spin polarization, i.e., no electron repulsion within the hybrid wavefunction, the amount of energy contributed by the transfer of the single spin will be influenced by the established alignment of the other spins occupying the orbital states involved. However, if a spin flip is necessary to complete the transfer in compliance with Hund's rule, the ionic transfer activation energy $U_{MM'}$ would be increased instead of reduced by U_{ex}, thereby resulting in a lowering of the transfer probability and ferromagnetic exchange stabilization, as diagrammed in Fig. 3.10c. The magnitudes of the respective delocalization exchange energies are given by

$$\begin{aligned}
\left|E_{\mathrm{ex}}^{\mathrm{deloc}}\right| &\approx \frac{b^2}{(U - U_{ex})} \quad \text{(ferromagnetic system)} \\
\left|E_{\mathrm{ex}}^{\mathrm{deloc}}\right| &\approx \frac{b^2}{(U + U_{ex})} \quad \text{(antiferromagnetic system)} \\
\left|\Delta E_{\mathrm{ex}}^{\mathrm{deloc}}\right| &\approx \frac{b^2}{(U - U_{\mathrm{ex}})} - \frac{b^2}{(U + U_{\mathrm{ex}})} = 2b^2 \left(\frac{U_{\mathrm{ex}}}{U^2 - U_{\mathrm{ex}}^2}\right) \\
&\approx \frac{2b^2}{U}\left(\frac{U_{\mathrm{ex}}}{U}\right) \quad \text{for } U_{\mathrm{ex}}^2 << U^2
\end{aligned} \tag{3.20}$$

in agreement with Goodenough's VB result [23]. As discussed in connection with real spin transfer, $\Delta E_{\mathrm{ex}}^{\mathrm{deloc}}$ represents an exchange trap energy for a mobile spin, e.g., a polaron with activation energy $U_{\mathrm{ex}} \neq 0$. Typical values of the relevant parameters are $b \sim -2\,\mathrm{eV}$, $U \sim -15\,\mathrm{eV}$, and $U_{\mathrm{ex}} \sim 2\,\mathrm{eV}$ (Slater integral), which permit the $U_{\mathrm{ex}}^2/U^2 << 1$ requirement to be satisfied. If the initial energy of $\mathrm{M_b}$ is restored in the process by an accommodation of the spin orientation, U_{ex} then represents the height of a barrier between two identical traps to be surmounted before a lossless transfer is complete. With only one electron involved in the stabilization energy, the MO orbital model can be used without loss of rigor.

In a special situation of Case 2 the combined energy of the initial and final states of mixed-valence $\mathrm{M_a}$ and $\mathrm{M_b}$ are unchanged in an adiabatic transfer event, with the ions essentially exchanging positions in the lattice. In this situation, the effect is called "double exchange" and occurs according to $\mathrm{M_a^{n+}} + \mathrm{M_b^{(n+1)+}} \leftrightarrow \mathrm{M_a^{(n+1)+}M_b^{n+}}$ that is basically ferromagnetic as diagrammed in Fig. 3.10c. However, the spin transfer still requires a mobility that is limited by an activation energy. In real situations, the spin transfer is between mixed-valence ions of the same atomic element, e.g., $\mathrm{Mn^{3+}}$ (d^4) and $\mathrm{Mn^{4+}}$ (d^3) or $\mathrm{Fe^{2+}}$ (d^6) and $\mathrm{Fe^{3+}}$ (d^5). The main interest in these combinations is the promotion of ferromagnetism through spin transfer (or vice-versa) by means of polaron carriers with mobility that is activated by thermal hopping of electrons across energy barriers of height (or traps of depth) generically defined as $U_{\mathrm{p}} = U_0 + U_{\mathrm{ex}}$. In special cases where exchange transfer is strong and the trap depth is small, these effects can also be realized by covalent tunneling through the bonding ground states. Here U_0 represents the trap energy of a small polaron that results from adjustments of the local electrostatic (elastic) environment [23].

For analytical purposes it is appropriate to introduce a polaron transfer integral b_{p} $(=b_{\mathrm{ab}})$ integral. For the case of $b_{\mathrm{p}} >> U_{\mathrm{p}}$, which is of interest for polarized spin transport, the magnitudes of the contributions can then be deduced from the one-electron molecular orbital solutions for double exchange from (2.46), noting that the energies $E_{\mathrm{a}} \approx E_{\mathrm{b}} = E_{\mathrm{F}}/2$ in this limit. Here the extent of the exchange stabilization that occurs because of the mutual ownership of the electron spin is determined by

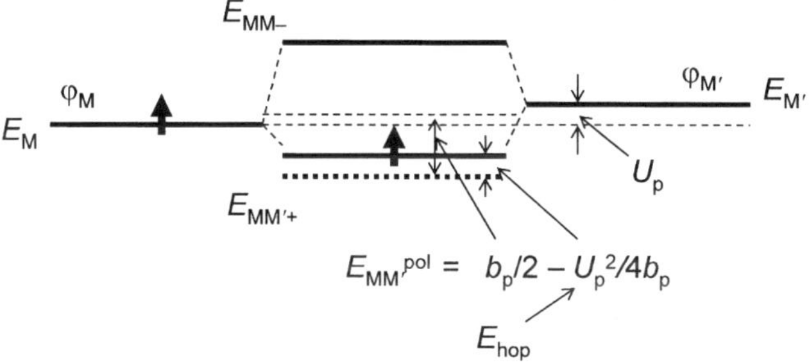

Fig. 3.11 Model of polaronic spin transfer indicating the origin of the E_{hop} activation energy

the degree of sharing. When U_{p} appears, both the sharing and stabilization decrease accordingly. For a pair of identical orbital states, the analysis in Sect. 3.1.1 can be applied to obtain the following relations

$$
\begin{aligned}
E_{\mathrm{ex}}^{\mathrm{pol}} &\approx \frac{b_{\mathrm{p}}}{2} && (\text{for } U_{\mathrm{p}} = 0),\\
E_{\mathrm{ex}}^{\mathrm{pol}} &\approx \frac{1}{2}\left(b_{\mathrm{p}}^2 + U_{\mathrm{p}}^2\right)^{1/2} && \left(\text{for } U_{\mathrm{p}} > 0\right),\\
E_{\mathrm{hop}} = \Delta E_{\mathrm{ex}}^{\mathrm{pol}} &\approx \frac{1}{2}\left[\left(b_{\mathrm{p}}^2 + U_{\mathrm{p}}^2\right)^{1/2} - b_{\mathrm{p}}\right] \\
&\approx \frac{U_{\mathrm{p}}^2}{4b_{\mathrm{p}}} && \left(\text{for } b_{\mathrm{p}} >> U_{\mathrm{p}}\right),
\end{aligned}
\tag{3.21}
$$

where E_{hop} is the effective polaron trap energy after U_{p} is reduced by the transfer integral b_{p} in proportion to the ratio $U_{\mathrm{p}}/b_{\mathrm{p}}$. The concept of polaronic electron spin transfer is diagrammed schematically in Fig. 3.11; for a filled to half-filled hole transfer, the mechanism is equivalent. This topic will be examined in greater detail in Chap. 8.

3.1.5 Goodenough–Kanamori Rules

On the basis of the exchange energies of individual molecular-orbital states, relations for the total ionic exchange constants can be deduced by applying (3.14) to combine the contributions from every occupied state. To this end, we define a resultant exchange constant between ions i and j:

$$
J_{ij} = -\frac{1}{2S_iS_j}\Delta E_{\mathrm{ex}} = -\frac{1}{2S_iS_j}\left(\sum \Delta E_{\mathrm{ex}}^{\mathrm{virtual}} - \sum \Delta E_{\mathrm{ex}}^{\mathrm{deloc}}\right), \tag{3.22}
$$

where $\Delta E_{\text{ex}}^{\text{virtual}}$ and $\Delta E_{\text{ex}}^{\text{deloc}}$ are for each individual orbital state and are positive quantities by definition. In most cases, the ferromagnetic influence of the orbital states linked by the single-electron transfers competes with the antiferromagnetism of any lower energy half-filled orbitals. In octahedral sites, where t_{2g} orbitals form weaker π bonds and e_g stronger σ bonds, the resultant magnetic alignment will depend on the particular ion combination. A more typical example is that of a d^4–d^3 coupling between ions of the same atomic element. Here the ferromagnetic stabilization of the single electron in the e_g–$p\sigma$ state is offset by the antiferromagnetic energies of the six t_{2g}–$p\pi$ electrons. The resultant spin alignments of these particular ions can assume parallel or antiparallel configurations depending on a variety of factors [3].

For any superexchange combination M_1–O–M_2, estimates of the J_{12} constant can be made if values can be assigned to the b_n integral and excitation energy U_n for each participating orbital state. In analytical terms, we express (3.15) as

$$J_{12} = -\left(\frac{1}{2S_1S_2}\right)\sum_n \frac{b_n^2}{U_n}. \tag{3.23}$$

This relation applies to any combination of $3d^n$ ions. To make calculations, it must first be realized that three of the five d orbitals are t_{2g}–$p\pi$ and the other two are e_g–$p\sigma$. The distinction is important because the overlap integral σ_π is significantly less than σ_σ. Estimates of these contributions that are made in Appendix 3A suggest that $b_\pi^2/U_\pi{:}b_\sigma^2/U_\sigma \sim 0.1$.

With this ratio one can construct models for individual ion combinations to predict the likely net superexchange interaction. As an example, the relation of (3.23) can be applied to the d^3–d^4 case $\left(t_{2g}^3e_g^0 \text{ to } t_{2g}^3e_g^1\right)$ occupying adjoining octahedral sites:

$$J_{12}^{\text{virtual}} \approx -\frac{1}{2S_1S_2}\left(\frac{3b_\pi^2}{U_\pi}\right) \quad \text{(antiferromagnetic)}, \tag{3.24}$$

or and for the single e_g spin

$$J_{12}^{\text{deloc}} \approx +\frac{1}{2S_1S_2}\left(\frac{b_\sigma^2}{U_\sigma}\right) \quad \text{(ferromagnetic)}, \tag{3.25}$$

which yields $J_{12}\,(\text{antiferro})\,/\,J_{12}\,(\text{ferro}) \approx -0.33$ if the above ratio of b_π^2/U_π : b_σ^2/U_σ is used.[4] Since the spin system can stabilize with either alignment, the greater stabilization energy will determine which one will dominate, i.e., which one will provide the larger spin-ordering temperature. The result for the above example suggests a condition of ferromagnetism dominated by the delocalization exchange from only one of the e_g–$p\sigma$ bonds.

[4] Note that if only one electron is stabilized in a semicovalent transfer, it could be argued that the $\Delta E_{\text{ex}}^{\text{virtual}}/\Delta E_{\text{ex}}^{\text{deloc}}$ is double at 0.2, which would result in $J_{12}^{\text{antiferro}}/J_{12}^{\text{ferro}} \approx -0.66$.

By pursuing this reasoning to evaluate the probable spin alignments of various combinations favored by particular cation–anion–cation linkages, a catalog of superexchange interactions among members of the $3d^n$ series can be created including relative strengths of the expected J_{12} constants. Such endeavors were undertaken independently by Goodenough and Kanamori with results that have become such valuable guidelines for the chemical design of magnetic compounds that the Goodenough–Kanamori rules have been added to the lexicon. In Table 3.2, a compilation of superexchange interactions based on Kanamori's review [24] is reproduced with some additions. The listing is restricted to some of the more standard combinations, as examined by Anderson [6], Goodenough [25, 26], Anderson and Hasegawa [18], and Slater [27]. Those with the same ionic valence are probably more credible than the others because of more consistent values anticipated for b and U.

Table 3.2 Summary of spin alignment estimates of d^n ion 180° superexchange interactions in octahedral sites (after Kanamori [24])

d^n–$d^{n'}$ combinations	Cations	Bond mechanism	Spin alignment	Probable result
d^3–d^3	Mn^{4+}–Mn^{4+}	σ, π-bonds		
	Cr^{3+}–Cr^{3+}	A, G, A–H, S	AF	AF
d^8–d^8	Ni^{2+}–Ni^{2+}	σ-bonds		
		A, G, A–H, S	AF	AF
d^5–d^5	Mn^{2+}–Mn^{2+}	σ-bonds		
	Fe^{3+}–Fe^{3+}	A, G, A–H, S	AF	AF
		π-bonds	AF(weak)	
		G, A–H, S	Uncertain	
		π-bonds	(Weak)	
		A		
d^8–d^3	Ni^{2+}–V^{2+}	σ, π-bonds		
	Ni^{2+}–Cr^{3+}	A, G, A–H, S	F	F
d^5–d^3	Fe^{3+}–Cr^{3+}	σ-bonds		
	Mn^{2+}–V^{2+}	A, G, A–H, S	F	F
	Fe^{3+}–V^{2+}	π-bonds	AF(weak)	
	Mn^{2+}–Cr^{3+}	G, A–H	Uncertain	
		π-bonds	(Weak)	
		A, S		
d^4–d^4	Mn^{3+}–Mn^{3+}	Bond angle		
	Mn^{3+}–Fe^{4+}	Dependent		
d^6–d^6	Fe^{2+}–Fe^{2+}	σ-bonds		
	(e.g., FeO)	A, G, A–H, S	AF	AF
	Co^{3+}–Co^{3+}	π-bonds	Uncertain	
	(e.g., $Co_2O_3^a$		(Weak)	
d^7–d^7	Co^{2+}–Co^{2+}	σ-bonds		
	(e.g., CoO)	A, G, A–H, S	AF	AF
		π-bonds	Bond-angle dependent	

A Anderson mechanism; *G* Goodenough mechanism; *A–H* Anderson-Hasegawa mechanism; *S* Slater mechanism

[a] Not 180° bond angles

Table 3.3 Spin ordering of d^n ions in octahedral oxygen cites, high-spin states, 180° bonds (after Goodenough [28])

	e_g / t_{2g}	d^0	d^1	d^2	d^3	d^4	d^5	d^6	d^7	d^8	d^9
	d^1	$S = 0$	NEGL	↑↓ W	↑↓ W	↑↑ ~	↑↑ M	↑↑ M	↑↑ M	↑↑ M	NEGL
	d^2			↑↓ W	↑↓ W	↑↑ ~	↑↑ M	↑↑ M	↑↑ M	↑↑ M	↑↑ W
	d^3				↑↓ W	↑↑ ~	↑↑ M	↑↑ M	↑↑ M	↑↑ M	↑↑ W
	d^4					↑↑ ~ M	↑↓ ~	↑↓ ~	↑↓ ~	↑↑ ~	↑↓ ~
	d^5						↑↓ S	↑↓ S	↑↓ S	↑↓ S	↑↓
	d^6							↑↓ S	↑↓ S	↑↓ S	↑↓
	d^7								↑↓ S	↑↓ S	↑↓
	d^8									↑↓ S	↑↓
	d^9										↑↓
	d^{10}										$S = 0$

S = Strong W = Weak NEGL = Negligible
M = Moderate ~ = Quasi-static

A more comprehensive study of cation combinations including those involving low-spin configurations is given by Goodenough [28]. In Table 3.3 results for 180° bonds between cations in octahedral sites are presented. It is important, however, to recognize that the spin ordering temperatures could be meaningful only where a homogeneous magnetic lattice is created. In most of these situations, particularly those involving combinations far from the diagonal or of widely differing cation valences, the actual materials structures may not be thermodynamically stable and may be fashioned only through artificial procedures such as molecular beam epitaxial layer growth. To estimate corresponding Néel or Curie temperatures, the proper relations between the spin values and the number of exchange coupled nearest neighbors must

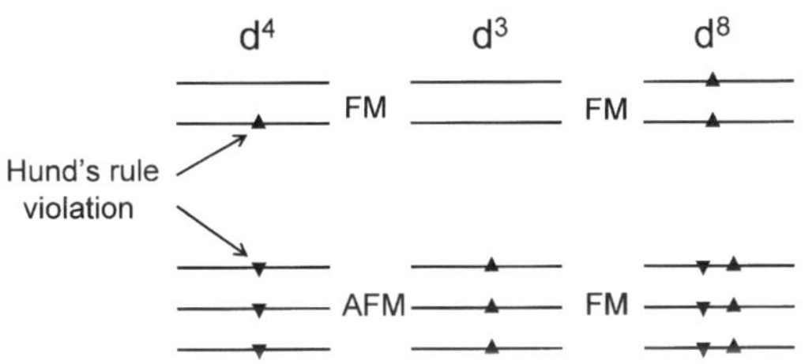

Fig. 3.12 Model of the Goodenough–Kanamori rules of spin alignment for d^4, d^3, and d^8 configurations in octahedral sites. E_g shell spin reversal in d^4 is shown intentionally to illustrate a Hund's rule violation

be selected for each pair of cations. An example of the application of these rules is offered by the Aufbau models of superexchange couplings between the aforementioned d^4–d^3 and d^8–d^3 combinations sketched in Fig. 3.12. As pointed out above, the former favors ferromagnetism in one state of the e_g shell that is opposed by all three antiferromagnetic virtual states of the t_{2g} shell. In a charge-ordered double perovskite lattice, e.g., $La^{3+}Ca^{2+}\left(Mn^{3+}Mn^{4+}\right)O_6$, the result is largely inconclusive. With Ni^{2+} $\left(d^8\right)$ in place of Mn^{4+}, however, all five orbital levels produce ferromagnetism that results in a Curie temperature near 300 K [29]. This subject will be addressed further in Sect. 3.3.

Before the topic of the effective magnetic fields created by superexchange is examined in the context of antiferromagnetism and ferrimagnetism, the implications of tetrahedral sites and non 180° bond angles should be mentioned. Recalling the discussion of the crystal field parameter $10Dq$ and how it reverses sign between octahedral and tetrahedral sites, we immediately recognize that the relative strengths of π and σ bonds should be expected to change since the locations of the negative anion charges are in the directions of the body diagonals in the tetrahedral case. Where the bonds form other than 180° angles, the influence of the t_{2g} orbitals is likely to increase relative to the e_g orbitals. These subtleties will be discussed further as the need arises in later chapters.

3.2 Antiferromagnetism

Because superexchange is generally associated with antiparallel spins, magnetic oxides usually feature antiferromagnetic ordering. At this point in our discussion, it is necessary to return to the subject of molecular fields introduced in Sect. 1.3 in preparation for a more detailed account of the opposing sublattice theories of antiferro and ferrimagnetism.

3.2.1 Superexchange and Molecular Fields

One of the most important results of the insights gained into the relation between covalent bonding and superexchange is the increased understanding of the origins of the molecular (or superexchange) field in spontaneous magnetism that was introduced in Sect. 1.3.1. Recalling (1.40), we express

$$H_{\text{eff}} = H + H_{\text{ex}} = H + N_W M, \tag{3.26}$$

where $N_{\text{W}} = \left(\frac{z}{n}\right)\frac{2J}{g^2 \text{m}_{\text{B}}^2}$ from (1.68) for the ferromagnetic case. For the two sublattices i and j,

$$\begin{aligned} N_{ij} &= \left(\frac{z_{ij}}{n_j}\right)\frac{2J_{ij}}{g_i g_j \text{m}_{\text{B}}^2}, \\ N_{ji} &= \left(\frac{z_{ji}}{n_i}\right)\frac{2J_{ji}}{g_j g_i \text{m}_{\text{B}}^2}, \end{aligned} \tag{3.27}$$

where n_i and n_j are the respective densities of spins of the i and j sublattices and z_{ij} and z_{ji} are the respective numbers of exchange-coupled neighbors.

From (3.15) we also see that the molecular field coefficient can be expressed in terms of the covalent stabilization energy

$$N_{ij} = -\frac{1}{g_i g_j \text{m}_{\text{B}}^2 S_i S_j}\left(\frac{z_{ij}}{n_j}\right)\sum_n \frac{b_n^2}{U_n}. \tag{3.28}$$

Since the molecular field in (3.26) from the j sublattice $H_{\text{ex}}^{(j)} = N_{ij} M_j$ acts on a magnetic moment $m_i = g_i m_{\text{B}} S_i$ in the i sublattice, the antiferromagnetic stabilization energy for the i sublattice is expressed as

$$E_m^{(i)} = -m_i H_{\text{ex}}^{(j)} = -g_i \text{m}_{\text{B}} S_i N_{ij} M_{\text{j}}. \tag{3.29}$$

Therefore, we can substitute for $M_j = n_j g_j \text{m}_B S_j$ and for N_{ij} from (3.28) to obtain

$$m_i H_{\text{ex}}^{(j)} = z_{ij} \sum_n \frac{b_n^2}{U_n}. \tag{3.30}$$

As a result, the Brillouin-Weiss function parameter a_{i} with the applied field $H = 0$ can be expressed as

$$a_i = \frac{m_i H_{\text{ex}}^{(j)}}{kT} = \frac{g_i \text{m}_{\text{B}} S_i N_{ij} M_j}{kT} = \frac{z_{ij}}{kT}\sum_n \frac{b_n^2}{U_n}. \tag{3.31}$$

Note that there is no direct indication that $m_i H_{\text{ex}}^{(j)}$ is of magnetic origin. In generic systems where orbital angular momentum and spin-orbit coupling effects are absent, i.e., for S-state ions such as Fe^{3+} or some crystal-field quenched systems of the d^n electron groups, the Pauli principle and Hund's rule are responsible for the superexchange stabilization. The influence of orbital angular momentum will be examined in Sect. 5.1.

3.2.2 Molecular Field Theory of Antiferromagnetism

To appreciate the importance of the molecular field concept in applying the theory of superexchange to magnetic oxides, we must first review Néel's extension of the Brillouin-Weiss theory of ferromagnetism to the case of opposing sublattices that are characteristic of antiferromagnetism and its more complex cousin ferrimagnetism [30]. For these situations, the spin values represent the number of orbital states of the two ions that are jointly participating in the superexchange covalent stabilization.

Although many instances of antiferromagnetism involve multiple sublattices, we limit this discussion to the simplest case of two sublattices comprising nearest-neighbor sites i and j occupied by ions with alternating spin directions. This is the case of a lattice in which the magnetic ions occupy the corners of a simple cubic structure, typical of a cubic perovskite to be examined in Sect. 3.3.2. There are also situations where the magnetic sublattices are formed between next-nearest neighbors, and these will be discussed later. In the ideal situation at $T = 0\,\text{K}$, the spin directions are assumed to be exactly parallel or antiparallel. Recalling the notions of the molecular field introduced in Sect. 1.3., we express the magnitudes of the resultant effective magnetic fields at the individual sites as

$$\begin{aligned} H_i &= H + N_{ii}M_i + N_{ij}M_j, \\ H_j &= H + N_{jj}M_j + N_{ji}M_i, \end{aligned} \tag{3.32}$$

where H is the applied magnetic field, $N_{ii}\left(= N_{jj}\right)$, and $N_{ij}\left(= N_{ji}\right)$ are the corresponding intra and intersublattice molecular field coefficients for the two sublattices of magnetizations M_i and M_j. Because the interaction between sublattices is antiferromagnetic, the N_{ij} coefficient is negative, while N_{ii} and N_{jj} could be positive or negative depending on the nature of the particular superexchange discussed in the previous section.

At thermal equilibrium, the individual sublattice magnetizations can be expressed by

$$M_i = n_i g \mathrm{m_B} S_i \mathcal{B}_{S_\mathrm{i}}\left(a_i\right), \tag{3.33}$$

where n_i is the volume density of spins S_i, $\mathcal{B}_{Si}$ is the Brillouin-Weiss function and $a_i = \frac{g_i \mathrm{m_B} S_i H_i}{kT}$ as defined previously from (3.31) and (3.32). As suggested by the introductory analysis of ferromagnetism in Sect. 1.3, a value of the threshold (Néel) temperature for spontaneous antiferromagnetic alignment will emerge from a solution of (3.33). To determine the behavior at or above the Néel temperature, we use the approximation for $\mathcal{B}_{S\mathrm{i}}\left(a_i\right) \to \left[\left(S_\mathrm{i}+1\right)/3S_\mathrm{i}\right] a_i$ near the limit where $a_i << 1$ from (1.37) to express the individual sublattice magnetizations as

$$\begin{aligned} M_i &= \frac{n_i g_i^2 \mathrm{m_B^2} S_i\left(S_i+1\right)}{3kT}\left(H + N_{ii}M_i + N_{ij}M_j\right), \\ M_j &= \frac{n_j g_j^2 \mathrm{m_B^2} S_j\left(S_j+1\right)}{3kT}\left(H + N_{ij}M_i + N_{jj}M_j\right), \end{aligned} \tag{3.34}$$

where the directions of H, M_i, and M_j can be treated as parallel in the paramagnetic regime above T_N. If we assume that each sublattice contains an equal division of the total population n, which comprises the same ionic species with spin S and spectroscopic factor g, $S_i = S_j = S$, $g_i = g_j = g$, and $n_i = n_j = n/2$, the resultant magnetization becomes

$$M = M_i + M_j = \frac{ng^2 \mathrm{m}_\mathrm{B}^2 S\,(S+1)}{3kT}\left[H + \left(N_{ii} + N_{ij}\right) M\right] \tag{3.35}$$

from which the susceptibility $\chi = M/H$ can be written in terms of the paramagnetic Néel temperature θ_N

$$\chi = \frac{C}{T + \theta_N}, \tag{3.36}$$

where $C = \frac{ng^2 \mathrm{m}_\mathrm{B}^2 S(S+1)}{3k}$ and $\theta_N = -\frac{C}{2}\left(N_{ij} + N_{ii}\right)$. Note that N_{ij} is assumed to be negative and of greater magnitude than N_{ii}. Therefore, θ_N is positive and consistent with (1.43). A graphical representation of the different situations of (3.36) is presented in Fig. 1.7. In situations where N_{ii} becomes equal to or greater than N_{ij}, a reordering of the sublattice spin alignments is expected to take place.

To obtain an expression for the actual Néel temperature T_N, (3.36) can be used in the limit of $H \to 0$,

$$\begin{aligned} M_i &= \frac{C}{2T}\left(N_{ii} M_i + N_{ij} M_j\right), \\ M_j &= \frac{C}{2T}\left(N_{ij} M_i + N_{ii} M_j\right). \end{aligned} \tag{3.37}$$

For M_i and M_j each to be nonzero, the determinant of the coefficients of M_i and M_j must be zero. Therefore, the solution for $T = T_N$ becomes

$$T_N = -\frac{C}{2}\left(N_{ij} - N_{ii}\right). \tag{3.38}$$

At this point, it is possible to compare the paramagnetic temperature with the Néel temperature by taking the ratio of (3.37) and (3.38):

$$\frac{\theta_N}{T_N} = -\frac{N_{ij} + N_{ii}}{N_{ij} - N_{ii}}. \tag{3.39}$$

There are limitations to the applicability of (3.39). In most cases N_{ii} and N_{ij} are both negative, and the ratio $\theta_N/T_N > 0$. Where N_{ii} is negligible compared with N_{ij}, $T_N \approx \theta_N$; where these two coefficients are comparable in size, instability in the ordering will occur; if N_{ii} were to dominate, the static spin ordering would assume a different pattern.

To complete the picture of the susceptibility as a function of temperature for a single-crystal antiferromagnet, we must examine the condition of the spin systems below the Néel temperature. Here the two sublattices tend to be antiparallel because of the superexchange stabilization, but are still influenced by an applied H. Because of the existence of crystalline anisotropy, the direction of the applied field relative to the preferred direction of the spins must be taken into account. The importance of magnetocrystalline anisotropy will be realized in our discussions of ferrite in Chap. 4. For this exercise, the anisotropy is considered to be uniaxial and our discussions will be limited to the cases of H parallel and perpendicular to the easy axis. Because of the complexity of the analytical procedure, only the results will be presented in this text. For details of the derivation, the reader is advised to consult texts such as Morrish [31]. The relations for the parallel and perpendicular susceptibility are stated as

$$\chi(T) = \frac{ng^2 \mathrm{m}_\mathrm{B}^2 S^2 \mathcal{B}'(a_0)}{kT - \frac{1}{2}\left(N_{ij} + N_{ii}\right) ng^2 \mathrm{m}_\mathrm{B}^2 S^2 \mathcal{B}'(a_0)},$$

$$\chi_\perp(T) = -\frac{1}{\left(N_{ij} + N_{ii}\right)}, \tag{3.40}$$

where $\mathcal{B}'(a_0)$ is the first derivative of the Brillouin-Weiss function and

$$a_0 = \frac{g\mathrm{m}_\mathrm{B} S}{kT}\left[\left(N_{ij} - N_{ii}\right) M_0\right], \tag{3.41}$$

where $M_0 = M_i = -M_j$ at $H = 0$. If it is assumed that $N_{ii} = 0$, (3.40) can be plotted as function of T/T_N with S as a variable parameter, without assigning a specific value to N_{ij}, as shown by Lidiard's $\chi(T)$ calculated graphs in Fig. 3.13 [32]. In the $1/\chi$ vs. T format where $N_{ii} \neq 0$, the curves appear as sketched in Fig. 3.14.

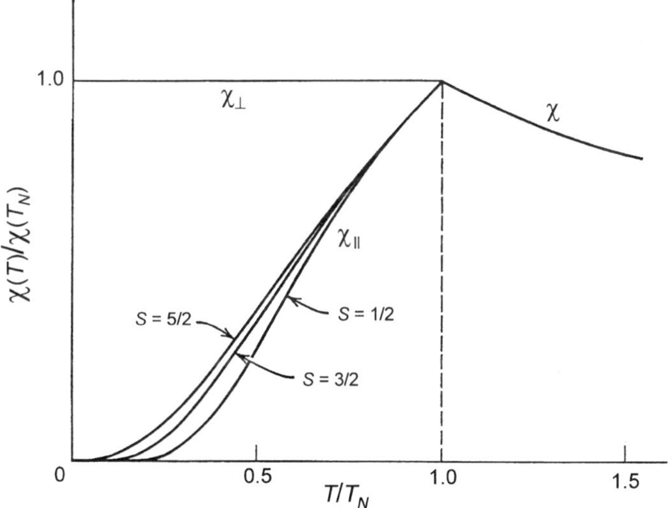

Fig. 3.13 Susceptibility of an antiferromagnet computed as a function of temperature in units reduced by the Neel temperature T_N. Image is based on computations of Lidiard [32]

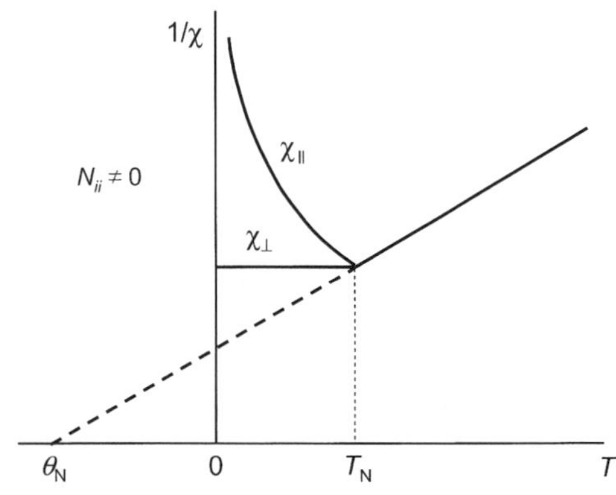

Fig. 3.14 Inverse antiferromagnetic susceptibility modeled from the Curie law as a function of temperature, indicating the asymptotic Neel temperature θ_N discussed in Sect. 1.4

For polycrystalline or powdered materials consisting of randomly oriented particles, this discussion can be extended by means of a simple argument. In general, the applied field will form an angle θ with the easy axis of a particular crystallite. Therefore, each single crystal grain will contribute a parallel and a perpendicular component to the magnetization, and thereby to the overall susceptibility which can be expressed as

$$\chi_{sc} = \frac{M_{\parallel}}{H}\cos\theta + \frac{M_{\perp}}{H}\sin\theta. \tag{3.42}$$

Since $\chi_{\parallel} = \frac{M_{\parallel}}{H\cos\theta}$ and $\chi_{\perp} = \frac{M_{\perp}}{H\sin\theta}$, (3.42) becomes

$$\chi_{sc} = \chi_{\parallel}\cos^2\theta + \chi_{\perp}\sin^2\theta, \tag{3.43}$$

and after averaging over all particle orientations, the susceptibility of a polycrystalline specimen is

$$\chi_p = \chi_{\parallel}\overline{\cos^2\theta} + \chi_{\perp}\overline{\sin^2\theta} = \frac{1}{3}\chi_{\parallel} + \frac{2}{3}\chi_{\perp}. \tag{3.44}$$

One additional estimate for the susceptibility should be added to this discussion. The ratio of the polycrystalline susceptibility at $T = 0$ to its value at $T = T_N$ can be obtained from (3.44) by setting $\chi_{\parallel} = 0$ at $T = 0$, and $\chi_{\parallel} = \chi_{\perp}$ (a constant value) at $T = T_N$ to yield

$$\frac{\chi_p(0)}{\chi_p(T_N)} = \frac{2}{3}. \tag{3.45}$$

This polycrystal susceptibility can be estimated from a simple interpolation suggested by (3.44) between the $\chi_{\parallel}$ and $\chi_{\perp}$ curves in Fig. 3.14.

Application of these theoretical results to specific materials systems has been expounded in a number of standard texts and comprehensive reviews. For our purposes, some generic examples will be given to illustrate the relation between covalent bonding, superexchange, and the magnetic sublattice arrangement.

3.2.3 Antiferromagnetic Spin Configurations

In nonmetallic compounds, antiferromagnetism reveals itself in a vast number of chemical structures that include, besides the oxides, sulfides, selenides, tellurides, and of course, halides of both simple and more complex combinations of cations in various crystal structures both natural and synthetic. To explore this ocean of possibilities in any depth would defeat the purpose of this text, but it is, nonetheless, important that the reader gain an appreciation of the background and the more basic facets of this most common type of magnetic interaction. As explained in Sect. 3.1, there are two overriding factors that determine the nature of the spin alignment in most situations: (1) direct overlap $e_g - p\sigma$ bonding is expected to dominate the exchange between octahedral sites and (2) the sign of the exchange constant J would be negative in situations where the cations have similar valence charges and electronic configurations.

For particular crystal lattices, the sites that make up the opposing sublattice, designated as j, are not always the crystallographic nearest neighbors. The determining factor in each case is the disposition of the anion (oxygen) relative to the cation neighbor, because it is the anion that provides the chemical bond which establishes the stabilization energy. There are three common magnetic structures that need to be distinguished: simple cubic (perovskites), body-centered cubic (rutile), and face-centered cubic (one-metal oxides). The first two are conventional in the sense that the j sublattice consists of nearest neighbor cations and the Néel temperature expressed by (3.38),

$$\begin{aligned} T_{\mathrm{N}} &= -\frac{C}{2}(N_{ij} - N_{ii}) && \text{for rutile and perovskite} \\ &\approx -\frac{C}{2}N_{ij} && \text{for } |N_{ij}| >> |N_{ii}| . \end{aligned} \tag{3.46}$$

The general relations between J, $\sum b_n^2/U_n$ and T_{N} for this antiferromagnet are

$$J = -\frac{1}{2S^2}\sum_n \frac{b_n^2}{U_n} = -q\frac{3kT_{\mathrm{N}}}{2zS(S+1)}, \tag{3.47}$$

from which we can express

$$\begin{aligned} T_{\mathrm{N}} &= \frac{z}{3k}\sum_n \frac{b_n^2}{U_n}\frac{S+1}{S} && (q = 2, \text{simple cube}), \\ T_{\mathrm{N}} &= \frac{z}{6k}\sum_n \frac{b_n^2}{U_n}\frac{S+1}{S} && (q = 4, \text{face-centered cube}), \\ T_{\mathrm{N}} &= \frac{2z}{3k}\sum_n \frac{b_n^2}{U_n}\frac{S+1}{S} && (q = 1). \end{aligned} \tag{3.48}$$

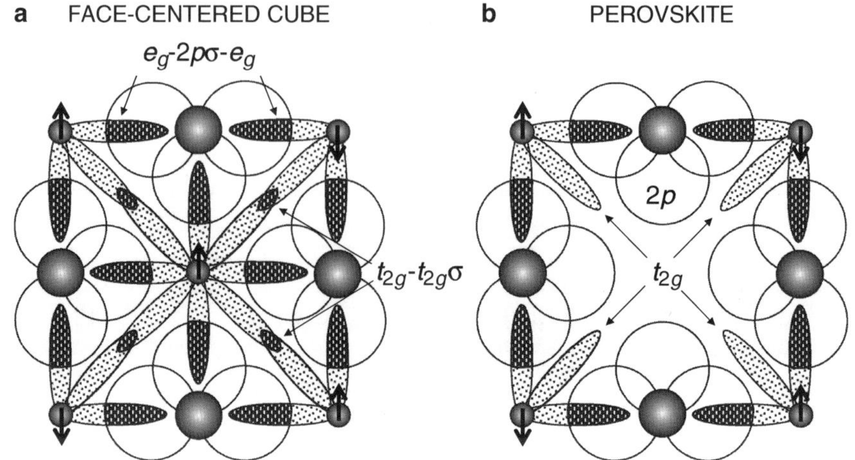

Fig. 3.15 Comparison of bonding linkages across cube faces for (**a**) a face-centered cube and (**b**) a perovskite. The difference is contribution of direct $t_{2g} - t_{2g}\sigma$ bonds in (**a**)

Note that $q = 2$ for antiferromagnetism and originates from the definition of (3.38); it is also double that of the ferromagnetic case ($q = 1$). A more interesting situation is the face-centered cube shown in Fig. 3.15a, for which $q = 4$. In this case, the opposing magnetic sublattice comprises next-nearest neighbors for reasons that can be seen from the diagram of the face. Along the unit cell edges 180°M–O–M linkages between cations occupying octahedral corner sites provide the strongest superexchange from e_g–$p\sigma$ bonds, and these next-nearest-neighbor interactions dominate over the diagonal interactions between the nearest-neighbor corner to face-center sites. If the intersublattice coefficient is greater than that of the intrasublattice coefficient in this structure, i.e., $|N_{ij}| > (3/4)\,|N_{ii}|$, and from [33]

$$T_N = -\frac{CN_{ij}}{4} \text{ (one-metal oxide)}, \tag{3.49}$$

and we see that only N_{ij} can be involved and the factor of 2 in the denominator of (3.38) is increased to 4. This occurs because of the sublattice configuration of layered <111> planes shown in Fig. 3.16 for this structure must have the 12 nearest neighbors occurring six in-plane of the same spin alignment and six in the adjacent plane of opposite spin alignment, thereby producing a cancellation of the N_{ii} contributions. However, coupling to nearest neighbors can still influence the value of N_{ij} through direct linkages in tandem. The relations corresponding to (3.38) apply here also, but with $q = 4$.

Experimental evidence of the spin configurations was obtained by the powerful tool of neutron diffraction, which can sort out planes of common spin orientations in a manner similar to that of X-ray diffraction with ordinary crystal lattices. An example of these data above and below the Néel temperature for MnO is shown in Fig. 3.17 [34]. Note the radical difference in the patterns for the two temperature regimes.

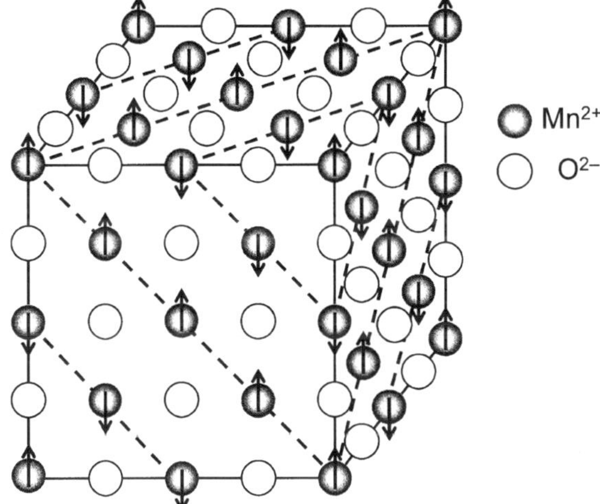

Fig. 3.16 Antiferromagnetic structure of a diatomic metal oxide of cubic symmetry

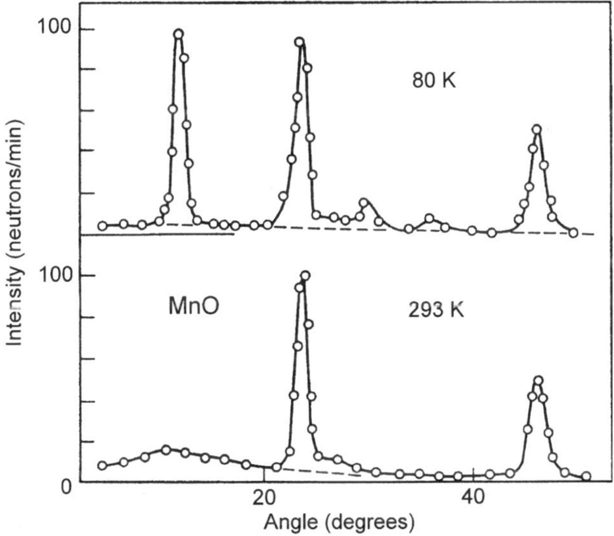

Fig. 3.17 Neutron-diffraction pattern from MnO, which is antiferromagnetic below the Neel temperature at 120 K. The upper trace was from the antiferromagnetic state; the lower one is for paramagnetism at room temperature. Image is adapted from Lax and Button presentation ([13], Fig. 3.22) of data by Shull et al. [34]

Figure 3.18 shows another common antiferromagnetic structure rutile, which is a tetragonal with magnetic cations centering octahedrally coordinated ligands with axes directed along twofold lattice symmetry axes. The local antiferromagnetic 180° cation–anion–cation arrangements produce the particular spin ordering shown in the figure. In this case the Néel temperature $T_{\mathrm{N}} = -(C/2)\left(N_{ij} - N_{ii}\right)$ follows (3.46).

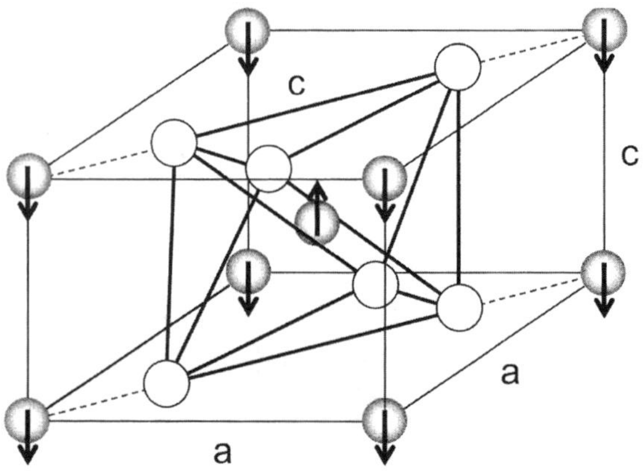

Fig. 3.18 Basic cell of rutile indicating the octahedrally coordinated ligands surrounding the internal cation with axes along a lattice axes of twofold symmetry

Table 3.4 Parameter data from selected antiferromagnetic compounds

Compound	Crystal structure	T_N (K)	θ_N (K)	θ_N/T_N	$\frac{\chi_p(0)}{\chi_p(T_N)}$	C_{mole} (cgs units)
MnO	fcc	122	610	5.0	0.69	4.40
FeO	fcc	198	570	2.9	0.77	6.24
CoO	fcc	293	280	0.96	–	3.0
NiO	fcc	520	–	–	0.67	–
CuO	monoclin.	~453	–	–	–	–
MnS	fcc	165	528	3.2	0.82	4.30
MnF_2	rutile	72	113	1.6	0.75	4.08
FeF_2	rutile	79	117	1.5	0.72	3.9
CoF_2	rutile	38	53	1.4	–	3.3
NiF_2	rutile	73	116	1.6	–	1.5
MnO_2[a]	rutile	84	–	–	0.93	–

References may be found in the review article by Nagamiya et al. [19]

[a] Mn is in the 4+ state, which means that only the t_{2g} orbitals are involved in the superexchange

Because of the relationship between spin alignments and bonding, simple metal oxides ($M^{2+}O^{2-}$) can form antiferromagnetic compounds of the face-centered cube (MO) or rutile (MO_2) types. Perovskites with magnetic structures that approximate simple cubic usually feature M^{3+} cations that have inactive e_g–$p\sigma$ bonds. When these 2+ ions are from the upper half of the series, the Néel temperatures are significant because of the availability of strong e_g–$p\sigma$ bonds. A comparison of the properties of these ions in the two crystal structures is informative. As listed in Table 3.4, T_N and θ_N values [19] are greater for the face-centered cubic structure. Part of the difference can be attributed to the 180°M–O–M bond angles, which provide the largest overlap integral and greatest stabilization energy. For rutile, the angle is less than 180°, which is more typical of ferrimagnetic oxides to be examined in Chap. 4

3.3 Antiferromagnetic Oxides

Chemical compounds with antiferromagnetic spin alignments are the most common of materials that exhibit magnetic properties. Even magnetically undiluted materials that are paramagnetic at room temperature usually reveal a Néel transition if the temperature is lowered far enough. Although the present discussion is restricted to oxides, halides and other compounds that incorporate ions of the $3d^{\mathrm{n}}$ series will be included wherever they can illustrate important features. The crystallographic systems that have been studied extensively both as vehicles for basic science investigations and for practical applications are the chemically simple one-metal compounds already introduced in the previous section and the more complex perovskites.

3.3.1 *One-Metal Oxides*

In the previous section, magnetocaloric properties of some of the divalent $3d^{\mathrm{n}}$ metal oxides were used as examples of the thermal effects that occur at the antiferromagnetic order–disorder transition. With the exception of CuO, which features a noncubic structure influenced by the Jahn-Teller distortions of the normally octahedral sites, each of them is of the face-centered cubic structure. The relation between the electron configurations and the Néel temperatures are summarized in Table 3.5. Exchange stabilization energies $z\sum b_n^2/U_n$ are deduced from the T_{N} values with the aid of (3.48).

The ions from the lower half of the series are typically trivalent, which precludes their occurrence in the $M^{2+}O^{2-}$ face-centered cubic oxides. As a result, lower symmetry molecular structures are formed without 180° bonds and strong antiferromagnetism does not appear. Since the ions from $n = 1, 2$, or 3 configurations have only t_{2g} electrons, bonding is achieved by means of t_{2g} orbitals with oxygen that can be stronger than the t_{2g}–$p\pi$ bonds of typical formations when 180° angles are available. In this case, the π overlaps are a combination of σ and π. As in all chemical compounds, the directionality of the t_{2g} orbital lobes in relation to their $2p$ lobe bonding partners of the oxygen ligands is the determining influence in establishing the particular stereochemistry of the molecular structure.

Table 3.5 Superexchange data of the $3d^n$ ions in one-metal oxides

Ion	Config.	S	T_N (K)	J (meV)	$z\sum b_n^2/U_n$ (meV)	ΔE_{hop} (meV)
$Mn^{2+}O$	$t_{2g}^3e_g^2$	5/2	122	−1.2	90	~100
$Fe^{2+}O$	$t_{2g}^3e_g^2$	2	198	−2.8	136	–
$Co^{2+}O$	$t_{2g}^5e_g^2$	3/2	293	−6.8	182	~300
$Ni^{2+}O$	$t_{2g}^6e_g^3$	1	520	−22.4	270	~600
$Cu^{2+}O$[a]	$t_{2g}^6e_g^3$	1/2	453	−52.0	156	~600

[a] CuO is of monoclinic structure probably because of the Jahn-Teller nature of the Cu^{2+} ion in an octahederal site

The magnitudes of the Néel temperatures presented in Table 3.5, which are correlated with the $z \sum b_n^2 / U_n$ through the relation of (3.48), were analyzed by considering the various covalent linkages that occur in the face-centered cubic system [35]. Inspection of these results immediately reveals the seemingly paradoxical conclusion that stability of the antiferromagnetic order varies inversely with the spin values in a systematic progression for which an explanation is not readily obvious. Insight can be gained by examining the nature of the chemical bonding in this structure. Recalling the implications of (3.41) when competing molecular field coefficients are present, we first recognize that the main e_g–$p\sigma$–e_g 180° antiferromagnetic couplings in Fig. 3.15a are between next-nearest neighbors along the cube edges. The nearest neighbor cation spins are along face diagonals with no intermediary oxygen ions. The issue, therefore, is whether competing ferromagnetic t_{2g}–t_{2g} direct exchange along the diagonals can offset the antiferromagnetism expected from the superexchange combinations of e_g–$p\sigma$–e_g and t_{2g}–$p\pi$–t_{2g} between next-nearest neighbors. Figure 3.15a indicates that such a mechanism could be t_{2g}–t_{2g} bonds acting consecutively across face diagonals. For this discussion, we refer to this direct exchange as $\left[t_{2g}–t_{2g}\right]^2$, with an individual stabilization energy designated as b_t.

Before we attempt to sort out these effects, an important point must be established. If the intersublattice coefficient is greater than that of the intrasublattice coefficient, i.e., $N_{ij} > (3/4)\, N_{ii}$, we recall from (3.49) that $T_N = -CN_{ij}/4$, and we see that only N_{ij} is involved. This arises from the sublattice configuration of layered <111> planes shown in Fig. 3.16. This structure has the 12 nearest neighbors occurring six in-plane of the same spin alignment and six in the adjacent plane of opposite spin alignment, thereby producing a cancellation of the N_{ii} contributions. However, the nearest neighbor couplings can still affect the value of N_{ij}. If we include the nearest-neighbor influence from two consecutive $t_{2g} - t_{2g}$ couplings that favor ferromagnetism, the computational models developed in Appendix 3A specifically for this problem can be used to deduce the values of $b_\sigma^2 / U_\sigma \approx 20\,\text{meV}$ and $b_\pi^2 / U_\pi \approx 2\,\text{meV}$, and the energy of the ferromagnetic coupling $b_t \approx -5\,\text{meV}$. Since this latter contribution represents the direct exchange J_{ij} term in (3.14), the negative sign must be used in this convention.

A summary of the values employed in the calculations of Appendix 3A is given in Table 3.6. From the resulting totals for the calculated stabilization energies, T_N values are determined and compared with the measured values listed in Tables 3.4–3.6 and plotted in Fig. 3.19. These data will be useful in interpreting the properties of certain perovskites to be described next.

3.3.2 ABO_3 and A_2BO_4 Perovskites

The early investigations of antiferromagnetic oxides logically focused on the one-metal systems. In later years, however, efforts were directed toward the perovskite family of compounds that also feature M–O–M 180° bonds along the cubic (or

Table 3.6 Magnetic exchange contributions to $M^{2+}O^{2-}$ antiferromagnetic compounds

$M^{2+}O^{2-}$	MnO	FeO	CoO	NiO	CuO
Half-filled orbitals	$2e_g$	$2e_g$	$2e_g$	$2e_g$	e_g
	$3t_{2g}$	$2t_{2g}$	t_{2g}	–	–
$z\sum b_\sigma^2/U_\sigma$	240	240	240	240	120
$z\sum b_\pi^2/U_\pi$	36	24	12	–	–
$z'\sum b_t$	−180	−120	−60	–	–
Net[a]	96	144	192	240	120
$z\sum b_n^2/U_n$	90	136	182	270	156
$T_N$calc. (K)	130	210	309	462	348
$T_N$exp. (K)	122	198	293	520	453

[a] The model employed here is based on the next-nearest neighbor values of $b_\sigma^2/U_\sigma = 20\text{meV}$, $b_\pi^2/U_\pi = 2\text{meV}$, and −5meV for the second-order direct exchange contribution from nearest neighbors b_t, which is designated as negative because of its destabilizing influence on the antiferromagnetic ordering. Note that $z' = 12$ for the nearest neighbors, which is double the $z = 6$ for the next-nearest neighbors

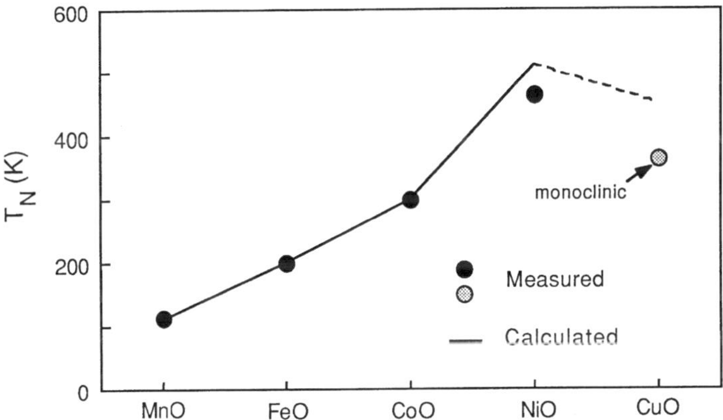

Fig. 3.19 Néel temperatures of simple transition-metal oxides. Data are listed in Tables 3.4–3.6

tetragonal) cell edges, but without the presence of direct exchange-coupled nearest neighbors (see Fig. 3.15b). One of the reasons for this emphasis is the availability of many of these compounds in stable ceramic or single-crystal form where their magnetic and dielectric properties could be utilized in practical applications. The discussion of this family will be confined to a description of two basic crystal structures and unique behavior that will be examined in more detail in later sections. For a comprehensive listing of this family and its various properties, the reader should consult Goodenough and Longo [36]. Reference to more recent developments will be mentioned in their proper context.

The basic perovskite (ABO_3) unit cell is sketched in Fig. 3.20, where A designates a large cation site with 12-fold dodecahedral oxygen coordination, often occupied by a trivalent member of the lanthanide (rare-earth) series, and B is an octahedral site that harbors ions of the $3d^n$ transition series. Néel temperatures are

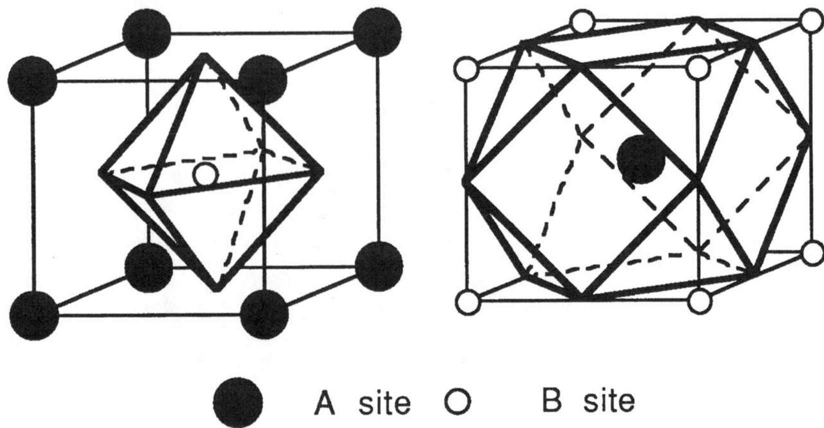

Fig. 3.20 Diagrams of A (octahedral) and B (dodecahedral) cation sites in generic perovskites.

generally low because e_g–$p\sigma$ bonds in the 3+ ions are not operative in most cases, and many properties of interest occur well below room temperature.[5]

Variations of the $A^{3+}B^{3+}O_3$ compounds, such as $A^{2+}B^{4+}O_3$ also occur (the original compound from which the term perovskite was coined is diamagnetic $Ca^{2+}Ti^{4+}O_3$), as well as solid solutions of multiple cations in each type of site that maintain the required 6+ total cation charge. The most frequently encountered example of a B^{4+} compound is the ferroelectric $Ba^{2+}Ti^{4+}O_3$ (BTO) with its relatives $Sr^{2+}Ti^{4+}O_3$ (STO), $Ca^{2+}Ti^{4+}O_3$, and $Pb^{2+}Ti^{4+}O_3$ in which the Ti^{4+}ion is d^0 and therefore diamagnetic.

The A_2BO_4 form, often referred to as the K_2MnF_4 or K_2NiF_4 structure sketched as part of Fig. 8.1, is tetragonal and therefore has a different crystal field symmetry at the octahedral (B) site. The elongation of the c axis results in an orbital state energy level structure of the type sketched in Fig. 2.29 for a tetragonal $c/a > 1$ distortion of the d^4 case. One implication of this situation is the splitting of the E_g doublet, which allows the d_{z^2} state to stabilize relative to its $d_{x^2-y^2}$ E_g partner. In cases where the e_g levels are only partially filled, i.e., with one electron or one hole and the exchange occurs by means of d_{z^2}–$p\sigma$ bonds as illustrated in Fig. 3.21. Where the B ion has a d^4 (e.g., Mn^{3+}) or d^9 (e.g., Cu^{2+}) configuration, the Jahn-Teller effect that would be expected to remove the E_g degeneracy from an undistorted cubic or rhombohedral site is preempted by the overriding D_{4h} lattice distortion.

In A_2BO_4 and synthetic variations of ABO_3 compounds, mixed-valence Cu ions in these tetragonally distorted sites can supply charge carriers for superconductivity. This subject will be examined in Chap. 8. The magnetic behavior of mixed-valence Mn ion combinations, however, lends credence to the notions of superexchange and charge transfer exchange explained in the foregoing sections.

[5] Ions of the transition elements with partially filled e_g orbitals can be formed from Mn, Fe, Co, Ni, and Cu in either the 2+ or 3+ states (in the case of Mn, the 4+ state as well). However, only Mn^{3+}, which has special magnetoelastic properties, and Fe^{3+} occur readily in the ABO_3 structure, which often departs from cubic symmetry.

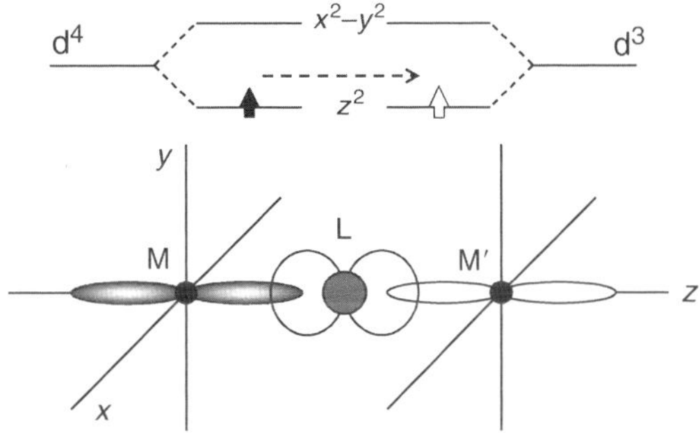

Fig. 3.21 Tutorial model of spin transfer in a tetragonally-split e_g doublet with an active lower z^2 state

3.3.3 The Mixed-Valence Manganite Anomaly

Jahn-Teller distortions from cubic (O_h) to tetragonal (D_{4h}) or orthorhombic (D_{2h}) are prevalent in oxygen coordinated octahedral sites containing Mn^{3+} (d^4) ions. In mixed-valence ABO_3 compounds, however, unexpected magnetic and crystallographic effects are observed. The typical manganite composition $La^{3+}Mn^{3+}O_3$ has been the subject of detailed investigations into its crystallography and magnetic behavior, particularly when Sr^{2+} or Ca^{2+} ions are blended with La^{3+} in the A sites to create a mixture of Mn^{3+} and Mn^{4+} ions in the B sublattice, according to the formula $\left(La^{3+}_{1-x}Ca^{2+}_x\right)Mn^{3+}_{1-x}Mn^{4+}_xO_3$[37].

In the generic $\left(A^{3+}A'^{2+}\right)MnO_3$ system, all three oxidation states of Mn can occur, offering a variety of 180° superexchange interactions. For the cubic octahedral coordination, crystal-field effects dictate that the t_{2g} orbital states are of lower energy and are half-filled to satisfy the Hund's rule spin polarization requirement for each of Mn^{2+} (d^5), Mn^{3+} (d^4), and Mn^{4+} (d^3). For all combinations of exchange pairs, the t_{2g} electrons favor antiferromagnetism via π-bonding as in the case of Mn^{4+}–O^{2-}–Mn^{4+}. For Mn^{2+}–O^{2-}–Mn^{2+}, the stabilization is determined by the stronger σ-bonding e_g states. For Mn^{3+}–O^{2-}–Mn^{3+}, a single electron in the e_g states (Mn^{3+} case) can be stabilized by a static Jahn-Teller (J-T) orthorhombic distortion that splits energy levels as shown in Fig. 2.29. Where the distortions are cooperative, a tetragonal/orthorhombic phase will appear with axis ratio $c/a > 1$ and the half-filled d_{z^2} orbital is stabilized relative to the empty d_{x2-y2} state. In the perovskites, however, Mn^{3+}–O^{2-}–Mn^{3+} $(d^4$–$d^4)$ couplings do not always follow these rules.

In (La,Ca)MnO_3, three types of exchange can occur: d^4–d^4, d^3–d^4, and d^3–d^3, as listed in Table 3.7. Based on the qualitative estimates of exchange energy summarized in Table 3.6 (including footnote), the d^4–d^4 couplings should be strongly antiferromagnetic with $3b_\pi^2/U_\pi + b_\sigma^2/U_\sigma$ (≈ 26 meV) stabilization energy, the d^3–d^4 (and d^4–d^3) couplings should be strongly ferromagnetic with b_σ^2/U_σ (≈ 20 meV) stabilization and the d^3–d^3 couplings should be antiferromagnetic with $3b_\pi^2/U_\pi$ (≈ 6 meV) stabilization. For a random distribution of the Ca^{2+} ions in the lattice, we would expect a corresponding randomization of the Mn^{3+}

Table 3.7 Magnetic exchange of d^n electrons in manganites

		$d_{x^2-y^2}$	d_{z^2}	
Ion pair	d_{xy}, d_{xz}, d_{yz}	$a-b$ Axes	c Axis	Net
$d^5 \leftrightarrow d^5$	↑↓ π (wk)	↑↓ σ (str)	↑↓ σ (str)	↑↓ (str)
$d^5 \leftrightarrow d^4$	↑↓ π (wk)	↑↑ σ (str)	–	↑↑ (str)
	–	–	↑↓ σ (str)	↑↓ (str)
$d^4 \leftrightarrow d^{4a}$	↑↓ π (wk)	↑↓ σ (str)	–	↑↓ (str)
$d^5 \leftrightarrow d^{3b}$	↑↓ π (wk)	↑↑ σ (mod)	↑↑ σ (mod)	↑↑ (mod)
$d^4 \leftrightarrow d^{3c}$	↑↓ π (wk)	↑↑ σ (str)	–	↑↑ (mod/str)
$d^3 \leftrightarrow d^3$	↑↓ π (wk)	–	–	↑↓ (wk)

Based on rules developed by J.B. Goodenough [28]

[a] J-T splitting of e_g orbitals producing $c/a > 1$ distortion

[b] Proposed quasi-static J-T version of $d^4 \leftrightarrow d^4$ that causes ferromagnetism through charge transfer into empty orbital states. It occurs in a rhombohedral structural phase that exists when the static cooperative distortions that cause orthorhombic phases are absent

[c] Conditions similar to those of footnote a, but featuring charge transfer superexchange that promotes ferromagnetism

and Mn^{4+} dispersal that would lead to antiferromagnetic ordering for almost every value of x, except possibly near $x = 0.5$, where the d^3–d^4 couplings could dominate with ideal charge ordering.

When the fraction of Mn^{4+} is in the range from 0.1 to 0.5, however, magnetization measurements indicate that ferromagnetism prevails [37] with Curie temperatures varying according to the data of Fig. 3.22. For ferromagnetism to occur at x values where the most abundant exchange combination is antiferromagnetic d^4–d^4, the ready explanation is that J-T effects of the d^4 ions could suppress the antiferromagnetic stabilization over this concentration range. Such an effect was proposed by Goodenough et al. to account for the observed ferromagnetic result [38]. The features of the phenomenon that are peculiar to this system will be described in connection with magnetoresistance properties in Sect. 8.3 as resulting from a vibronic breathing mode of the octahedral ligands that occurs when the symmetry is cubic or rhombohedral (trigonal) with a degenerate E_g term. This latter condition was supported by experiment as indicated by the correspondence between crystallographic and magnetic states in Fig. 8.10 [39].

In Appendix 3B, a mathematical model utilizing empirical exponential functions is constructed to represent the peculiar variations in superexchange between d^4 and d^4 ions as a function of the d^3 ion fraction x. Apparently, the transition from antiferromagnetism to ferromagnetism near $x = 0.1$ and back to antiferromagnetism above $x = 0.5$, is influenced by the concentration of Mn^{4+} ions on crystallographic symmetry, which in turn allows for an E_g term degeneracy and the vibronic J-T effect that occurs only in this limited x range. The results of the model applied to the calculation of Curie temperatures fitted to the data of Figs. 3.22 and 8.10 are plotted in Fig. 3.24. For this example, a random dispersal of valence states is assumed and the previously determined stabilizations energies of $b_\sigma^2/U_\sigma = 20\,\text{meV}$ and $3b_\pi^2/U_\pi = 6\,\text{meV}$ are used to compute values effective exchange constants

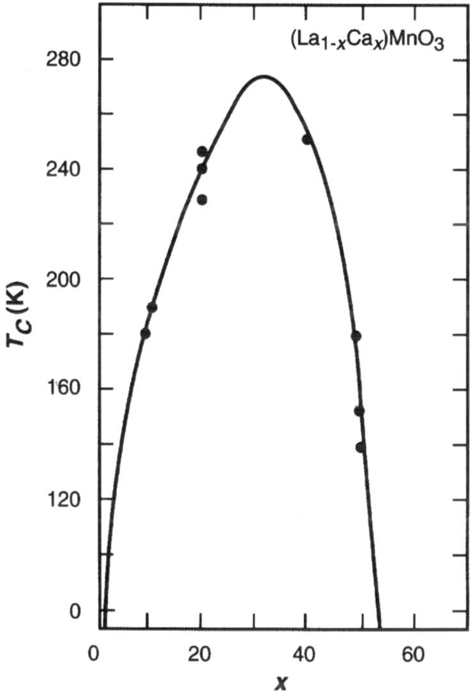

Fig. 3.22 Classic plot of Curie temperature vs. x for $(La^{3+}_{1-x}Ca^{2+}_x)Mn^{3+}_{1-x}Mn^{4+}_xO_3$. Data are from Jonker and Van Santen [40]. See also **Fig. 8.10**

$J'_{33} = 2.2\,\text{meV}$, J_{34} (and J_{43}) $= 3.3\,\text{meV}$, and $J_{44} = -1.2\,\text{meV}$. The resultant antiferromagnetic stabilization energy for end-member $CaMnO_3$ (at $x = 1$) becomes 36 meV, which equates to $T_N = 116\,\text{K}$ as computed from the appropriate relation in (3.49), in reasonably good agreement with the measured value of 130 K. For the opposite end-member $LaMnO_3$ (at $x = 0$), a net stabilization of 156 meV would produce an expected $T_N = 453\,\text{K}$, far from an experimental value of only ~60 K. At $x = 0.5$, the combined d^3–d^4 and d^4–d^3 couplings with random charge ordering would have a total ferromagnetic stabilization energy of 240 meV capable of producing a $T_C = 364\,\text{K}$, if acting independently. Clearly the data and model constructed to fit them indicate that the d^4–d^4 couplings do not behave in the conventional manner particularly over the lower half of the Mn^{4+} concentration range.

At a point in the regime $x \leq 0.1$, the crystallographic phase changes to rhombohedral (trigonal symmetry with no J-T splitting of the e_g levels) and remains such until $x \approx 0.5$ is reached. With half the Mn ions in the 4+ state, the vibronic J-T condition breaks down, and the remaining Mn^{3+}–O^{2-}–Mn^{3+} couplings revert to antiferromagnetism, combining with the growing population of antiferromagnetic Mn^{4+}–O^{2-}–Mn^{4+} couplings to produce mainly antiferromagnetism in the range from $0.5 \leq x \leq 1.0$. At higher temperatures, changes in the cation charge distribution could enable the rhombohedral phase to extend beyond $x = 0.5$, to possibly account for a second antiferromagnetic/ferromagnetic transition first reported by Jonker and Van Santen [40] for $\left(La^{3+}_{0.3}Sr^{2+}_{0.7}\right)MnO_3$.

Appendix 3A Analysis of $M^{2+}O^{2-}$ Exchange Interactions

If we consider that the nearest neighbor influence on N_{ij} arises from two consecutive t_{2g}–t_{2g} direct couplings that favor ferromagnetism according to the direct exchange J_{ij} term of (3.14), the following model can be constructed: We begin by assuming that the values of $z\sum b_{\mathrm{n}}^2/U_{\mathrm{n}}$ deduced from measured T_{N} values listed in Table 3.4 are the arithmetic sum of contributions from b_σ^2/U_σ and b_π^2/U_π superexchange and b_{t} direct exchange. Simultaneous linear equations are then formulated from the data according to

$$\begin{aligned}
&\text{MnO: } 2zb_\sigma^2/U_\sigma + 3zb_\pi^2/U_\pi + 3z'b_t = 90\,\text{meV},\\
&\text{FeO: } 2zb_\sigma^2/U_\sigma + 2zb_\pi^2/U_\pi + 2z'b_t = 136\,\text{meV},\\
&\text{CoO: } 2zb_\sigma^2/U_\sigma + zb_\pi^2/U_\pi + z'b_t = 182\,\text{meV},\\
&\text{NiO: } 2zb_\sigma^2/U_\sigma + 0 + 0 = 270\,\text{meV},\\
&\text{CuO: } zb_\sigma^2/U_\sigma + 0 + 0 = 156\,\text{meV},
\end{aligned} \tag{3.50}$$

where z and z' are the respective numbers of neighbors in each case. Solution of these equations produces nearly uniform values that are averaged to $zb_\sigma^2/U_\sigma \approx$ 120 meV and $zb_\pi^2/U_\pi + z'b_t = -48$ meV.

Further dissection of these effects can lead to estimates of the relative values of b_π^2/U_π and b_σ^2/U_σ bonding components. Estimates of these superexchange contributions can be obtained from measurements of the Néel temperatures of Mn^{4+} (d^3) in the cubic perovskite $CaMnO_3$ and Ni^{2+} (d^8) in face-centered cubic NiO because there is no possibility of ferromagnetic effects from some form of t_{2g}–t_{2g} bonding. The d^3 configuration of VO shown in Fig. 3.23 involves superexchange of the three half-filled t_{2g} states, with no contributions from the e_g shell, i.e., $t_{2g}^3e_g^0$ in conventional nomenclature. On the contrary, the d^8 configuration shown in Fig. 3.23b has a filled t_{2g} shell, but provides full superexchange from its two half-filled e_g states, i.e., $t_{2g}^6e_g^2$. As a consequence, these two situations are good examples for comparing the superexchange effects originating from individual π and σ-bonding orbitals in this

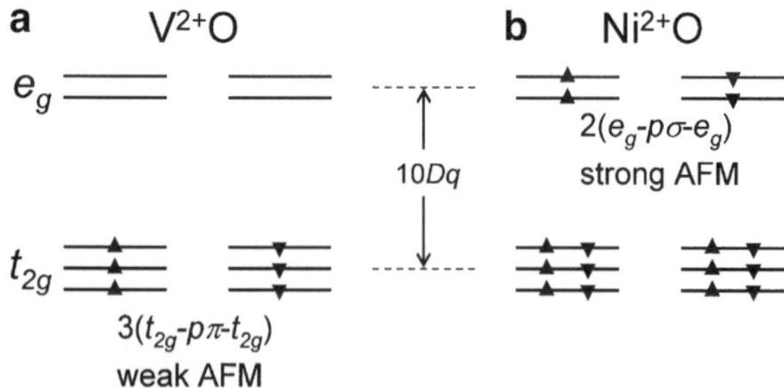

Fig. 3.23 Model comparing the Aufbau cation spin occupation for V^{2+} $\left(t_{2g}^3\right)$ and Ni^{2+} $\left(t_{2g}^6e_g^2\right)$ that determines the superexchange energy and the ultimate values of T_{N}

lattice geometry. In both cases, the cations occupy octahedral sites and are linked covalently through 180° bonds. From (3.48),

$$\begin{aligned} z\frac{3b_\pi^2}{U_\pi} &= 2\frac{3}{2}kT_\mathrm{N}\frac{S}{S+1} \quad \text{for CaMnO}_3, \\ z\frac{2b_\sigma^2}{U_\sigma} &= 4\frac{3}{2}kT_\mathrm{N}\frac{S}{S+1} \quad \text{for NiO} \end{aligned} \tag{3.51}$$

and the relative magnitudes for the average b_π^2/U_π and b_σ^2/U_σ ratio, weighted accordingly, can be expressed by

$$\frac{3b_\pi^2/U_\pi}{2b_\sigma^2/U_\sigma} \approx \frac{1}{2}\frac{T_\mathrm{N}\,(\mathrm{Mn})}{T_\mathrm{N}\,(\mathrm{Ni})}\left(\frac{S_\mathrm{Mn}}{S_\mathrm{Mn}+1}\right)\left(\frac{S_\mathrm{Ni}+1}{S_\mathrm{Ni}}\right), \tag{3.52}$$

where $S_\mathrm{Mn} = 3/2$ and $S_\mathrm{Ni} = 1$. Measurements have indicated that T_N (Mn) $\sim$ 130 K [37] and T_N (Ni) $\sim$ 520 K [19]. If $U_\pi \approx U_\sigma$ (an average value for the oxides is about 8 eV, with variations due to differing ionization potentials, crystal-field splittings and overlap integrals), we arrive at $b_\pi^2/U_\pi : b_\sigma^2/U_\sigma = 0.1$. If the calculation is extended further, with the aid of (3.26) for nearest neighbors number $z = 6$ and the Boltzmann constant $k = 8.625\times10^{-5}$ eV/K, we estimate magnitudes of $b_\sigma^2/U_\sigma \approx 20$ meV from the total of 120 meV listed in Table 3.4, and we can now deduce $zb_\pi^2/U_\pi \approx 12$ meV, or the individual $b_\pi^2/U_\pi \approx 2$ meV.

From this result for b_π^2/U_π, we can estimate the total stabilization energy for the ferromagnetic contribution as 60 meV, i.e., 48 + 12 meV. Since there are twelve nearest neighbors ($z' = 12$), the magnitude of the individual $b_t \approx 5$ meV.

Appendix 3B Curie Temperature Model for (La,Ca) MnO_3

If the spatial distribution of Mn^{4+} ions is assumed to be random, the gain in stabilization energy per covalent coupling $E_\mathrm{ex} = z\sum b_n^2/U_n$ due to magnetic ordering may be approximated from the exchange Hamiltonian $\mathcal{H}_\mathrm{ex} = -2\sum J_{ij}\boldsymbol{S}_i\cdot\boldsymbol{S}_j$. By setting $E_\mathrm{ex} = -\mathcal{H}_\mathrm{ex}$, we can work with positive energies and define further

$$\begin{aligned} E_\mathrm{ex} &= \sum E_{ij} = E_{33} + E_{34} + E_{43} + E_{44} \\ &= 2z\left(p_3^2 J_{33} S_3^2 + p_3 p_4 J_{34} S_3 S_4 + p_4 p_3 J_{43} S_4 S_3 + p_4^2 J_{44} S_4^2\right)\cos\theta, \end{aligned} \tag{3.53}$$

where z is the number of nearest neighbors, J_{33}, J_{34} ($= J_{43}$), and J_{44} are the respective exchange constants of the individual magnetic pair interactions, and p_3 and p_4 are the probabilities of occurrence of S_3 and S_4 spins in any given site, and θ is the angle between spins, and is assumed to be uniform throughout the spin system with a value of 0 or π depending on whether the ordering is parallel or antiparallel.

Since $p_3 = 1 - x$ and $p_4 = x$ for a random distribution, (3.53) may be expressed as a function of x:

$$E_{ex} = 2z\left[(1-x)^2 J_{33}S_3^2 + 2x(1-x)J_{34}S_3S_4 + x^2 J_{44}S_4^2\right]\cos\theta$$
$$= 2zJS^2\cos\theta, \tag{3.54}$$

where $S = (1-x)S_3 + xS_4$ is a weighted average spin value, and J is the corresponding average J_{ij} coefficient.

The variation in E_{ex} with x may be inferred directly from the Curie/Néel temperature data of Fig. 3.23, which register a peak $T_C > 300\,K$ at $x \approx 0.3$. A mathematical model is helpful in sorting out the contributions of the individual terms in (3.54). To this end, the following empirical relation for J_{33} was constructed to simulate the vibronic J-T effects of the Mn^{3+}–O^{2-}–Mn^{3+} interactions as a function of x:

$$J_{33} = J'_{33}\{1 - \exp[-(x - x_0)/d_0]\}\{1 - \exp[-(x_1 - x)/d_1]\}, \tag{3.55}$$

where x_0, d_0, x_1, and d_1 are parameters that may be structurally and temperature dependent. To relate E_{ex} to the Curie temperature the relation $kT_C = E_{ex}(S+1)/3S$ is employed.

Figure 3.24 shows plots of the calculations based on (3.53) through (3.55) over the full range of x. To obtain a good fit with experiment the parameters were chosen

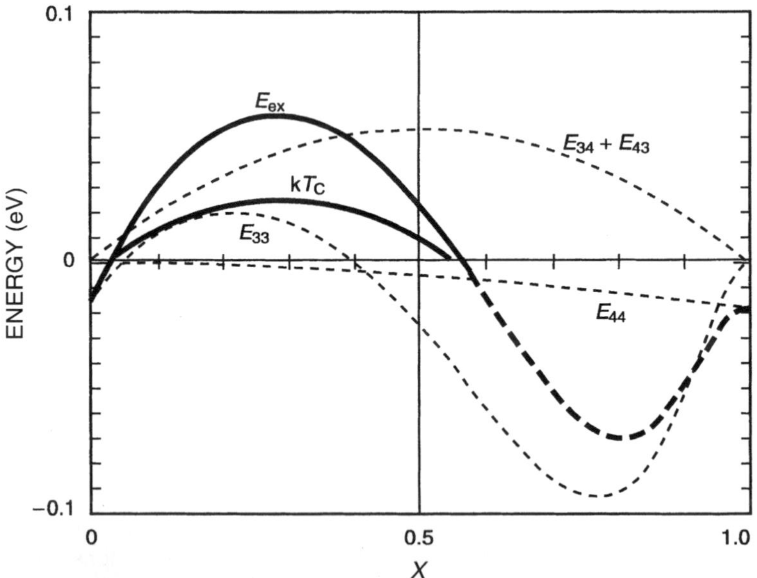

Fig. 3.24 Calculated plot of E_{ex} and kT_C vs. x, including estimated contributions of E_{33}, E_{44}, $E_{34} + E_{43}$. Parameter values used with exponential functions of (3.55) are $x_0 = 0.1$, $d_0 = 0.4$, $x_1 = 0.7$, and $d_1 = 0.14$

to produce a $kT_C \approx 25\,\text{meV}$ ($T_C = 300\,\text{K}$) with a peak at $x = 0.3$. From this exercise, $J'_{33} = 2.2$ and $J_{34} = 3.3\,\text{meV}$. The value of $J_{44} = -1.2\,\text{meV}$ is based on the reported Néel temperature $T_N \sim 130\,\text{K}$ for $La^{3+}Ca^{2+}Mn^{4+}O_3$ [38]. (J_{33} values above $x = 0.5$ are shown as a dashed line that is an artifact of the mathematical function of (3.55) and are not intended to represent any quantitative physical reality in this regime.) It is, therefore, concluded that the relative importance of the terms in (3.53) is heavily weighted toward the charge transfer $E_{34} + E_{43}$ components in the vicinity of $x = 0.3$. The source of ferromagnetism in this system lies both in the strength of the J_{34} interaction and in the J-T assisted ferromagnetism of J_{33} in the range $0 \leq x \leq 0.5$.

References

1. W. Heitler and F. London, Z. Physik **44**, 455 (1927)
2. L.E. Orgel, *Introduction to Transition-Metal Chemistry: Ligand-Field Theory*, (John Wiley, New York, 1959)
3. G.F. Dionne, *Magnetic Interactions and Spin Transport*, A. Chtchelkanova, S. Wolf, and Y. Idzerda, eds., (Springer, New York, 2003), Chapter 1
4. C.J. Ballhausen, *Molecular Electronic Structures of Transition Metal Complexes*, (McGraw-Hill International, Chatham, Great Britain, 1979), pp. 84–89
5. E. Cartmell and G.W.A. Fowles, *Valency and Molecular Structures*, (Butterworths, London, 1961), Chapter 8
6. P.W. Anderson, *Phys. Rev.* **115**, 2 (1959)
7. J.B. Goodenough, *Prog. Solid State Chem.* **5**, 145 (1972), Section IID
8. A.H. Morrish, *The Physical Principles of Magnetism*, (John Wiley, New York, 1965), p. 279
9. K. Yosida, *Theory of Magnetism*, (Springer, New York, 1996), p. 54
10. R.M. White, *Quantum Theory of Magnetism*, (Springer-Verlag, New York, 1983), Chapter 2
11. P.W. Anderson, *Solid State Phys.* **14**, 99 (1969)
12. J. Hubbard, Proc. Roy. Soc. (*London*) **A276**, 238 (1962); **A277**, 237 (1964); **A281**, 401 (1964); **A285**, 542 (1965); **A296**, 82, 100 (1966)
13. B. Lax and K.J. Button, *Microwave Ferrites and Ferrimagnetics*, (McGraw-Hill, New York, 1962), p. 65
14. H.B. Kramers, *Proc. Amsterdam Acad. Sci.* **33**, 959 (1930); also H.B. Kramers, Physica **1**, 182 (1934)
15. J.B. Goodenough, *Magnetism and the Chemical Bond*, (Wiley Interscience, New York, 1963), Chapter 3
16. C. *Zener, Phys. Rev.* **82**, 403 (1951)
17. P.-G. de Gennes, *Phys. Rev.* **118**, 141 (1960)
18. P.W. Anderson and H. Hasegawa, *Phys. Rev.* **100**, 675 (1955)
19. T. Nagamiya, K. Yosida, and R. Kubo, *Adv. Phys.* **4**, 1 (1955)
20. M.A. Ruderman and C. Kittel, *Phys. Rev.* **96**, 99 (1954)
21. T. Kasuya, *Prog. Theor. Phys.* **16**, 45 (1959)
22. K. Yosida, *Phys. Rev.* **106**, 893 (1957)
23. J.B. Goodenough, *New Developments in Semiconductors*, P.R. Wallace, R. Harris, and M.J. Zuckermann, eds., (Nordhoff International Publishing, Leyden, 1973), pp. 145–151
24. J. Kanamori, *J. Phys. Chem.* Solids **10**, 87 (1959)
25. J.B. Goodenough, *Magnetism and the Chemical Bond*, (Wiley Interscience, New York, 1963), Chapter 3, p. 213
26. J.B. Goodenough and A.L. Loeb, *Phys. Rev.* **8**, 391 (1955)
27. J.C. Slater, *Phys. Rev.* **35**, 509 (1930)

28. J.B. Goodenough, *Magnetism and the Chemical Bond*, (Wiley Interscience, New York, 1963), Chapter 3, Table XII; also J.B. Goodenough, *Phys. Rev.* **117**, 1442 (1960)
29. N.S. Rogado, J. Li, A.W. Sleight, and M.A. Subramanium, *Adv. Mater. (Weinheim, Ger.)* **17**, 2225 (2005)
30. L. Néel, *Ann. Phys.* (Paris) **17**, 64 (1932)
31. A.H. Morrish, *The Physical Principles of Magnetism*, (John Wiley, New York, 1965), Chapter 8
32. A.B. Lidiard, *Rept. Prog. Phys.* **17**, 201 (1954)
33. A.H. Morrish, *The Physical Principles of Magnetism*, (John Wiley, New York, 1965), p. 457
34. C.G. Shull, W.A. Strausser, and E.O. Wollan, *Phys. Rev.* **83**, 333 (1951)
35. R.K. Nesbet, *Phys. Rev.* **122**, 1497 (1961)
36. J.B. Goodenough and J.M. Longo, *Crystallographic and Magnetic Properties of Perovskite and Perovskite-Related Compounds*, Landolt-Bornstein, Volume 4a (Springer-Verlag, New York, 1970) pp. 126–314
37. G.H Jonker and J.H. Van Santen, *Physica* **XVI**, 337 (1950)
38. J.B. Goodenough, A. Wold, N. Menyuk, and R.J. Arnott, *Phys. Rev.* **124**, 373 (1961)
39. J.B. Goodenough and J.M. Longo, *Crystallographic and Magnetic Properties of Perovskite and Perovskite-Related Compounds*, Landolt-Bornstein, Volume 4a (Springer-Verlag, New York, 1970), Fig. 39
40. J.H. Van Santen and G.H Jonker, *Physica* **XVI**, 599 (1950)

Chapter 4
Ferrimagnetism

In the previous chapters, the origins of spontaneous magnetism for parallel (ferromagnetism) and antiparallel spin alignments (antiferromagnetism) have been reviewed. In their pristine forms, the former occurs through direct exchange in metals and alloys, and the latter in nonmetallic ionic compounds comprising oxygen or other elements from the right-hand side of the Periodic table as the anion lattice. Utilitarian applications of ferromagnets are self-evident to even the most casual observer of physical phenomena, but the situation is much less so in the case of antiferromagnetism. For the most part, antiferromagnetism has been a portal to fundamental research in materials, particularly involving the diagnostic methods of neutron and more recently, muon diffraction and scattering.

There are, however, select groups of transition-metal oxides that combine the magnetic properties of ferromagnetic metals with the electrically insulating characteristics of the antiferromagnetic compounds described in the previous section. These magnetic insulators are termed *ferrimagnets*, and the phenomenon that characterizes their magnetic properties is called *ferrimagnetism*. Ferrimagnetic oxides have also served as rich sources of knowledge about the fundamental physics of materials, but unlike the antiferromagnetic oxides, they continue to add to their already widespread uses in modern electronics technology. For these reasons, the properties of *ferrites*, as they are commonly designated, will be treated generously for the remainder of this book.

4.1 Ferrimagnetic Order

In the previous chapter, the concept of multiple magnetic sublattices was introduced to explain the phenomenology of antiferromagnetism. From this starting point, we may define a ferrimagnet as an antiferromagnet with unbalanced magnetic sublattices due to either differing populations of similar spins or sublattices with ions of different spin values altogether. For this to occur, it is apparent that some kind of crystallographic selection must be involved to distinguish the sublattices. In the common oxide systems where ferrimagnetism occurs, the sublattices are defined by cation sites of different oxygen coordinations: octahedral, tetrahedral,

G.F. Dionne, *Magnetic Oxides*, DOI 10.1007/978-1-4419-0054-8_4,

and dodecahedral, as described in Sect. 2.2.2. In a crystal lattice, these sites form interspersed but ordered arrays that enable each to be occupied by entirely different ionic species, or ions of the same atomic species but of different valence, or various combinations of both. In the simplest case, a two-sublattice molecular field theory can be applied, but now requiring three coefficients instead of the two for the antiferromagnet – one each for the ions in the same sublattice (intrasublattice), and a third (intersublattice) linking ions between the sublattices.

4.1.1 Generic Ferrimagnetic Systems

In Fig. 4.1, a one-dimensional sketch is offered to clarify the difference between a ferromagnet, an antiferromagnet, and a ferrimagnet. In practical terms, a ferrimagnet behaves magnetically as a ferromagnet and can be analyzed in terms of the Brillouin–Weiss theory from Sect. 1.3.2. It must be recognized, however, that all of the bonding linkages produce principally superexchange and therefore favor antiparallel spin alignments. In an ideal ferrite with collinear moments, the final ordering of the magnetic moments in the sublattices is therefore the result of competing exchange fields that are generally dominated by the antiparallel influence from intersublattice coupling. The resultant magnetic moment then becomes the arithmetic difference of the two opposing sublattice moments.

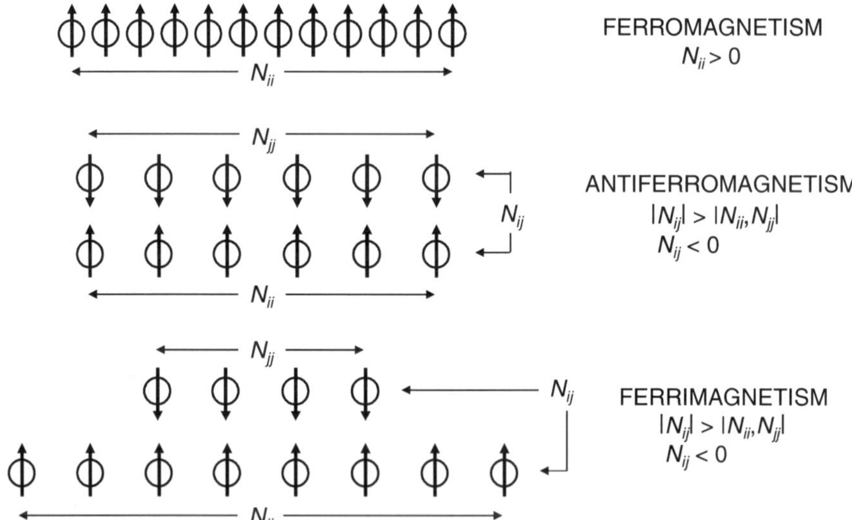

Fig. 4.1 One-dimensional exchange models of spontaneous spin alignment: (**a**) for a single magnetic lattice ferromagnetism: $N_{ii} > 0$; (**b**) for two equal sublattices with N_{ij} as the dominant coefficient, antiferromagnetism: $N_{ij} < 0$, $N_{ii} <$ or > 0, (**c**) for two unbalanced sublattices with N_{ij} dominant, ferrimagnetism: $N_{ij} < 0$, N_{ii}, $N_{jj} <$ or > 0

Although there are exotic chemical compounds in which ferrimagnetic sublattice spin arrangements exist, our discussion of ferrimagnetism will be limited to three families of transition-metal oxides that have become the foundation of this branch of magnetism. The first of these systems to be recognized is the spinels designated by the generic formula $A\,[B_2]\,O_4$, where A is the tetrahedral site with O_4 coordination and B is the octahedral site with O_6 coordination. The brackets around B serve to indicate the octahedral sites in the actual chemical formulae. Later the magnetic garnets with the generic formula $\{c_3\}\,[a_3]\,(d_3)\,O_{12}$ were synthesized and have become equally important particularly in microwave and optical applications. The bracket/site designations are $\{c\}$ for dodecahedral with O_{12} coordination, $[d]$ for tetrahedral, and (a) for octahedral. It is important to note that unlike the spinels in which the octahedral sites dominate the tetrahedral sites by a ratio of 2:1, in the garnets the tetrahedral sites dominate by a ratio of 3:2.

A third family of ferrimagnetic compounds is the magnetoplumbites, named after the naturally occurring of $PbFe_{19}O_{12}$ (lead ferrite). These compounds are commonly referred to as hexagonal ferrites or hexaferrites because of their sixfold symmetrical uniaxial crystallographic structures. In many respects, hexaferrites resemble spinels because they feature the same ratio of octahedral to tetrahedral sites, but also include one trigonal bipyramid (O_5) site (see Fig. 2.7) that contains an iron ion, and one large site to house the usually divalent Pb, Ba, or Sr. These compounds are important for permanent magnet applications and will be described in more detail along with the spinels and garnets in Sect. 4.3.

4.1.2 Molecular Field Theory of Ferrimagnetism

Recalling the exposition of the molecular field model of antiferromagnetism from Sect. 3.2.2, we can now apply this formalism in the manner of Néel to the case of a ferrimagnet [1]. For the individual sublattices of a two-site system, (3.35) with $N_{ii} \neq N_{jj}$ can be used to express the individual sublattice magnetizations as

$$\begin{aligned} M_i &= \frac{C_i}{T}\left(H + N_{ii}M_i + N_{ij}M_j\right), \\ M_j &= \frac{C_j}{T}\left(H + N_{ji}M_i + N_{jj}M_j\right), \end{aligned} \tag{4.1}$$

where

$$\begin{aligned} C_i &= \frac{n_i g_i^2 m_B^2 S_i\left(S_i+1\right)}{3k}, \\ C_j &= \frac{n_j g_j^2 m_B^2 S_j\left(S_j+1\right)}{3k}. \end{aligned} \tag{4.2}$$

As in the discussion of antiferromagnetism, a Curie temperature relation can be extracted from solutions of (4.1) in the paramagnetic region above T_C and in the magnetically ordered region below T_C. Some of the details of the mathematical manipulations will be left to the readers, who can also consult standard text books, e.g., Morrish [2]. The analytical results, however, can serve as helpful approximations as well as providers of physical insight into the stability of the ferromagnetic state.

In the paramagnetic region, (4.1) can be solved simultaneously to produce individual relations for M_i and M_j as a function of H. A susceptibility can then be defined as $\chi = \left(M_i + M_j\right)/H$, which then leads to a rather complicated expression for $1/\chi$ vs. T that takes the form of a hyperbolic function displayed graphically in Fig. 4.2 for which the asymptote as $T \to \infty$ is given by

$$\frac{1}{\chi} = \frac{T}{C_i + C_j} - \frac{1}{\chi_0}, \tag{4.3}$$

where

$$\frac{1}{\chi_0} = \frac{1}{\left(C_i + C_j\right)^2}\left(C_i^2 N_{ii} + C_j^2 N_{jj} + 2C_i C_j N_{ij}\right). \tag{4.4}$$

The asymptotic or paramagnetic Curie temperature is the intercept with the T axis, given by

$$\theta_C = \frac{C_i + C_j}{\chi_0}. \tag{4.5}$$

By the convention adopted so far in this text, all of the molecular field coefficients are treated as negative quantities, so that χ_0 and therefore θ_C will be negative. Equation (4.3) can be expressed as the Curie-Weiss law

$$\chi = \frac{C_i + C_j}{T + \theta_C}. \tag{4.6}$$

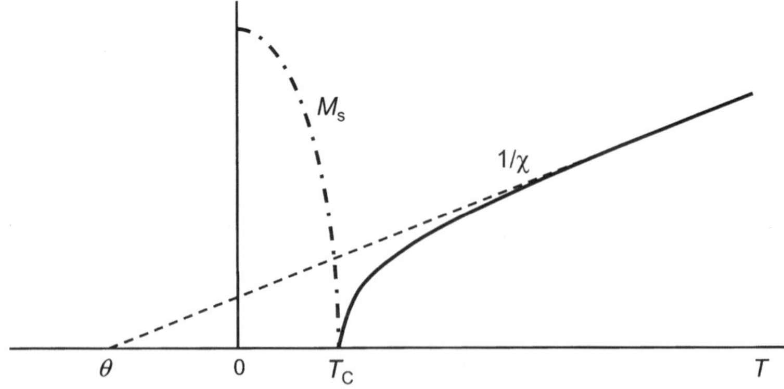

Fig. 4.2 Inverse susceptibility of a ferrimagnet above the Curie temperature, including the asymptote of the hyperbola. Magnetization curve is added to illustrate the behavior at $T < T_C$

The above analytical method can be used to determine values of N_{ii}, N_{jj}, and N_{ij} in various situations that are explained in Sect. 4.2.

In the magnetically ordered region, an expression for the true Curie temperature T_C can be derived from (4.1) by allowing $H = 0$ and solving the determinant of the coefficients of M_i and M_j to yield

$$T_C = \frac{1}{2}\left(C_i N_{ii} + C_j N_{jj}\right) + \frac{1}{2}\sqrt{\left(C_i N_{ii} - C_j N_{jj}\right)^2 + 4C_i C_j {N_{ij}}^2}. \qquad (4.7)$$

In spinel and garnet ferrites, it will be shown that $C_i N_{ii} \sim C_j N_{jj}$ and since N_{ij} is usually the dominant coefficient, $4C_i C_j {N_{ij}}^2 >> \left(C_i N_{ii} - C_j N_{jj}\right)^2$, allowing (4.7) to be simplified to

$$T_C \approx \frac{1}{2}\left(C_i N_{ii} + C_j N_{jj}\right) + \sqrt{C_i C_j {N_{ij}}^2}. \qquad (4.8)$$

A first-order approximation that is often used with generally disappointing results is $T_C \approx \sqrt{C C_j {N_{ij}}^2}$. Because N_{ii} and N_{jj} are negative, N_{ij} estimates arrived at by this approximation can be substantially larger than the true values.

The most important application of the Néel molecular field theory of ferrimagnetism is the computation of the spontaneous magnetization characteristic of a given chemical composition as a function of temperature (thermomagnetization). In the case the resultant magnetization of the opposing sublattices is given by

$$M = \left|M_i - M_j\right|. \qquad (4.9)$$

The procedure once again involves the Brillouin–Weiss function, which is applied to each sublattice according to

$$\begin{aligned} M_i(T) &= M_i(0) B_{S_i}(T), \qquad (4.10) \\ M_i(T) &= M_i(0) B_{S_i}(T), \end{aligned}$$

where

$$\begin{aligned} \mathcal{B}_{a_i}(T) &= \frac{m_i H_{\text{ex}}^{(i)}}{kT} = \frac{g_i m_B S_i}{kT}\left(N_{ii} M_i + N_{ij} M_j\right), \qquad (4.11) \\ \mathcal{B}_{a_j}(T) &= \frac{m_j H_{\text{ex}}^{(j)}}{kT} = \frac{g_j m_B S_j}{kT}\left(N_{ji} M_i + N_{jj} M_j\right), \end{aligned}$$

and $M_i(0) = n_i g_i m_B S_i$ and $M_j(0) = n_j g_j m_B S_j$. Solution of (4.11) cannot be done in closed form, but can be accomplished by self-consistent iteration procedures involving multiple sublattices simultaneously. Computer programs for this

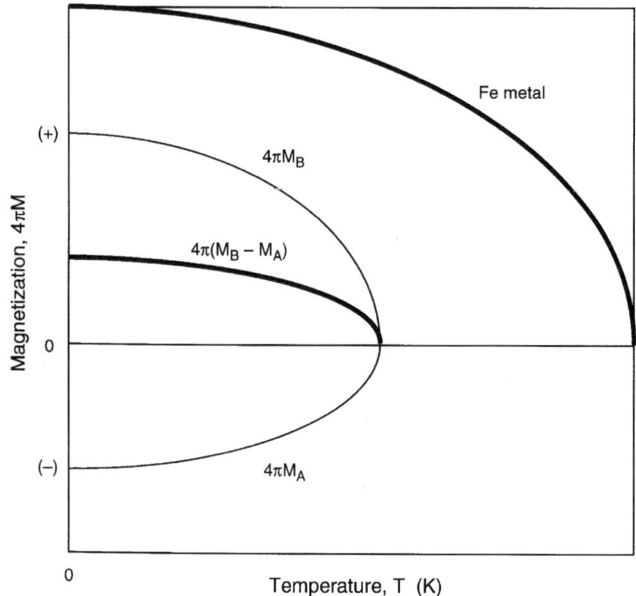

Fig. 4.3 Thermomagnetic characteristics of a ferromagnetic metal and a two-sublattice ferrite. Note greater magnetization expected from the itinerant magnetism metal, for which no oxygen occupies lattice sites and there are no opposing sublattices

purpose are listed in MIT Lincoln Laboratory Technical Reports ([3] (garnet), [4] (RE garnet), [5] (spinel)).[1] A sketch of a typical thermomagnetization computation is given in Fig. 4.3. The usefulness of the molecular field model has proven to be enormous over the past four decades, particularly the refined versions of it that have made possible the explanation and prediction of thermomagnetism behavior of compounds in which the sublattice moments are diluted with diamagnetic substitutions for the purpose of tailoring magnetic properties to specific applications.

In most of these situations, dilution of the magnetic sublattices has been accompanied by departures from ideal magnetic spin alignments, commonly referred to as "spin canting." These reductions in the effective magnetic moments occur beyond the normal disruptions of the magnetic ordering arising from the thermal randomization accounted for in the application of the Brillouin–Weiss function. Before the more general theory of thermomagnetization is discussed, some background on canting effects must be reviewed.

[1] These published documents can be readily obtained from the U.S. National Technical Information Service (NTIS).

4.1.3 Magnetic Frustration and Spin Canting

In ferrimagnetism, the molecular fields comprise contributions from magnetically opposing sublattices. As a consequence, there exists the possibility that breakdown of the long-range magnetic order by local cancellation (or even reversal through overcompensation) of the magnetic moments can occur through variations in spatial ordering of these individual moments from inhomogeneous site distributions. Another cause for cancellation is magnetic dilution, i.e., the replacement of the magnetic ions by diamagnetic $S = 0$ substitutes (which could include actual lattice vacancies). For a site in the i sublattice that is missing even one of its nearest neighbor spins, there is a finite probability that the ion does not participate in the exchange stabilization. Such an occurrence is called magnetic frustration, which effectively renders the ion paramagnetic at the site in question. The resulting spin canting is a departure from collinearity of the spin directions independent of the thermal lattice vibrational disruptions that accompany rising temperatures. The general subject of spin canting has been examined from various approaches and will be reviewed in the approximate chronological order in which they were reported.

The first examination of noncollinearity of spins in magnetically ordered systems was reported by Yafet and Kittel (Y-K) [6] as an attempt to explain the magnetic behavior of spinel ferrites with A sites diluted by zinc. In the example of Fig. 4.4

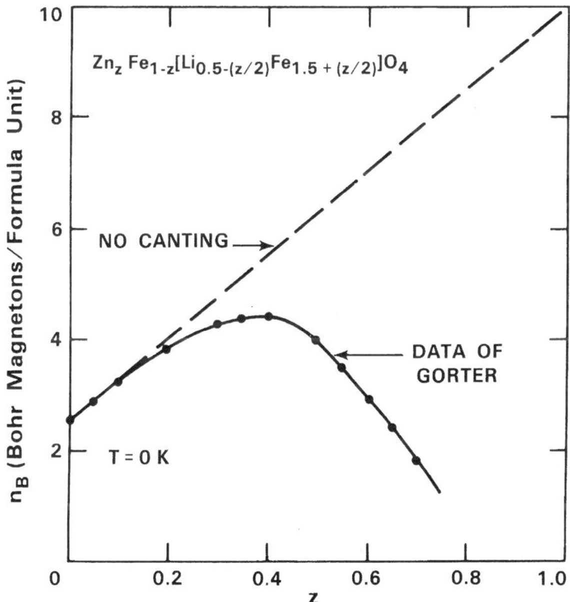

Fig. 4.4 Magnetic moment of lithium zinc spinel ferrite at $T \rightarrow 0$ K, showing the canting effects from Zn^{2+} dilution of the minority tetrahedral A sublattice. Note departure from the linear Néel model and peak in data at $z \approx 0.4$. Data are from Gorter [7]. Figure reprinted from G.F. Dionne, *J. Appl. Phys.* **45**, 3621 (1974) with permission. © 1974 by the American Institute of Physics

for a lithium ferrite host [7], i.e., $Fe_{1-z}Zn_z\left[Li_{0.5-z/2}Fe_{1.5+z/2}\right]O_4$, the saturation moment expressed in Bohr magnetons per formula unit n_B is shown to fall below the collinear Néel model and reach a peak as the zinc content $z \rightarrow 0.4$. The basic notion of the Y-K model was that the i–i and $j-j$ interactions under the right conditions could form antiferromagnetic spin alignments within the their own sublattices, thereby breaking up the main $i-j$ antiferromagnetic ordering to produce four sublattices and causing radical changes in the net magnetization. Although the Y-K model failed to fit the data of the ferrite anomalies, it called attention to the existence of canted spins that was later confirmed in the spinel $CuCr_2O_4$ by neutron diffraction [8]. Another important result of this concept was the reasoning by de Gennes that the canting of a sublattice would be principally the result of dilution of the *opposing* sublattice [9].

There have been several attempts to devise a theory of canting that could place the concept on a more quantitative basis. The seminal work was carried out by Gilleo in an analysis of magnetic moment departures from the Néel model for the magnetic garnets. In his most successful model, he assumed that Fe^{3+} ions linked to no more than one nearest neighbor nonmagnetic ions would not contribute to the spontaneous magnetization [10]. This condition can be more stringent than that of frustration, which can occur by means of a cancellation of exchange fields and therefore requires less dilution, but the results of the model nonetheless provided some degree of satisfaction. In this model, the net magnetic moment per molecule in the garnet system is

$$n_B\left(k_d, k_a\right) = n_{Bd}\left(k_d, k_a\right) - n_{Ba}\left(k_d, k_a\right), \tag{4.12}$$

where $n_{Bd} = 15\left(1-k_d\right)\left[1-E_d\left(k_a\right)\right]$ and $n_{Ba} = 10\left(1-k_a\right)\left[1-E_a\left(k_d\right)\right]$. (Note that each Fe^{3+} ion carries a magnetic moment of $5m_B$.) The parameters $E_a\left(k_d\right) = 6k_d{}^5 - 5k_d{}^6$ and $E_d\left(k_a\right) = 4k_a{}^3 - 3k_a{}^4$ are the respective sublattice canting probabilities from opposite sublattice dilution as determined from random probability theory. From these relations both the magnetic moment at $T = 0\,\mathrm{K}$ and the Curie temperature can be estimated. In this attempt to fit measurement data, only the intersublattice interactions were taken into account. Nonetheless, the results as applied to the magnetic garnet compositions with a-site dilution by Sc^{3+} in the forms of $\{Y_3\}\left[Fe_{2-x}Si_x\right]\left(Fe_3\right)O_{12}$ or by Zr^{4+} with charge compensating c-site Ca^{2+} in the form of $\{Y_{3-x}Ca_x\}\left[Fe_{2-x}Zr_x\right]\left(Fe_3\right)O_{12}$ have given reasonable qualitative agreement with experiment, as shown in Fig. 4.5. For d-site dilution by Ge^{4+} or Si^{4+} with charge compensating c-site Ca^{2+} in the form of $\{Y_{3-x}Ca_x\}\left[Fe_2\right]\left(Fe_{3-x}Ge_x\right)O_{12}$ or $\{Y_{3-x}Ca_x\}\left[Fe_2\right]\left(Fe_{3-x}Si_x\right)O_{12}$, the model has also given similar agreement with experiment, as shown in Fig. 4.6.

By the same procedure, Gilleo also deduced a model for the ferrimagnetic spinel system AB_2O_4, with

$$n_B\left(k_B, k_A\right) = n_{BB}\left(k_B, k_A\right) - n_{BA}\left(k_B, k_A\right), \tag{4.13}$$

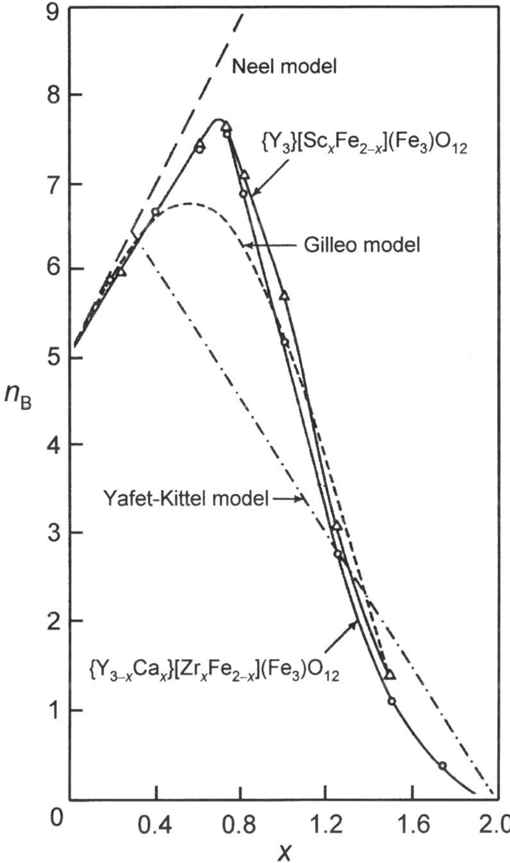

Fig. 4.5 Initial canting effects on magnetic moment of a-sublattice diluted yttrium-iron garnet at $T \approx 0$ K. Data of $\{Y_{3-x}Ca_x\}[Fe_{2-x}Si_x](Fe_3)O_{12}$ and $\{Y_{3-x}Ca_x\}[Fe_{2-x}Zr_x](Fe_3)O_{12}$ are from Geller [15]. Figure reprinted from G.F. Dionne, *J. Appl. Phys.* **41**, 4874 (1970) with permission.

where $n_{BB} = n_{BB}{}^0 (1 - k_B)[1 - E_B(k_A)]$, $n_{BB} = n_{BB}{}^0 (1 - k_A)[1 - E_A(k_B)]$, and $E_A(k_B) = 12k_B{}^{11} - 11k_B{}^{12}$, $E_B(k_A) = 6k_A{}^5 - 5k_A{}^6$. In this case the respective undiluted Bohr magnetons per molecule are left as variables $n_{BB}{}^0$ and $n_{BA}{}^0$ because of the greater likelihood of varying ionic spin values, i.e., other than $S = 5/2$, in the spinel system.

In fashioning a physical description of spin canting in the magnetic garnets, Geller realized the importance of the intrasublattice exchange fields and proceeded to picture the evolution of the antiferromagnetic ground state from its initial ferrimagnetism by examining the influence of the intrasublattice a–a and d–d exchange fields on the stronger intersublattice a–d interaction [11]. The reasoning proceeds as follows: When dilution of the a sublattice, for example, is made, the initial effect is a net increase of n_B which would proceed initially as a linear function of k_a; however,

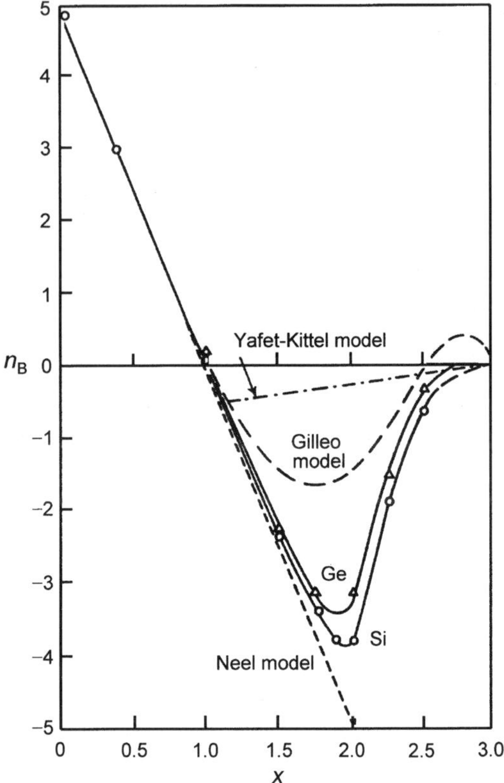

Fig. 4.6 Initial canting effects on magnetic moment of d-sublattice diluted yttrium-iron garnet at $T \approx 0\,K$. Data of $\{Y_{3-x}Ca_x\}\,[Fe_2]\,(Fe_{3-x}M_x)\,O_{12}$, where $M = Ge^{4+}$ and Si^{4+}, are from Geller [15]. Figure reprinted from G.F. Dionne, *J. Appl. Phys.* **41**, 4874 (1970) with permission. © 1970 by the American Institute of Physics

spins of the d sublattice experience a weakened a–d exchange field, thereby causing them to react more strongly to their own antiferromagnetic d–d exchange field. This event leads to canting in the d sublattice that causes a lowering of n_B. If the dilution is allowed to continue, n_B will reach a peak at some critical value of k_a and will eventually drop to zero when the a sublattice is depleted of spins and an antiferromagnetic ground state takes over in the d sublattice. When the same logic is applied to dilution of the d sublattice with the more abundant spins, n_B will pass through zero and changes sign when the a sublattice moment becomes dominant. The net moment then peaks and reverses back towards zero, indicating cancellation of the a sublattice spins at the point of spin depletion of the d sublattice. These events are seen in Figs. 4.5 and 4.6.

There have been other diligent attempts to interpret the experimental results of garnet and spinel spin canting by Borghese [12], Nowik [13], and Rosencwaig [14] following in general terms the above concepts set down by Gilleo and Geller. A more comprehensive discussion of spin canting may be found in review articles by Geller [15] and Gilleo [16]. So far in our discussion, the effects of spin departures from collinearity have been confined to the state at $T = 0\,K$. More important from a practical standpoint and also for an understanding of the overall magnetic state of the ferrimagnetic material is the variation of the magnetization or magnetic moment as a function of temperature, i.e., thermomagnetization over the range $0 \leq T \leq T_C$. For this analysis, the molecular field concepts of the Néel theory must be revisited.

4.2 Theory of Superexchange Dilution

During the time period of Gilleo and Geller's work on spin canting, precise calculated fits to thermomagnetic data were reported by Anderson [17] for yttrium iron garnet and by Rado and Folen [18] for lithium spinel ferrite. These results produced accurate values of the molecular field coefficients that would serve as the basis for the magnetic dilution analyses carried out by Dionne [19]. The refinement to Néel's theory emerged from a mathematical representation of Geller's reasoning on the effects of selective replacement of magnetic cations by diamagnetic substitutes. Unlike the earlier attempts to develop spin frustration models described in the previous section that were limited to the interpretation of magnetization as a function of diamagnetic substitutions observed at $T = 0\,\mathrm{K}$, the Dionne theory deals directly with the influence of dilution on the molecular field stabilization energy that is the essence of the Brillouin–Weiss thermomagnetization formalism.

4.2.1 Superexchange Energy Stabilization

Since small changes in the exchange energy due to spin canting or frustration will affect the Curie temperature first, the net magnetic moment at $T = 0\,\mathrm{K}$ will be less sensitive to the frustration effects of dilution at low levels of diamagnetic substitutions. As a consequence, probabilities of canting can initially be represented by simple linear approximations. From this approach an analytical formalism for use in thermomagnetic computations was made possible. Although the correct form of the effective molecular coefficient relations as functions of dilution was originally deduced semiempirically, subsequent theoretical underpinning has been developed to explain the results. For instructional purposes, we exercise the luxury of discussing the theoretical basis first.

By manipulating (3.27) through and (3.30), we can express the undiluted i and j sublattice exchange energies in terms of exchange field H_{ex}, according to

$$
\begin{aligned}
E_{\mathrm{ex}}{}^{i} &= -m_i\left(H_{\mathrm{ex}}{}^{ii} + H_{\mathrm{ex}}{}^{ij}\right) = -2S_i\left(z_{ii}{}^{0}J_{ii}S_i + z_{ij}{}^{0}J_{ij}S_j\right), \qquad (4.14)\\
E_{\mathrm{ex}}{}^{j} &= -m_j\left(H_{\mathrm{ex}}{}^{ji} + H_{\mathrm{ex}}{}^{jj}\right) = -2S_j\left(z_{ji}{}^{0}J_{ji}S_i + z_{jj}{}^{0}J_{jj}S_j\right),
\end{aligned}
$$

where $S_i = -S_j$ to observe the proper signs of the opposing i and j sublattices, and $z_{ij}{}^{0}$ is the number of S_j nearest neighbors of spin S_i. The removal of a spin, e.g., one isolated S_j, will cause a series of magnetic frustration events in the vicinity of the missing S_j spin as depicted in the sketch of Fig. 4.7.

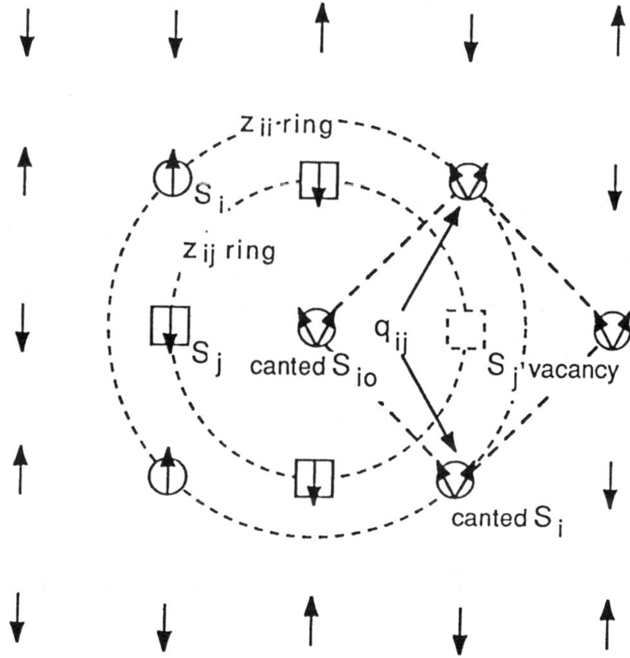

Fig. 4.7 Two-dimensional model of canting of i-site spins surrounding a j-site vacancy

Direct Reduction of Exchange Fields: If a dilution fraction k_j exists among the spins in the j sublattice, the effective number $z_{ij}{}^0$ and $z_{jj}{}^0$ are reduced to $z_{ij} = z_{jj}{}^0 \left(1 - k_j\right)$ and $z_{jj} = z_{jj}{}^0 \left(1 - k_j\right)$. Therefore, the exchange field at random i sites, e.g., S_{i0}, produced by the j sublattice will also decrease by the factor $\left(1 - k_j\right)$. This first-order dilution effect is seen as a direct reduction in the $H_{\text{ex}}{}^{ij}$ component of (4.14).

Intersublattice Spin Canting: Because the intersublattice $S_i \leftrightarrow S_j$ interaction is the dominant antiferromagnetic interaction that establishes the antiparallel spin alignments between the sublattices, the removal of spins from the j sublattice would have the indirect effect of causing partial frustration of neighboring spins of the i sublattice by spin canting. For small dilutant concentrations, we first characterize the canting effect in the i sublattice as the probability k_j of a dilutant appearing at a j site weighted by a canting probability c_j, thereby creating a frustration fraction $c_j k_j$. To account for the canting of the i-site spin, we then reduce S_{i0} by the inverse probability $\left(1 - c_j k_j\right)$. Since S_{i0} has $z_{ij}{}^0$ equivalent j sites that can be diluted, the $\left(1 - c_j k_j\right)$ factor must be applied z_{jj} times to determine the probability of S_{i0} making a full contribution to the exchange energy. The canting reduction factor expressed in the form of $\left(1 - c_j k_j\right)^{z_{ij}}$ can be expanded binomially to $\left(1 - z_{ij} c_j k_j\right)$ for small values of $c_j k_j$.

Intrasublattice Spin Canting: A third frustration effect occurs in the ring of S_i nearest neighbors surrounding S_{i0} in the i sublattice. The contribution from each of these S_i neighbors that are also neighbors of the missing S_j spin would also be reduced by the $(1 - c_j k_j)$ canting factor because they are crystallographically and therefore magnetically equivalent in the small dilution limit. If q_{ij} of the j sites in the z_{ij} ring shown in Fig. 4.7 are common neighbors to any one of the spins in the z_{ii} ring surrounding S_{i0}, their contribution to the $H_{ex}{}^{ii}$ intrasublattice exchange field term inside the bracket of (4.14) must be reduced by the factor $(1 - c_j k_j)^{q_{ij}}$, which can be expanded binomially to $(1 - q_{ij} c_j k_j)$ for low k_j levels. Furthermore, this canting factor must be applied collectively to all of the spins of the S_i ring, i.e., z_{ii} times, so that the complete canting factor becomes $(1 - q_{ij} c_j k_j)^{z_{ii}}$, which is the probability that none of the S_i ring is canted. In the example of Fig. 4.7, $z_{ij}{}^0 = 4$, $q_{ij} = 2$, and $z_{ii}{}^0 = 4$. Equation (4.14) then becomes

$$E_{\text{ex}}^j \approx -2S_i \left(1 - z_{ij} c_j k_j\right) \left[z_{ii}{}^0 J_{ii} S_i \left(1 - z_{ii} q_{ij} c_j k_j\right) + z_{ij}{}^0 \left(1 - k_j\right) J_{ij} S_j\right],$$

$$E_{\text{ex}}^j \approx -2S_j \left[z_{ji}{}^0 J_{ji} S_i \left(1 - z_{ij} c_j k_j\right) + z_{jj}{}^0 \left(1 - k_j\right) J_{jj} S_j\right]. \tag{4.15}$$

Note that the canting factor $(1 - z_{ij} c_j k_j)$ of the i-sublattice spins is also reflected in the intersublattice term of E_{ex}^j.

If dilution of the i sublattice also occurs, this procedure can be repeated and the result of these manipulations is the formation of the general relations

$$\begin{aligned} E_{\text{ex}}^i &\approx -2S_i \left(1 - z_{ij} c_j k_j\right) \left[z_{ii}^0 \left(1 - k_i\right) J_{ii} S_i \left(1 - z_{ii} q_{ij} c_j k_j\right) \right. \\ &\quad \left. + z_{ij}^0 \left(1 - k_j\right) J_{ij} S_j \left(1 - z_{ji} c_i k_i\right)\right], \\ E_{\text{ex}}^j &\approx -2S_j \left(1 - z_{ji} c_i k_i\right) \left[z_{ji}^0 \left(1 - k_i\right) J_{ji} S_i \left(1 - z_{ij} c_j k_j\right) \right. \\ &\quad \left. + z_{jj}^0 \left(1 - k_j\right) J_{jj} S_j \left(1 - z_{jj} q_{ji} c_i k_i\right)\right]. \end{aligned} \tag{4.16}$$

To a first approximation, the parameter c_j is treated as a semiempirical constant that is proportional to the magnitude of the intersublattice exchange field coupling to spin S_i relative to the magnitude of the net exchange field of the inter and intrasublattice contributions, i.e.,

$$c_j \propto \left| \frac{z_{ij}^0 J_{ij} S_j \left(1 - k_j\right)}{z_{ii}^0 J_{ii} S_i \left(1 - z_{ii}^0 q_{ij} c_j k_j\right) + z_{ij}^0 J_{ij} S_j \left(1 - k_j\right)} \right|. \tag{4.17a}$$

and by symmetry for the opposing sublattice

$$c_i \propto \left| \frac{z_{ji}^0 J_{ji} S_i (1 - k_i)}{z_{ji}^0 J_{ji} S_i (1 - k_i) + z_{jj}^0 J_{jj} S_j \left(1 - z_{jj}^0 q_{ji} c_i k_i\right)} \right|. \tag{4.17b}$$

A test for the validity of the relations of (4.15) and (4.16) can be found from the corrections to the molecular field coefficients required for the fitting of magnetic moment versus temperature data.

4.2.2 Molecular Field Coefficients

For application to the Brillouin–Weiss theory, (4.16) can be converted back to the molecular field format by first defining for the undiluted state

$$\begin{aligned} M_i^0 &= n_i{}^0 g_i m_B S_i, \\ M_j^0 &= n_j{}^0 g_j m_B S_j, \end{aligned} \tag{4.18a}$$

and

$$N_{ii}^0 = \frac{z_{ii}^0}{n_i{}^0} \frac{2J_{ii}}{g_i{}^2 m_B{}^2},$$

$$N_{jj}^0 = \frac{z_{jj}^0}{n_j^0} \frac{2J_{jj}}{g_j^2 m_B{}^2},$$

$$N_{ij}^0 = \frac{z_{ij}^0}{n_j^0} \frac{2J_{ij}}{g_i g_j m_B{}^2}; \quad N_{ji}^0 = \frac{z_{ji}^0}{n_i^0} \frac{2J_{ji}}{g_j g_i m_B{}^2},$$

where $n_i{}^0$ and $n_j{}^0$ are the number densities of magnetic ions in the respective sublattices.

After appropriate substitution for S_i, S_j and the different J_{ij} factors in the bracketed exchange fields, (4.16) can now be expressed as

$$\begin{aligned} E_{ex}^i &\approx -g_i m_B S_i \left(N_{ii} M_i + N_{ij} M_j\right), \\ E_{ex}^j &\approx -g_j m_B S_j \left(N_{ji} M_i + N_{jj} M_j\right), \end{aligned} \tag{4.19}$$

where

$$\begin{aligned} M_i &= M_i^0 (1 - k_i) \\ M_j &= M_j^0 (1 - k_j) \end{aligned} \tag{4.20}$$

and

$$
\begin{aligned}
N_{ii} &\approx N_{ii}^{0}\left[1-\left(1+\frac{z_{ii}q_{ij}}{z_{ij}}\right)z_{ij}c_j k_j\right], \\
N_{jj} &\approx N_{jj}^{0}\left[1-\left(1+\frac{z_{jj}q_{ji}}{z_{ji}}\right)z_{ji}c_i k_i\right], \\
N_{ij} = N_{ji} &\approx N_{ij}^{0}\left(1-z_{ij}c_j k_j\right)\left(1-z_{ji}c_i k_i\right), \\
&\approx N_{ij}^{0}\left(1-z_{ij}c_j k_j - z_{ji}c_i k_i\right).
\end{aligned}
\tag{4.21}
$$

In this conversion to molecular field format, it should be noted that the dilution factors $(1-k_i)$ and $\left(1-k_j\right)$ have been absorbed into M_i and M_j, thereby reflecting the appropriate reduction in sublattice magnetizations. For convenience as much as any other reason, the canting or frustration factors have been included in the definitions of the molecular field coefficients for computational convenience with the Brillouin–Weiss function, as will be demonstrated for the case of the magnetic garnet system.

4.2.3 Solution for Yttrium Iron Garnet

For d and a sublattices of the garnet system (e.g., $Y_3Fe_5O_{12}$), the relevant parameters are as follows: $z_{dd}{}^{0}=4$, $z_{aa}{}^{0}=6$, $z_{da}{}^{0}=4$, $z_{ad}{}^{0}=6$; $q_{da}=q_{ad}=2$. If these values are applied in (4.21) the molecular field coefficients become

$$
\begin{aligned}
N_{dd} &\approx N_{dd}{}^{0}\left(1-12c_a k_a\right), \\
N_{aa} &\approx N_{aa}{}^{0}\left(1-18c_d k_d\right), \\
N_{ad} &\approx N_{ad}{}^{0}\left(1-4c_a k_a-6c_d k_d\right).
\end{aligned}
\tag{4.22}
$$

In Dionne's original work [19], the above molecular field relations were applied to the yttrium iron garnet (YIG) system. For the convenience of relating the results to chemical formulae, the sublattice magnetic moments were expressed per formula unit or molecule ($\mathcal{M}$) rather than per unit volume (M for magnetization). In the text that follows, wherever the discussion involves this particular model,

$$
\begin{aligned}
\mathcal{M}_d(T) &= \mathcal{M}_{\mathrm{d}}(0)\,B_{S_{\mathrm{d}}}(a_d), \\
\mathcal{M}_a(T) &= \mathcal{M}_{\mathrm{a}}(0)\,B_{S_{\mathrm{a}}}(a_a),
\end{aligned}
\tag{4.23}
$$

where

$$
\mathcal{M}_d(0) = 3g_d m_{\mathrm{B}} S_d N_{\mathrm{A}}\left(1-k_d\right)\left(1-0.1k_d\right)
$$

and

$$
\mathcal{M}_a(0) = 2g_a m_{\mathrm{B}} S_a N_{\mathrm{A}}\left(1-k_a\right)\left(1-k_a{}^{5.4}\right),
\tag{4.24}
$$

where Avogadro's number N_{A} is required to convert per molecule to per mole. The additional factors are small adjustments that were created empirically to provide a close fit to experiment and bear no relation to the present frustration model or those discussed in the previous section. The corresponding parameters a_d and a_a are

$$a_d(T) = \frac{m_d H_{\mathrm{ex}}^{(d)}}{kT} = \frac{g_d m_{\mathrm{B}} S_d}{kT}\left[N_{dd} M_d + N_{da} M_a\right],$$
$$a_a(T) = \frac{m_a H_{\mathrm{ex}}^{(a)}}{kT} = \frac{g_a m_{\mathrm{B}} S_a}{kT}\left[N_{ad} M_d + N_{aa} M_a\right]. \quad (4.25)$$

Because the exchange fields (inside brackets) are in units of oersteds, the N_{ij} coefficients will be dimensionless if M is expressed in gauss (or emu/cm^3). As a consequence, the conversion factor between M $\left(\mathrm{emu/cm}^3\right)$ and $\mathcal{M}$ (emu/mol) must be absorbed in the $\mathcal{N}_{ij}$ coefficients that are now expressed in units of mol/cm^3. For these purposes, the units of $H_{\mathrm{ex}}^{(i)}$ is that of the factor multiplying the magnetic moment $m_i = g_i m_{\mathrm{B}} S_i$, i.e.,

$$H_{\mathrm{ex}}^{(i)} = \sum_j \frac{2z_{ij} J_{ij} S_i S_j}{g_i m_{\mathrm{B}} S_i} = \sum_j \mathcal{N}_{ij} \mathcal{M}_j = \sum_j N_{ij} M_j, \quad (4.26)$$

in emu/cm^3.

By fitting theory to data compiled from magnetic moment measurements as a function of temperature, the following relations for the molecular field coefficients were deduced as functions of dilution:

$$\mathcal{N}_{dd} \approx -30.4\,(1 - 0.87k_a) \approx -30.4\left(1 - \tfrac{7}{8}k_a\right),$$
$$\mathcal{N}_{aa} \approx -65.0\,(1 - 1.26k_d) \approx -65.0\left(1 - \tfrac{5}{4}k_d\right), \quad (4.27)$$
$$\mathcal{N}_{da} = \mathcal{N}_{ad} \approx +97.0\,(1 - 0.25k_a - 0.38k_d) \approx +97.0\left(1 - \tfrac{1}{4}k_a - \tfrac{3}{8}k_d\right).$$

for dilution limits $k_d \le 0.65$ and $k_a \le 0.35$.[2] Comparison of theory with experimental data of Geller et al. [20] is shown in Fig. 4.8 for the $\{Y_3\}[Fe_{2-x}Mg_x](Fe_{3-x}Si_x)O_{12}$ system that includes both d and a sublattice dilution. To convert among the $\mathcal{N}_{ij}$'s, N_{ij}'s and J_{ij}'s, the following relation based on (3.28) between their magnitudes can be used.

$$J_{ij} = \frac{n_j g_i g_j m_{\mathrm{B}}{}^2}{2z_{ij}^0} N_{ij} = \frac{n_j g_i g_j m_{\mathrm{B}}{}^2}{2z_{ij}^0}\left(\frac{a_0{}^3 N_{\mathrm{A}}}{8}\right)\mathcal{N}_{ij}, \quad (4.28)$$

[2] Note that the sign of $\mathcal{N}_{ad}{}^o$ is designated as positive. This serves to account for the sign reversal in the sublattice moments. An alternative convention would be to leave all of the coefficients negative and change the sign in the expression for the exchange field to indicate the opposing contributions of the antiferromagnetically aligned spins.

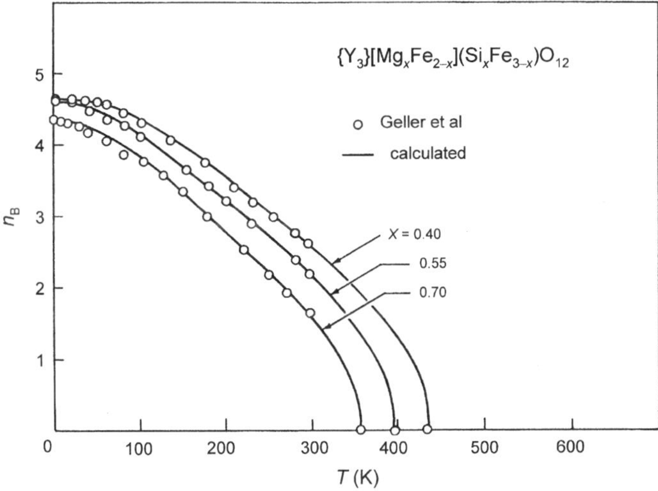

Fig. 4.8 Comparison of theory with experiment for three compositions of $\{Y_3\}$ $[Mg_xFe_{2-x}](Si_xFe_{3-x})O_{12}$, which feature substitutions in both sublattices. Figure reprinted from G.F. Dionne, *J. Appl. Phys.* **41**, 4874 (1970) with permission. © 1970 by the American Institute of Physics

Table 4.1 Molecular and exchange field parameters of yttrium iron garnet and lithium spinel ferrite calculated from theory

		$z_{ij} : n_j$	$\mathcal{M}_j$ (emu/mol × 10^4)	$\mathcal{N}_j$ (mol/cm^3)	N_{ij} $(\times 4\pi)$[a]	J_{ij} (ergs× 10^{-15})[b]	$H_{ex}^{(i)}$ (T) $(T = 0\,K)$
$Y_3Fe_5O_{12}$							
$Fe_d \rightarrow$	Fe_d	4 : 3	8.4	−30.4	−347	−2.4	288
	Fe_a	4 : 2	5.6	97.0	1107	−4.9	
$Fe_a \rightarrow$	Fe_a	8 : 2	5.6	−65.0	−742	−1.6	451
	Fe_d	6 : 3	8.4	97.0	1107	−4.9	
$Li_{0.5}Fe_{2.5}O_4$							
$Fe_B \rightarrow$	Fe_B	4.5[c] : 1.5[c]	4.2	−60	−212	−2.1	510
	Fe_A	6 : 1	2.8	273	969	−4.8	
$Fe_A \rightarrow$	Fe_A	4 : 1	2.8	−150	−533	−3.8	720
	Fe_B	9[c] : 1.5[c]	4.2	273	969	−4.8	

[a]In the determination of these parameter values, the lattice parameter of the garnet $a_o \approx 12.4$ Å, and of the spinel $a_o \approx 8.4$ Å

[b]By convention, all J_{ij} values are negative

[c]These values reflect the 25% initial dilution of the B sublattice by Li^{1+} ions

where a_0 is the lattice cubic cell dimension (containing eight molecules for garnets and spinels) and $g_i = g_j = 2$ for the $3d^n$ group. In Table 4.1, where a summary of these parameter values is given, the sign convention mentioned above for the N_{ij} and $\mathcal{N}_{ij}$ has been adopted. Note that the values of $H_{ex}^{(i)}$ computed from (4.26) for each sublattice of the garnets and spinels is in the range of 10^2–10^3 T.

As predicted by (4.22), the relations of (4.27) are linear and the coefficients of k_d and k_a are in the approximate 3:1 ratio as they appear in the intra and intersublattice factors. Moreover, the values of the frustration parameters c_d and c_a computed from these results are 0.070 and 0.0725, respectively, or about 7% for both sublattices. By means of the values from Table 4.1 applied to (4.15) with $S_i = S_j = 5/2$ for Fe^{3+}ions, computations show that the ratio of c_d and c_a are approximately equal, confirming the equality of this low dilution limit frustration factor between the two sublattices.

Other attempts to model the thermomagnetic properties of the iron garnets based on this seminal work produced refinements for specific ionic dilutants. Noteworthy among these efforts was the work of Röschmann and Hansen in support of research into the magneto-optical properties of the diluted garnets to be discussed in Chap. 7 [21].

In addition to the YIG-based system and the rare-earth iron garnet system [3, 4, 22] to be examined in Sect. 4.3.3, Dionne also analyzed the lithium spinel ferrite family [5, 23] and later included high-permeability nickel-zinc and manganese-zinc spinels commonly used for inductor cores and in magnetic recording applications [24]. Although the superexchange interactions are more complicated than in the simple garnet system because of the presence of multiple species of cations such as Ni^{2+} $(3d^8)$ and Mn^{2+} $(3d^5)$, the principles of magnetic dilution apply in the same manner. Details of the diluted spinel system lithium-zinc-titanium ferrite that is particularly important for microwave applications are summarized in Appendix 4A.

In the present analysis, higher order canting effects within the diluted sublattice have been ignored at low dilution levels. Part of these effects appear as second-order terms in ${k_d}^2$ and ${k_a}^2$ that enter through the $z_{ij} = {z_{ij}}^0 \left(1 - k_j\right)$ dependencies in (4.16).

The dilution theory developed from elementary probability arguments confirms the original experimental findings for yttrium-iron garnet (YIG). The earlier work led to the conclusions that (1) the dilution relations are at least initially linear for each of the molecular-field coefficients, (2) the dilution of one sublattice does not influence its own intrasublattice coefficient ($\mathcal{N}_{dd}$ or $\mathcal{N}_{aa}$) to first order, and (3) the reduction of the intrasublattice coefficient is three to four times greater than that of the intersublattice coefficient ($\mathcal{N}_{da}$ or $\mathcal{N}_{ad}$). Note that the reduction in exchange energy E^i_{ex} is not directly caused by the J_{ij} exchange constants, which are fixed by covalent bonding between individual magnetic ion pairs. The effect is manifested in the molecular-field coefficients because it is the dilution of z_{ij} spin neighbors (see (4.21)) that causes exchange frustration between the opposing sublattices, generally referred to as spin canting.

4.3 Ferrimagnetic Oxides

In the previous section, some of the basic concepts peculiar to ferrimagnetism were described. In particular, the behavior of the spontaneous magnetism and its relation to the magnetic exchange was explained. The actual magnetic properties of the various ferrimagnetic systems are important in themselves, however. Although they have been documented in other volumes, failure to include at least the basic properties of standard ferrite compounds in this text would be a conspicuous omission. The following subsections will attempt to define the spinel, garnet, and hexagonal ferrites of general interest to workers in the fields of magnetics. Some of the more unusual or anomalous issues will be treated as the subjects arise in later parts of the book.

4.3.1 Spinel Ferrites $A\,[B_2]\,O_4$

The 2:1 octahedral *B* to tetrahedral *A* site ratio in the spinels lattice is determined by the crystal lattice structure sketched in Fig. 4.9. Unlike the garnets that will be discussed next, this cell features congruity between the overall crystal axes to those of the individual sites. In its generic form the spinel can be (1) normal, with the divalent ions occupying only *B* sites, i.e., $A^{3+}\left[B^{3+}B^{2+}\right]O_4$, or (2) inverse, in which the divalent ion resides in *A* sites. In most practical cases, the ferrite is a mixture between normal and inverse. A number of excellent references document the properties of spinel ferrites: Smit and Wijn [25], von Aulock [26], Gorter [7], Blasse [27], von Grenou [28], Folen [29], Griefer [30], and Schieber [31]. The magnetization properties of common spinel families are described in the following paragraphs

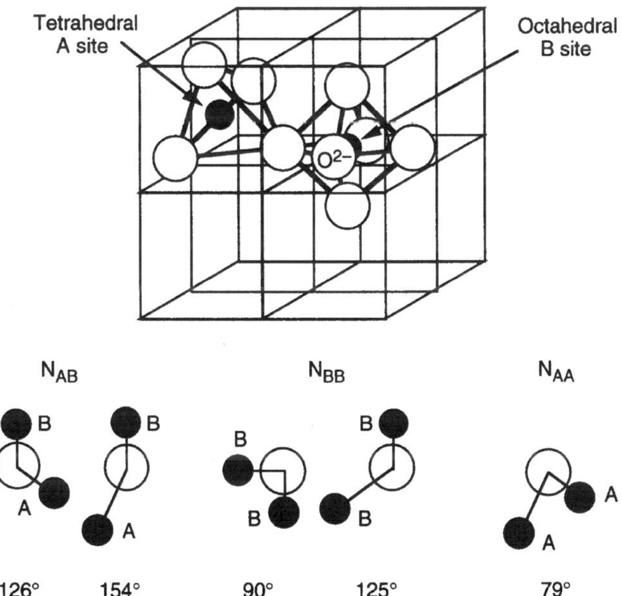

Fig. 4.9 Spinel crystal structure with bond angle diagrams

Table 4.2 Molecular field coefficients of nickel and manganese spinel ferrite (determined semiempirically)

	$Fe^{3+}[Ni^{2+}Fe^{3+}]O_4$			$Mn_{0.8}{}^{2+}Fe_{0.2}{}^{3+}[Mn_{0.2}{}^{2+}Fe_{1.8}{}^{3+}]O_4$		
	Fe–Fe (mol/cm³)	Fe–Ni (mol/cm³)	Ni–Ni (mol/cm³)	Fe–Fe (mol/cm³)	Fe–Mn (mol/cm³)	Mn–Mn (mol/cm³)
$\mathcal{N}_{AA}$	–200	–	–	−200	−180	−160
$\mathcal{N}_{BB}$	–60	–58	–60	−60	−59	−58
$\mathcal{N}_{AB}$	312	276	–	312	187	62

Table 4.3 Magnetic parameters of common spinel and garnet cubic ferrites

Compound	n_B (tet)	n_B (oct)	n_B (theor.)	$4\pi M_s$ (0) (G)	$4\pi M_s$ (300 K) (G)	T_C (exp.) (K)
SPINEL						
Fe_3O_4	5	5 + 4	4	6,400	6,000	858
γFe_2O_3[a]	5	8.3	3.3	~5,000	~4,500	948
$ZnFe_2O_4$	0	5 − 5	0 (antiferro)	paramag	paramag	–
$CdFe_2O_4$	0	5 − 5	0 (antiferro)	paramag	paramag	–
$MnFe_2O_4$[b]	5	5 + 5	5	7,000	5,000	573
$CoFe_2O_4$	5	5 + 3	3	6,000	5,300	793
$NiFe_2O_4$	5	5 + 2	2	3,800	3,400	858
$CuFe_2O_4$	5	5 + 1	5	7,000	5,000	728
$MgFe_2O_4$[c]	5	5 + 0	0	1,800	1,500	713
$Li_{0.5}Fe_{1.5}O_4$	5	5 + 2.5	2.5	4,200	3,900	943
GARNET						
$Y_3Fe_5O_{12}$	15	10	5	2,400	1,800	560

Typical spinel lattice parameter $a_o \approx 8.4$ Å; density $\approx$ 4.5–5.5 gm/cm³; typical garnet (YIG) lattice parameter $a_o \approx 12.4$ Å; density ≈ 5.17 gm/cm³

[a]A defect spinel structure that is called maghemite and written as $Fe^{3+}\left[\square_{1/3}Fe^{3+}{}_{2/3}\right]O_4$, where $\square$ represents a cation site vacancy. A more common antiferromagnetic nonspinel form is known as hematite with designation αFe_2O_3

[b]Mn^{2+} ions occupy the tetrahedral sublattice in the amount of ~20%. The anomalously low Curie temperature is caused by very weak $\mathcal{N}_{AB}$ interaction involving Mn^{2+}

[c]The experimentally observed magnetic moment is attributed to upwards of 10% of the Mg^{2+} ions occupying the tetrahedral sublattice thereby placing more Fe^{3+} in the octahedral sublattice

based on the experimental parameter values listed in Tables 4.2 and 4.3. In Fig. 4.10, thermomagnetization characteristics from measurements are compared with the results of computations in which the molecular orbital exchange concepts developed in Sect. 3.1 were applied to the spinel ferrite family [32]. The corresponding parameter values used for the calculations are listed in Table 4.4.

Magnetite (lodestone) is a naturally occurring normal spinel of chemical formula $Fe^{3+}[Fe^{3+}Fe^{2+}]O_4$. Because the electrical conductivity introduced by the full complement of octahedral-site Fe^{2+} ions that can transfer electrons among the Fe^{3+} neighbors (a process called polaronic electron hopping), magnetite is more a vehicle for studying the electrical properties of oxides than a practical magnetic insulator

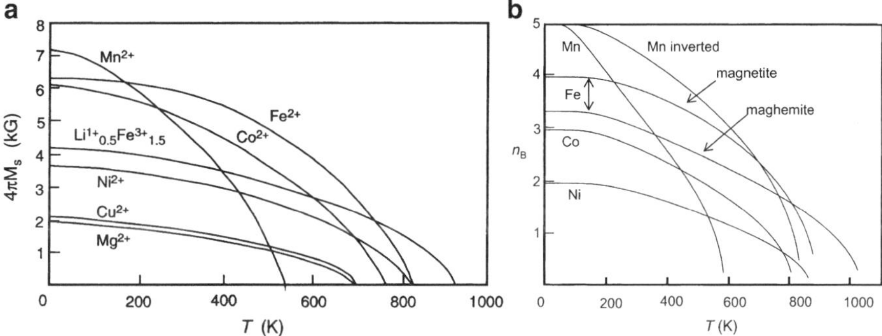

Fig. 4.10 Thermomagnetism characteristics of basic spinel ferrites: (**a**) measured curves inspired by Fig. 32.7 in reference [24], and (**b**) computations by a molecular-orbital model [32]

Table 4.4 Spinel ferrite molecular field coefficients (determined from molecular-orbital theory)

B-site Ion-S_M	$\mathcal{N}_{AA}$ $(\mathrm{mol/cm^3})$	$\mathcal{N}_{BB}$ $(\mathrm{mol/cm^3})$	$\mathcal{N}_{AB}$ $(\mathrm{mol/cm^3})$
$Li^{1+}Fe^{3+} - 5/2$	−150	−60	+273
$Fe^{3+} - 5/2$	−200	−60	+300
γ-$Fe^{3+} - 5/2$	−170	−60	+282
$Mn^{2+} - 5/2$[a]	−135	−51	+178
$Mn^{2+} - 5/2$[b]	−200	−32	+222
$Fe^{2+} - 2$	−200	−25[c]	+240
$Co^{2+} - 3/2$	−200	−58[c]	+269
$Ni^{2+} - 1$	−200	−90[c]	+330

[a]Standard site distribution $Fe_{0.2}{}^{3+}Mn_{0.8}{}^{2+}\left[Mn_{0.2}{}^{2+}Fe_{1.8}{}^{3+}\right]O_4$
[b]Fully inverted site distribution $Fe^{3+}\left[Mn^{2+}Fe^{3+}\right]O_4$
[c]Values are corrected for t_{2g}–t_{2g} ferromagnetic direct exchange which decreases monotonically as the shell fills with paired spins. For $Fe^{3+}\left[Fe^{2+}Fe^{3+}\right]O_4$ (magnetite), $\mathcal{N}_{BB}$ was adjusted to −25 from $-44\,\mathrm{mol/cm^3}$ to account for ferromagnetic double exchange

[33]. It is also one of the best hosts for studying the magnetic contribution of the Fe^{2+} ion, which can cause important magnetoelastic and dielectric effects. One interesting feature of this compound is the order-disorder transition of Fe^{2+}–Fe^{3+} ions in the B sublattice that occurs at $T \approx 120\,\mathrm{K}$. Below this temperature, the crystal structure undergoes a phase transition from cubic to orthorhombic symmetry, lending further credence to the proposition that the magnetoelastic Fe^{2+} $(3d^6)$ ions undergo a J–T or S–O condensation that stabilizes the transfer electrons in their polaronic traps and causes an anomalous decrease in conductivity. This general subject is discussed further in Chap. 8.

Lithium ferrite contains the most iron ions next to magnetite and is the ferrite compound with the largest Curie temperature ($T_C \approx 940\,\mathrm{K}$). As can be recognized immediately from the chemical formula $Fe\,[Li_{0.5}Fe_{1.5}]\,O_4$, 25% of the B sublattice is diluted by the Li^{1+} ions. Nonetheless, the net number of Bohr magnetons is 2.5, which is sufficient to produce a room-temperature $4\pi M \sim 3{,}500\,\mathrm{G}$. To lower the moment, magnetic dilution can be accomplished by direct Al^{3+} in the B sublattice or more commonly by substitutions of $0.5Li^{1+} + Ti^{4+} \rightarrow 1.5Fe^{3+}$ with chemical formula $Fe_{1-t/2}Li_{t/2}\,[Li_{0.5}Fe_{1.5-t}Ti_t]\,O_4$. As the site distributions

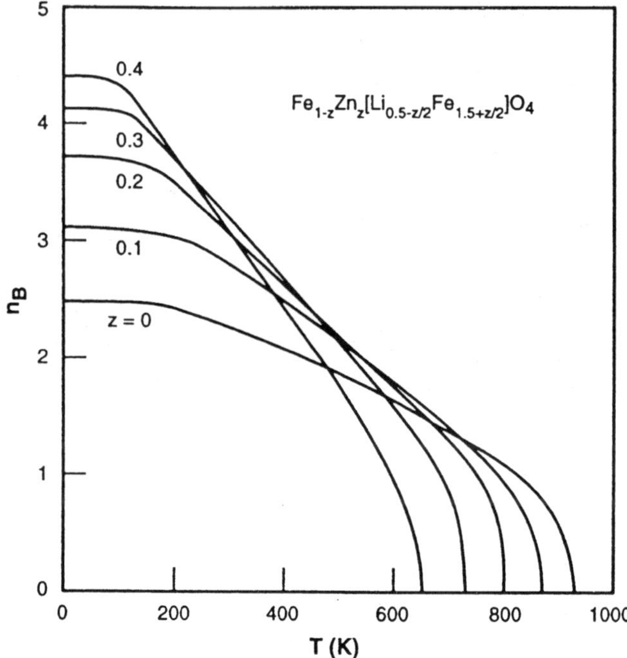

Fig. 4.11 Calculated lithium-zinc ferrite thermomagnetization curves

indicate, when the *B* sublattice is diluted with Ti^{4+} ions, the extra Li^{1+} ions required for electrical charge neutralization replace Fe^{3+} ions in the *A* sites. To increase the moment, $Zn^{2+} \rightarrow 0.5Fe^{3+} + 0.5Li^{1+}$ are substituted into the *A* sublattice according to $Fe_{1-z}Zn_z\left[Li_{0.5-z/2}Fe_{1.5+z/2}\right]O_4$. This modification also increases the Fe^{3+} content of the *B* sublattice, producing the initial rise in magnetization shown in Fig. 4.4, but it also causes a reduction in T_C because of the decreased net number of ${Fe^{3+}}_A$–${Fe^{3+}}_B$ interactions shown in the curves of Fig. 4.11. The results are from a molecular field analysis [23] and are also listed in Table 4.1 and discussed in Appendix 4A. The uses of Li ferrite have become mainly in microwave applications because of its high T_C made possible by the larger amounts of Fe^{3+} in both sublattices and generally good dielectric properties in the microwave bands. Its higher $4\pi M$ capabilities make it particularly attractive in the millimeter-wave bands (above 35-GHz frequencies). The development of Li ferrite of good ceramic quality was delayed by the high volatility of Li. When these materials are subjected to temperatures above 1,000°C, Li_2O is lost and the compound becomes iron-rich with the result that diamagnetic α-Fe_2O_3 hematite phase precipitates and Fe^{3+} is reduced to Fe^{2+} in the B sublattice, causing charge transfer that lowers the resistivity. To make dense microwave-quality Li ferrite, bismuth oxide (Bi_2O_3) is added as a flux to lower sintering temperatures [34]. Unfortunately, the tendency of Bi_2O_3 to segregate along grain boundaries also causes a deterioration in mechanical integrity.

Nickel ferrite $Fe^{3+}\left[Ni^{2+}Fe^{3+}\right]O_4$ is an inverted and generally well-behaved spinel with Ni^{2+} occupying almost half of the octahedral B sublattice. A small amount (<5%) can be expected to find its way into the tetrahedral A sites under thermal equilibrium conditions, causing magnetoelastic and spin–lattice relaxation effects. Because Ni^{2+} carries a spin value $S = 1$ in addition to an effective Landé g factor greater than 2 (see Sect. 5.1) nickel ferrite can offer higher magnetizations. Since Ni^{3+} occurs only occasionally and in special situations (e.g., Li^{1+}-substituted NiO) [35, 36], the incidence of Fe^{2+} is usually negligible, and dielectric properties are excellent for microwave applications. Because the e_g orbitals are half filled, this ion presents a significant covalent transfer integral b and consequently a strong superexchange coupling to neighbors of its own kind and to the adjacent Fe^{3+} ions of both sublattices. Similar to Li ferrite its magnetic moment can be reduced, in this case typically by direct substitutions of Al^{3+} for Fe^{3+} ions, which initially favor B-site dilution according to the theoretical analysis reported by Borghese [12].

The magnetic moment can also be raised by the Zn^{2+} ions diluting the A sublattice (similar to the LiZn ferrite case shown in Fig. 4.4, with the attendant increase in the sensitivity of $4\pi M$ to temperature resulting from the decrease in T_C, as shown in Fig. 4.12 [37]. The peak maximum magnetization of ~5,000 G at room temperature can be obtained with Zn replacing Ni, but occupying 35% of the tetrahedral sites, i.e., $Fe^{3+}{}_{0.65}Zn^{2+}{}_{0.35}\left[Ni^{2+}{}_{0.65}Fe^{3+}{}_{1.35}\right]O_4$. An additional benefit of Zn for materials preparation is a reduction in the solid-state reaction (sintering) temperature. As examined in Chap. 5, a drawback to nickel ferrite can be the large negative λ_{100} magnetostriction constant that renders the anisotropy and hysteresis-loop properties

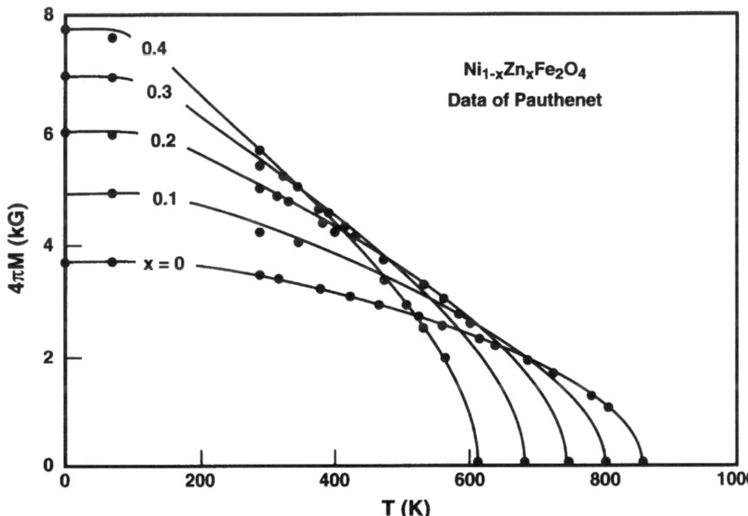

Fig. 4.12 Thermomagnetism curves of nickel-zinc spinel ferrite, showing the effects of Zn^{2+} dilution of the minority tetrahedral A sublattice. Note decrease in Curie temperatures. Data are from Pauthenet [37]

of the material highly stress-sensitive. The origin of this property could be the unquenched orbital momentum of the A-sublattice Ni^{2+} ground state, which is discussed in Sect. 5.3. Manganese in the 3+ state helps to compensate magnetostriction through its local Jahn-Teller <100> axial expansion effect [38], and should be used as an additive where square hysteresis loops are desired. Although the Curie temperature of Ni ferrite is not as high as that of Li ferrite, the material in ceramic form is generally considered to be of better quality in a microstructural sense, i.e., it is denser and has more uniform grain size. Because of these properties, as well as because of its wide range of magnetization values, Ni ferrite ranks as one of the most-used ferrite families.

Manganese ferrites are mainly normal with site distributions indicated by the formula $Mn^{2+}{}_{0.8}Fe^{3+}{}_{0.2}\left[Mn^{2+}{}_{0.2}Fe^{3+}{}_{1.8}\right]O_4$. The Mn^{2+} ion has a larger radius than most of the ions of the $3d^n$ series, 0.80 Å instead of ~0.65 Å. To accommodate this larger cation, the lattice parameter is increased with the result that some of the b^2/U factors of the covalent bonding are reduced, thereby increasing the $\mathcal{N}_{BB}$ coefficient leading to a Curie temperature that is the lowest of all of the basic spinel ferrite families, as shown in Fig. 4.10. This effect is reflected in the values of the molecular field coefficients $\mathcal{N}_{ij}$, which reveal weaker interactions that involve Mn^{2+}, as shown in Table 4.2 where large reductions occur in the $\mathcal{N}_{AB}$ coefficients for Fe-Mn and Mn-Mn [24]. These results support the conclusions of Simsa and Brabers [39] that a canting angle of 54° among the Mn^{2+} ions in the B sublattice accounts for an initial deficiency in the magnetic moment at $T = 0\,K$.

For microwave applications, dielectric losses due to poor dielectric properties stem from the charge transfer mechanism that stabilizes Fe^{3+} in B sites, i.e., $Fe^{2+} + Mn^{3+} \rightarrow Fe^{3+} + Mn^{2+}$. There are also two possible polaronic conduction opportunities here: $Fe^{2+} \leftrightarrow Fe^{3+} + e^-$ and $Mn^{2+} \leftrightarrow Mn^{3+} - e^-$. In general, the Curie temperatures of this family are unacceptably low for applications with power dissipation effects that are likely to increase the temperature. Similar to Li and Ni ferrites, T_C is reduced further when Zn^{2+} is added to increase $4\pi M$.

Magnesium ferrite is mainly inverted with a typical formula $Mg^{2+}{}_{0.1}Fe^{3+}{}_{0.9}\left[Mg^{2+}{}_{0.9}Fe^{3+}{}_{1.1}\right]O_4$. The magnetization and permeability are too low for most low frequency applications, but the family has proven to be useful for certain microwave regimes in situations where the high temperature sensitivity is tolerable. In these cases, manganese is added for the purpose of suppressing any formation of Fe^{2+} through the charge transfer equation $Fe^{2+} + Mn^{3+} \rightarrow Fe^{3+} + Mn^{2+}$ [40,41]. As a consequence, this family is usually referred to as the magnesium-manganese ferrites. Final properties of ceramic versions are sensitive to the firing schedule because of the uncertain ionic site dispositions. The principal virtues of this hybrid family are its square hysteresis loops and insensitivity to stress made possible by the presence of Mn^{3+} ions. The compensation of magnetostriction is believed to result from a combination of J–T effects in B sites and a minority of S–O effects in A sites.

Cobalt ferrite is also an inverted spinel of formula $Fe^{3+}\left[Co^{2+}Fe^{3+}\right]O_4$, with Co occupying the B sites. For reasons that will be explained in subsequent chapters, this compound is of interest primarily for its magnetoelastic properties from which the phenomena of magnetocrystalline anisotropy and magnetostriction arise. Like Ni^{2+}, it can also feature a g factor greater than 2 in octahedral sites. Divalent cobalt in a B site is a classic spin-orbit stabilized ion that favors a compressive distortion along the c axis of the ligand octahedron, analogous to Ni^{2+} in tetrahedral sites. Unfortunately, these properties are generally undesirable for many conventional applications. Moreover, the strong coupling of Co^{2+} orbital states to the lattice is responsible for large power losses when this ferrite is used as a microwave propagation medium. In small amounts, however, Co^{2+} ions can be beneficial for adjusting anisotropy and magnetostriction, and in controlling relaxation effects that influence peak power limits.

Copper ferrite is another spinel that features a magnetoelastically active constituent Cu^{2+}, which like Mn^{3+} is a classic Jahn-Teller ion in an octahedral site. Also similar to Mn^{3+}, magnetoelastic and relaxation effects from a small fraction of Cu^{2+} in tetrahedral sites would also be expected from unquenched L in the upper T_2 triplet. The formula is generally inverted, in the form of $Fe^{3+}\left[Cu^{2+}Fe^{3+}\right]O_4$, but can in theory assume a monovalent state in ${Cu^{1+}}_{0.5}{Fe^{3+}}_{2.5}O_4$ analogous to Li ferrite, but with part of the Cu^{1+} ions occupying tetrahedral A sites [31]. As in the case of Ni^{2+} and Co^{2+} in octahedral sites, the g factor of the Cu^{2+} ion can be greater than 2, but because of its low spin value $S = 1/2$, Cu ferrites are not of high permeability and are rarely considered for microwave applications because their chemistry and microstructure in ceramic form are difficult to control. At high temperatures copper has a volatility comparable to that of lithium. Sintering temperatures are lower than typical for spinels, and the attainment of good stoichiometry is a challenge. Furthermore, because of the strong Jahn-Teller tendencies, the Cu^{2+} ion can drive the lattice symmetry tetragonal, as it does in $Fe^{3+}\left[Cu^{2+}Fe^{3+}\right]O_4$ (and also the superconducting cuprate perovskite host compounds such as ${La^{3+}}_2Cu^{2+}O_4$), and will also produce strong positive magnetostrictive effects when used as an additive to cubic lattices. Another drawback is the polaronic conduction mechanisms similar to those of Mn ferrite ($Cu^{2+} \leftrightarrow Cu^{3+} + e^-$ and $Cu^{2+} \leftrightarrow Cu^{1+} - e^-$) [42]. One interesting possibility is the use of monovalent copper in A sites to produce higher magnetization compounds for microwave applications. This subject is discussed in Appendix 4B. Table 4.5 lists a qualitative summary of spinel ferrite magnetization properties as well as magnetoelastic effects that are discussed in Chap. 5.

4.3.2 Garnet Ferrites $\{c_3\}\,[a_2]\,(d_3)\,O_{12}$

The garnet family is the best-behaved chemically and structurally of all ferrimagnetic oxides. From a magnetic standpoint, it is also the most thoroughly investigated. Following the first synthesis of the iron garnets by Bertaut and Forrat [43], careful analysis of the crystal structure was carried out by Geller and Gilleo [44, 45]. The three cation sites dodecahedral c, octahedral a, and tetrahedral d are shown in relation to a section of the crystallographic unit cell in Fig. 4.13. It can be readily seen that the principal symmetry axes of the a and d sites do not conform to the cubic cell axes <100>, <111>, and <110>, in marked contrast to the spinels which have the A and B sites neatly packed into the cubic cell. Only the octahedral site

Table 4.5 Initial effects of dilutant ions in inverted spinel ferrite $A^{3+}\left[B^{3+}B^{2+}\right]O_4$

Ions	Site	M_s	$\lvert K_1\rvert$	$\lvert K_1/M_s\rvert$	$\lvert \lambda_s/K_1\rvert$
Al^{3+}, Ga^{3+} [a]	B (A)	⇓	⇓	–	–
Mg^{2+} [b]	B	⇓	⇓	–	–
$Li^{1+}+2Ti^{4+}$ [c]	$A+2B$	⇓	⇓	–	–
Zn^{2+}	A	⇑	–	⇓	–
Co^{2+} [d]	B	–	⇓→⇑	⇓→⇑	⇑→⇓
Mn^{3+} [e]	B	–	–	–	⇓→⇑
Fe^{2+} [f]	B	–	⇑⇑	⇑⇑	–

[a] Al^{3+} and Ga^{3+} occupy both octahedral and tetrahedral sites, but opposite to the garnets, they initially have strong preference for the octahedral sites in the inverted spinels
[b] Mg^{2+} is the divalent-ion basis for the magnesium ferrite family
[c] $Li^{1+}+2Ti^{4+}$ combination is used in Li ferrite. The added Li^{1+} replaces Fe^{3+} in the tetrahedral sites, but the net result is a lowering of M_s by the $2Ti^{4+}$ ions diluting the octahedral sublattice
[d] Co^{2+} is used in very small amounts to compensate K_1. It will cause K_1 to pass from negative to positive at about 0.01–0.02 ions per formula unit. Because it is spin-orbit stabilized, it is also a fast-relaxing ion
[e] Mn^{3+} is a Jahn-Teller ion that influences magnetostriction by adding a positive contribution to the large λ_{100} constant that makes the average value λ_s pass through zero at a concentration of about 0.2 ions per formula unit
[f] Fe^{2+} is a spin-orbit stabilized ion that is a natural constituent of spinel ferrites. It enhances the negative anisotropy constant by favoring the $\langle 111\rangle$-axis extensions, and thereby increases the λ_{111} magnetostriction constant. In combination with host Fe^{3+} ions, it can provide a hopping electron conduction mechanism that is detrimental to high-frequency applications

has one of its trigonal axes aligned with a lattice <111> direction [46]. An even more important difference is that the tetrahedral sites are more numerous than the octahedral sites in the ratio of 3:2, thereby making the *d* sublattice dominant in contributing to the magnetization. As will be shown in Chap. 5, this feature allows the magnetoelastic properties to be varied over a wider range.

Dielectric properties of the garnets are generally superior to those of the spinels, due in part because all of the cations are naturally in the 3+ state, thereby affording less opportunity for Fe^{2+} to form. Magnetocrystalline anisotropy and magnetostriction are also manageable quantities. Moreover, there are a plethora of substitutions available for all three sites that may be used to tailor magnetic, microwave, and optical properties. Because the system is more refractory than the spinels, chemical stability is generally higher among conventional compounds, i.e., those based on yttrium or lanthanum series ions in the *c* sublattice. Garnet systems with lower sintering temperatures are the families with cation combinations {Y, Ca} [Fe, Zr] (Fe, V) and {Bi, Ca} [Fe, Zr] (Fe, V), and are used where lower magnetizations are acceptable and the high sublimation rate of Bi at solid reaction or sintering temperatures can be tolerated.

Yttrium iron garnet is the basic magnetic host of this family, and can be altered by a variety of substitutions. Al^{3+} and Ga^{3+} ions are used to reduce the magnetization by forming solid solutions of $Y_3Fe_5O_{12}$ (YIG) with either $Y_3Al_5O_{12}$ (YAG) or $Y_3Ga_5O_{12}$ (YGG). With these two ions, the dilution begins in the *d* sublattice and gradually spills over to the *a* sites [47,48]. For this reason Al^{3+} (or Ga^{3+}) are used to reduce the magnetization, as shown in Gilleo's data [16] of Fig. 4.14. To increase

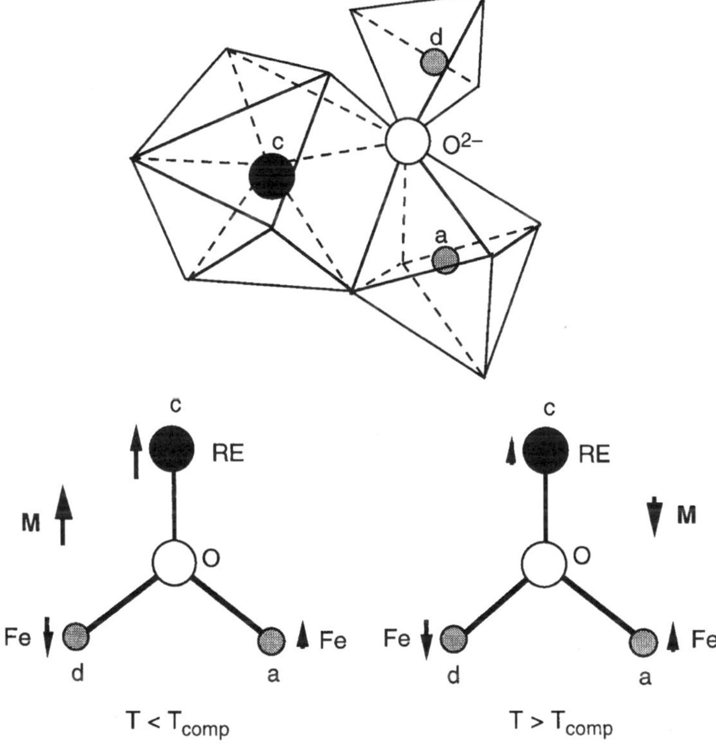

Fig. 4.13 Garnet crystal structure with bond angle diagrams, below and above the compensation temperature T_{comp} that occurs when the c-sublattice is occupied by heavy rare-earth ions of the $4f^n$ series

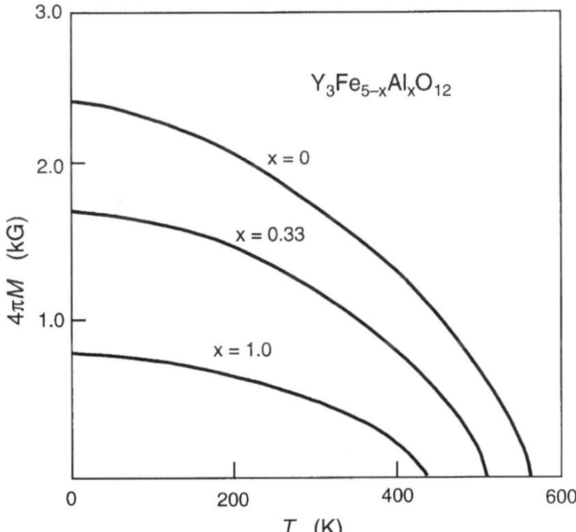

Fig. 4.14 Thermomagnetism curves of yttrium aluminum iron garnet ferrite, showing the effects of Al^{3+} net dilution of the majority tetrahedral d sublattice. Data are from Gilleo and Geller [16]

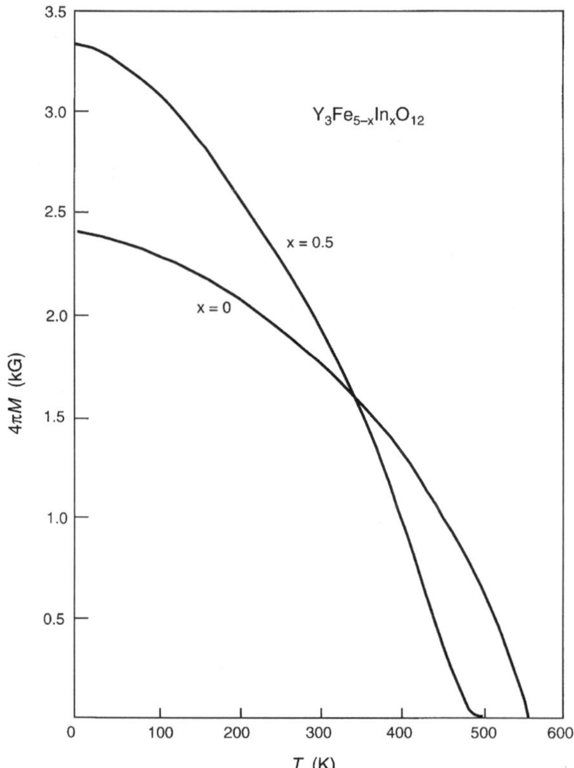

Fig. 4.15 Thermomagnetism curves of yttrium indium iron garnet ferrite, showing the effects of In^{3+} dilution of the minority octahedral a sublattice. Compare with Fig. 4.14. Data are from Gilleo and Geller [16]

magnetization, the larger ions In^{3+} (or Sc^{3+}) can be used up to a concentration level of about 0.5 per formula unit because they occupy *a* sites almost exclusively. With the greater reduction in the number of J_{ad} exchange couplings by *a*-site dilution; however, the Curie temperature drops quickly with dilution, similar to the case of Zn in the *A* sites of spinels. Examples of these effects are shown in the data of Fig. 4.15.

Yttrium-calcium iron garnet is a common modification of the basic YIG composition that has been considered for use in special microwave and magneto-optical applications with vanadium diluting *d* sites or zirconium *a* sites. The advantages of this family lie in two features: (1) V^{5+} and Zr^{4+} occupy exclusively their respective sites, which means that the adjustments in magnetization can be achieved without unwanted dilution of the opposite sublattice that would cause a further decrease in Curie temperature; and (2) calcium, vanadium, and zirconium are less expensive chemicals than the alternative yttrium, lanthanum, indium, or scandium. A further benefit of the use of calcium actually comes from the reduced amounts of yttrium

Table 4.6 Initial effects of dilutant ions in garnet ferrites $\left\{c_3^{3+}\right\}\left[a_2^{3+}\right]\left(d_3^{3+}\right)O_{12}$

Ions	Site	M_s	$\lvert K_1\rvert$	$\lvert K_1/M_s\rvert$	$\lvert \lambda_s/K_1\rvert$
Al^{3+}, Ga^{3+} [a]	d (a)	$\Downarrow$	$\Downarrow$	$\Downarrow$, $\Uparrow$	–
$2Ca^{2+} + V^{5+}$ [b]	$2c + d$	$\Downarrow$	$\Downarrow$	$\Downarrow$	–
$Ca^{2+} + Ge^{4+}$ [b]	$c + d$	$\Downarrow$	$\Downarrow$	$\Downarrow$	–
$Ca^{2+} + Si^{4+}$ [b]	$c + d$	$\Downarrow$	$\Downarrow$	$\Downarrow$	–
In^{3+}, Sc^{3+} [b]	a				
$Ca^{2+} + Zr^{4+}$ [b]	$c + a$	$\Uparrow$	$\Downarrow$	$\Downarrow\Downarrow$	$\Uparrow$
$Ca^{2+} + Sn^{4+}$ [b]	$c + a$	$\Uparrow$	$\Downarrow$	$\Downarrow\Downarrow$	$\Uparrow$
$Co^{2+} + Si^{4+}$ [c]	$a + d$	$\Uparrow$	$\Downarrow$	$\Downarrow\Downarrow$	$\Uparrow$
Mn^{3+} [d]	a	–	–	–	$\Downarrow_{\rightarrow}\Uparrow$
Bi^{3+}, Pb^{2+} [e]	c	–	$\Downarrow_{\rightarrow}\Uparrow$	$\Downarrow_{\rightarrow}\Uparrow$	$\Uparrow^{\rightarrow}\Downarrow$
Fe^{2+} [f]	a	$\Uparrow\Uparrow$	$\Uparrow\Uparrow$	–	$\Uparrow\Uparrow$

[a] Al^{3+} and Ga^{3+} occupy both octahedral and tetrahedral sites, but initially have strong preference for the tetrahedral sites in the garnets. Ga^{3+} reduces M_s more efficiently and can raise K_1/M_s

[b] Sr^{2+} can be used instead of Ca^{2+} in c sites, as well as La^{3+} instead of Y^{3+} for lattice parameter adjustments

[c] Co^{2+} is used in very small amounts to compensate K_1. It will cause K_1 to pass from negative to positive at about 0.02 ions per formula unit. Because it is spin-orbit stabilized, it is also a fast-relaxing ion

[d] Mn^{3+} causes a Jahn-Teller distortion that adds a positive contribution to the magnetostriction constants and makes the average value pass through zero at a concentration between 0.05 and 0.1 ions per garnet formula unit

[e] Bi^{3+} and Pb^{2+} are important for magnetooptical applications but can increase microwave losses

[f] Fe^{2+} is a spin-orbit stabilized ion that occurs as a natural impurity. It enhances the negative anisotropy constant by favoring the $\langle 111\rangle$-axis extensions, and thereby increases the λ_{111} magnetostriction constant. In combination with host Fe^{3+} ions, it can provide a hopping electron conduction mechanism that is detrimental to high-frequency applications

or lanthanum because these rare-earth elements can contain fast-relaxing impurities that are magnetically lossy to microwaves. One other ion that is used particularly for magneto-optical application is Bi^{3+} for reasons that will be explained in Chap. 7. This ion, however, is extremely large and does not freely substitute for Y^{3+} over the full range. In fact, the compound $Bi_3Fe_5O_{12}$ (BIG) may exist only as films of submicron thickness scale [49].

Table 4.6 presents a qualitative summary of magnetic and magnetoelastic properties of YIG-based garnets similar to that given for the spinels in Table 4.5. It is appropriate to mention at this point that in all of the magnetic garnets and spinels, magnetoelastic properties can be tailored by select ions identified previously. To reduce the negative anisotropy, very small amounts of Co^{2+} paired with a tetravalent diamagnetic ion can be substituted. To offset the negative magnetostriction, Mn^{3+} ions can be substituted directly for Fe^{3+}. This subject will be discussed in Chap. 5.

Other ions of the lanthanide group can be used generously in the c sublattice to form what amounts to a distinct set of ferrimagnetic compounds. Because the ionic spins that are introduced to the c sublattice exchange couple to the iron in the d and a sublattices, a wide variety of thermomagnetic, magnetoelastic, microwave, and optical properties can be created in these rare-earth iron garnets.

4.3.3 Rare-Earth Garnet Ferrites

When rare-earth ions with unfilled $4f^n$ shell are substituted into the c sublattice to form solid solutions of $(RE)_3Fe_5O_{12}$ and $Y_3Fe_5O_{12}$, the magnetic properties of the iron garnets take on a remarkably different character. The earliest studies on the magnetization behavior of this family were reported by Bertaut and Pauthenet [50, 51] whose data are replotted in Fig. 4.16 for the series from $Gd_3Fe_5O_{12}$ to $Yb_3Fe_5O_{12}$, with Y^{3+} included as a diamagnetic reference ion in the c sites. The most distinctive feature of these curves is the magnetization reversal that takes place at a compensation temperature. This occurs when the magnetic moment of the c sublattice equals and then surpasses the net moments of the opposing d and a sublattices, causing $\mathcal{M}_c$ to switch its alignment from $\mathcal{M}_d$ to $\mathcal{M}_a$ in the simple three-ion model of Fig. 4.17, according to

$$\mathcal{M} = |(\mathcal{M}_d - \mathcal{M}_a) - \mathcal{M}_c| . \tag{4.29}$$

To explain this effect analytically, the Néel theory must be expanded to three or more sublattices with the number of molecular field coefficients increasing from four to nine or more, according to

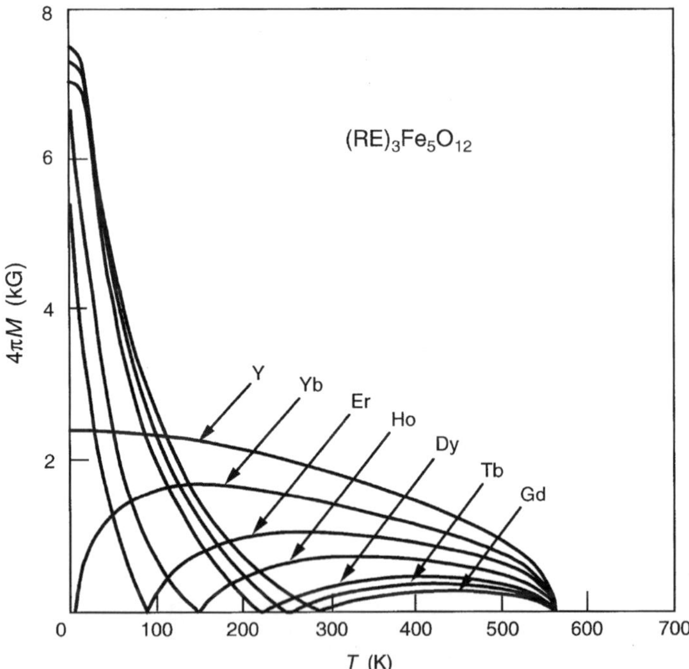

Fig. 4.16 Thermomagnetism curves of the rare-earth series occupying c sites in the iron garnet system $(RE)_3Fe_5O_{12}$. Curves are replotted from data of Bertaut and Pauthenet [50]

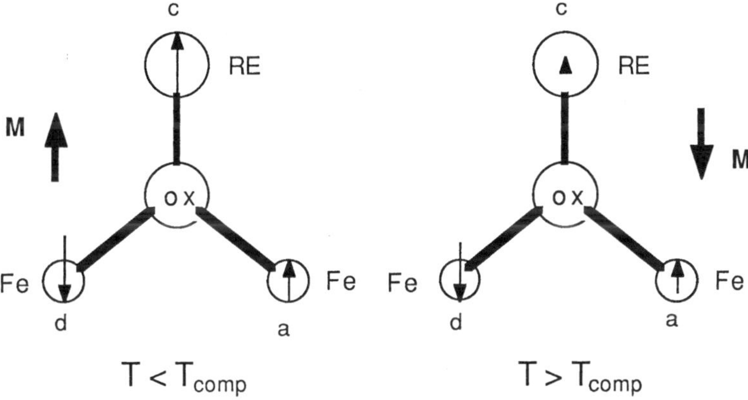

Fig. 4.17 The anatomy of the sublattice magnetic moment cancellation that causes the compensation temperature in rare-earth iron garnets

$$\begin{aligned}\mathcal{M}_d(T) &= \mathcal{M}_d(0)\,\mathcal{B}_{S_d}(a_d)\\ \mathcal{M}_a(T) &= \mathcal{M}_a(0)\,\mathcal{B}_{S_a}(a_a),\\ \mathcal{M}_c(T) &= \mathcal{M}_c(0)\,\mathcal{B}_{J_c}(a_c)\end{aligned} \tag{4.30}$$

where

$$\begin{aligned}\mathcal{M}_d(0) &= 3g_d m_{\mathrm{B}} S_d N_{\mathrm{A}}(1-k_d),\\ \mathcal{M}_a(0) &= 2g_a m_{\mathrm{B}} S_a N_{\mathrm{A}}(1-k_a),\\ \mathcal{M}_c(0) &= 3g_c m_{\mathrm{B}} J_c N_{\mathrm{A}}(1-k_c).\end{aligned} \tag{4.31}$$

The complete expressions for the Boltzmann energy ratios are given by

$$\begin{aligned}a_a(T) &= \frac{m_a H_{\mathrm{ex}}^{(a)}}{kT} = \frac{g_a m_{\mathrm{B}} S_a}{kT}\left[\mathcal{N}_{ad}\mathcal{M}_d + \mathcal{N}_{aa}\mathcal{M}_a + \mathcal{N}_{ac}\mathcal{M}_c\right],\\ a_c(T) &= \frac{m_c H_{\mathrm{ex}}^{(c)}}{kT} = \frac{g_c m_{\mathrm{B}} J_c}{kT}\left[\mathcal{N}_{cd}\mathcal{M}_d + \mathcal{N}_{ca}\mathcal{M}_a + \mathcal{N}_{cc}\mathcal{M}_c\right],\\ a_d(T) &= \frac{m_d H_{\mathrm{ex}}^{(d)}}{kT} = \frac{g_d m_{\mathrm{B}} S_d}{kT}\left[\mathcal{N}_{dd}\mathcal{M}_d + \mathcal{N}_{da}\mathcal{M}_a + \mathcal{N}_{dc}\mathcal{M}_c\right].\end{aligned} \tag{4.32}$$

Figure 4.18 illustrates the thermomagnetic contribution of the rare-earth ions from the c sites, which are seen to exchange couple only to the a and d sites and not to each other.

In the relations involving the c sublattice, the magnetic moment is determined by the total angular momentum J_c instead of only the spin component S_c.

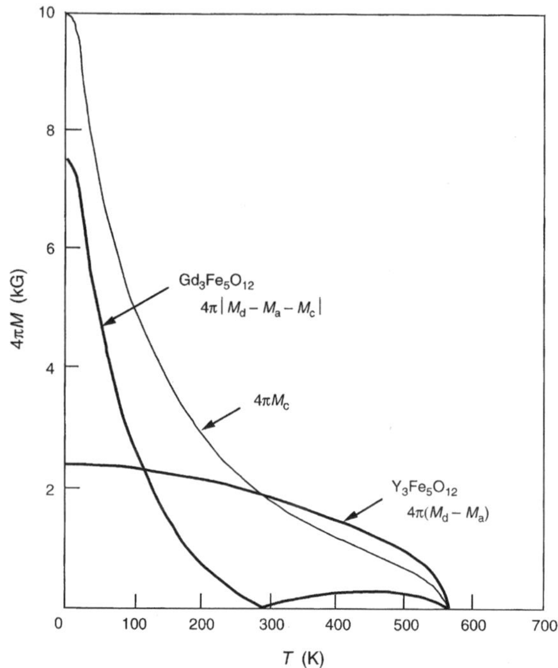

Fig. 4.18 Thermomagnetism curves of gadolinium-iron garnet ($Gd_3Fe_5O_{12}$) and yttrium-iron garnet ($Y_3Fe_5O_{12}$), showing the contribution of the Gd^{3+} sublattice and indicating the formation of the compensation temperature T_{comp}. The curve for the Gd sublattice was obtained by a simple arithmetic subtraction, which is not strictly accurate but adequate for illustrative purposes

Because the rare-earth ions of the c sublattice have the unpaired spins in the $4f^n$ shell, the orbital angular momentum is largely unquenched by the crystal field, as discussed in Sect. 2.3. Unlike the $3d^n$ series where the crystal field strength $10Dq \sim 1\,\text{eV}$, the crystal fields to which the unpaired $4f$ electrons are exposed are only $\sim 10^{-2}\,\text{eV}$ $\left(\sim 100\,\text{cm}^{-1}\right)$. This means that the vector sum $\boldsymbol{J}_c$ remains a "good" quantum number. For this reason, it must be used in the expression for the magnetic moment of the c sublattice, and also in the expression for g_c based on (1.19)

$$g_c = 1 + \frac{J_c\,(J_c+1) + S_c\,(S_c+1) - L_c\,(L_c+1)}{2J_c\,(J_c+1)}. \tag{4.33}$$

For the rare-earth ions of the upper half of the $4f^n$ series (heavy rare earths, $n > 7$), the Russell-Saunders coupling scheme for the multiplet structure dictates that the ground state $J_c = |L_c + S_c|$ and (4.33) conveniently reduces to

$$g_c = \frac{L_c + 2S_c}{L_c + S_c}. \tag{4.34}$$

It is instructive to point out that while the $\boldsymbol{L}_c$ and $\boldsymbol{S}_c$ vectors are collinear for the heavy rare-earth ions, they oppose each other in the lighter ($n < 7$) ions, with $J_c = |L_c - S_c|$ according to the required multiplet ordering. Because it is only $\boldsymbol{S}_c$ that must align antiparallel with the superexchange-coupled sublattices, $\boldsymbol{L}_c$ will automatically tend to be parallel to the neighboring spins, thereby reversing the alignment of $\boldsymbol{J}_c$ to $\boldsymbol{S}_d$ and $\boldsymbol{S}_a$, if $L_c > S_c$. In other words, the sign of $\mathcal{M}_c$ in (4.31) will change to positive. Since this condition always applies in the rare-earth group, we can substitute $J_c = |L_c - S_c|$ into (4.33) to obtain

$$g_c = \frac{L_c - 2S_c + 1}{L_c - S_c + 1}. \tag{4.35}$$

Equations (4.33)–(4.35) apply well to the rare-earth series, but have no validity in the $3d^n$ group because of the breakdown in Russell-Saunders coupling due to crystal field lifting the L_c degeneracy. This topic will be discussed in Chap. 5.

The fact that $\boldsymbol{S}_c$ alone interacts with the exchange fields raises a conceptual question when J_c is introduced to $\mathcal{M}_c$ in the molecular field terms $\mathcal{N}_{dc}\mathcal{M}_c, \mathcal{N}_{ac}\mathcal{M}_c$, and $\mathcal{N}_{cc}\mathcal{M}_c$. If the model is formulated to maintain the spin-only rigor [52], the coefficients will scale to larger values by the ratio of J_c/S_c in order to be consistent with the relations of (4.32). For academic interest, a spin-only version of the above model is outlined in Appendix 4C.

The earliest attempt to determine the molecular field coefficients for the rare-earth series was reported by Aléonard from reduction of paramagnetic susceptibility data above the Curie temperature [53] employing the method described in Sect. 4.1.2. Although efforts to apply these results to fit the magnetization vs. temperature measurement data below the Curie temperature were unsuccessful, Aléonard's values served as useful starting points for Dionne's solution [22, 54] that was carried out in the manner described previously for diluting the $Y_3Fe_5O_{12}$ host system. In Table 4.7 molecular field values are listed, including a set reported by Brandle and Blank [55] who also used the approach of Dionne.

At the lowest temperatures, there remains a controversy in the initial ($T = 0$ K) magnetization of $Tm_3Fe_5O_{12}$ and to a lesser extent that of $Yb_3Fe_5O_{12}$. Because the steepness of the slope of $\mathcal{M}_c$ vs. T could have allowed the overlooking of compensation points very close to the lowest measurement temperature of 4.2 K, misinterpretation of the data is a distinct possibility. This problem is illustrated in the data Fig. 4.19. The issue is whether the c-sublattice contribution to the measured $\mathcal{M}$ is parallel or antiparallel to $\mathcal{M}_c$, i.e., whether $\mathcal{M}_c$ takes a positive or negative value in (4.29). A summary of molecular field coefficients derived from various values of $\mathcal{M}_c$ for $Tm_3Fe_5O_{12}$, including a proposed case that might fit if the measurements were extended down to $T = 0$ K. A similar effort is made for $Yb_3Fe_5O_{12}$. For a more comprehensive discussion of this question the reader again is directed to the review article by Geller [15]. This subject will be continued in the next section as part of the analysis of another peculiar effect.

Table 4.7 Molecular field coefficients of rare-earth ions in garnet ferrites*

	Dionne [54]		Brandle-Blank [55]		Aléonard [53]	
Ion	$\mathcal{N}_{dc}$	$\mathcal{N}_{ac}$	$\mathcal{N}_{dc}$	$\mathcal{N}_{ac}$	$\mathcal{N}_{dc}$ [a]	$\mathcal{N}_{ac}$ [a]
Gd^{3+}	6.00	−3.44	3.40	−1.20	–	–
Tb^{3+}	6.50	−4.20	4.60	−4.40	3.40	−1.80
Dy^{3+}	6.00	−4.00	3.60	−3.20	3.95	−3.35
Ho^{3+}	4.00	−2.10	2.40	−4.00	1.50	−0.75
Er^{3+}	2.20	−0.20	1.00	−0.60	1.25	−0.75
Tm^{3+} [b]	17.0	−1.00	–	–	8.00	−1.00
Tm^{3+} [c]	10.4	−0.61	–	–	–	–
Tm^{3+} [d]	9.2	−0.54	–	–	–	–
Tm^{3+} [e]	6.1	−0.36	–	–	–	–
Yb^{3+} [b]	8.0	−4.00	8.80	−1.00	2.00	−1.70
Yb^{3+} [e]	6.8	−3.4	–	–	–	–

*All coefficients are expressed in units of mol/cm^3
[a]Derived from paramagnetic susceptibility measurements [53]
[b]Dionne values based on data of Geller et al. [56]; Aléonard values were reduced from his own data [53]
[c]Dionne values based on reinterpretation of data of Geller et al. [56]
[d]Dionne values based on data of Bertaut and Pauthenet [50]
[e]Proposed values; see footnotes to Table 4.8

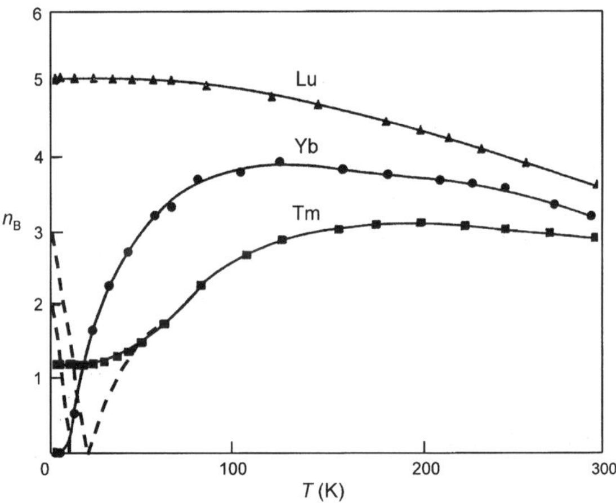

Fig. 4.19 Details thermomagnetism curves of Lu, Tm, and Yb iron garnet systems below $T = 300$ K. Dashed curves indicate alternative interpretations. Data are from Geller [15]

4.3.4 *Rare-Earth Canting Effect*

Before leaving this topic, we must pay attention to an important effect that emerges from the analysis of the rare-earth magnetic garnets. As can be noted in the $J_{c'}/J_c$ column of Table 4.8, Dionne's original reduction of Geller et al. data [56] indicates values of the total angular momentum that are less than the nominal rare earth

Table 4.8 Rare-earth ion angular momentum in garnet ferrites (based on data at $T = 4.2\,\mathrm{K}$)

Ion	L_c	S_c	J_c	g_c	$J_{c'}$	$J_{c'}/J_c$	ϕ (deg.)	γ	g_c''	J_c''
Gd^{3+}	0	3.5	3.5	2	3.5	1.0	0	–	2.0	3.5
Tb^{3+}	3	3	6	3/2	4.6	0.84	40.0	0.32	1.73	4.0
Dy^{3+}	5	2.5	7.5	4/3	5.3	0.71	45.0	0.41	1.54	4.6
Ho^{3+}	6	2	8	5/4	5.0	0.62	51.5	0.37	1.49	4.2
Er^{3+}	6	1.5	7.5	6/5	4.6	0.61	52.0	0.42	1.38	4.0
Tm^{3+} [a]	5	1	6	7/6	1.1	0.18	79.6	–	–	–
Tm^{3+} [b]	5	1	6	7/6	1.77	0.30	72.5	0.01	1.95	1.06
Tm^{3+} [c]	5	1	6	7/6	2.0	0.33	70.7	0.07	1.75	1.33
Tm^{3+} [d]	5	1	6	7/6	3.0	0.50	60.0	0.30	1.40	2.5
Yb^{3+} [e]	3	0.5	3.5	8/7	1.5	0.43	64.8	0.24	1.42	1.21
Yb^{3+} [f]	3	0.5	3.5	8/7	1.75	0.50	60.0	0.33	1.33	1.5

[a] Values based on $n_B = 1.2$ from data of Geller et al. [56]; This moment assumes no compensation point. Corresponding g_c'', J_c'', and L_c'' values were interpreted as negative or unrealistic, suggesting a possible misjudgment in the reduction of the data
[b] Values based on $n_B = -1.2$ from reinterpretation of Geller et al. data. In this case the same $\mathcal{M}$ value is used but a compensation point is assumed
[c] Values based on $n_B = -2$, deduced by Bertaut and Pauthenet [50]
[d] Proposed values based on $n_B = -5$ at $T = 0\,\mathrm{K}$ to provide better fit to the monotonic trends
[e] Values based on $n_B \sim 0$ from data of Geller et al. [56]
[f] Proposed values based on $n_B = -3$ at $T = 0\,\mathrm{K}$ to provide better fit to the monotonic trends

free-ion values, i.e., $\boldsymbol{J}_{c'} < \boldsymbol{L}_c + \boldsymbol{S}_c$. For a decrease to be realized in the effective value of J_c in the iron garnet lattice, orbital degeneracy would be lifted through direct coupling of the $4f$ lobes to the $2p$ orbitals of the oxygen ligands and the $3d$ lobes of the Fe^{3+} ions in the d and a sites. The net result of these interactions would provide a crystal field and molecular orbital stabilization of the ground J_c term sufficient to partially quench the orbital magnetic moment, as described in Sect. 2.3.

We consider two ways to view this phenomenon: (1) semiclassically with the magnetic moments associated with the rare-earth ions $\mathcal{M}_c = g_c m_B J_c$ simply canted as in the case of iron spins in diluted sublattices described previously in Sect. 4.2 where the canting is characterized by $\cos\phi = J_c'/J_c$, and (2) quantum mechanically with the orbital angular momentum partially quenched, leaving a reduced orbital component $\boldsymbol{L}_c''$, such that $\boldsymbol{J}_c'' = \boldsymbol{L}_c'' + \boldsymbol{S}_c$.

In the semiclassical case, we assume that the orbital angular momentum is not quenched because spin-orbit coupling is the dominant interaction. The canting of $\boldsymbol{J}_c$ will then be determined by the relative strengths of the exchange interaction between $\boldsymbol{S}_c$ with the net spin moment of the iron sublattices and the Stark coupling between $\boldsymbol{L}_c$ and the c-site crystal field. In the model diagrammed in Fig. 4.20, the exchange field $\boldsymbol{H}_{\mathrm{ex}}$ competes with the crystal field vector $\boldsymbol{E}_{\mathrm{cf}}$ (assumed to be orthogonal to $\boldsymbol{H}_{\mathrm{ex}}$ for illustrative purposes) for the $\boldsymbol{J}_c$ vector. Stabilization of the combined energies as a function of canting angle ϕ is related by

$$\begin{aligned} E(\phi) &= e\boldsymbol{E}_{\mathrm{cf}}\cdot\boldsymbol{L}_c + g_c m_B \boldsymbol{S}_c\cdot\boldsymbol{H}_{\mathrm{ex}} + g_c m_B \boldsymbol{J}_c\cdot\boldsymbol{H} \\ &= eE_{\mathrm{cf}}L_c\sin\phi + g_c m_B\left(S_c H_{\mathrm{ex}} + J_c H\right)\cos\phi, \end{aligned} \tag{4.36}$$

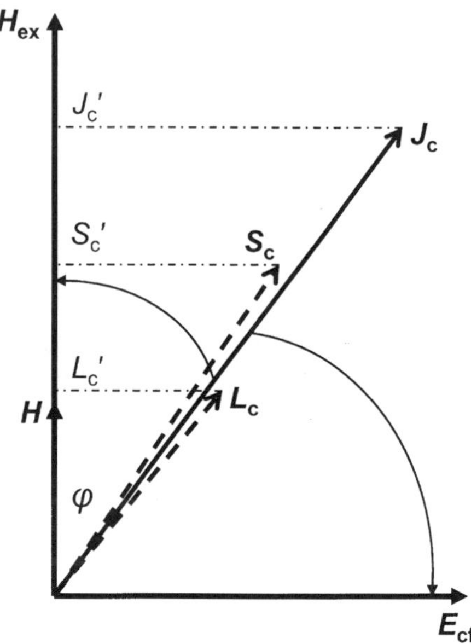

Fig. 4.20 The J_c' effect and the meaning of the canting angle ϕ

where the exchange field term is from the spin-only version defined in Appendix 4C and $\boldsymbol{H}$ is an applied field that couples both $\boldsymbol{L}_c$ and $\boldsymbol{S}_c$, thereby tending to offset the canting effects of $\boldsymbol{E}_{\text{cf}}$ on $\boldsymbol{L}_c$. A value of ϕ then follows from minimization of $E(\phi)$ according to $\partial E(\phi)/\partial\phi = 0$:

$$\cos\phi \approx \frac{g_c m_{\text{B}}(S_c H_{\text{ex}} + J_c H)}{\sqrt{[g_c m_{\text{B}}(S_c H_{\text{ex}} + J_c H)]^2 + (eE_{\text{cf}} L_c)^2}}. \tag{4.37}$$

For $eE^{\text{cf}} \approx g_c m_{\text{B}} H_{\text{ex}}$ (experiments indicate that both crystal fields and exchange fields produce energy stabilizations of about 10^{-2} eV), and $H_{\text{ex}} >> H$, (4.37) reduces to

$$\cos\phi \approx \frac{S_c}{\sqrt{S_c^2 + L_c^2}}. \tag{4.38}$$

For $H_{\text{ex}} << H$, $\cos\phi \to 1$ because the applied $\boldsymbol{H}$ field overrides the crystal field and realigns $\boldsymbol{L}_c$ to $\boldsymbol{S}_c$. For energies $\sim 10^{-2}$ eV, H would have to approach values of 10 T.

As a physical rationale for the reduction of J_c, this simplified picture can be useful. It cannot, however, take into account the crystalline anisotropy of $\boldsymbol{E}_{\text{cf}}$ that would manifest itself through a variation of J_c with the direction of an applied magnetic field $\boldsymbol{H}$. Moreover, according to the above model J_c' should equal J_c with $\boldsymbol{H}$ parallel to $\boldsymbol{E}_{\text{cf}}$, a result that has not been observed in measurements with single crystals. To provide a better interpretation for the actual measurements, we proceed to examine the crystal field influence in more detail.

For a quantum mechanical approximation, angular momentum quenching discussed in Sect. 2.3.6 can be considered. With the $3d^n$ orbitals, the crystal field Hamiltonian energy dominates spin-orbit coupling or $\mathcal{H}_{\text{cf}} >> \mathcal{H}_{LS}$, and $\boldsymbol{L}_c$ has a minor role in magnetic phenomena. With the rare-earth $4f^n$ ions, however, $\mathcal{H}_{LS} > \mathcal{H}_{\text{cf}}$ and $\boldsymbol{L}_c$ is fully active magnetically as part of $\boldsymbol{J}_c$, particularly where the ion is isolated as in paramagnetic systems. When placed in a ferrimagnetic lattice, however, the $\boldsymbol{J}_c$ vector and its $\boldsymbol{L}_c$ and $\boldsymbol{S}_c$ components are subjected to additional influences, according to the perturbation Hamiltonian

$$\begin{aligned} \mathcal{H}_1 &= \mathcal{H}_{LS} + \mathcal{H}_{\text{cf}} + \mathcal{H}_h + \mathcal{H}_{\text{ex}} \\ &= \lambda \boldsymbol{L}_c \cdot \boldsymbol{S}_c + e\boldsymbol{E}_{\text{cf}} \cdot \boldsymbol{J}_c + \boldsymbol{g}_c m_{\text{B}} \boldsymbol{J}_c \cdot H + \boldsymbol{g}_c m_{\text{B}} \boldsymbol{S}_c \cdot \boldsymbol{H}_{\text{ex}} \\ &= \lambda L_c S_c + \mathcal{V}_{\text{cf}} J_c'' + g_c'' m_{\text{B}} \left(L_c'' + S_c \right) H + g_c'' m_{\text{B}} S_c H_{\text{ex}} \end{aligned} \tag{4.39}$$

Equation (4.39) mirrors (4.36), except for the $\mathcal{V}_{\text{cf}} J_c''$ term that represents a splitting of the angular momentum degeneracy instead of a classical precession of the $\boldsymbol{J}_c$ vector. Here the angular momentum is designated as $\boldsymbol{J}_c''$ instead of the J_c' of Dionne's original model because we now consider an actual reduction in $\boldsymbol{J}_c$ rather than simply a trigonometric component. Thus it is appropriate to also redefine to g_c'' such that $g_c'' m_{\text{B}} J_c'' = g_c m_{\text{B}} J_c'$, which represents the magnetic moment component along the z direction of measurement. Since $\boldsymbol{S}_c$ is not influenced by the crystal field, in reality L_c is reduced to L_c''. We restate that the $\mathcal{H}_{\text{ex}}$ term is expressed in the spin-only format of Appendix 4C to emphasize that $\boldsymbol{S}_c$ is the only relevant momentum vector (although the b^2/U covalent stabilization probably influences the magnitude of the $\boldsymbol{J}_c$ quenching as part of the crystal field effects). Finally, the interaction between the applied magnetic field $\boldsymbol{H}$ and the total magnetic moment now embodied in the $J_c'' = L_c'' + \boldsymbol{S}_c$ vectors is expressed by $\mathcal{H}_{\text{h}}$.

Because of the complexity of the matrix solution involved in carrying out degenerate perturbation theory with operations that will mix the eigenstates at each stage of the procedure, no attempt will be made at a formal solution. However, a qualitative projection can still be made by examining the influence of the different perturbation terms.

In a crystalline environment, the rare-earth ion undergoes a lifting of its angular momentum degeneracy in a manner determined by the analysis of Lea, Leask, and Wolf [57], from which the ground state terms are listed in Table 4.9. An example of the competing perturbations in a rare-earth ion occupying a cubic oxygen site is given by the sketches of the energy level structures for the Ho^{3+} ion with ground term $^5\text{I}_8$ in Fig. 4.21. For this model, we first recognize from (4.39) that the largest term is spin-orbit coupling, typically on the order of 10^{-1} eV, which splits the J_c degeneracy into levels running from $|L_c + S_c|$ up to $|L_c - S_c|$ for the group from $4f^7$ to $4f^{13}$. The next three terms are smaller by at least an order of magnitude and each affects a different part of the J_c degeneracy. The $\mathcal{H}_{\text{cf}}$ stabilization will lift the $^5\text{I}_8$-state degeneracy ($J_c = 8$) by creating states of lower angular momentum, in this case a ground triplet $T_{2\text{g}}$ and various higher energy terms. Despite the small energy level separations between these crystal field splittings ($\mathcal{V}_{\text{cf}} \sim 10^{-2}$ eV or 100 cm^{-1}),

Table 4.9 Rare-earth ion ground terms of quenched J angular momentum in O_h crystal fields

Rare-earth ion	$\lambda L \cdot S$ term	$J_c = \lvert L_c - S_c \rvert$	Cubic ground term Mulliken	Cubic ground term Bethe
Ce^{3+}	$^2F_{3/2}$	3/2	G_g	Γ_8
Pr^{3+}	$^3H_4$	4	A_{1g}	Γ_1
Nd^{3+}	$^4I_{9/2}$	9/2	$E_{1/2g}$	Γ_6
Pm^{3+} [a]	$^5I_4$	4	A_{1g}	Γ_1
Sm^{3+}	$^6H_{5/2}$	5/2	G_g	Γ_8
Eu^{3+}	$^7F_0$	0	–	–
		$J_c = \lvert L_c + S_c \rvert$		
Gd^{3+}	$^8S_{7/2}$	$S_c = 7/2$	–	–
Tb^{3+}	$^7F_6$	6	T_{2g}	Γ_5
Dy^{3+}	$^6H_{15/2}$	15/2	$E_{5/2g}$	Γ_7
Ho^{3+}	$^5I_8$	8	T_{2g}	Γ_5
Er^{3+}	$^4I_{15/2}$	15/2	$E_{5/2g}$	Γ_7
Tm^{3+}	$^3H_6$	6	T_{2g}	Γ_2
Yb^{3+}	$^2F_{7/2}$	7/2	$E_{1/2g}$	Γ_6

Ground terms taken from the computations of Lea, Leask, and Wolf [57]
[a]Synthetic element

the net result is a significantly reduced effective orbital angular momentum $J_c'' < J_c$ that carries with it anisotropy consistent with the lattice c-site symmetry. As stated above, the $\mathcal{H}_{ex}$ perturbation stabilizes only $\boldsymbol{S}_c$, while $\mathcal{H}_h$ stabilizes both $\boldsymbol{L}_c$ and $\boldsymbol{S}_c$, thereby restoring the integrity of $\lambda \boldsymbol{L}_c \boldsymbol{S}_c$.

To account for reduction of L_c due to crystal field interactions, Van Vleck first reasoned that $L_c'' = \gamma L_c$ and $J_c'' = \gamma L_c + S_c$, where the orbital quenching parameter $0 \leq \gamma \leq 1$, and then derived the relation for a modified g_c according to [58]

$$\begin{aligned} g_c'' &= \gamma + (2-\gamma)\frac{J_c''\left(J_c''+1\right) + S_c\left(S_c+1\right) - L_c''\left(L_c''+1\right)}{2J_c''\left(J_c''+1\right)} \\ &= \gamma + (2-\gamma)\frac{S_c}{\gamma L_c + S_c}. \end{aligned} \tag{4.40}$$

After recalculations from the original data, γ, g_c'', and J_c'' entries can be made to Table 4.8 based on the relation

$$\gamma = \frac{(L_c + 2S_c)\cos\phi - 2S_c}{L_c} = \frac{(L_c + 2S_c)\left(J_c'/J_c\right) - 2S_c}{L_c}. \tag{4.41}$$

These results reveal a decreasing monotonic trend for J_c'/J_c, J_c''/J_c and a fairly narrow range of γ values ($0.3 \leq \gamma \leq 0.4$) through the upper half of the $4f^n$ series. From (4.41) an expression for $\cos\phi$ can be deduced as

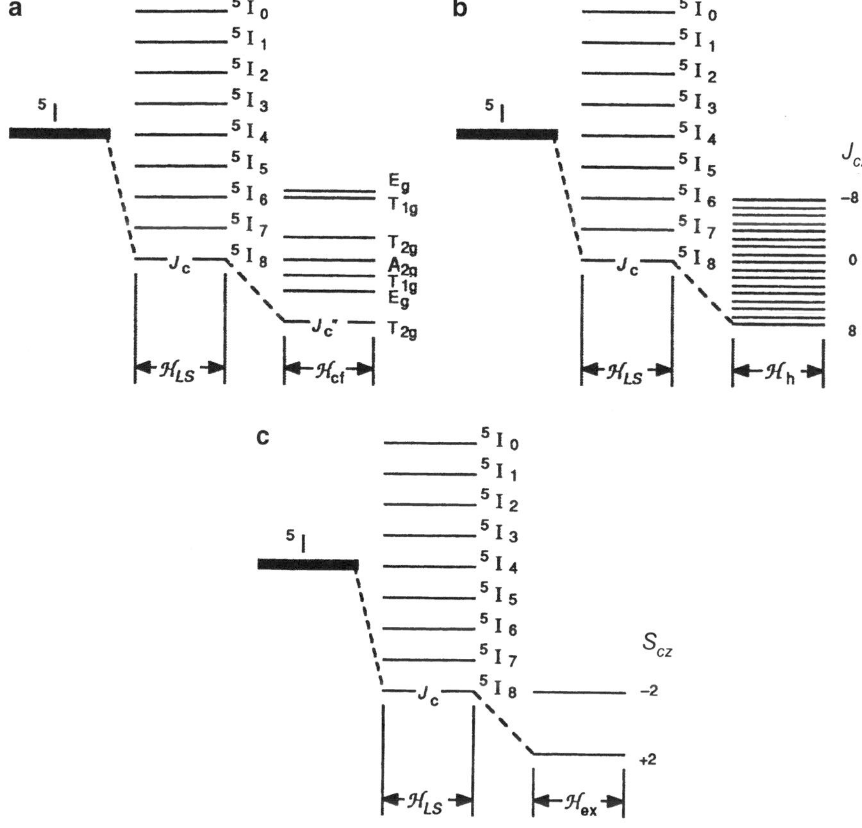

Fig. 4.21 Energy-level model of Ho^{3+} in iron garnet, isolating the three principal perturbations that follow the spin-orbit coupling $\mathcal{H}_{LS}$, i.e., the multiplet structure. (**a**) the crystal field $\mathcal{H}_{cf}$, (**b**) the Zeeman effect $\mathcal{H}_h$ on the total J_c in an magnetic field; and (**c**) the exchange splitting of the spin only bonding and antibonding states. In this diagram, $\mathcal{H}_{ex}$ is treated as a scalar energy so that $S_{cz} = \pm 2$ for a complete ionic spin flip consistent with the discussion in Sect. 7.1

$$\cos\phi = \frac{\gamma L_c + 2S_c}{L_c + 2S_c}. \tag{4.42}$$

If we compare (4.42) with the previous expression derived from (4.38), we note that both relations point to a increasing ϕ trend with decreasing S_c that supports the experiment-derived values in Table 4.8.

Experiments with $Ho_3Fe_5O_{12}$ compounds in high magnetic fields have confirmed that the J_c'/J_c ratio expressed in terms of canting angle $\phi = 51.5°$ is a plausible explanation [59]. In applied magnetic fields exceeding 10 T, J_c' (or J_c'') returned to its uncanted value as the applied $\boldsymbol{H}$ offset the crystal field quenching of $\boldsymbol{L}_c$ and restored $\boldsymbol{J}_c$ as a good quantum number. In addition to the decoupling effects on $\lambda \boldsymbol{L}_c \cdot \boldsymbol{S}_c$, H_{cf} introduce anisotropy effects in $\boldsymbol{J}_c'$ that correspond to the lattice symmetry. Anisotropy of $\boldsymbol{J}_c'$ following the [100, 111], and [110] axes were also reported [60].

Another important finding from these experiments that is consistent with the sublattice canting discussion of Sect. 4.2 was that the molecular field coefficients $\mathcal{N}_{dc}$ and $\mathcal{N}_{ac}$ are reduced by a frustration factor $(1 - \alpha k_a)$, where $\alpha = 0.185$ for Sc^{3+} ions substituted into the a sublattice. The value of $\phi = 51.5°$ at low H was found to be insensitive to k_a, thereby confirming the above contention that the exchange field from the iron sublattices does not contribute to the J_c reduction and that the effect is caused by crystal field quenching of the orbital component of the total angular momentum. It is also noteworthy that the value of ϕ decreased with applied $\boldsymbol{H}$ and declined more sharply at fields above 10 T, beginning with the larger values of k_a (representing the smallest $\mathcal{H}_{ex}$), as expected.

When one considers the breadth of the practical applications of these versatile materials, it is not surprising that the studies of the magnetic garnets have become exhaustive, as documented in the book by Winkler [61]. In subsequent chapters, these materials will be examined in terms of their magnetoelastic, microwave, and magneto-optical properties for which they have found use in their technology of information storage, radar, and communications systems. Before the focus is shifted to these magnetic related phenomena, the basic structural and magnetization properties of the magnetically hard hexagonal ferrites will be included.

4.3.5 Hexagonal Ferrites

The spinel and garnet ferrites discussed in the previous sections have proven to be rich in physics because their cubic crystal structure has made both measurement and theoretical analysis easy compared with another technologically important class of ferrimagnetic oxides that features a crystallographic axis of sixfold symmetry (designated as the c axis and usually selected as the z direction in a Cartesian system). These hexagonal ferrites, "hexaferrites" as they are commonly called, comprise the same Fe^{3+} sites as the cubic ferrites: octahedral and tetrahedral. In addition, they also contain a low density of Fe^{3+} trigonal bipyramid sites (see Fig. 2.7) that enhance the uniaxial or planar anisotropy that is strongly reflected in their magnetic properties.

Although this family has not been a vehicle of choice for the exploration of basic physical knowledge, and in some respects remains uncharted territory with regard to the origins of its properties, at least an introductory treatment is mandated by this book because of their widespread application. In this section, we shall outline the salient features of the family and how it contrasts with the cubic systems.

The hexaferrite family consists of several structures comprising a stack of alternating building blocks shown in Fig. 4.22 for a type-M lattice structure, beginning with a spinel cell, labeled S, but aligned with the c axis along a <111> body diagonal. As a consequence, the ferromagnetic spin arrangements of the spinel are in place automatically, but with only one easy (or hard) magnetic direction instead of four in a cubic structure. Moreover, the symmetry is already threefold in this

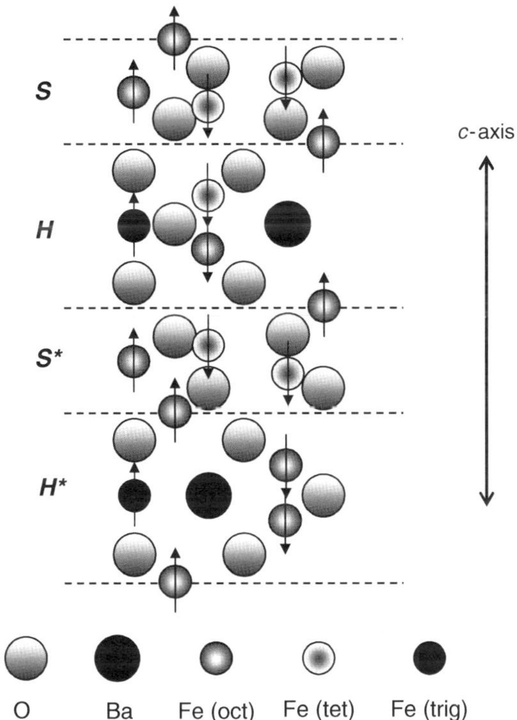

Fig. 4.22 Cross section of a unit cell of the magnetoplumbite M-type hexagonal ferrite structure $BaFe_{12}O_{19}$. The *H* blocks are hexagonal and *S* are spinel. As presented, H* represents a partial rotation about the *c* axis and S* indicates a rotation orthogonal to the *c* axis. The spheres are as labeled on the image. Note that the Fe^{3+} ions with vertical hatching occupy a ligand coordination that forms a trigonal bipyramid. These sites, which are due to the absence of an O^{2-} ion, occur as only one among the 12 iron sites, but are likely responsible for at least part of the sixfold axial symmetry of the magnetic properties. Image is adapted from Braun [66]

direction. To obtain the sixfold symmetry, the S blocks are systematically rotated by 180° relative to one another with the rotated designation given by S*. In the stack are also blocks or layers that contain large cations, Ba^{2+} (radius 1.36 Å), Sr^{2+} (1.16 Å), Pb^{2+} (1.18 Å) or combinations thereof. These blocks can also contain the fivefold oxygen-coordinated trigonal bipyramid sites which occur because only one Fe^{3+} ion is available for two tetrahedral sites due to the presence of the large cation in these layers. Additional blocks in the figure feature hexagonal symmetry and are labeled accordingly as H (or H*).

The most common hexaferrite barium (M-type) is a solid solution $BaO \cdot 6Fe_2O_3$ written in molecular form $BaFe_{12}O_{19}$ and derived from the mineral magnetoplumbite $PbFe_{7.5}Mn_{3.5}Al_{0.5}Ti_{0.5}O_{19}$. For technical applications Sr is regularly used with Ba because it provides slightly better hysteresis loop properties. Crystallographically, this compound is described as HSH*S* with 10 atomic layers per unit cell with a c-axis length of 23.2 Å. Magnetically, the 12 Fe^{3+} ions per formula unit consist of the single trigonal-site ion[3] in the H block directed up and the 2

[3] As pointed out by Smit and Wijn [62], the source of very large magnetocrystalline anisotropy in Ba ferrite (type M) is a not explained by a conventional spin dipole-dipole alignment mechanism. Because all of the magnetic ions are S-state ($L = 0$) Fe^{3+}, there should neither be first-order spin-orbit or Jahn-Teller stabilizations. A curious suggestion regarding the single highly distorted ligand coordination is that the trigonal crystal field might stabilize the d^5 electrons into a low-spin

Table 4.10 Hexagonal ferrite structural and magnetic data

Composition	Type	Block order	$4\pi M_s$ (300 K)[a] (kG)	H_K (300 K)[b] (kOe)
$(Ba,Sr)Fe_{12}O_{19}$	M	RSR^*S^*	4.8	17–19
$BaMe_2Fe_{16}O_{27}$	W (= MS_2)	$RS_2R^*S_2^*$	4.0– 4.8	2.2–19
$BaMe_2^{3+}Fe_{14}O_{23}$	X (= MS)	$3(RS_2R^*S^*)$		
$BaMeFe_6O_{11}$	Y	$(ST)_3$	1.5–3.8	9–28 (Co^{2+})
$Ba_2Me_2Fe_{18}O_{30}$	U (= MY)	$(RSR^*S_2^*)_3$	∼3.8 (Zn^{2+})	∼11 (Zn^{2+})
$Ba_3Me_2Fe_{24}O_{41}$	Z	$RSTSR^*S^*T^*S^*$	3.1– 3.9	13 (Co^{2+})
$(Ba,Ca)Fe_2O_4$	F	–	1.2 (Ca^{2+})	2.2 (Ca^{2+})

H.P.J. Wijn [62]

[a]Curie temperatures are typically in the vicinity of 700 K, and are accordingly lower when the room temperature magnetization is higher, e.g., where Me = Zn^{2+}

[b]H_K can be lowered by substituting Sc^{3+} or In^{3+} for Fe^{3+} in S-block octahedral sites which contribute most of the anisotropy energy. Conversely, Al^{3+} substitutions will raise H_K, but mainly by reducing the magnetization

octahedral-site ions down; in the spinel S block, 7 octahedral ions are up and 2 tetrahedral ones are down. For each Fe^{3+} ion contributing $5m_B$, the theoretical magnetic moment is $20m_B$, which has been confirmed by measurements at low temperatures.

Although other forms of hexagonal ferrites exist and occasionally find use, e.g., W, X, Y, and Z, [63] according to the listing presented in Table 4.10, only the M structure has proven to be of important technical value. The reasons for this lie principally in its uniaxial magnetic direction, as opposed to others that feature a planar magnetic orientation. Because of the comparatively high magnetic moment relative to ferrites in general Ba and Sr ferrite have become permanent magnet materials. As in the case of the spinel and garnet ferrites, dilution of the magnetic moment can be accomplished by substitutions of diamagnetic ions that favor octahedral sites. Experimental work by DeBitetto has shown that Al^{3+} ions substituting for Fe^{3+} will lower the magnetization [64]. Similar work by Roschmann et al. [65] and Wilber et al. [66] with In^{3+} and Sc^{3+} ions will not only reduce the magnetization but also lower the magnetic anisotropy that is responsible for the magnetic hardness property of the material. This topic will be examined in Sect. 5.4 on hysteresis properties.

For a more complete treatment of the hexagonal ferrite family, the reader is directed to seminal works by Braun [67], Went et al. [68], Jonker et al. [69], subsequent texts by Smit and Wijn [62], Lax and Button [70], and the comprehensive review by Wijn [63].

In our discussions up to this point, the origin of the magnetization has been the primary emphasis. Equally important is the process of magnetization, i.e., the initial permeability of a material and the factors that influence it. Prominent among these effects is the magnetoelasticity that enhances the anisotropy of the magnetic moment vector relative to the symmetry axes of the crystal in which it occurs. From these effects come many of the utilitarian properties of magnetic oxides, including magnetostriction, magnetic domains, and hysteresis.

($S = 1/2$) configuration, rendering some of the Fe^{3+} cations highly magnetoelastic with a degenerate spin-orbit stabilized ground state.

Table 4.11 Room-temperature magnetization of rare-earth orthoferrites

RE[b]	Y	La	Ce	Pr	Nd	Sm	Eu	Gd	Tb	Dy	Ho	Er	Tm	Yb	Lu
J_c	0	0	3/2	4	9/2	5/2	0	7/2	6	15/2	8	15/2	6	7/2	0
M_s (G)[c]	8.4	6.6	–	5.7	4.9	6.7	6.6	7.5	10.9	10.2	7.3	6.5	11.2	11.4	9.5

Data are from Bobeck et al. [73]
[b]All ions are 3+ valence state
[c]M_s values are in Gauss units

4.3.6 Orthoferrites

Perovskites represent another oxide structure compositionally similar to the garnets that can be made magnetically active. An advantage over the garnets is that they can be grown as films on a range of substrates. With 180° cation–anion–cation bonds, perovskites are also preferred lattices for superexchange investigations. Magnetic perovskites, with the generic formula ABO_3 (see Fig. 3.21, where A is an twelve-coordinated oxygen site occupied commonly by magnetic ions of the lanthanide (rare-earth) series, and B houses a transition metal ion, usually Fe^{3+}), have found applications in liquid-phase epitaxial film form as cylindrical-domain (bubble) memories. In later studies, magneto-optical properties of $CeFeO_3$ [71] and $Ba(Ti,Fe)O_3$ [72] have been reported.

For rare-earth orthoferrites $(RE)FeO_3$ [73], the basic ferrimagnetism that arises from a canted (∼0.5°) antiferromagnetic spin ordering of the Fe^{3+}–O^{2-}–Fe^{3+} interactions in octahedral sites, a net magnetization on the order of 100 emu/cm^3 (about a factor of 10 less than the corresponding garnet) is created. A summary of magnetization values for this series is given in Table 4.11. More importantly, the Néel (Curie) temperatures exceed 600 K, which is more than sufficient to provide collinear alignment of orbital and spin angular momentum in the A sublattice for room-temperature operation. If the frustration of the $4f$ spins of RE ions leads to such alignment with the resultant magnetization of the canted Fe^{3+} lattices, as sketched in Fig. 4.23, a Faraday rotation effect could occur in proportion to the net Fe^{3+} exchange fields. The subject of polarization rotation in magnetic oxides is discussed further in Chap. 7.

Appendix 4A Molecular Field Analysis of LiZnTi Ferrite

For the lithium zinc titanium spinel system $Li_{0.5-z/2+t/2}Fe_{1.5-z/2-3z/2}Zn_zTi_tO_4$, there are two cation site distribution regimes to consider, $z \geq t$ and $z \leq t$ [22]. In terms of generic dilution fractions, a set of general relations for the coefficients can be expressed as

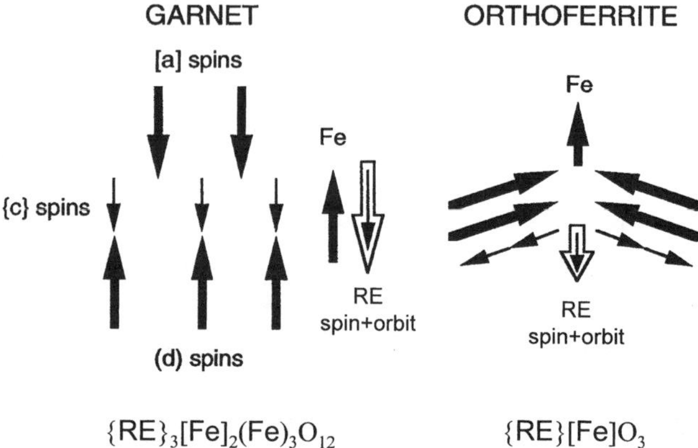

Fig. 4.23 Two-dimensional schematic diagram comparing the likely spin ordering of $(RE)_3\,Fe_5O_{12}$ garnet and $(RE)_4\,Fe_4O_{12}$ orthoferrite (expressed as four molecules). Note that the magnetically opposing tetrahedral (*d*) and octahedral (*a*) sites of the Fe sublattices in the garnet are in the ratio of 3:2, while the net moment of the orthoferrite arises from the canting of spins equally divided between the two antiferromagnetic lattices that comprise only octahedral sites. Figure reprinted from M. Bolduc, A.R. Taussig, A. Rajamani, G.F. Dionne, and C.A. Ross, *IEEE Trans. Magn.* **42**, 3093 (2006) with permission; © 2006 IEEE

$$\begin{aligned} \mathcal{N}_{AA} &\approx -150\left(1-\frac{4}{3}k_B^t-\frac{4}{3}k_B^l\right), \\ \mathcal{N}_{BB} &\approx -60\left(1-k_A^l-\frac{1}{3}k_A^z\right), \\ \mathcal{N}_{BA}=\mathcal{N}_{AB} &\approx +273\left[1-\frac{4}{5}k_B^t\left(k_A^z+k_A^l\right)-\frac{1}{5}k_B^t\right], \end{aligned} \tag{4.43}$$

where the superscripts t, l, and z refer to titanium, lithium and zinc dilutions.

For $t \geq z$, the chemical formula is

$$Fe_{1-t/2-z/2}Zn_zLi_{t/2-z/2}\,[Li_{0.5}Fe_{1.5-t}Ti_tO_4]$$

and the coefficient relations are

$$\begin{aligned} \mathcal{N}_{AA} &\approx -150\left(1-\frac{2}{3}t\right) \\ \mathcal{N}_{BB} &\approx -60\left(1-\frac{1}{2}t-\frac{1}{6}z\right) \\ \mathcal{N}_{AB} &\approx +273\left(1-\frac{1}{5}t(t+z)-\frac{1}{5}z\right) \end{aligned} \tag{4.44}$$

For $z \geq t$, $Fe_{1-z}Zn_z\left[Li_{0.5-z/2+t/2}Fe_{1.5+z/2-3t/2}Ti_tO_4\right]$, and the molecular field coefficient relations are

$$\begin{aligned} \mathcal{N}_{AA} &\approx -150\left(1 - t - \frac{1}{3}z\right), \\ \mathcal{N}_{BB} &\approx -60\left(1 - \frac{1}{3}z\right), \\ \mathcal{N}_{AB} = N_{BA} &\approx +273\left(1 - \frac{2}{5}tz - \frac{1}{5}z\right). \end{aligned} \tag{4.45}$$

Appendix 4B High-Magnetization Limits

The search for ferrites with magnetizations significantly greater than 5,000 G is motivated by the desire for higher permeability media in general and for device applications at millimeter wavelengths, which is discussed in Chap. 6. The basic challenge, apart from sublattice spin canting that is unavoidable, is to retain as much Fe as possible. Ideally, the maximum magnetization would be obtained if every magnetic ion were Fe^{3+} according to the formula ${Fe^{3+}}_{1-x}{Q^{n+}}_x\left[{Fe_2}^{3+}\right]O_4$, where Q represents a combination of A-site ions (including vacancies) with an effective valence of $n = 3 - 1/x$ for the range $0.33 \leq x \leq 1$. In practical cases, this has never been achieved mainly because the required dilutants have not been found. Figure 4.24 shows computed thermomagnetization curves in the range of $x < 0.5$ [74]. Note that only the $x = 0.4$ curve is realizable if stoichiometry is to be preserved; a curve for $x = 0.33$ (not shown) would represent the case of one-third of the A-sublattice lost to vacancies, similar to maghemite $Fe^{3+}\left[{Fe^{3+}}_{2-0.33}\square_{0.33}\right]O_4$, but with all of the vacancies in the B sublattice (see computed curve in Fig. 4.10).

Generally, the A sublattice accepts divalent ions more readily, and the closest one can come to reaching the high magnetization goals is to use Zn^{2+}. Solid solutions of magnetite $Fe^{3+}\left[Fe^{2+}Fe^{3+}\right]O_4$ and zinc ferrite $Zn^{2+}\left[{Fe_2}^{3+}\right]O_4$ will form the series ${Zn^{2+}}_x{Fe^{3+}}_{1-x}\left[{Fe^{2+}}_{1-x}{Fe_{1+x}}^{3+}\right]O_4$ has produced room-temperature magnetization values well over 7,000 G [75]. Unfortunately, these compositions are not useful at microwave frequencies because the large Fe^{2+} content renders the material dielectrically lossy. A possible solution to this impediment could be the monovalent cation system with generic formula ${A_x}^{1+}{Fe_{1-x}}^{3+}\left[{B_{0.5-x}}^{1+}{Fe_{1.5+x}}^{3+}\right]O_4$, where A and B represent any combination of monovalent ions occupying the tetrahedral and octahedral sites, respectively. The computed results of a thermomagnetization analysis shown in Fig. 4.25 predicted that a room-temperature $4\pi M \approx$ 6,500 G could be obtained with x in the 0.3–0.4 range [74]. This would require that at least 60% of the monovalent cations, Li^{1+}, Na^{1+}, or Cu^{1+} be forced into the A sites.

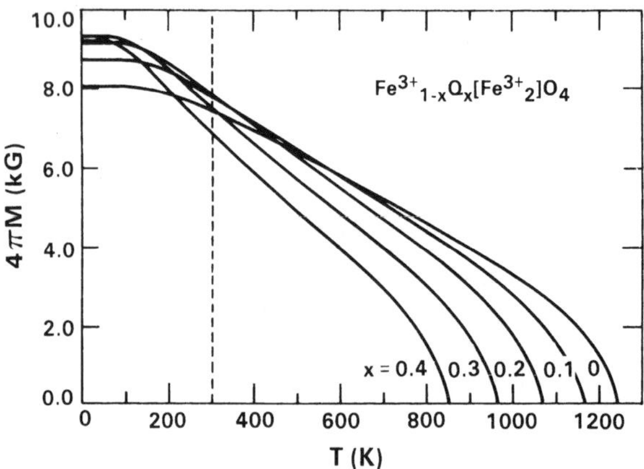

Fig. 4.24 Thermomagnetism curves for fictional inverse spinel $\left(Fe_{1-x}{}^{3+}Q_x\right)^{2+}\left[Fe_2{}^{3+}\right]O_4$. The range of values $x = 0, 0.1, 0.2, 0.3$, and 0.4 cannot be realized in practice, but the exercise serves to illustrate the challenge of the search for higher magnetization ferrites, particularly for microwave applications where Fe^{2+} ions cannot be used. Figure reprinted from [74] with permission. © 1987 by the American Institute of Physics

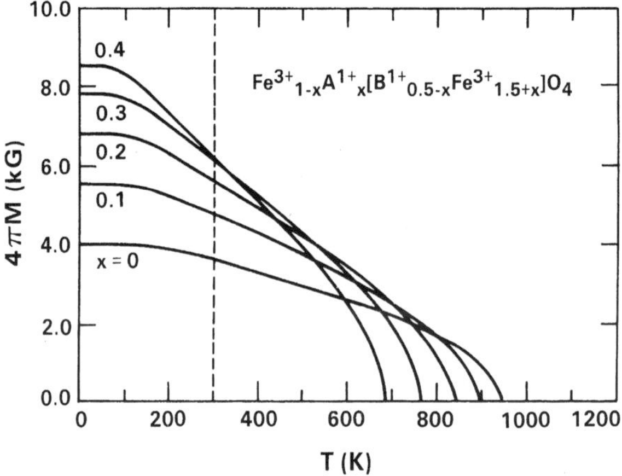

Fig. 4.25 Thermomagnetism curves of monovalent cation diluted spinel $Fe_{1-x}{}^{3+}A_x{}^{1+}\left[B_{0.5-x}{}^{1+}Fe_{1.5+x}{}^{3+}\right]O_4$ for $x = 0, 0.1, 0.2, 0.3$, and 0.4. Figure reprinted from [74] with permission. © 1987 by the American Institute of Physics

Appendix 4C Brillouin Functions in Exchange Energy Format

If the Brillouin functions are expressed in terms of exchange constants instead of molecular fields coefficients in the relations, the spin parameters are used in (4.30) according to

$$S_d\left(T\right) = S_d\left(0\right)\mathcal{B}_{S_d}\left(a_d\right),$$

$$S_a(T) = S_a(0)\,\mathcal{B}_{S_a}(a_a)\,,$$
$$S_c(T) = S_c(0)\,\mathcal{B}_{S_c}(a_c)\,, \tag{4.46}$$

where $S_d(T)$, $S_a(T)$, and $S_c(T)$ are average spin values of ions in the respective sublattices that are randomized by lattice thermal energy. Employing the expressions for the sublattice exchange energy from (4.32) (without the canting and dilution factors), we define

$$a_d(T) = \frac{E_{\text{ex}}^{(d)}}{kT} = -\frac{2S_d}{kT}\left[z_{dd}J_{dd}S_d + z_{da}J_{da}S_a + z_{dc}J_{dc}S_c\right],$$
$$a_a(T) = \frac{E_{\text{ex}}^{(a)}}{kT} = -\frac{2S_a}{kT}\left[z_{ad}J_{ad}S_d + z_{aa}J_{aa}S_a + z_{ac}J_{ac}S_c\right], \tag{4.47}$$
$$a_c(T) = \frac{E_{\text{ex}}^{(c)}}{kT} = -\frac{2S_c}{kT}\left[z_{cd}J_{cd}S_d + z_{ca}J_{ca}S_a + z_{cc}J_{cc}S_c\right].$$

To convert back to magnetic moments, the relations listed previously in Sect. 4.3.2 can be applied. However, for the case of rare-earth ions from the upper half of the $4f^n$ series in the c sublattice two adjustments must be made: first, the g_c factor must be that defined by (4.33) to account for the fact that the contribution to the magnetic moment M_c comes from the relation

$$\mathcal{M}_{\text{c}} = 3g_c m_{\text{B}} J_c N_{\text{A}} = 3g_c m_{\text{B}} N_{\text{A}}(L_c + S_c)\,, \tag{4.48}$$

and second, J_c must replace S_c in the determination of the magnetic moment $\mathcal{M}_c$. In addition, the molecular field coefficients $\mathcal{N}_{dc}$ and $\mathcal{N}_{ac}$ must be multiplied by the factor S_c/J_c to offset the change in $\mathcal{M}_c$ and keep the $\mathcal{N}_{dc}\mathcal{M}_c$ and $\mathcal{N}_{dc}\mathcal{M}_c$ products constant. A redefinition of $\mathcal{N}_{ij}$ in terms of J_{ij} from that of (4.28) to

$$N_{ij} = \frac{z_{ij}2J_{ij}}{n_{ij}g_i g_j m_{\text{B}}^2}\left(\frac{S_i S_j}{J_i J_j}\right) \tag{4.49}$$

is necessary to take into account that the L_c contribution to the moment does not take part in the exchange stabilization. This spin-only format can be used as an alternative to the conventional molecular field approach that is described in the text, but would require vigilance to make certain that J_c is used in $\mathcal{M}_c$ where necessary and that the $\mathcal{N}_{ij}$ coefficients are adjusted according to (4.49).

References

1. L. Néel, *Ann. Phys. (Paris)* **3**, 137 (1948)
2. A.H. Morrish, *The Physical Principles of Magnetism*, (John Wiley, New York, 1965)
3. G.F. Dionne, *Magnetic Moment versus Temperature Curves of Ferrimagnetic Garnet Materials*, (MIT Lincoln Laboratory Techn. Rept. **480**, 1970), ADA715284
4. G.F. Dionne, *Magnetic Moment versus Temperature Curves of Rare-Earth Iron Garnets*, (MIT Lincoln Laboratory Techn. Rept. **534**, 1979), ADA0773564

5. G.F. Dionne, *Magnetic Moment versus Temperature Curves of LiZnTi Ferrites*, (MIT Lincoln Laboratory Techn. Rept. **502**, 1974), ADA7824212
6. Y. Yafet and C. Kittel, *Phys. Rev.* **87**, 1203 (1955)
7. E.W. Gorter, *Philips Res. Rept.* **9**, 295 (1954)
8. E. Prince, *Acta Cryst.* **10**, 554 (1957)
9. P.-G. deGennes, *Phys. Rev.* Lett. **3**, 209 (1959)
10. M.A. Gilleo, *J. Phys. Chem. Solids* **13**, 33 (1960)
11. S. Geller, *J. Appl. Phys.* **37**, 1408 (1966)
12. C. Borghese, *J. Phys. Chem. Solids* **28**, 2225 (1967)
13. I. Nowik, *Phys. Rev.* **171**, 550 (1968); also I. Nowik, *J. Appl. Phys.* **40**, 5184 (1969)
14. A. Rosencwaig, *Can. J. Phys.* **48**, 2857 and 2868 (1970)
15. S. Geller, *Physics of Magnetic Garnets, Proc. Int'l School Phys. "Enrico Fermi," Course LXX*, (North-Holland Publishing Co., New York, 1978), p. 1
16. M.A. Gilleo, *Ferromagnetic Materials* **2**, W.P. Wohlfarth, ed., (North-Holland, New York, 1980), Chapter 1; also M.A. Gilleo and S. Geller, *Phys. Rev.* **110**, 73 (1958)
17. E.E. Anderson, *Phys. Rev.* **134**, A1581 (1964)
18. G.T. Rado and V.J. Folen, *J. Appl. Phys.* **31**, 62 (1960)
19. G.F. Dionne, *J. Appl. Phys.* **41**, 4874 (1970); *for FORTRAN program Magnetic Moment versus Temperature Curves of Ferrimagnetic Garnet Materials*, (MIT Lincoln Laboratory Technical Report TR-480, 1970), AD-715284
20. S. Geller, H.J. Williams, R.C. Sherwood, and G.P. Espinosa, *J. Appl. Phys.* **36**, 88 (1965)
21. P. Roschmann and P. Hansen, *J. Appl. Phys.* **52**, 6257 (1981)
22. G.F. Dionne, *J. Appl. Phys.* **46**, 4220 (1976); *for FORTRAN program Magnetic Moment versus Temperature Curves of Rare-Earth Iron Garnets*, (MIT Lincoln Laboratory Technical Report TR-588, 1981), AD-A107898/9
23. G.F. Dionne, *J. Appl. Phys.* **45**, 3347 (1974); *for FORTRAN program Magnetic Moment versus Temperature Curves of LiZnTi Ferrites*, (MIT Lincoln Laboratory Technical Report TR-502, 1974), AD-782421/2
24. G.F. Dionne, *J. Appl. Phys.* **63**, 3777 (1988)
25. J. Smit and H.P.J. Wijn, *Ferrites*, (Wiley, New York, 1959)
26. W.H. von Aulock, *Handbook of Microwave Ferrite Materials*, (Academic Press, New York, 1965)
27. G. Blasse, *Philips Research Repts.* Suppl. 3 (1964)
28. A. Broese van Groenou, P.F. Bongers, and A.L. Stuijts, *Mater. Sci. Eng.* **3**, 317 (1968)
29. V.J. Folen, *Landolt-Bornstein III/4b*, (Springer-Verlag, New York, 1970), p. 315
30. A.P. Greifer, *IEEE Trans. Magn.* **5**, 774 (1969)
31. M.M. Schieber, *Experimental Magnetochemistry*, (Wiley, New York, 1967)
32. G.F. Dionne, *J. Appl. Phys.* **99**, 08M913 (2006)
33. M.I. Klinger and A.A. Samokhvalov, *Phys. Stat. Sol.(b)* **79**, 9 (1977)
34. P.D. Baba, G.M. Argentina, W.E. Courtney, G.F. Dionne, and D. H. Temme, *IEEE Trans. Magn.* **8**, 83 (1972)
35. G.F. Dionne, *J. Appl. Phys.* **67**, 4561 (1990)
36. R.R. Heikes and W.D. Johnston, *J. Chem. Phys.* **26**, 582 (1957)
37. R. Pauthenet, *Ann. Phys.* **7**, 710 (1952)
38. G.F. Dionne and R.G. West, *Appl. Phys. Lett.* **48**, 1488 (1986); also *J. Appl. Phys.* **61**, 3868 (1987)
39. Z. Simsa and V.A.M. Brabers, *IEEE Trans. Magn.* **11**, 1303 (1975)
40. L.G. Van Uitert, *J. Chem. Phys.* **24**, 306 (1956)
41. F.K. Lotgering, *J. Phys. Chem. Solids* **25**, 95 (1964)
42. G.F. Dionne, *J. Appl. Phys.* **79**, 5172 (1996)
43. F. Bertaut and F. Forrat, *Comptes Rend.* **242**, 382 (1956)
44. S. Geller and M.A. Gilleo, *J. Phys. Chem. Solids* **3**, 30 (1957)
45. S. Geller and M.A. Gilleo, *J. Phys. Chem. Solids* **9**, 235 (1959)
46. G.F. Dionne, *J. Appl. Phys.* **40**, 1839 (1969)
47. E.R. Czerlinsky and R.A. MacMillan, *Phys. Stat. Sol.* **41**, 333 (1970)

48. E.R. Czerlinsky, *Phys. Stat. Sol.* **34**, 483 (1969)
49. A. Thavendrarajah, M. Pardavi-Horvath, P.E. Wigen, and M. Gomi, *IEEE Trans. Magn.* **25**, 4015 (1989)
50. F. Bertaut and R. Pauthenet, *Proc. IEE (London)* **104B**, 261 (1956)
51. R. Pauthenet, *Ann. Phys. (Paris)* **3**, 424 (1958)
52. J.H. Van Vleck and M.A. Gilleo, private communications urging the author to rework the model in terms of spin only. Regrettably, the opportunity to collaborate with these giants of magnetism ended prematurely when both passed away within months of the conversations. The revised model is presented in Appendix 4C
53. R. Aléonard, *J. Phys. Chem. Solids* **15**, 167 (1960)
54. G.F. Dionne and P.L. Tumelty, *J. Appl. Phys.* **50**, 8257 (1979)
55. C.D. Brandle and S.L. Blank, *IEEE Trans. Magn.* **12**, 14 (1976)
56. S. Geller, J.P. Remeika, R.C. Sherwood, H.J. Williams, and G.P. Espinosa, *Phys. Rev.* **137**, A1034 (1965)
57. K.R. Lea, M.J.M. Leask, and W.P. Wolf, *J. Phys. Chem. Solids* **23**, 1381 (1962)
58. J.S. Griffith, *The Theory of Transition-Metal Ions*, (Cambridge University Press, London, 1961)
59. J. Ostoréro and M. Guillot, *J. Appl. Phys.* **75**, 6792 (1994)
60. J. Ostoréro and M. Guillot, *J. Appl. Phys.* **81**, 4797 (1997)
61. G. Winkler, *Magnetic Garnets*, (Friedr. Vieweg & Sohn, Braunschweig/Wiesbaden, 1981)
62. J. Smit and H.P.J. Wijn, *Ferrites*, (Wiley, New York, 1959), Chapter IX
63. H.P.J. Wijn, *Landolt-Bornstein III/4b*, (Springer-Verlag, New York, 1970), p. 547
64. D.J. De Bitetto, *J. Appl. Phys.* **35**, 3482 (1964)
65. P. Roschmann, M. Lemke, W. Tolksdorf, and F. Welz, *Mater. Res. Bull.* **19**, 385 (1984)
66. W.D. Wilber, L.E. Silber, and A. Tauber, *Hexagonal Ferrites for Millimeter-Wave Control Devices*, US Army Laboratory Command Research and Development Technical Report No. SLCET-TR-87–4, 1987
67. P.B. Braun, *Nature* **170**, 708 and 1123 (1952)
68. J.J. Went, G.W. Rathenau, E.W. Gorter, and G.W. van Oosterhout, *Philips Tech. Rev.* **13**, 194 (1952)
69. G.H. Jonker, J.H. van Santen, H.P.J. Wijn, and P.B. Braun, *Philips Tech. Rev.* **18**, 145 (1956)
70. B. Lax and K.J. Button, *Microwave Ferrites and Ferrimagnetics*, (McGraw-Hill, New York, 1962), Section 3.3
71. D.S. Schmool, N. Keller, M. Guyot, R. Krishnan, and M. Tessier, *J. Appl. Phys.* **86**, 5712 (1999)
72. A. Rajamani, G.F. Dionne, D. Bono, and C.A. Ross, *J. Appl. Phys.* **98**, 063907 (2005)
73. A.H. Bobeck, R.F. Fischer, A.J. Perneski, J.P. Remeika, and L.G. Van Uitert, *IEEE Trans. Magn.* **5**, 544 (1969)
74. G.F. Dionne, *J. Appl. Phys.* **61**, 3865 (1987)
75. D. Stopples, P.G.T. Boonen, U. Enz, and L.A.H. van Hoof, *J. Magn. Magn. Matls.* **37**, 116 (1983)

Chapter 5
Anisotropy and Magnetoelastic Properties

In this chapter, we discuss the local origins of the two measurable macroscopic effects that occur from interactions between the ionic magnetic moments and the lattices in which they reside: magnetocrystalline anisotropy and magnetostriction. In the preceding chapters, the focus has been on the molecular origin of the magnetic moments in crystal lattices. For the $3d^n$ transition group in particular, the disposition of spin alignments as determined by covalent-induced superexchange and the randomizing effect of temperature has been reviewed. The spin system is also influenced by geometrical shape of the specimen in which it resides (described in Chap. 1) and the symmetry of the lattice itself and its elastic properties, each of which contribute to the anisotropy that influences the magnetization process and other magnetic properties. In addition, large anisotropic magnetic effects can result from asymmetry of the local crystal fields and their interactions with magnetoelastic cations. In this sense, magnetoelasticity refers to the coupling between the magnetic moment of the cation and local crystal field of the anion coordination. All of these mechanisms, however, involve interaction between the spins and the elastic properties of the lattice, which can be collective, as in the case of dipole–dipole interactions in fixed array of lattice sites, or individual through orbital angular momentum coupling to the crystal field. The conventional macroscopic phenomenological model is presented later in this chapter, but it is the molecular origins of these properties where our initial attention will be focused.

Following the context established by the preceding chapters, we begin by examining the local origins of the local anisotropy. In particular, self-induced anisotropy in the form of crystal-field distortions derived from spin–orbit coupling and the Jahn–Teller effect will be emphasized. The underlying physics is reviewed first through the properties of individual ions. With the single-ion concepts in hand, we then examine the ions in an exchange-coupled ferromagnet (or ferrimagnet) to determine how the macroscopic anisotropy and magnetostriction effects influence the collective magnetization statically, and then dynamically in Chap. 6.

G.F. Dionne, *Magnetic Oxides*, DOI 10.1007/978-1-4419-0054-8_5,

5.1 Quantum Paramagnetism of Single Ions

In the previous chapter, the concept of magnetic anisotropy was introduced somewhat incidentally in order to explain the variation in magnetic moment of the rare-earth ion Ho^{3+} in iron garnet through the effects of crystal fields on the total angular momentum. Although crystal-fields influence the magnetic properties of rare-earth ions, they are critically important in the $3d^n$ transition iron group series. Because the coupling to the electrostatic fields of the crystalline environment is with the orbital angular momentum, the reason why the iron group ions are so sensitive to their surroundings is the unshielded $3d$ shell. Not only do the d electrons sense the electric fields of the ligands, but they also participate in the chemical bonding by sharing their orbital states with the oxygen anion $2p$ electrons. Much of the underlying theory is reviewed in Chap. 2, but the actual relation between the electron–lattice effects and the properties of the oxides (and other families of transition-metal compounds) was not discussed. The role of spin–orbit coupling as the intermediary between spin and lattice and the interaction between orbit and spin angular momentum with a magnetic field must now be examined.

5.1.1 Theory of Anisotropic g Factors

The classical approach to paramagnetism from a collection of isolated magnetic moments $\boldsymbol{m}$ was treated in Sect. 1.2.1. From this model Curie's law of $\chi \propto 1/T$ was derived with versions that employed the Langevin or semiclassical Brillouin functions. In either case, the applicable variable is $\boldsymbol{m} \cdot \boldsymbol{H}/kT$ (or $gm_{\mathrm{B}}HJ/kT$ for the Brillouin function). In later developments spearheaded by Van Vleck, the theory was refined to take into account an angular momentum energy level structure above the ground state [1]. Because these excited levels have differing values of $\boldsymbol{m}$, a Boltzmann probability distribution can be used to determine a weighted average susceptibility shown schematically in Fig. 5.1 for a five-level orbital group (with spin implied) split in a crystal field. The general expression for the paramagnetic susceptibility of a molecule with quantum numbers L, S, and J can be written as

$$\chi = \frac{\sum\limits_{J=|L-S|}^{L+S} (2J+1)\,\chi\,(LSJ) \exp\left[-\lambda\gamma J\,(J+1)/2\mathrm{k}T\right]}{\sum\limits_{J=|L-S|}^{L+S} (2J+1) \exp\left[-\lambda\gamma J\,(J+1)/2\mathrm{k}T\right]}, \tag{5.1}$$

where $\chi\,(LSJ)$ is the susceptibility contribution from the energy level designated as ${}^{2S+1}L_J$. Details of the derivation of (5.1), including an expression for $\chi(LSJ)$, may be found in Griffith's book [2].

The most successful application of this theory was carried out by Van Vleck for the $4f^n$ lanthanide series. The results are shown in Fig. 5.2, where the effective

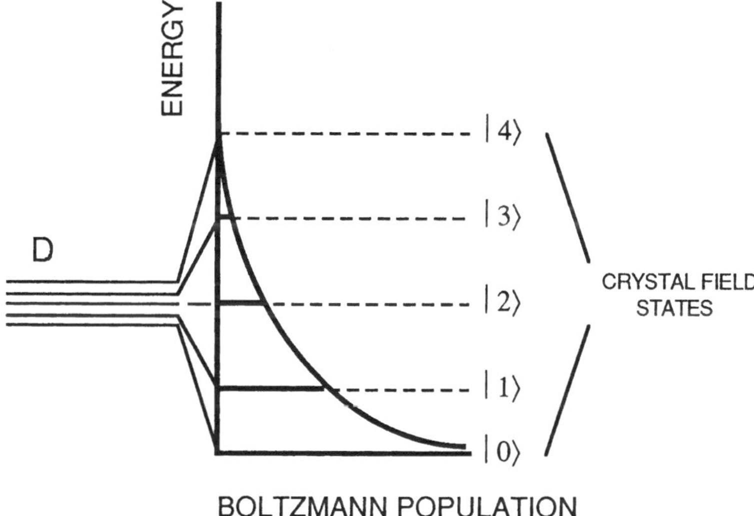

Fig. 5.1 Schematic diagram of Boltzmann population distribution function showing the relative occupancies of the five d orbital states controlled by the exp $(-E/kT)$

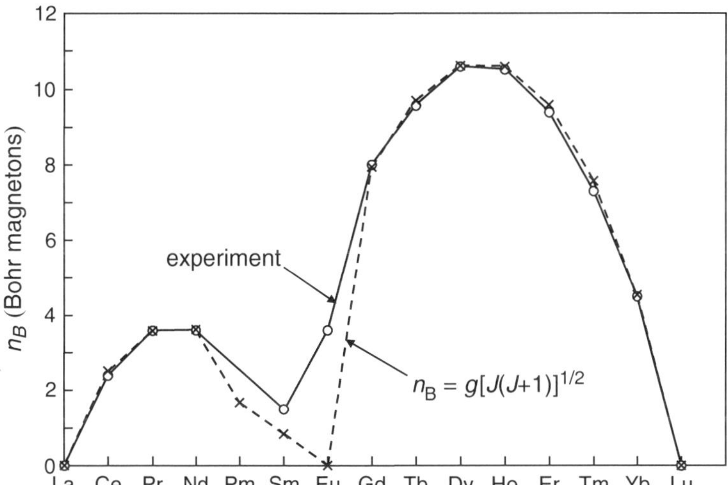

Fig. 5.2 Rare-earth ion magnetic moments as a function of the number of $4f^n$ electrons. The lanthanide series is used as the x-axis coordinate, which extends from 0 to 14. The experimental data and calculated fit were reported by Van Vleck and Frank [1]. Note that the lack of agreement with the simple theory for Eu^{3+} and Sm^{3+} indicates that the more complex model of (5.1) is needed

moments are graphed in terms of Bohr magnetons $n_B = (\chi 3kT/N_A m_B{}^2)^{1/2}$, where N_A is Avogadro's number. These data may be found in tabular form in most textbooks on magnetism. Despite intense efforts to extract information from susceptibility measurement data, results of analysis based on this theory have generally fallen short of expectations. The difficulty in sorting out the individual contributions from the ladder of energy levels with varying separations, compounded by the uncertainty in the values of the Landé g factors that are implicit within the $\chi(LSJ)$ of individual states led researchers to adopt more sophisticated approaches in both measurement and theory. Because this necessary spectral information is so sensitive to the effects of the crystal field, more powerful tools were developed to probe the lowest energy states and directly measure g factors by the Zeeman effect in many of the transition metal ions. In these situations, the analysis applied to electron paramagnetic resonance (EPR) spectroscopy proved to be fertile ground for theorists with skill in the application of group theory and quantum mechanical perturbation theory.

If we now return to the general perturbation Hamiltonian for an ion in a crystalline environment introduced as (4.38) for the lanthanide series, $\mathcal{H}_1 = \mathcal{H}_{LS} + \mathcal{H}_{\mathrm{cf}} + \mathcal{H}_{\mathrm{h}} + \mathcal{H}_{\mathrm{ex}}$, we can examine the effects of the remaining interactions that influence the magnetic moment of an isolated ion. For the individual ions of the $3d^n$ transition series $\mathcal{H}_{\mathrm{ex}}$ does not apply, and the relation is changed to

$$\begin{aligned} \mathcal{H}_1 &= \mathcal{H}_{\mathrm{cf}} + \mathcal{H}_{\mathrm{LS}} + \mathcal{H}_{\mathrm{h}} \\ &= \mathcal{H}_{\mathrm{cf}} + \lambda \boldsymbol{L} \cdot \boldsymbol{S} + m_{\mathrm{B}} \left(\boldsymbol{L} + g_{\mathrm{e}} \boldsymbol{S} \right) \cdot \boldsymbol{H}, \\ &= \mathcal{H}_{\mathrm{cf}} + \lambda \boldsymbol{L} \cdot \boldsymbol{S} + m_{\mathrm{B}} g_{ij} S_i H_j, \end{aligned} \tag{5.2}$$

where $g_{\mathrm{e}} = 2$ is the g factor for an electron and g_{ij} is the tensor that results from the orbital quenching and its attendant anisotropy effects from the crystal field symmetry. Since the effects of $\mathcal{H}_{\mathrm{cf}}$ on the orbital states have already been dealt with in Chap. 2 for D and F terms of the iron group, we can now continue with an examination of the next perturbations for this series: spin–orbit coupling $\lambda \boldsymbol{L} \cdot \boldsymbol{S}$ and the combined effect of the orbital and spin moments in a magnetic field $m_{\mathrm{B}} \left(\boldsymbol{L} + 2\boldsymbol{S} \right) \cdot \boldsymbol{H}$. Smaller energy terms involving spin–spin and electron–nuclear spin interactions that produce hyperfine spectral structure will be omitted from this discussion.

For the $3d^n$ series, we recall from Chap. 2 that an undistorted octahedral crystal field will stabilize either a triplet (from a free-ion D-state d^1 and d^6, from an F-state d^2 and d^7), a doublet (from a D-state d^4 and d^9), or a singlet (from an F-state d^3 and d^8). However, in cases where the ground state is expected to be degenerate based on the host lattice symmetry, local ligand distortions usually occur spontaneously to lower the symmetry. In the context of magnetocrystalline anisotropy, it is the extent to which the crystal field lifts the orbital degeneracy that determines the magnetoelastic properties. To this end, we will review the various electronic structures that exist in the transition metal ions residing in ligand fields of initially cubic symmetry.

Analysis of any of these electronic configurations is solvable by conventional degenerate perturbation theory with the use of high-speed digital computation technology once the appropriate Hamiltonian matrices are set up. To demonstrate the origin of single-ion anisotropy as reflected in the variation of the g factor with magnetic field direction, a detailed examination of the simplest case will be presented to sensitize the reader to the important physics issues involved in these critical spin–lattice interactions.

5.1.2 Conventional Perturbation Solutions

For the general case of a single d electron in an octahedral crystal field with an orthorhombic distortion dominated by a z-axis contraction,[1] the energy level structure is shown in Fig. 5.3. For cases where the free-ion term is a $^{2S+1}D_J$, the appearance of the crystal-field energy levels in the ground-state occupancy diagram conforms to the actual energy states of a multiple d-electron ion (as opposed to the $^{2S+1}F_J$ cases to be reviewed in subsequent sections). The ground term is indicated by $^2D_{3/2}$ and the eigenvectors of the crystal-field states are listed as $|0\rangle$, $|1\rangle$, $|2\rangle$, $|3\rangle$, and $|4\rangle$, where from (2.12)

$$|4\rangle = d_{z^2} = d_0 = Y_2^0 = \frac{1}{2}\left(3z^2 - r^2\right),$$

$$|3\rangle = d_{x^2-y^2} = \frac{\sqrt{3}}{\sqrt{2}}\left(d_2 + d_{-2}\right) = \frac{\sqrt{3}}{\sqrt{2}}\left(Y_2{}^2 + Y_2^{-2}\right) = \frac{\sqrt{3}}{2}\left(x^2 - y^2\right),$$

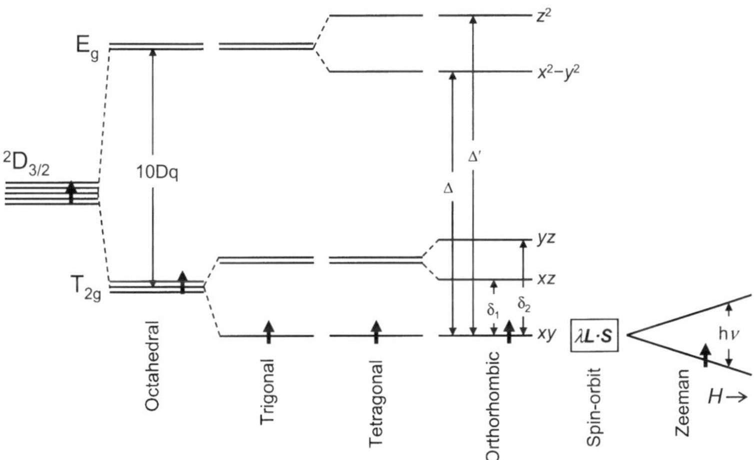

Fig. 5.3 $3d^1$ energy-level model in an orthorhombic crystal field, showing spin stabilization in a Zeeman splitting of the ground-state Kramers doublet

[1] In other situations, the distortion can be an expansion, which is manifested by the orbital functions with a z-dependence lying below the in-plane x- or y-dependent states. Note also that a trigonal [111]-axis expansion is compatible with a [100]-axis contraction.

$$|2\rangle = d_{yz} = \mathrm{i}\frac{\sqrt{3}}{\sqrt{2}}(d_1 + d_{-1}) = \mathrm{i}\frac{\sqrt{3}}{\sqrt{2}}\left(Y_2{}^1 + Y_2^{-1}\right) = \sqrt{3}yz, \tag{5.3}$$

$$|1\rangle = d_{xz} = -\frac{\sqrt{3}}{\sqrt{2}}(d_1 - d_{-1}) = -\frac{\sqrt{3}}{\sqrt{2}}\left(Y_2{}^1 - Y_2^{-1}\right) + \sqrt{3}xz,$$

$$|0\rangle = d_{xy} = -i\frac{\sqrt{3}}{\sqrt{2}}(d_2 - d_{-2}) = -\mathrm{i}\frac{\sqrt{3}}{\sqrt{2}}\left(Y_2^2 - Y_2^{-2}\right) = \sqrt{3}xy.$$

The solutions of these perturbation effects will be carried out in two steps: first, the eigenstates of $\mathcal{H}_{LS}$ term will be determined and a new ground state will be found; then the Zeeman term will be applied to determine the anisotropy of the magnetic moment which will appear in the inequality of the elements of the g tensor. The problem will be analyzed in terms of the Ti^{3+}-substituted hydrated alum salt[2] $Rb^{1+}\left(Al^{3+}, Ti^{3+}\right)(SO_4)^{2-}{}_2 \cdot 12H_2O$ from three approaches, beginning with the most general [3]. First, degenerate theory will be applied whereby a full perturbation matrix is established for the elements $\langle k|\mathcal{H}_{\text{cf}} + \lambda \boldsymbol{L} \cdot \boldsymbol{S}|n\rangle$, where $\langle k|$ and $|n\rangle$ are eigenfunctions from the list in (5.3) with the spin states included. Since each orbital state has a twofold spin degeneracy ($m_s = \pm 1/2$) this exercise involves a total of ten wavefunctions, requiring the diagonalization of a 10×10 matrix. For the spin–orbit contributions a convenient form of the operator is $\boldsymbol{L} \cdot \boldsymbol{S} = L_z S_z + (1/2)(L_+ S_- + L_- S_+)$.

Details of the solution are given in Appendix 5A, where the secular equation is shown to be separable into two identical functions that yield degenerate energy states, each representing a twofold spin degeneracy (known as Kramers doublets, which occur in ions with odd numbers of d electrons). The resulting ground state eigenfunctions from (5.72) are

$$|e_0+\rangle = a\left|0, \tfrac{1}{2}\right\rangle + b\left|1, -\tfrac{1}{2}\right\rangle + c\left|2, -\tfrac{1}{2}\right\rangle + d\left|3, \tfrac{1}{2}\right\rangle, \tag{5.4}$$
$$|e_0-\rangle = a\left|0, -\tfrac{1}{2}\right\rangle + b\left|1, \tfrac{1}{2}\right\rangle - c\left|2, \tfrac{1}{2}\right\rangle - d\left|3, -\tfrac{1}{2}\right\rangle,$$

where the highest energy $|4\rangle$ level has been dropped to simplify the calculation with only a small loss in accuracy, are then used for the 2×2 ground state Zeeman matrix according to $\langle k|m_{\text{B}}(\boldsymbol{L} + 2\boldsymbol{S}) \cdot \boldsymbol{H}|n\rangle$.[3]

To compute the Zeeman splittings for a particular direction of $\boldsymbol{H}$, we determine the matrix elements by applying the appropriate operators, recalling that $L_x = (1/2)(L_+ + L_-)$ and $L_y = -(i/2)(L_+ - L_-)$ and likewise for the spin operators. In each case the Zeeman matrix is diagonalized and the expressions for the energies of the split spin doublet are subtracted to give $\Delta E_i = g_i m_{\text{B}} H_i$, from

[2] In this lattice, the ligands of the Ti^{3+} ions are an octahedron of H_2O molecules that mimic O^{2-} ions.

[3] For this exercise we consider only the lowest Kramers doublet (Zeeman state) of the three originating from the T_{2g} of the cubic field. Implicit in this choice is that the ground state is a singlet that results from either an axial field component of the appropriate sign or a Jahn–Teller effect.

which expressions for g_z, g_x, and g_y in terms of the a, b, c, and d coefficients can be deduced, as shown in Appendix 5A. To solve for the g factors, values are assigned to the crystal field splittings δ_1, δ_2, and Δ; a value for λ must also be assigned, usually the free-ion value. Conversely, if g factors are found from measurement, paramagnetic resonance at microwave frequencies in many cases, the reverse procedure could be followed to determine the energy splittings. This method was used to compute energy-level splittings from measured g factors for a series of Ti^{3+}-substituted alum salts starting with $Rb^{1+}Al^{3+}(SO_4)^{2-}{}_2 \cdot 12H_2O$, and the results are included in Table 5.1. Among these compounds are two types of crystal field: for Rb, K, Tl, and Na alum, orthorhombic with complete splitting of all five d orbitals that produces three g factors [3–6], and for the Cs alum-based compounds, a trigonal field directed along the lattice <111> axes produces a ground state singlet and two excited doublets as depicted in Figs. 2.13 and 5.3 (for $\alpha > 60°$) [7], [8]. In each case the degenerate method outlined in Appendix 5A was employed. In the Cs alum cases, accuracy of fits between theory and experiments were limited by the approximations to the theory because the values of the lower energy splittings was on the order of 200 cm^{-1} so that the assumption that $\lambda/\delta < 1$ could not be justified, i.e., off-diagonal elements are as large as the diagonal ones. Among the orthorhombic alums, variations in the g factors and their interpreted lower symmetry orbital splittings were discussed by Dionne and MacKinnon [6] in terms of the possible effect on the crystal fields by differing locations of the $(SO_4)^{2-}$ radicals. The ionic radius of the large monovalent cation relative to that of the Al^{3+} host ion was correlated

Table 5.1 Anisotropic g factors and crystal field splittings of Ti^{3+} (d^1) in octahedral sites. $(\lambda_{50} - 154\,cm^{-1})$

Ti^{3+} host	Radius (Å)	Theory	g_z	g_x	g_y	δ_1 (cm^{-1})	δ_2 (cm^{-1})	Δ (cm^{-1})	Refs.
Rb alum	1.48	SpinHam	1.895	1.715	1.767	1,070	1,310	11,500	[4]
		nondeg.	"	"	"	1,050	1,320	17,000	[4]
		degen.	"	"	"	1,050	1,320	20,300[a]	[4]
Tl Alum	1.40	degen.	1.938	1.790	1.834	1,462	1,843	20,300	[5]
K Alum	1.33	degen.	1.975	1.828	1.897	1,780	2,950	20,300	[6]
Na Alum	0.95	degen.	~2.00	~1.86	~1.86	>2,000	>2,000	20,300	[6]
Ti^{3+} host	Radius (Å)	Theory		$g_\parallel$	$g_\perp$	δ (cm^{-1})	Δ (cm^{-1})		Refs.
CsTi Alum	1.69	Degenerate		1.24	0.93	~200	20,300		[8]
		"		1.19	0.70	"	"		[8]
		"		1.17	0.23	"	"		[8]
CsAlum	1.69	Degenerate		1.25	1.14	~300	20,300		[7]
Al_2O_3	–	–		1.067	<0.1				5.19

[a] The value $\Delta = 20{,}300\,cm^{-1}$ was measured by optical absorption in an aqueous solution [10], where the ligands are octahedra of water molecules. This situation is close to that of oxygen ligands because of the relative location of the O in the H_2O molecule. To obtain the measured value of $g_z = 1.895$ by the (corrected) degenerate method, $\Delta = 22{,}000\,cm^{-1}$

with the magnitude of the apparent orthorhombic departures of the crystal field from its normal alum trigonal symmetry that is seen only in the Cs alum case. One result that supported this contention was the observation that Cs(TiAl) alum crystals could be grown from aqueous solution over the full range of solid solution, while the rest of the family with the apparent orthorhombic distortions would accept only small concentrations of Ti as substitutions for Al.

If we assume that lower symmetry exists from local site distortions to produce a singlet ground state as depicted in Fig. 5.3, we can also treat the problem as a conventional nondegenerate case whereby the matrix elements of interest become the off-diagonal ones $\langle k|\lambda \boldsymbol{L}\cdot\boldsymbol{S}|n\rangle$ for use in the standard relation

$$\left|n'\right\rangle = |n\rangle - \sum_{k\neq n} |k\rangle \frac{\langle k|\,\lambda \boldsymbol{L}\cdot\boldsymbol{S}\,|n\rangle}{E_k - E_n}, \tag{5.5}$$

where $|n\rangle$ refers to $|0,+1/2\rangle$ and $|0,-1/2\rangle$ and $|k\rangle$ is any of the other eigenfunctions in (5.2) with the $|+1/2\rangle$ and $|-1/2\rangle$ spin functions added. Applying the matrix elements from Appendix 5A, we obtain for the ground states of (5.5)

$$\begin{aligned}
|e_0+\rangle &= \left|0,\tfrac{1}{2}\right\rangle - \mathrm{i}\,(\lambda/2\delta_1)\left|1,-\tfrac{1}{2}\right\rangle - (\lambda/2\delta_2)\left|2,-\tfrac{1}{2}\right\rangle + \mathrm{i}\,(\lambda/\Delta)\left|3,\tfrac{1}{2}\right\rangle, \\
|e_0-\rangle &= \left|0,-\tfrac{1}{2}\right\rangle - \mathrm{i}\,(\lambda/2\delta_1)\left|1,\tfrac{1}{2}\right\rangle + (\lambda/2\delta_2)\left|2,\tfrac{1}{2}\right\rangle - \mathrm{i}\,(\lambda/\Delta)\left|3,-\tfrac{1}{2}\right\rangle.
\end{aligned} \tag{5.6}$$

To calculate the Zeeman energy splitting of the Kramers doublet $|e_0\pm\rangle$ we follow the same procedure for the degenerate solution above to produce matrix elements for the z direction

$$\begin{aligned}
\mathcal{H}_{11} &= \left[1 - \frac{4\lambda}{\Delta} - \frac{\lambda^2}{2\delta_1\delta_2} - \frac{\lambda^2}{4}\left(\frac{1}{{\delta_1}^2} + \frac{1}{{\delta_2}^2}\right) + \frac{\lambda^2}{\Delta^2}\right] m_{\mathrm{B}} H_z, \\
\mathcal{H}_{22} &= -\mathcal{H}_{11}, \\
\mathcal{H}_{12} &= \mathcal{H}_{21} = 0.
\end{aligned} \tag{5.7}$$

By solving the secular equation and relating $\Delta E = 2\mathcal{H}_{11} = g_z m_{\mathrm{B}} H_z$, and then repeating the procedure for the x and y directions, the diagonal elements of the g tensor can be expressed as

$$\begin{aligned}
g_z &= 2 - \frac{8\lambda}{\Delta} - \frac{\lambda^2}{\delta_1\delta_2} - \frac{\lambda^2}{2}\left(\frac{1}{{\delta_1}^2} + \frac{1}{{\delta_2}^2}\right) + \frac{2\lambda^2}{\Delta^2}, \\
g_x &= 2 - \frac{2\lambda}{\delta_1} + \frac{\lambda^2}{2}\left(\frac{1}{{\delta_1}^2} - \frac{1}{{\delta_2}^2}\right) + \frac{2\lambda^2}{\delta_2\Delta} - \frac{2\lambda^2}{\Delta^2}, \\
g_y &= 2 - \frac{2\lambda}{\delta_2} + \frac{\lambda^2}{2}\left(\frac{1}{{\delta_2}^2} - \frac{1}{{\delta_1}^2}\right) + \frac{2\lambda^2}{\delta_1\Delta} - \frac{2\lambda^2}{\Delta^2}.
\end{aligned} \tag{5.8}$$

As listed in Table 5.1 for Rb alum, splitting energies for the measured $g_z = 1.895$, $g_x = 1.715$, and $g_y = 1.767$ are $\delta_1 = 1{,}050$, $\delta_2 = 1{,}320$, and $\Delta = 17{,}000\,\text{cm}^{-1}$ [3,4].

5.1.3 The Spin Hamiltonian for 3d^n Ions

To complete this instructional example, it is important that we introduce albeit briefly the "spin" Hamiltonian that was developed by Pryce and Abragam as an approximation to the nondegenerate model outlined in Sect. 5.1.2 [9]. This method was used to analyze the paramagnetic resonance spectrum of multiple-d-electron ions with an orbital *singlet* ground state, in particular the ^4F-state Cr^{3+} ion that was the key to the early ruby maser and laser demonstrations. If the excited orbital states of the crystal field splittings are high enough to be ignored in following calculations, the approach differs from those of the two conventional methods in that the perturbation matrix elements are computed for a combination of spin–orbit and Zeeman interactions, $\langle k|\mathcal{H}_1|n\rangle = \langle k|\lambda \boldsymbol{L}\cdot\boldsymbol{S} + m_{\text{B}}(\boldsymbol{L}+2\boldsymbol{S})\cdot\boldsymbol{H}|n\rangle$, with the result that the spin **S** is treated as the only angular momentum quantum number and the contribution of L becomes absorbed in the g factors, consistent with (5.8) but with less precision. Since the $|\text{e}_0\rangle$ states are singlets that carry no orbital angular momentum, i.e., S-state behavior, the only first-order nonzero elements will be $\langle \text{e}_0|2m_{\text{B}}\boldsymbol{S}\cdot\boldsymbol{H}|\text{e}_0\rangle$ and the second-order correction will simplify to

$$E^{(1)} = -\sum_{n\neq 0} \frac{|\langle \text{e}_0|\,\lambda \boldsymbol{L}\cdot\boldsymbol{S} + m_{\text{B}}\boldsymbol{H}\cdot\boldsymbol{L}\,|\text{e}_n\rangle|^2}{E_n - E_0}. \tag{5.9}$$

After collecting all first- and second-order terms that are linear in $\lambda m_{\text{B}} H$, (5.9) can be reduced to

$$E^{(1)} = 2m_{\text{B}}\left(\delta_{ij} - \lambda\Lambda_{ij}\right) H_i S_j, \tag{5.10}$$

where

$$\Lambda_{\text{ij}} = \sum_{n\neq 0} \frac{\langle \text{e}_0|\,L_i\,|\text{e}_n\rangle\,\langle \text{e}_n|\,L_j\,|\text{e}_0\rangle}{E_n - E_0}. \tag{5.11}$$

From (5.9) we see that the matrix elements used are only those linked to the ground state, thereby substantially simplifying the labor involved. With this approximation only the terms to first order in λ appear, and

$$\begin{aligned} g_z &= 2 - \frac{8\lambda}{\Delta}, \\ g_x &= 2 - \frac{2\lambda}{\delta_1}, \end{aligned} \tag{5.12}$$

$$g_y = 2 - \frac{2\lambda}{\delta_2}.$$

For Rb alum measured g factors are fitted to splitting energies of $\delta_1 = 1{,}070$, $\delta_2 = 1{,}310$, and $\Delta = 11{,}500\,\mathrm{cm}^{-1}$, as listed in Table 5.1. Note that only the larger cubic field component Δ deviates significantly from the measured reference value of $20{,}300\,\mathrm{cm}^{-1}$ [10].

For ions with more complicated spin multiplets, i.e., $S > 1/2$, such as Cr^{3+} and Ni^{2+}, higher-order terms in λ become important for spectral analysis and a more general form of (5.10) can be expressed as [9]

$$E^{(2)} = 2m_\mathrm{B}\left(\delta_{ij} - \lambda\Lambda_{\mathrm{ij}}\right) H_i S_j - \lambda^2 \Lambda_{ij} S_i S_j, \tag{5.13}$$

leading to the general form of the spin Hamiltonian expressed to second order in λ as

$$H_\mathrm{s} = m_\mathrm{B}\boldsymbol{g}\cdot\boldsymbol{H}\cdot\boldsymbol{S} + DS_z{}^2 + E\left[S_x{}^2 - S_y{}^2\right] + \tfrac{1}{6}a\left[S_z{}^4 + S_x{}^4 + S_y{}^4\right], \tag{5.14}$$

where the anisotropic $\boldsymbol{g}$ is now expressed in tensor format. The zero-field splitting parameters D and E (both functions of Λ_{ij}) can be determined from experiment and can be used to deduce information about the electronic structure of the ions in their crystal field. Included in this version of the spin Hamiltonian is the fourth-order term characterized by the a parameter which becomes significant in S-state ions $3d^5$ (Fe^{3+}, Mn^{2+}) and $4f^7$ (Gd^{3+}, Eu^{2+}) for which the first-order orbital angular momentum contributions are absent. The subject of S-state ions will be reviewed further in Sect. 5.2.5.

Equations (5.8) and (5.12) expose the role of spin–orbit coupling in determining the magnitude and anisotropy of the magnetic moments in these transition-metal ions. The ratios of the constant λ to the various orbital level splittings can dictate not only the degree of anisotropy of the magnetic moment as the magnetic field is applied to different directions, but also whether the moment increases or decreases as in this case of Ti^{3+} ion. According to Table 5.2, the λ constants increase in magnitude and reverse sign in the upper half of the $3d^n$ series, thereby alerting us to the fact that g factors greater than 2 as well as increased anisotropy is to be expected from the ions with larger d-electron populations. Through the rest of the series, the Aufbau principle is employed to show how the peculiarities of the orbital ground states of the various ions are related to their magnetoelastic properties.

5.1.4 *The Crystal-Field Hamiltonian for* $4f^n$ *Ions*

For the lanthanide series, analysis of isolated ion magnetic properties in solids follows the same perturbations approach as that for the iron group. The hierarchy of perturbations is totally different, however. In the $4f^n$ ions spin–orbit coupling exceeds that of the $3d^n$ group by at least a factor of 3. Moreover, crystal field effects

Table 5.2 $3d^n$ transition group data[a]

	d^1	d^2	d^3	d^4	d^5	d^6	d^7	d^8	d^9
Free ion	Ti^{3+}	V^{3+}	Cr^{3+}	Mn^{3+}	Fe^{3+}	Fe^{2+}	Co^{2+}	Ni^{2+}	Cu^{2+}
		Ti^{2+}	V^{2+}	Cr^{2+}	Mn^{2+}	Co^{3+}	Ni^{3+}	Cu^{3+}	
$\lambda\ (\mathrm{cm}^{-1})$	154	104	87	85	–	–100	–180	–335	–852
			55	57					
Hund term	${}^2D_{3/2}$	${}^3F_2$	${}^4F_{3/2}$	${}^5D_0$	${}^6S_{5/2}$	${}^5D_4$	${}^4F_{9/2}$	$3F_4$	${}^5D_{3/2}$
High spin	–	–	–	$(e_g)^1$	$(e_g)^2$	$(e_g)^2$	$(e_g)^2$	$(e_g)^2$	$(e_g)^3$
	$(t_{2g})^1$	$(t_{2g})^2$	$(t_{2g})^3$	$(t_{2g})^3$	$(t_{2g})^3$	$(t_{2g})^4$	$(t_{2g})^5$	$(t_{2g})^6$	$(t_{2g})^6$
Low spin	–	–	–	–	–	–	$(e_g)^1$	$(e_g)^{2b}$	$(e_g)^{3b}$
	$(t_{2g})^1$	$(t_{2g})^2$	$(t_{2g})^3$	$(t_{2g})^4$	$(t_{2g})^5$	$(t_{2g})^6$	$(t_{2g})^6$	$(t_{2g})^6$	$(t_{2g})^6$
S(hs)	1/2	1	3/2	2	5/2	2	3/2	1	1/2
(ls)				1	1/2	0	1/2	0[b]	
$\langle g \rangle$	0 to $\sim$1.9	$\sim$1.9	1.9–2	1.9–2	$\sim$2	$> 3 - \sim 7$	> 4	> 2.25	2–2.5

[a]Data obtained from W. Low, "Paramagnetic Resonance in Solids," (*Academic Press*, New York 1960) Table XIX [13]

[b]Low-spin states can occur in lower symmetry crystal fields that split the E_g degeneracy sufficiently to cause a violation of Hund's rule by creating a spin pair in the lower e_g state

are only a small fraction of those in the iron group with the result that free-ion states can now include the $\lambda \boldsymbol{L} \cdot \boldsymbol{S}$ interaction and the perturbation calculation employs J as the quantum operator and begins with the spin–orbit multiplet functions. The shielding of the $4f$ electrons by the filled $5s$ and $5p$ shells is contrasted with the exposed $3d$ shell in the iron group by comparison of the calculated radial distributions [11] shown in Fig. 5.4. As a consequence, the $\boldsymbol{S}$ operator alone does not enter the picture until exchange effects are present as discussed in Sect. 4.3. The Hamiltonian operator of concern for the single-ion analysis of rare-earths is therefore that of the point-charge crystal field model of Stevens [12] and Hutchings [13], which is expressed in general terms of operator equivalents as

$$H_{\mathrm{cf}} = \sum_{l,m} B_l^m O_l^m, \tag{5.15}$$

where the coefficients B_l^m are related physically to the atomic number of the cation, its ionic radius, and the interionic distances. There is additional mathematical formalism that relates B_l^m to another set of coefficients designated as $A_l^m \langle r^l \rangle$ multiplied by constants α_J, β_J, and γ_J for respective values of $l = 2, 4$, and 6, which may be computed but are usually determined semiempirically for the particular situation. A partial list of operator equivalents is given in Table 5.3 for some of the more significant crystal field components. More comprehensive lists are found in the papers by Stevens and Hutchings cited above, as well as the book by Low [14].

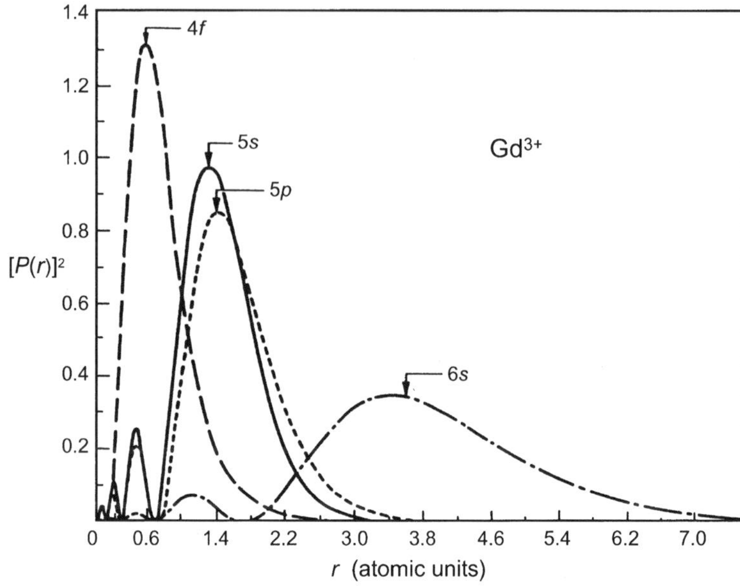

Fig. 5.4 Radial densities of Gd^{3+} Hartree–Fock wavefunctions. Figure reprinted from [11] with permission. http://link.aps.org/doi/10.1103/PhysRev.127.2058

Table 5.3 Partial list of lower-order crystal field operator equivalents[a]

Electrostatic potential	Operator equivalent	Notation
$\sum(3z^2 - r^2)$	$\alpha_J\langle r^2\rangle[3J_z^2 - J(J+1)]$	$\alpha_J\langle r^2\rangle O_2^0$
$\sum(x^2 - y^2)$	$\alpha_J\langle r^2\rangle\frac{1}{2}\left[J_+^2 - J_-^2\right]$	$\alpha_J\langle r^2\rangle O_2^2$
$\sum(35z^4 - 30r^2z^2 + 3r^4)$	$\beta_J\langle r^4\rangle\begin{bmatrix}35J_z^4 - 30J(J+1)J_z^2\\ +25J_z^2 - 6J(J+1)\\ +3J^2(J+1)^2\end{bmatrix}$	$\beta_J\langle r^4\rangle O_4^0$
$\sum(7z^4 - r^2)(x^2 - y^2)$	$(BH)_{\max} = \frac{1}{4}(4\pi M_s)^2$	$\beta_J\langle r^4\rangle O_4^2$
$\sum z(x^3 - 3xy^2)$	$\beta_J\langle r^4\rangle\frac{1}{4}\left[J_z\left(J_+^3 + J_-^3\right) + \left\{J_+^2 + J_-^2\right\}J_z\right]$	$\beta_J\langle r^4\rangle O_4^3$
$\sum(x^4 - 6x^2y^2 + y^4)$	$\beta_J\langle r^4\rangle\frac{1}{2}\left[J_+^4 + J_-^4\right]$	$\beta_J\langle r^4\rangle O_4^4$
$\sum(231z^6 - 315z^4r^2 + 105z^2r^4 - 5r^6)$	$\gamma_J\langle r^4\rangle\frac{1}{2}\begin{bmatrix}231J_z^6 - 315J(J+1)J_z^4 +\\ 735J_z^4 + 105J^2(J+1)^2J_z^2\\ -525J(J+1)J_z^2 + 294J_z^2 -\\ 5J^3(J+1)^3 + 40J^2(J+1)^2\\ -60J(J+1)\end{bmatrix}$	$\gamma_J\langle r^6\rangle O_6^0$

[a]Based on theory developed by Stevens [11]

5.2 Anisotropy of Single Ions

The foregoing discussion included the case of a single d electron occupying a singlet ground state stabilized from the T_{2g} triplet term by an axial tetragonal field or possible Jahn–Teller distortion of an octahedral site. This situation represents a type of building block for the remainder of the d^n series, and we shall now review the essential features of each member, first with regard to the g factors and the influence of partially quenched orbital angular momentum on anisotropy, and then in relation to magnetostriction. Although most of the discussion will focus on octahedral ligand coordinations, the contributions from tetrahedral sites will be pointed out where appropriate.

5.2.1 $3d^1$ and $3d^6$ *D*-State Triplet

In the previous section, the example of the d^1 orbital singlet ground state was analyzed to illustrate theoretical tools available for determining the single-ion paramagnetic anisotropy. In most situations, the electronic structure is not so amenable to analysis because the orbital ground state is not a singlet or involves an even-integer spin state ($S = 1$ or 2) that does not produce Kramers doublets, which are spin degeneracies that cannot be removed by a crystalline field and occur in odd numbered d-electron systems [15]. Furthermore, when the spin–orbit coupling operator is introduced to the calculation, energy multiplicities increase to the extent that the one-electron approximation must give way to the complete analysis in order to gain a more accurate picture of the electronic structure. However, unless detailed spectroscopy issues arise, we shall work with the simpler models.

To continue the discussion of the previous section, we first consider the case where the threefold orbital degeneracy of the lower T_{2g} state of the free-ion D term is not fully quenched. This occurs with Ti^{3+} $\left(d^1\right)$ and Fe^{2+} $\left(d^6\right)$ as described by the one-electron ground state diagrams and the corresponding multiple electron energy level structure of Fig. 5.5. For the d^6 case, it is convenient to view the degeneracy as arising from the single electron that begins the second half of the d shell, recalling that the first five are locked into collinear spin polarization dictated by Hund's rule with $\boldsymbol{L} = 0$ that therefore contribute nothing to the orbital degeneracy. [A similar approach can be made for the case of two electrons in the T_{2g} states (d^2 or d^7) by treating the source of degeneracy as spin vacancies or "holes" in the T_{2g} shell]. For both d^1 and d^6 ions, the triplet can be further stabilized as a singlet (analyzed in the previous section) or as a doublet, which can have an important influence on all magnetoelastic properties because the remaining degeneracy carries orbital angular momentum into the ground state, and hence a strong spin–orbit link to the crystal lattice.

Despite the success in interpreting the data for the orthorhombic crystal field of Ti^{3+} as an impurity substituted for Al^{3+} in certain hydrated alum salts [generic formula $(R \cdot 6H_2O)^{1+}$ $(Al \cdot 6H_2O)^{3+}$ $(SO_4)^{2-}{}_2$, where R^{1+} is a large monovalent (alkali metal) ion] described above, there remains uncertainty about the origins of the lower symmetry crystal field across the various members of the alkali metal members of this family. An examination of the data in Table 5.1 reveals a curious correlation between the strengths of lower symmetry crystal fields and the radius of the monovalent ions. The smaller the radius, the smaller the lattice parameter

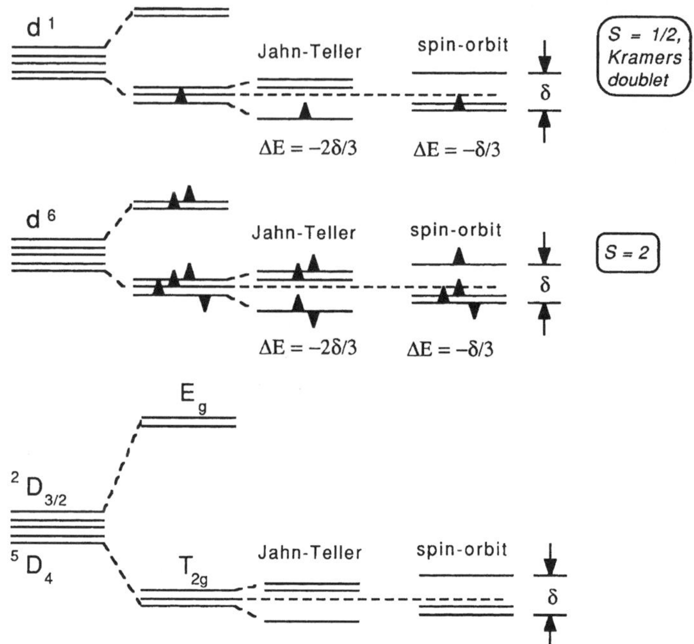

Fig. 5.5 Electronic structure of d^1 and d^6 configurations: one-electron and combined electron models

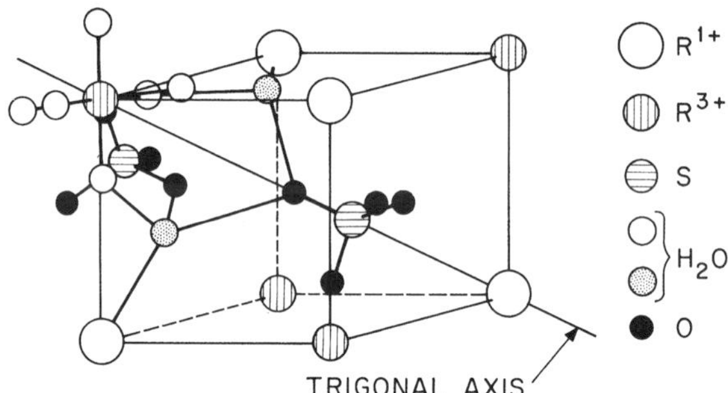

Fig. 5.6 Structure details of one octant of an α-type alum lattice, with trigonal symmetry axis indicated. Figure reprinted from [6] with permission. © 1968 by the American Physical Society. http://link.aps.org/doi/10.1103/PhysRev.172.325

(Vegard's law), and the greater is the departure from cubic symmetry. This recognition led to the local crystallographic model [6] that began with an axial field along a ⟨111⟩ axis of each $(Ti \cdot 6H_2O)^{3+}$ complex that is influenced by a trigonal distortion of a tetrahedral grouping of sulfate radicals $(SO_4)^{2-}$ illustrated in Fig. 5.6. Since the monotonic reductions in lattice parameter from Cs, Rb, Tl, K, through to Na correlate with increases in the T_{2g} splittings, it is reasonable to presume that the trigonal field would increase accordingly, preserving the condition $\delta > \lambda_{so}$ that is necessary to stabilize the orbital singlet in the T_{2g} shell, as discussed in Sect. 2.4.

The appearance of a further distortion as the smaller alkali ions were introduced was attributed to local lattice adjustments needed to accommodate the larger Ti^{3+} ion (0.76 Å radius) into the smaller host Al^{3+} (0.51 Å) site. Another mechanism would stabilize the d_{xy} singlet by a Jahn–Teller effect that causes a tetragonal z-axis expansion of the H_2O octahedron that is set up along the $\langle 100 \rangle$ cubic lattice axes. The combined distortions then account for the lower than axial symmetry that gives rise to the expected twelve equivalent paramagnetic complexes reported for RbAlTi alum [3,4].

Another concern is the concentration-dependent factor of spin–orbit stabilization that can occur in magnetically ordered situations induced by local exchange fields due to clustered Ti^{3+} impurity sites. A perturbation analysis of the T_{2g} splitting from a $\langle 100 \rangle$ axial distortion, spin–orbit coupling, and an exchange field is presented for the d^1 case in Appendix 5B. From this model, it is concluded that (2.34) for the threshold condition of doublet stabilization should be modified according to [16]

$$\frac{\lambda_{\text{so}}}{\delta} \geq \frac{2}{3}\left(1 + \frac{2}{3}\frac{\delta}{g m_{\text{B}} H_{\text{ex}}}\right), \tag{5.16}$$

for $\delta < g m_{\text{B}} H_{\text{ex}}$. This approximation loses validity when H_{ex} decreases with temperature as described by a Brillouin function, reaching zero at the Curie or Néel temperature. Based on the results from Ti^{3+} in Rb alum with $\lambda_{\text{so}} = 154\,\text{cm}^{-1}$ and $\delta \sim 10^3\,\text{cm}^{-1}$ deduced from the anisotropy of the measured g factors, this effect will produce S–O stabilization only if δ is significantly less than λ_{so} regardless of the value of H_{ex}. Moreover, the observation of paramagnetic resonance (EPR) signals from regions dominated by antiferromagnetic ordering is questionable. Nonetheless, any unquenched orbital momentum in a ground-state doublet could increase magnetoelastic effects that would enhance EPR line broadening through spin–lattice relaxation, as discussed in Chap. 6.

The earliest work on undiluted CsTi alum by Bleaney et al. [7] reported resonances with trigonal field Zeeman splitting factors $g_{\parallel} = 1.25$ and $g_{\perp} = 1.14$ and very large homogeneous linewidths even at the lowest temperature of 1.2 K. Such a result gave credence to the argument that strong spin–orbit–lattice interactions could be rendering $\delta < \lambda_{\text{so}}$ and allowing the line broadening to take place from a rapid Orbach relaxation process (see Chap. 6). A small axial strain that allows an orbital doublet ground state could be the result of a trigonal axis contraction from a J–T effect in the T_{2g} term, encouraged further by interionic Ti^{3+} exchange at $T \leq 4\,K$. Subsequent experiments by Woonton and MacKinnon [8] with dilute specimens of CsTiAl alum tended to confirm the Bleaney results. These later results, however, indicated that a smaller splitting $\left(\sim 10^2\,\text{cm}^{-1}\right)$ of the T_{2g} state likely exists and could explain a stronger than usual influence of the $l_z = \pm 1$ doublet. Later, work reported by Dubicki et al. [17] and Pigott et al. [18] have also suggested that spontaneous orbit–lattice deformation within the T_{2g} manifold could contribute to the observed peculiarities in the CsTi case.

Another occurrence of the Ti^{3+} octahedral water complex worthy of mention is as a dopant in sapphire Al_2O_3. In low concentrations this cation creates the blue-colored gem, but its technical importance lies in the 5,000 Å wavelength, i.e., $10Dq \sim 2.5\,eV$, of its optical transition which is useful for laser applications. The crystal field is trigonal, with a ligand coordination depicted in Fig. 2.6. Magnetically, the g-factors $g_{\parallel} = 1.067$ and $g_{\perp} < 0.1$ reported by [19] as S–O stabilized doublet ground state modified by dynamic Jahn–Teller effects, which suggests a field opposite in sign to that of CsTi alum. This interpretation was proposed by Ham [20] and later analyzed by Macfarlane et al. [21], Bates and Bentley [22], Stevens [23] and Abou-Ghantous et al. [24]. The question of how the degeneracy of the T_{2g} state is lifted can be viewed in the context of Jahn–Teller (singlet) vs. spin–orbit (doublet) stabilizations introduced in Sect. 2.4. In either case an axial (or lower) symmetry crystal field must occur by distorting of the ligands, if an appropriate lower symmetry field is not already present. Even if there is a lower symmetry component, a spontaneous distortion to enhance (or reduce) the existing field is always a possibility, as appears to occur in ferrimagnetic spinels and garnets.

The principal $3d^6$ ions of interest are Fe^{2+} and Co^{3+}, usually occupying an octahedral site.[4] Despite the similarities of the ground state, this configuration differs from the $3d^1$ case in various ways. With an even number of spins, there are no Kramers doublets to enable microwave spectroscopy; the value of Dq can be smaller if the cation valence charge is lower; the sign of the multielectron spin–orbit coupling constant λ_{so} is negative because the contributing $\boldsymbol{L}$ and $\boldsymbol{S}$ vectors of all d^n ions with $n > 5$ (upper half of the shell) are in opposite directions. This latter condition causes the g factor to increase to values greater than 2 in the perturbation analysis result given in (5.8) and (5.12), therefore causing an increased magnetic moment and greater paramagnetic anisotropy. Despite the absence of Kramers doublets in this even-electron system, microwave resonance measurements were carried out on Fe^{2+} in MgO because the negative λ_{so} combined with small-to-moderate δ splittings of the T_{2g} term likely caused by local random J–T effects that produced isotropic $g = 3.47$ and 6.83 [25].

An even more important effect for magnetoelastic properties is the occurrence of low and intermediate spin states with the higher valence Co^{3+} ion because the E_g shell is occupied. As pointed out in Sect. 2.4 (Fig. 2.19b), because of the intra exchange energy U_{ex} that compels the alignment of spins, the occupation can assume various configurations, depending on the relative values of $10Dq$ and U_{ex}. In Fig. 5.7, various situations are depicted. For the high-spin (hs) case of $S = 2$ with $U_{ex} \gg 10Dq$, the E_g shell is half-filled, and all of the magnetoelastic effects are the result of T_{2g} occupancies. Similar J–T stabilizations of an orbital singlet will occur for either a trigonal or tetragonal ligand distortion. However, inspection of the orbital lobes in relation to the ligand charges indicates that the local-site magnetostriction will be $\lambda_{111} > 0$ in the trigonal case (which has been confirmed with Fe^{2+}) and $\lambda_{100} < 0$ from a tetragonal contraction, typical of Co^{3+}. As a consequence, the singlet is stabilized with energy defined as $\Delta E = (2/3)\delta_t$ in either case. At the extreme where $U_{ex} \ll 10Dq$, all six spins condense into a low-spin (ls) $S = 0$ state

[4] See footnote at end of Sect. 2.4 for comment on Fe^{2+} in a tetrahedral site.

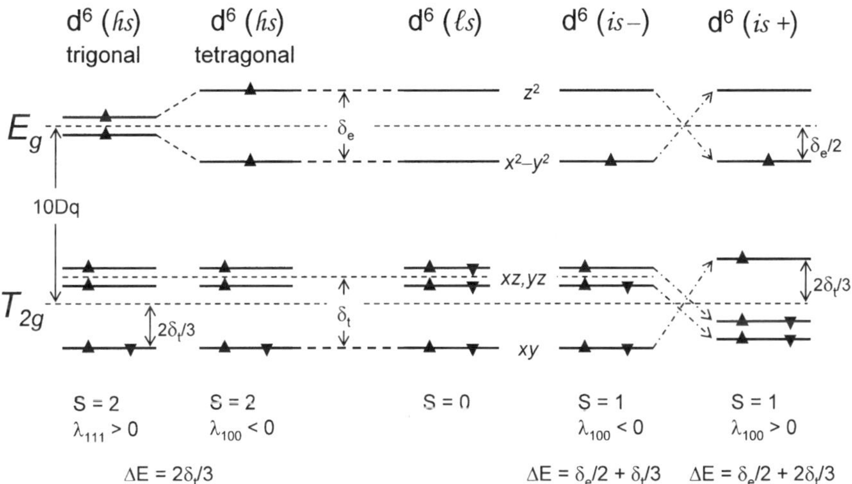

Fig. 5.7 Aufbau diagram of various octahedral-site distortions induced by S–O or J–T stabilizations of d^6 spin configurations. Symbols hs, ls, ±is indicate high, low, and intermediate spin states

to fully occupy the T_{2g} shell, thereby creating a diamagnet. The most interesting situation arises where $U_{ex} \sim 10Dq$, allowing the formation of intermediate-spin states (is-) and (is+) that are diagrammed in Fig. 5.7. In these two cases both the E_g and T_{2g} shells are occupied, such that the conventional J–T splitting of the e_g states provides a strong stabilization $\delta_e/2$ that is augmented by either $(1/3)\delta_t$ or $(2/3)\delta_t$, depending on the sign of the axial distortion. The implications of these competing possibilities become evident in experiments where the magnetostrictive strains reverse from $\lambda_{100} < 0$ to $\lambda_{100} > 0$ [31] of Chap. 2.

Most significant are the properties associated with magnetic exchange. Because of the partially filled e_g states that allows for strong exchange coupling, S–O stabilization of the T_{2g} term can occur in ferro- or ferrimagnetic systems such as magnetite Fe_3O_4, enhancing anisotropy and magnetostriction effects discussed in Sect. 5.3. In tetrahedral sites, d^1 and d^6 become pure Jahn–Teller ions with the lower E_g doublet term split to stabilize a singlet ground state. An example is Fe^{2+} in the uncommon spinel $Fe^{2+}\left[Al^{3+}\right]_2 O_4$ [26].

5.2.2 3d^4 and 3d^9 D-State Doublet (J–T Effect)

For the classic Jahn–Teller configurations, the g factors are not strongly influenced by the ligand environment for the simple reason that the ground state E_g doublet described by the diagrams in Fig. 5.8 contains no orbital angular momentum. The influence of the upper $|xz\rangle$ and $|yz\rangle$ states therefore occurs through mixing by the $\lambda \boldsymbol{L} \cdot \boldsymbol{S}$ operator, but as the perturbation theory dictates, the effect on g_z will be reduced to terms on the order of $\lambda \boldsymbol{L} \cdot \boldsymbol{S}/10Dq \sim 10^{-2}$ in most situations – significant

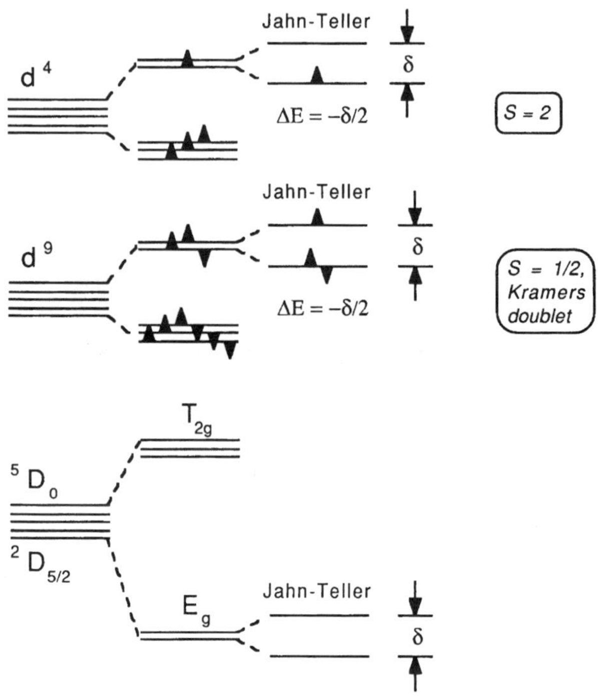

Fig. 5.8 Electronic structure of d^4 and d^9 configurations: one-electron and combined electron models

but not substantial for spin–lattice interaction considerations. Moreover, the d^4 configuration does not yield Kramers doublets for convenient microwave spin resonance analysis of the g factors. A multielectron calculation of the electronic structure is described by Low [27] that reveals the expected five-fold ladder of levels split according to λ^2/Δ multiples. Regrettably, the effect of a strong Jahn–Teller splitting of the E_g doublet was not found in the literature. The most common example of this configuration is Mn^{3+} (although Fe^{4+} makes an occasional appearance in nonstoichiometric garnets and other magnetic oxides). It almost universally occupies octahedral sites in oxygen coordinations[5] and therefore brings with it structural perturbations associated with the attendant J–T tetragonal distortions of the ligands that manifest themselves as magnetostriction effects or actual crystallographic phase changes when they become cooperative. In mixed-valence situations with Mn^{2+} or Mn^{4+}, charge transfer mechanism can provide interesting and sometimes anomalous electrical conductivity behavior, as will be examined in Chap. 8.

[5] There is evidence that Mn^{3+} (and Cu^{2+}) in ferrites enters tetrahedral sites in ferrimagnetic oxides where it can stabilize a doublet ground state in the T_{2g} shell with unquenched orbital momentum through spin–orbit coupling. The implications of this possibility are examined in relation to magnetostriction and spin–lattice relaxation in Sect. 5.3.6.

The paramagnetism of d^9 is seen most commonly in Cu^{2+} which served as the vehicle for the discovery of the J–T effect discussed in Sect. 2.4.3, and in that respect parallels the magnetoelastic behavior of the d^4 configuration. In early EPR work with the cupric salts, g-factors in the range of 2–2.5 were found, consistent with the expectation for $\lambda < 0$ [27] and theoretical discussions have been reported by Pryce [28] and Bates and Chandler [29]. The importance of this ion, however, may reside more in its charge transfer capability with Cu^{3+} and Cu^{1+} oxidation states for high-temperature superconductivity (also discussed in Chap. 8 than in its fundamental magnetism. The low spin value contributes little to magnetic systems and although, like Mn^{3+}, its magnetoelastic capability can assist in tailoring magnetostriction in magnetic oxides, cooperative J–T effects on lattice structural phases can sometimes be more of an irritant than an asset.

5.2.3 $3d^2$ and $3d^7$ F-State Triplet

In octahedral sites, the two d-electron case produces a ground-state triplet similar to that of the d^1 and d^6 cases, and can be treated in a similar manner by considering the T_{2g} state as occupied by a single spin vacancy or hole for simple approximations. However, an important distinction must be made. The total $\boldsymbol{L}$ value for these configurations is 3 instead of 2 (Fig. 2.3). This means that the free-ion term is F, not D, and there are seven orbital states instead of only five. The distinction becomes particularly of interest for optical transitions because of the different orbital splitting energies. For F states, the first expected orbital term above a ground triplet is the other triplet which is separated by $8Dq$. Nonetheless, the ground state can still be treated by the single electron model for our present purposes. Inspection of the diagrams of Fig. 5.9 indicate that J–T and S–O stabilizations can take place in a manner similar to the d^1 and d^6 cases, but with inverted electronic energy structures. This feature will be examined more critically in relation to magnetostriction and spin–lattice relaxation in ordered spin systems. The splitting of the lower triplet, however, will contribute significantly to the single-ion anisotropic g factors because the δ splittings are small enough to render denominator of the Λ_{ij} factor in the spin Hamiltonian in (5.11) small enough to cause a significant departure of the g-factor from the isotropic spin-only value of 2. For this reason, g factors for V^{3+} (d^2) in Al_2O_3 and Co^{2+} (d^7) in MgO have been measured, respectively, as 1.92 [30] with a positive $\lambda = 104\,\text{cm}^{-1}$ and 4.28 [31] with a negative $\lambda = -180\,\text{cm}^{-1}$.

When the d^7 case is analyzed in the multielectron ^4F term format, the degenerate T_{1g} state is lowest and its threefold degeneracy can be approximated by a pure P state with $L = 1$ [32, 33]. This approach simplifies the perturbation calculation by the adoption of an effective spin–orbit coupling operator to produce a set of orbital states with Kramers spin degeneracies included. For our purposes, however, the one-electron model is sufficient to gain a physical understanding of how spin–orbit coupling can further stabilize an unquenched orbital doublet by enhancing the effects of an exchange field. This effect can account for the large Co^{2+} anisotropy

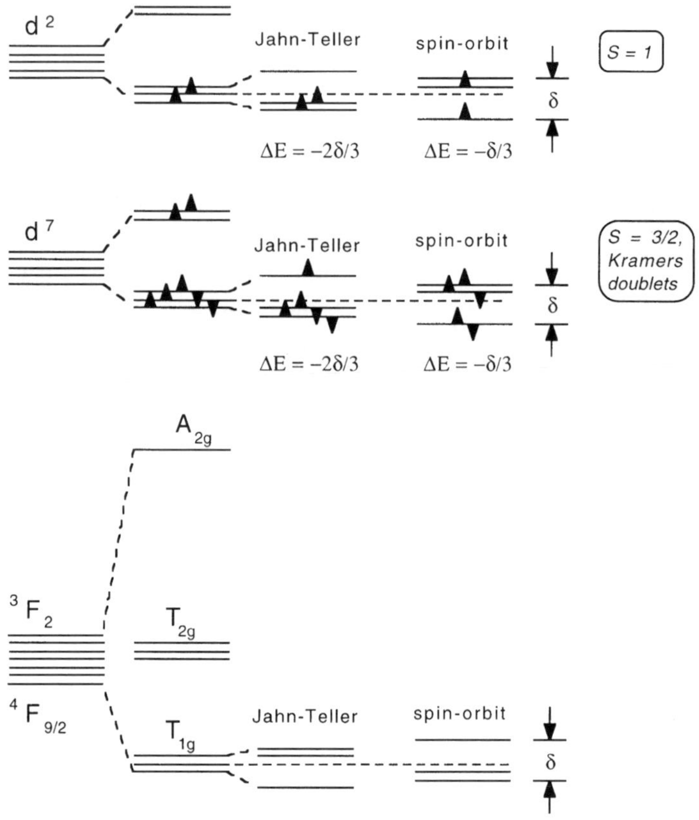

Fig. 5.9 Electronic structure of d^2 and d^7 configurations: one-electron and combined electron models

contributions in magnetically ordered systems. It can also help to explain the source of the small orbital contribution to the paramagnetic anisotropy and magneto-optical effects of the S-state Fe^{3+} ion in cubic crystal fields typical of spinel and garnet ferrimagnets.

5.2.4 *$3d^3$ and $3d^8$ F-State Singlet*

The influence of orbital angular momentum on the paramagnetism and anisotropy of the g factors is felt less in the F state cases, where the A_{1g} singlet in cubic field is lowest, as shown in Fig. 5.10 for d^3 and d^8 configurations in an octahedral site. With the first excited orbital state separated in energy from the ground state by $10Dq$, a smaller $8\lambda/10Dq$ correction to g in (5.12) can usually be expected. The most studied ion of this type in an octahedral site is Cr^{3+} (d^3), with a spin

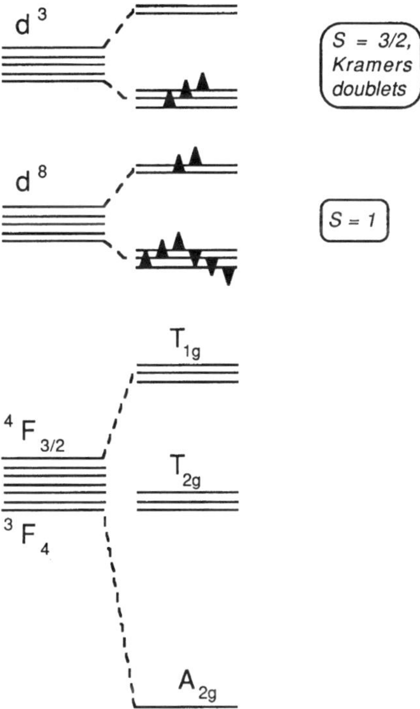

Fig. 5.10 Electronic structure of d^3 and d^8 configurations: one-electron and combined electron models

$S = 3/2$ that provides three Kramers doublets. Its EPR spectrum has been studied in paramagnetic salts [34] and also in Al_2O_3 (*ruby*) [35] where it not only served as the vehicle for the invention of the *maser* [36] by utilization of its comparatively large trigonal D parameter that required a millimeter-wave frequency, but also produced the earliest *laser* utilizing the $10Dq$ $(\approx 17{,}000\,\text{cm}^{-1})$ red optical transition. To analyze these singlet ground states, the expanded spin Hamiltonian of (5.14) is employed, with emphasis on the second-order D and E terms. In octahedral sites, neither Cr^{3+} nor its d^8 counterpart Ni^{2+} are expected to be active in a spin–lattice sense because of the $l_z = 0$ singlet ground state.[6] However, despite the absence of J–T or S–O stabilization possibilities in the octahedral sites that they almost always occupy, the Ni^{2+} ions have exhibited an isotropic $g \approx 2.25$ in MgO [37]. The absence of anisotropy is consistent with the first-order cubic crystal field, and the origin of the apparent larger spin–lattice contribution can be found in a consideration of the relative magnitudes of λ and Dq in the first-order approximation given by (5.12), $g = 2 - 8\lambda/10Dq$. If we apply $\lambda = -335\,\text{cm}^{-1}$ from Table 5.2 and assume

[6] In nickel spinel ferrite, a small fraction of Ni^{2+} enters the tetrahedral sublattice, where a triplet similar to that of the Co^{2+} case is stabilized.

$Dq \approx 1{,}000\,\text{cm}^{-1}$ for divalent cations taken from optical absorption measurements of Ni^{2+} in octahedral fields such as $Ni(H_2O)_6{}^{2+}$ [38–41], we arrive at g values ≈ 2.26 without even taking into account the influence of covalent bonding on reducing orbital contributions. For Cr^{3+} with $\lambda = 87\,\text{cm}^{-1}$ and $Dq \approx 1{,}700\,\text{cm}^{-1}$ for trivalent cations, this calculation yields $g \approx 1.96$ in good agreement with the measured value. However, the EPR spectrum of this ion features large angular variations because of the D and E terms.

5.2.5 3d⁵ S-State Singlet

There remains to be discussed the d^5 configuration exemplified most commonly by Fe^{3+} and Mn^{2+}, which is the least sensitive to its ligand environment while providing the strongest superexchange interaction of all the transition-metal ions. To explain this apparent paradox, we must begin by pointing out that although five d electrons in the high-spin state produce an $S = 5/2$ total spin, the addition of orbital angular momentum leaves an $L = 0$ state and a ${}^6S_{5/2}$ free-ion ground term. As a consequence, orbit–lattice effects are nonexistent in the pure ground state, but the $\Sigma b^2/2U$ exchange term involves all five half-filled orbital states to produce a large $\boldsymbol{JS} \cdot \boldsymbol{S}$ energy. Small contributions to the anisotropy are introduced through the excited free-ion terms that are discussed in Sect. 2.3 as part of the exposition of the "strong-field" approach to crystal field theory. In the $L = 0$ case, any orbital angular momentum influence must be derived from coupling between the ground ${}^6S$ term and the excited terms from the Russell–Saunders coupling listed in Table 5.4.

As indicted by the sketches in Fig. 5.11, ${}^4G$, ${}^4P$, ${}^4D$, ${}^4F$, and others are sequentially laddered above the ground ${}^6A$ singlet, each with its own set of degeneracies. The states that are sensitive to the cubic crystal field parameter are shown with limiting-case electron occupancies. These energy levels were computed by Tanabe and Sugano using the Slater integrals and A, B, and C Racah parameters also defined in Sect. 2.3. The procedure employed is outlined in Appendix 5C.

In a cubic field, the first excited state with orbital angular momentum becomes the ${}^4T_{1g}$ from the ${}^4G$ term for both octahedral and tetrahedral sites. Orbital influence felt in the ground state through spin–orbit coupling that links the two separate terms instead of states within a J manifold, can be analyzed by the theoretical tools outlined by Griffith [2], with the result that a hybrid ground state φ_o is formed according to

Table 5.4 d^5 orbital energy terms subject to splitting in a cubic crystal field

Russell–Saunders (free-ion) term	Cubic crystal-field states
^{2}I	${}^2A_{1g} + {}^2A_{2g} + {}^2E_g + {}^2T_{1g} + {}^2T_{2g}$
^{4}F	${}^4A_{2g} + {}^4T_{1g} +^4 T_{2g}$
^{4}D	${}^4E_g + {}^4T_{2g}$
^{4}P	${}^4T_{1g}$
^{4}G	${}^4A_{1g} + {}^4E_g + {}^4T_{1g} + {}^4T_{2g}$
^{6}S	${}^6A_{1g}$

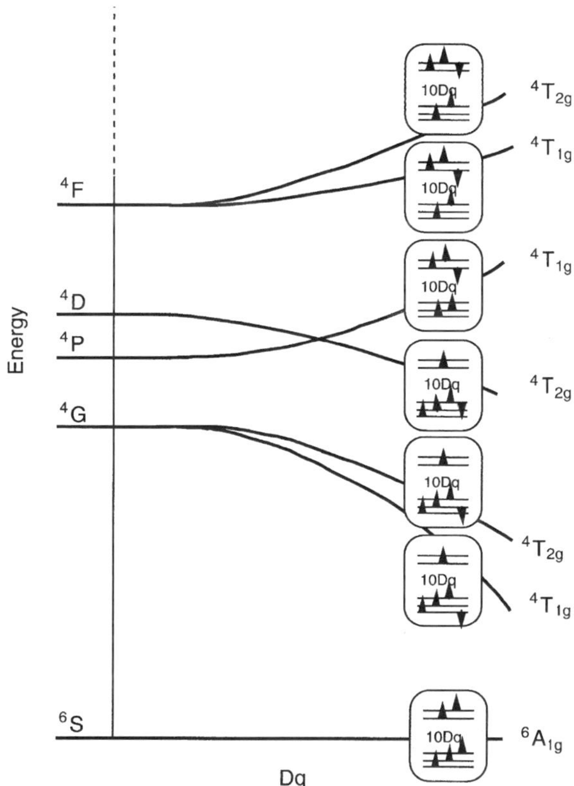

Fig. 5.11 Variation of d^5 energy terms as a function of cubic field strength $10Dq$, showing proposed electron configurations

$$|\varphi_{\mathrm{o}}\rangle \approx \left\langle {}^6A_{1\mathrm{g}}, m_{\mathrm{s}}\right| + \varepsilon \left|{}^4T_{1\mathrm{g}}, m_{\mathrm{s}}\right\rangle, \tag{5.17}$$

where

$$\varepsilon \approx \frac{\sum_n \left\langle A_{1\mathrm{g}}, m_{\mathrm{s}}\right|\lambda L \cdot S \left|T_{1\mathrm{g},n}, m_{\mathrm{s}}\right\rangle}{\left|E_{T_{1\mathrm{g}}} - E_{A_{1\mathrm{g}}}\right|}. \tag{5.18}$$

where n corresponds to any of the three orbital states within the $T_{1\mathrm{g}}$ triplet (analogous to a P state with $l_z = 1, 0, -1$). The approximations in (5.18) and (5.18) represent only part of a more comprehensive computation of ε for the spin–orbit interaction with ${}^4P_{5/2}$ excited state that was reported by Watanabe [42] who produced the result for the g-factor reduction as

$$g = \left(1 - \varepsilon^2\right) g_{\mathrm{s}} + \varepsilon^2 g_{\mathrm{p}}, \tag{5.19}$$

where g_{S} and g_{P} are the g factors for the ${}^6S_{5/2}$ and ${}^4P_{5/2}$ states, and $\varepsilon = \sqrt{2}\lambda\,(E_{\mathrm{P}} - E_{\mathrm{S}})$, with $(E_{\mathrm{P}} - E_{\mathrm{S}})$ representing the denominator of (5.18) [43]. The ${}^4P_{5/2}$ excited state is featured here as the logical mixing state because it has

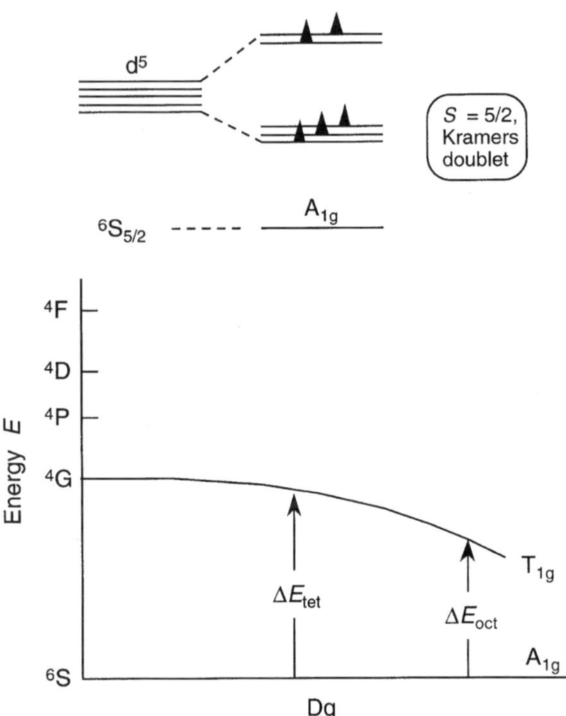

Fig. 5.12 Electronic structure of d^5 configurations: one-electron and combined electron models, plus the more complete diagram showing the relation of the A_{1g} ground state and the ^4G term

the same $J(=5/2)$ and would therefore blend readily with $^6S_{5/2}$ under a $\lambda \boldsymbol{L} \cdot \boldsymbol{S}$ perturbation. As demonstrated by Watanabe, a complete treatment would require a summation over the full ladder of spin states, rather than the single excited state approximation given by (5.19) shown here to illustrate how the excited terms can affect the orbital angular momentum of an S state in a crystal field. The influence of the cubic crystal field can be seen in Fig. 5.12 where the inverse dependence of ΔE on Dq is illustrated. As a consequence, the degree of mixing of the excited state into the ground singlet would be greater for an octahedral site than a tetrahedral site because $Dq_{\text{tet}} = -(4/9)Dq_{\text{oct}}$. This conclusion is supported by observations of the behavior of magnetic anisotropy in the dilution of ferrimagnetic spinels and garnets discussed in Sect. 4.3.

Paramagnetic anisotropy of single d^5 ions, therefore, is a higher order effect than for the other members of the series. g-factors are, in effect, equal to the pure spin value of 2. However, paramagnetic resonance measurements reveal anisotropic effects in the angular dependence of the microwave spectra that are attributed to the a parameter. These data have been interpreted by the spin Hamiltonian of the $^6S_{5/2}$ state in a cubic crystal field with lower symmetry components given by

$$H_s = m_B \boldsymbol{g} \cdot \boldsymbol{H} \cdot \boldsymbol{S} + \frac{1}{6} a \left[S_z^4 + S_x^4 + S_y^4 - \frac{1}{5} S(S+1)\left(3S^2 + 3S - 1\right) \right]$$
$$+ D\left[S_z^2 - \frac{1}{3} S(S+1) \right] + E\left[S_x^2 - S_y^2 \right] \tag{5.20}$$
$$+ F\left[35S_z^4 - 30S(S+1)S_z^2 + 25S_z^2 - 6S(S+1) + 3S^2(S+1)^2 \right],$$

where the x, y, and z axes of the spin system are chosen here to coincide with those of the crystal-field coordinates.

In most cases, the cubic parameter a can be treated as negligible. D and E, however, have influence where significant lower symmetry fields are present. Such a situation occurs when Fe^{3+} ions occupy the trigonal bipyramid sites (Fig. 2.7) in magnetoplumbites, e.g., M-type Ba hexaferrite, which exhibit very large collective anisotropies. In general, the a term of (5.20) causes the ^6S state to split into a Kramers doublet, $m_s = \pm 5/2$ and a quartet with $m_s = \pm 3/2$ and $\pm 1/2$ [44]. In an axial field the spin state group divides further into three Kramers doublets, $m_s = \pm 5/2$, $\pm 3/2$, and $\pm 1/2$ [32, 33]. These Zeeman states are depicted in Fig. 5.13. Experimental results and their analysis have been reported by a number of authors, and a review of this work can be found in Bleaney and Stevens [45] and Low [13]. More relevant to the present discussion perhaps is the work of Folen on Fe^{3+} in lithium spinel ferrite with Al^{3+} and Ga^{3+} dilutions [46].

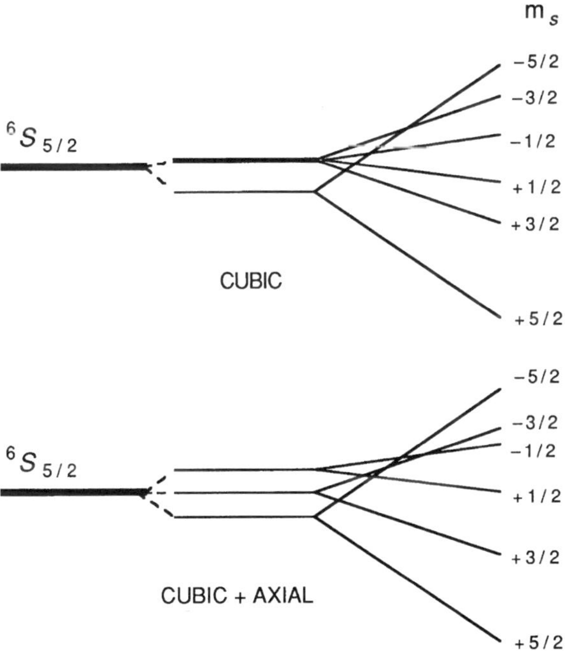

Fig. 5.13 Higher-order energy level stabilizations of the $^6S_{5/2}$ ground term of Fe^{3+} $(3d^5)$ in cubic and cubic+ axial crystal fields (based on discussions in Section 19 of Low [13]

Throughout the ions of the transition series, competition between crystal fields and spin–orbit coupling for control of the orbital angular momentum dictates the crystal magnetocrystalline anisotropy and magnetostriction, which in turn determines the relative ease (or difficulty) in magnetizing a material and the degree to which the magnetized state is retained once the magnetic field is removed, i.e., the hysteresis properties. A strong spin–orbit coupling will also provide a ready path for electromagnetic energy to flow into the lattice from the spin system when the magnetic material becomes a microwave frequency transmission medium. We can now begin to examine these phenomena in the context of these discussions.

5.2.6 $4f^n$ Ion Anisotropy

To illustrate the difference between the crystal-field effects on the exposed $3d$ electrons of the iron-group ions and the shielded $4f$ electrons of the lanthanide group, we consider the energy-level diagrams of the respective three-electron cases in Fig. 5.14. These examples represent idealized situations for typically Cr^{3+} $(3d^3)$ and Nd^{3+} $(4f^3)$. Note that the exchange field is indicated as acting after spin–orbit coupling and possibly the crystal field as well in the $4f^n$ case because of the reduced covalency, i.e., smaller b exchange integrals. An example of crystal-field splittings of the lowest multiplets for rare earths in chlorides is given in reference [15]. Table 5.5 lists the prominent crystal field parameters for operator equivalent use in cubic, C_{3v}, and C_{3h} symmetries. It should also be commented that rare-earth ions condense into sites of lower symmetry in many compounds, although in magnetic oxides they are found in cubic coordinations such as in the garnets and perovskite families.

The absence of quenching of the orbital angular momentum leaves spin–orbit coupling as the dominant interaction in the $4f^n$ group. For this reason, only a small crystal field stabilization energy is sufficient to create large anisotropy of the g factors in single ions. The strong coupling between the spin and the lattice that results will be examined in Chap. 6 in connection with high-frequency properties. Microwave spectroscopy has provided data on the g-factor anisotropy for the various ions of the rare-earth series and some of these are listed in Table 5.6 as compiled by Elliott and Stevens [47]. For more details of individual ion behavior through the series, the reader is directed to reviews by Orton [48], Low [13], and Bowers and Owen [49].

Among the $4f^n$ group we have seen in Chap. 4 how the heavy rare earths ($n = 8$–13 with $J = |L + S|$) participate in the extraordinary magnetization properties of the iron garnets. These ions also strongly influence the anisotropy and magnetostriction behavior when exchange fields are significant. The light rare earths ($n = 1$–6, with $J = |L - S|$) do not offer strong magnetism because of their lower J values, but certain paramagnetic and optical properties have proven to be important in laser and magneto-optical applications.

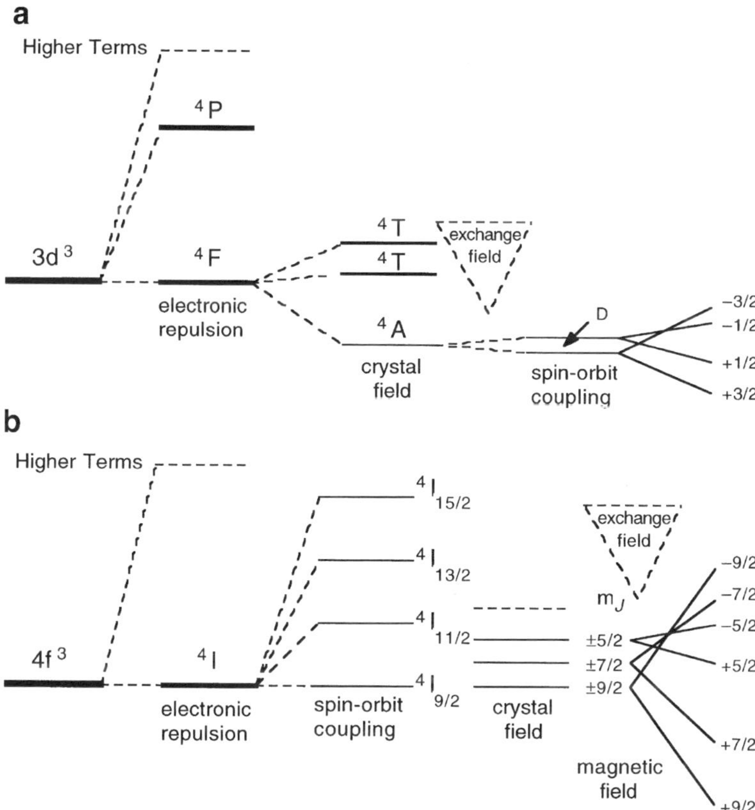

Fig. 5.14 Ground-term energy-levels: **(a)** the $3d^3$ configuration, typical of the Cr^{3+} ^4F-state ion in point-charge cubic crystal field with an axial distortion characterized by the parameter D. The Zeeman splitting energy $h\nu = gm_B H$ for $\Delta S = 1$ can occur at three values of H (not indicated). In this case $g \approx 2$ and is reduced only slightly by the multiplet interaction because the ground term after the crystal field splitting is an orbital singlet (^4A with $L = 0$); and **(b)** the $4f^3$ configuration, typical of the light rare-earth ion Nd^{3+} ^4I-state ion. For this series, $J = |L - S|$ in the lower half (light rare earths) and $|L + S|$ in the upper half (heavy rare earths). In the manifold of energy levels induced by a magnetic field, the signs of J_z are reversed between the light and heavy rare earth ions

Table 5.5 Crystal field symmetries and operator equivalent terms common to $4f^n$ ions in solids[a]

Compound	Site symmetry	$\mathcal{V}_{cf}$ terms
CaF_2, ThO_2	Cubic	$O_2^0, O_4^4, O_6^0, O_6^4$
Double Nitrate	C_{3v}	$O_2^0, O_4^4, O_6^0, O_6^4, O_6^6$
Ethyl Sulfate; Chloride	C_{3h}	$O_2^0, O_4^4, O_6^0, O_6^6$ (O_6^{-6})

[a]Compiled by Taylor and Darby [14]

Table 5.6 Rare-earth ion anisotropy parameter data for ethyl sulfate hosts[a]

$4f$ Ion	n	λ	$g_{\parallel}$	$g_{\perp}$	$A_2^0\langle r^2\rangle$	$A_4^0\langle r^4\rangle$	$A_6^0\langle r^6\rangle$	$A_6^6\langle r^6\rangle$	Ground state
	–	cm^{-1}	–	–	10^{-4} cm^{-1}	10^{-4} cm^{-1}	10^{-4} cm^{-1}	10^{-4} cm^{-1}	
Ce^{3+}	1	640	0.995 3.725	2.185 0.20	−15	−40	–92	1,150	One/Two Kramers doublets
Pr^{3+}	2	800	1.525		−50	−100	−48	660	Split doublet
Nd^{3+}	3	900	3.535	2.072	−15	−35	−60	640	Kramers doublet
Pm^{3+}	4	(1,070)							
Sm^{3+}	5	1,200	0.596	0.604	0	−30	−54	590	Kramers doublet
Eu^{3+}	6	1,410							
Gd^{3+}	7	1,540	1.991	1.991					
Eu^{2+}			1.991	1.991					
Tb^{3+}	8	1,770	17.72	<0.3	37	−20	−20	220	Split doublet
Dy^{3+}	9	1,860							
Ho^{3+}	10	2,000	15.36		0	−18	−22	170	
Er^{3+}	11	2,350	1.47	8.85	0	–40	–30	330	Kramers doublet
Tm^{3+}	12	2,660							
Yb^{3+}	13	2,940							

[a]Data compiled by Elliott and Stevens [47]

The single-ion anisotropy of the $3d^n$ and $4f^n$ ions seen in paramagnetic systems can also have a strong influence on spontaneous magnetism to produce magnetocrystalline anisotropy that is responsible for the directional magnetic capabilities that are essential to many applications.

5.3 Magnetocrystalline Anisotropy and Magnetostriction

The foregoing discussions have focused on magnetization established spontaneously through the medium of covalent bonding by the mechanism called magnetic exchange. As the molecular-orbital models showed, the exchange creates an energy stabilization of ionic spins in parallel or antiparallel ordering that we initially assumed to be insensitive to the symmetry of the crystal lattice. When a ferro- or ferrimagnetically ordered system is placed in an external magnetic field, the anisotropic tendencies created by the interaction of magnetic ions (collectively or as individuals) with the lattice are acquired by the resultant magnetic moment to create preferred directions of magnetization called "easy" axes. The corresponding magnetically "hard" directions alternate with the easy directions in any crystallographic plane of the lattice. Magnetocrystalline anisotropy energy is the magnetostatic work necessary to rotate the magnetic moment from an easy to hard direction, and the field required is the anisotropy field. Magnetostriction is the corresponding reaction by the lattice in the form of an elastic strain that occurs along any axis, but is usually characterized by the deformation along the magnetic field direction. Although both of these effects are manifestations of a magnetoelastic interaction, the relation between them can vary widely, depending on the origin of the crystalline source of the anisotropy and the strength of the spin–orbit coupling of the magnetic ion.

5.3.1 Phenomenological Anisotropy Theory

The beginnings of magnetocrystalline anisotropy theory were phenomenological, resulting from interpretation of measurements and based on symmetry arguments. This early history will not be reviewed here. However, the fundamental relations and parameter definitions are essential elements of the dialogue. For background and derivations the reader is directed to accounts found in standard magnetism texts by Chikazumi [50], Smit and Wijn [51], Bozorth [52], and Morrish [53].

Referring to the coordinate-axe/$\boldsymbol{M}$-vector sketches of Fig. 5.15a for the case of uniaxial anisotropy, we can express the magnetocrystalline anisotropy energy as a series expansion according to

$$E_{\mathrm{Ku}} = K_{\mathrm{u}1}\sin^2\theta + K_{\mathrm{u}2}\sin^4\theta + \cdots, \tag{5.21}$$

where $K_{\mathrm{u}1}$ and $K_{\mathrm{u}2}$ are the first- and second-order uniaxial anisotropy constants, and θ is the polar angle made by $\boldsymbol{M}$ with the z-axis. Here the situation is straightforward

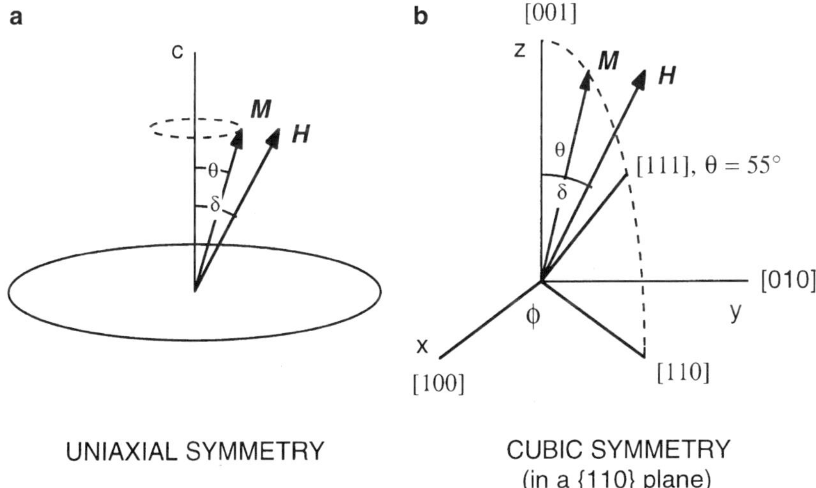

Fig. 5.15 Vector diagrams of the magnetocrystalline anisotropy **(a)** the uniaxial case, and **(b)** the cubic case, with rotation confined to a {110} plane, i.e., $\phi = 45°$

because the analysis is two-dimensional, involving only the polar angle θ between $\boldsymbol{M}$ and the axis of symmetry. The cubic symmetry case is more complex because of the anisotropy in the azimuthal plane. In polar coordinates (depicted for rotation of $\boldsymbol{M}$ in a $\{110\}$ plane in Fig. 5.15b), in-plane variation of E_{Kc} is characterized by the azimuthal angle ϕ, according to

$$\begin{aligned} E_{\mathrm{Kc}} = {} & K_1 \left(\sin^4\theta \sin^2\varphi \cos^2\phi + \sin^2\theta \cos^2\theta\right) \\ & + K_2 \sin 4\theta \cos^2\theta \sin^2\varphi \cos^2\varphi, \end{aligned} \tag{5.22}$$

where K_1 and K_2 are the respective first- and second-order cubic anisotropy constants.

A static method for determining the anisotropy constants in a particular plane is to measure the mechanical torque imposed on the specimen when it is rotated from an equilibrium position where $\boldsymbol{M}$ is initially aligned with $\boldsymbol{H}$ along an easy axis. After rotation of the specimen easy axis through angle δ away from $\boldsymbol{H}$, $\boldsymbol{M}$ assumes an equilibrium angle θ with respect to the easy axis. The new equilibrium is established when the anisotropy torque balances the torque from the magnetostatic energy $E_{\mathrm{HM}} = -HM\cos(\delta\text{–}\theta)$. The offsetting torques as function of θ are determined by differentiation, according to

$$\tau_{\mathrm{Ku}}(\theta) = -\frac{\partial E_{\mathrm{Ku}}(\theta)}{\partial\theta} = -K_{\mathrm{u}1}\sin 2\theta - 4K_{\mathrm{u}2}\sin^3\theta\,\cos\theta, \tag{5.23}$$

and

$$\tau_{HM}(\theta) = -\frac{\partial E_{HM}(\theta)}{\partial\theta} = -HM\,\sin(\delta-\theta)\,. \tag{5.24}$$

If the specimen is magnetically saturated, $\theta = \delta$ and measurement of τ_{Ku} as a function of θ will provide the means for measurement of τ_{Ku} as a function of $K_{\mathrm{u}1}$ and $K_{\mathrm{u}2}$.

For the cubic case

$$\tau_{\mathrm{Kc}}(\theta) \approx -K_1\left(2\sin\theta\cos^3\theta - \sin^3\theta\cos\theta\right) - K_2\left(\cos 4\theta\cos^2\theta - \frac{1}{4}\sin^4\theta\sin 2\theta\right). \tag{5.25}$$

In these models, $\boldsymbol{H}$ sets the azimuthal plane ($\theta = 0$) and $\boldsymbol{M}$ and $\boldsymbol{H}$ share the same value of ϕ.

If the specimen is constrained from rotating, the effect of magnetization rotation away from an easy direction as the magnetic field is increased and can be observed by studying the change in the $\boldsymbol{M}$ component along the direction of $\boldsymbol{H}$. The equilibrium positions of the $\boldsymbol{M}$ vector at each value of H is given by the value of θ for which the torque τ_{K} is balanced by τ_{H} as obtained from (5.23, 5.24, and 5.25), according to

$$\begin{aligned} \frac{\partial(\tau_{\mathrm{Ku}} + \tau_{\mathrm{H}})}{\partial\theta} = {} & -2K_{\mathrm{u}1}\cos 2\theta - K_{\mathrm{u}2}\left(4\sin^4\theta + 12\sin^2\theta\cos^2\theta\right) \\ & - H_{\mathrm{Ku}}M\cos(\delta-\theta) = 0, \end{aligned} \tag{5.26}$$

and

$$\begin{aligned}\frac{\partial\left(\tau_{\mathrm{Kc}}+\tau_{\mathrm{H}}\right)}{\partial\theta} = & -2K_1\left(2-\sin^2\theta-3\sin^2 2\theta\right)\\ & -K_2\left[\frac{1}{2}\sin^2\theta\left(6\sin^4\theta-11\sin^2\theta\cos^2\theta+\sin^4\theta\right)\right]\\ & -H_{\mathrm{Kc}}M\cos\left(\delta-\theta\right)=0.\end{aligned} \tag{5.27}$$

Examples of the magnetization as a function of field where the fields required to magnetize the specimens vary according to material and crystal orientation with respect to the applied field $\boldsymbol{H}$ can be found in [54]. Calculations of these measured curves can be carried out by finding the roots of angle θ for a fixed value of δ by solving the equation or by numerical computation to determine $M\cos\theta$ for each value of H, given values of M, angle δ, and the appropriate K constants. The field required to fully magnetize by rotating $\boldsymbol{M}$ between symmetry axes in a particular crystallographic plane is used to characterize the difficulty in rotating $\boldsymbol{M}$ and is treated as an intrinsic *anisotropy* field. This effective internal "bias" field equates to the rotational "stiffness" of the anisotropy torque in the limit where $\theta \rightarrow \delta$, according to

$$\begin{aligned}H_{\mathrm{Ku}} &\approx \frac{2K_{\mathrm{u}}}{M}\cos 2\delta\\ H_{\mathrm{Kc}} &\approx \frac{2K_1}{M}\left(1-\frac{1}{2}\sin^2\delta-\frac{3}{2}\sin^2 2\delta\right)\\ &\quad+\frac{K_1}{2M}\left[\sin^2\delta\left(6\sin^4\delta-11\sin^2\delta\cos^2\delta+\sin^4\delta\right)\right].\end{aligned} \tag{5.28}$$

The above example for H_{Kc} was originally derived by Bickford [55], and applies to locations of the vector in a plane of the $\{110\}$ family, with angle δ referred to a $\langle 100\rangle$ axis. The values of E_{K} and H_{K} expressed in terms of anisotropy constants along principal axes of symmetry are summarized in Table 5.7 for rotation in the three most important families of planes in a cubic lattice: $\{100\}$, $\{110\}$, and $\{111\}$. The axes of minimum and maximum E_{K}, i.e., easy and hard directions, are (100) and (110) in the $\{100\}$ planes, (100), (110), and (111) in the $\{110\}$ planes, and (111) and (112) in the $\{111\}$ planes. In Appendix 5D, the complete relations for H_{K} as functions of angle are listed for torques both in-plane and out-of-plane. The H_{K} out-of-plane is necessary in ferromagnetic resonance, which is reviewed in Chap. 6.

5.3.2 *Phenomenological Magnetostriction Theory*

The second part of the magnetoelastic phenomena is the reaction of the lattice to the internal stress imparted by the torque when the magnetization is rotated from its minimum energy or easy axis position. The result is called magnetostriction and is manifested by a measurable deformation strain $\Delta l/l$ of the lattice, usually in the

Table 5.7 Magnetocrystalline anisotropy energies and anisotropy fields in cubic lattices

{Plane}	⟨Axis⟩ (δ)	E_K	H_K (in plane)	H_K (out of plane)
{100}	⟨100⟩ (0°, 90°)	0	$2K_1/M_s$	$2K_1/M_s$
	⟨110⟩ (45°)	$K_1/4$	$-2K_1/M_s$	$K_1/M_s + K_1/2M_s$
{110}	⟨100⟩ (0°)	0	$2K_1/M_s$	$2K_1/M_s$
	⟨112⟩ (35°)	$K_1/4 + K_2/54$	$-K_1/M_s + K_1/18M_s$	$-K_2/3M_s$
	⟨111⟩ (55°)	$K_1/3 + K_1/27$	$-4K_1/3M_s$ $-4K_1/9M_s$	$-4K_1/3M_s$ $-4K_1/9M_s$
	⟨110⟩ (90°)	$K_1/4$	$K_1/M_s + K_1/2M_s$	$-2K_2/M_s$
{111}	⟨112⟩ (0°, 60°)	$K_1/4 + K_2/54$	$-K_2/3M_s$	$-K_1/M_s + K_1/18M_s$
	⟨110⟩ (30°, 90°)	$K_1/4$	$-K_2/3M_s$	$-K_1/M_s + K_1/2M_s$
{112}	⟨111⟩ (0°)	$K_1/3 + K_1/27$	$-4K_1/3M_s$ $-4K_1/9M_s$	$-4K_1/3M_s$ $-4K_1/9M_s$
	⟨110⟩ (90°)	$K_1/4$	$-K_2/3M_s$	$-K_1/M_s + K_2/6M_s$

range of 10^{-5}–10^{-6}, depending on the amount of anisotropy energy involved and the elasticity of the lattice. The formal treatment of the origins of magnetostriction will not be included here because their mathematical details would unnecessarily burden the text. This information can be found in most standard texts on magnetism, and in particular, the review article by Lee [56] and the book by du Trémolet de Lacheisserie [57]. For our purposes, the discussion will focus on the results of these analyses from two standpoints: (a) the phenomenology of magnetostriction and its relation to magnetocrystalline anisotropy and (b) the effect on the anisotropy from applied stress, known as *inverse* magnetostriction.

The lattice strain resulting from the magnetization of a single crystal of cubic symmetry is expressed as

$$\begin{aligned}\frac{\Delta l}{l} &= \frac{3}{2}\lambda_{100}\left(\alpha_x{}^2\beta_x{}^2 + \alpha_y{}^2\beta_y{}^2 + \alpha_z{}^2\beta_z{}^2 - \frac{1}{3}\right) \\ &\quad + 3\lambda_{111}\left(\alpha_x\alpha_y\beta_x\beta_y + \alpha_y\alpha_z\beta_y\beta_z + \alpha_z\alpha_x\beta_z\beta_x\right), \qquad (5.29)\end{aligned}$$

where λ_{100} and λ_{111} are the magnetostriction constants, and α_1, α_2, and α_3 are the direction cosines of the $\boldsymbol{M}$ vector, while β_1, β_2, and β_3 are the direction cosines of a strain Δl. Inspection of (5.29) reveals that $\Delta l/l = \lambda_{100}$ or λ_{111} when $\boldsymbol{M}$ and Δl are collinear and directed along the corresponding ⟨100⟩ and ⟨111⟩ axes. Further inquiry into the meaning of λ_{100} and λ_{111} shows that they depend on the elastic constants of the crystal with standard designations of c_{11}, c_{12}, and c_{44}, and on the magnetoelastic constants B_1 and B_2 which are dependent on the configuration of nearest neighbors, according to

$$\begin{aligned}\lambda_{100} &= -\frac{2}{3}\frac{B_1}{c_{11} - c_{12}} \\ \lambda_{111} &= -\frac{1}{3}\frac{B_2}{c_{44}}. \qquad (5.30)\end{aligned}$$

An additional relation that is useful for polycrystalline materials defines an average magnetostriction constant

$$\lambda_s = \frac{2}{5}\lambda_{100} + \frac{3}{5}\lambda_{111}. \tag{5.31}$$

The merging of the anisotropy and magnetostriction concepts into a single theory was first formalized by Kittel [58]. The relation that combines the interactive effects between strain and anisotropy in the free energy is given by

$$E_K = \left(K_1^0 + \Delta K_1\right)\left(\alpha_x{}^2\alpha_y{}^2 + \alpha_y{}^2\alpha_z{}^2 + \alpha_z{}^2\alpha_x{}^2\right), \tag{5.32}$$

where $\Delta K_1 = \frac{9}{4}\left[(c_{11} - c_{12})\,\lambda_{100}{}^2 - 2c_{44}\lambda_{111}{}^2\right]$. Note that ΔK_1 can enhance or reduce the total K_1 depending on the values of the magnetostriction and elastic constants.

The influence of magnetostriction on anisotropy is particularly important where external stress is involved. With a uniaxial compressive stress σ applied to a magnetized body, the combined energy of anisotropy and inverse magnetostriction

$$\begin{aligned} E_{K\sigma} = {} & K_1\left(\alpha_x{}^2\alpha_y{}^2 + \alpha_y{}^2\alpha_z{}^2 + \alpha_z{}^2\alpha_x{}^2\right) + \frac{3}{2}\sigma\lambda_{100}\left(\alpha_x{}^2\gamma_x{}^2 + \alpha_y{}^2\gamma_y{}^2 + \alpha_z{}^2\gamma_z{}^2\right) \\ & + 3\sigma\lambda_{111}\left(\alpha_x\alpha_y\gamma_x\gamma_y + \alpha_y\alpha_z\gamma_y\gamma_z + \alpha_z\alpha_x\gamma_z\gamma_x\right), \end{aligned} \tag{5.33}$$

where γ_x, γ_y, and γ_z are the direction cosines of the uniaxial stress vector. If the $\boldsymbol{M}$ and σ are coincident, (5.33) can be simplified to

$$E_{K\sigma} = \left[K_1 + 3\sigma\left(\lambda_{111} - \lambda_{100}\right)\right]\left(\alpha_x{}^2\alpha_y{}^2 + \alpha_y{}^2\alpha_z{}^2 + \alpha_z{}^2\alpha_x{}^2\right) + \frac{3}{2}\sigma\lambda_{100}. \tag{5.34}$$

Special cases have been analyzed for situations where the magnetostriction is comparable to the anisotropy. In particular, the variation of the easy axis direction when $\boldsymbol{H}$ and σ are applied collinearly is of interest in forecasting the effects of stress on the magnetization process, more specifically, the hysteresis loop. An example of this is shown in Fig. 5.16 for a single crystal of YIG with stress applied along a <100> hard axis ($K_1 < 0$) [59]. Note how the anisotropy field has become smaller and the remanence ratio $R\,(= M_r/M_s)$ has increased through the action of the uniaxial anisotropy component caused by the stress. In this case,

$$\begin{aligned} H_{K\sigma} &= -\frac{2K_1}{M_s} + \frac{3\sigma\lambda_{100}}{2M_s}, \\ R &= \frac{M_r}{M_s} = \sqrt{\frac{1}{3} + \frac{\sigma\lambda_{100}}{K_1}}, \end{aligned} \tag{5.35}$$

for which the room-temperature saturation magnetization $4\pi M_s = 1{,}780$ G, $K_1 \approx -6{,}000\,\mathrm{erg/cm^3}$, and $\lambda_{100} = -1.3 \times 10^{-6}$.

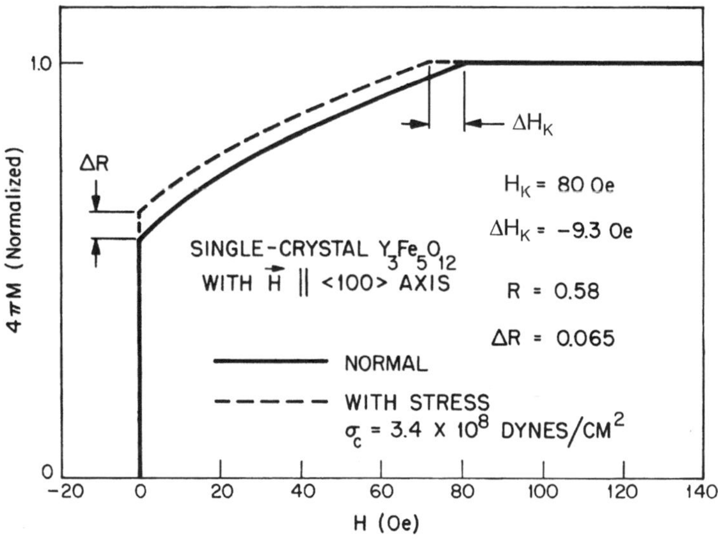

Fig. 5.16 Trace of oscilloscope photograph showing remanence and approach to saturation regions from a hysteresis loop of a $Y_3Fe_5O_{12}$ single-crystal with magnetic field and uniaxial compressive stress applied along [001] "hard" axis. Effects of stress on remanence ratio (R) and anisotropy field (H_K) are indicated with their measured values (after Dionne [59]). Figure reprinted from [59]. © 1969 by the IEEE

If the anisotropy induced by external stress is large enough to dominate the unstressed anisotropy, it is possible to create a uniaxial component that can produce an easy axis normal to the plane of a magnetic thin film, i.e., $K_{u\sigma} \gg K_1$, even one in which a demagnetizing field of magnitude $4\pi M$ of a single domain must be overcome [60].

With this brief phenomenological background, we can proceed to examine sources of the anisotropy and magnetostriction parameters in magnetic oxides.

5.3.3 Dipolar Pair Model of Magnetic Anisotropy

Among the earliest attempts to analyze the molecular origins of magnetocrystalline anisotropy was the dipole pair model of Van Vleck [61]. In two dimensions, the spin configurations in an ordered lattice plane are depicted in Fig. 5.17, where the angle ϕ is the angle between the aligned spin direction and a direction set by the bonds connecting the spins. The pair energy can be expressed by a Legendre polynomial expansion

$$E_{\text{dipole}}(\cos\varphi) = E_0 + E_1\left(\cos^2\varphi - \frac{1}{3}\right) + E_2\left(\cos^4\varphi - \frac{6}{7}\cos^2\varphi + \frac{3}{35}\right) + \ldots \tag{5.36}$$

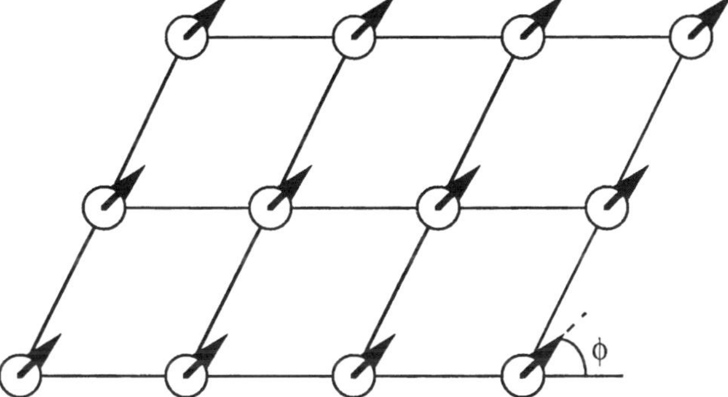

Fig. 5.17 Two-dimensional model of ferromagnetic ordering of magnetic dipoles in a lattice plane with a non-orthogonal bonding arrangement

Fig. 5.18 Three-dimensional model of ferromagnetic ordering of magnetic dipoles in a cubic cell

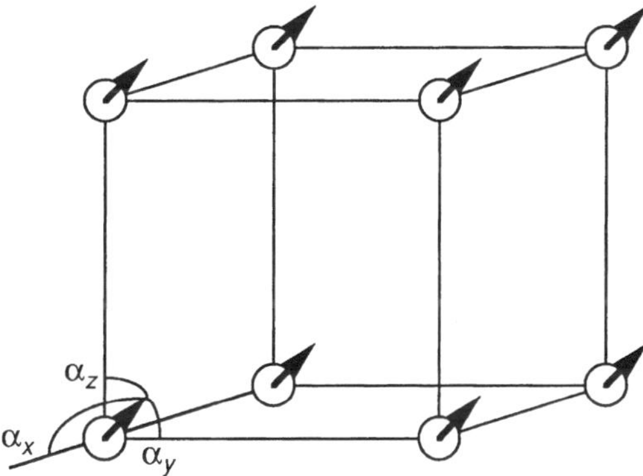

The zero-order term would include the isotropic exchange energy; the first and second-order terms are the phenomenological contributors to the anisotropy that would comprise spin–orbit–lattice interactions. For N ions per unit volume in a lattice of cubic symmetry shown in Fig. 5.18, direction cosines α_x, α_y, and α_z of the three-dimensional structure are introduced and the relation of (5.36) becomes

$$E_{\text{dipole}} = N \begin{bmatrix} E_0 + E_1\left({\alpha_x}^2 - \frac{1}{3}\right) + E_2\left(\alpha_x^4 - \frac{6}{7}{\alpha_x}^2 + \frac{3}{35}\right) \\ +E_1\left({\alpha_y}^2 - \frac{1}{3}\right) + E_2\left(\alpha_y^4 - \frac{6}{7}{\alpha_y}^2 + \frac{3}{35}\right) \\ +E_1\left({\alpha_z}^2 - \frac{1}{3}\right) + E_2\left(\alpha_z^4 - \frac{6}{7}{\alpha_z}^2 + \frac{3}{35}\right) \end{bmatrix}. \tag{5.37}$$

Recalling the identity ${\alpha_x}^2 + {\alpha_y}^2 + -{\alpha_z}^2 = 1$, we can reduce (5.37) to

$$E_{\text{dipole}} = N\left[E_0 + E_2\left(\alpha_x^4 + \alpha_y^4 + \alpha_z^4 + \frac{9}{35}\right)\right], \tag{5.38}$$

and then write the angle-dependent anisotropy energy as

$$E_K \approx NE_2 \left(\alpha_x^4 + \alpha_y^4 + \alpha_z^4\right) = K_1 \left(\alpha_x^2\alpha_y^2 + \alpha_y^2\alpha_z^2 + \alpha_z^2\alpha_x^2\right) + K_2 \text{ (higher order factors)}. \tag{5.39}$$

By inspection of (5.21) and (5.22) we recognize that K_{1u} and K_1 correspond to the E_1 and E_2 energies of the Legendre series and that the former terms correspond to an axially symmetric component that cancels out in perfect cubic symmetry.

The dipolar model of ferromagnetic anisotropy has produced in some cases reasonable estimates of actual measured values, and spawned much of the early analysis of this phenomenon. One useful and often-cited relation between the anisotropy and magnetization as a function of temperature comes from a calculation by Zener [62] that produced the equation

$$\frac{K_1(T)}{K_1(0)} = \left[\frac{M(T)}{M(0)}\right]^{10}. \tag{5.40}$$

This result was found to be accurate for metallic iron over the entire range from zero to the Curie temperature, but not so appropriate for cobalt and nickel for which the exponent is variable over the full range due to the larger unquenched orbital angular momentum and spin–orbit coupling.

For further details on this aspect of the subject and the derivation and implications of (5.39), the reader is directed to Chikazumi [63]. For the present discussion, however, the origins of the anisotropy constants are of greater interest and the focus will be placed on the microscopic theory of single-ion anisotropy as it contributes to the macroscopic value of K_1.

5.3.4 Single-Ion Model of Ferrimagnetic Anisotropy

In magnetic oxides, the microscopic origins of K_1 can be extrapolated from the single-ion properties discussed above. When the individual magnetic moments become collective under the stabilizing influence of magnetic exchange, the effects of anisotropy with respect to the crystal lattice are retained, often as competing influences between opposing sublattices in ferrites or between individual sites where different ion species with their own particular anisotropy and magnetostriction contributions, i.e., spin–orbit or Jahn–Teller ties to the lattice, become significant. The most important yet least dramatic of these cases is logically Fe^{3+} $(3d^5)$. With $L = 0$, it is an S-state ion with no first-order spin–orbit coupling and therefore a small single-ion interaction with the lattice, as discussed previously. Among the more active ions are Co^{2+} $(3d^7)$, Fe^{2+} $(3d^6)$, Ni^{2+} $(3d^8)$, Mn^{3+} $(3d^4)$, and Cu^{2+} $(3d^9)$, and they will be examined in the next section. However, because of the practical importance of low anisotropy ferrites, the analysis of the single-ion contributions of Fe^{3+} will be summarized to serve as background to the reader's awareness.

For the $g \approx 2$ ions, we shall describe the single-ion mechanism set forth independently by Yosida and Tachiki [64], and Wolf [65]. Wolf began with the general spin-Hamiltonian based on (5.20) with no first-order anisotropy terms

$$\mathcal{H}_s = g m_B \boldsymbol{H} \cdot \boldsymbol{S} + \frac{1}{6} a \left(S_z^4 + S_x^4 + S_y^4\right) + D S_\alpha^2 + F S_\alpha^4, \tag{5.41}$$

where subscript α designates the direction of a local axial distortion. For this procedure, Wolf began with the assumptions that the exchange energy dominates the lower symmetry crystal fields, i.e, $a < D \ll g m_B H_{ex}$, and the F contribution from equivalent sites averaged over the cubic symmetry axes is absorbed into the a term. Equation (5.41) is then applied to the $S = 5/2$ case to compute the energy levels of a single $3d^5$ ion:

$$\begin{aligned}
E_{\pm\frac{5}{2}} &= \pm\frac{5}{2} g m_B H_{ex} + \frac{1}{2} a (1 - 5\varphi) + \frac{5}{3} D \left(3\cos^2\theta - 1\right) \\
&\quad \pm \frac{D^2}{4 g m_B H_{ex}} \left(5 + 70\cos^2\theta - 75\cos^4\theta\right) \\
E_{\pm\frac{3}{2}} &= \pm\frac{3}{2} g m_B H_{ex} - \frac{3}{2} a (1 - 5\varphi) + \frac{1}{3} D \left(3\cos^2\theta - 1\right) \\
&\quad \pm \frac{D^2}{4 g m_B H_{ex}} \left(9 - 66\cos^2\theta + 57\cos^4\theta\right) \\
E_{\pm\frac{1}{2}} &= \pm\frac{1}{2} g m_B H_{ex} + a (1 - 5\varphi) - \frac{4}{3} D \left(3\cos^2\theta - 1\right) \\
&\quad \pm \frac{D^2}{4 g m_B H_{ex}} \left(4 - 40\cos^2\theta + 36\cos^4\theta\right),
\end{aligned} \tag{5.42}$$

where θ is the angle that H_{ex} makes with the direction of the distortion, l, m, and n are the direction cosines of H_{ex} with respect to the cubic axes, and $\phi = l^2m^2 + m^2n^2 + n^2l^2$. The anisotropy energy is extracted from the free energy of each sublattice individually and then combined in a manner similar to that for the solution of thermomagnetization properties. Boltzmann statistics are used to account for the population distributions over the energy levels of (5.42), with the partition function

$$Z = \left[\sum_{m=-5/2}^{m=+5/2} \exp\left(-E_m/kT\right)\right]^N, \tag{5.43}$$

where N is the number of ions, and the free energy $\mathcal{F} = -(kT)\log_e(Z)$. Wolf expanded these exponential terms in $(-E_m/\mathrm{k}T)$ that contain the crystal field energy parameters which are dominated by the exchange energy at lower temperatures and small compared with $\mathrm{k}T$ except at the higher temperatures. The following general result was obtained for the i sublattice:

$$F_i = N \left\{ \begin{array}{l}
F_{0i}(y) + D\cos^2\theta p(y) + \frac{D^2}{g m_B H_{ex}} \cos^4\theta q(y) \\
+a\left(l^2m^2 + m^2n^2 + n^2l^2\right) r(y) - \frac{1}{2}\frac{D^2}{kT}\cos^4\theta s(y) \\
+\frac{1}{2}\frac{D^2}{\mathrm{k}T}\cos^4\theta \left[p(y)\right]^2
\end{array} \right\}, \tag{5.44}$$

where $\mathcal{F}_{0i}(y)$ is the angle independent part of the free energy. The anisotropy terms are given by

$$\begin{aligned}
p(y) &= (1/Z_0)\left(5 - y - 4y^2 - 4y^2 - y^4 + 5y^5\right),\\
q(y) &= (1/4Z_0)\left(75 - 57y - 36y^2 - 36y^2 - 57y^4 + 75y^5\right),\\
r(y) &= (5/2Z_0)\left(-1 + 3y - 2y^2 - 2y^2 + 3y^4 - y^5\right),\\
s(y) &= (1/Z_0)\left(25 + y + 16y^2 - 4y^2 - y^4 + 5y^5\right),\\
Z_0 &= 1 + y + y^2 + y^3 + y^4 + y^5,\\
y &= \exp\left(-gm_B H_{ex}/kT\right).
\end{aligned} \tag{5.45}$$

Equation (5.44) for the free energy can be condensed into

$$F_i = N\left[F_{0i}(y) + D\cos^2\theta p(y) + a\left(l^2m^2 + m^2n^2 + n^2l^2\right)r(y) - \frac{D^2}{kT}\cos^4\theta t(y)\right], \tag{5.46}$$

where $t(y) = -q(y)/\ln(y) - \frac{1}{2}s(y) + \frac{1}{2}[p(y)]^2$. The value of y as a function of temperature can be found indirectly from the Brillouin function of the relative sublattice magnetization M_i/M_{i0}, which can be obtained from the relation

$$\frac{M_i}{M_{i0}} = \mathcal{B}\left(gm_B H_{eff}/kT\right) = \sum_{m=+S}^{m=-S} my^{-m}/S\sum_{m=+S}^{m=-S} y^{-m} \tag{5.47}$$

either graphically or computed by a numerical iterative procedure of the kind used by Dionne [19] of Chap. 4. For each value of M_i/M_{i0} the functions of (5.44) can be determined in generic form for each sublattice and for different values of S. To apply these results to a specific situation such as yttrium–iron garnet with two iron sublattices, values of the different sublattice magnetizations must be found from the molecular field analysis discussed in Chap. 4. Regrettably, this accurate refinement of the thermomagnetization could have extended Wolf's model to garnet compounds with varying levels of magnetic dilution if it had been available for use in (5.47).

The first-order cubic anisotropy for N ions can be reduced from (5.46) as

$$K_1 = N\left[ar(y) + \gamma\left(D^2/kT\right)t(y)\right]. \tag{5.48}$$

where $\gamma = -2/3$ or 4/9 for axial distortions along <100> or <111> axes, respectively. The single-ion theory of anisotropy has been applied to analyze magnetic garnets. For two sublattices, e.g., the d and a of a garnet, the resultant anisotropy is

$$K_1 = K_{1d} + K_{1a} = N_d\left[\left(a_d + \frac{7}{12}F_d\right)r(y_d) + \gamma\left(D_d^2/kT\right)t(y_d)\right] + N_a\left[\left(a_a + \frac{7}{12}F_a\right)r(y_a) + \gamma\left(D_a^2/kT\right)t(y_a)\right], \tag{5.49}$$

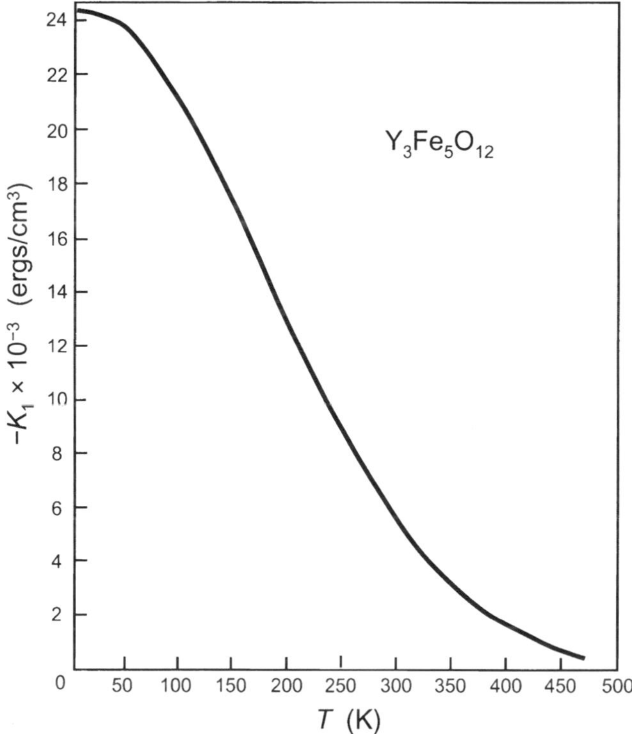

Fig. 5.19 Variation of $-K_1$ with T for yttrium–iron garnet. Image was traced from data reported by Pauthenet ([51] of Chap. 4) and later fitted closely by calculations of Rodrigue et al. [66]

Higher-order F contributions have been added to the a terms. Rodrigue et al. [66] fitted K_1 vs. T data with the Wolf model, shown in Fig. 5.19, and produced expressions for the bracketed weighting factors of $r(y_i)$ of the Fe^{3+} contributions from the a and d sites in yttrium–iron garnet (YIG) using values of a and F reported by Geschwind [67]: for site d, $(0.791a_d + 0.389F_d)$; for site a, $(0.335a_a - 0.259F_a)$.

The comparison of their results given in Table 5.8 raises questions. From the interpretation of Rodrigue et al., the more numerous (by a factor 3/2) tetrahedral d sites appear to produce larger contributions to K_1 than the octahedral a sites. The results of Geschwind are even more surprising in that the a-site contributions are of opposite sign to Rodrigue et al., contrary to experiment. Rodrigue et al.'s result was supported by studies of d-sublattice dilution by Ga^{3+} substitutions in YIG [68]. Measurements of Ga^{3+} dilution of YIG by Hansen that included a complete range of temperature dependence also indicated a monotonic reduction of K_1 with concentration [69]. Adding to the contradictions, a corresponding study of a-sublattice dilution by large radii ($\approx$0.81 Å) Sc^{3+} ions revealed an even more dramatic reduction in K_1 [70]. In both cases of Ga^{3+} and Sc^{3+}, the anisotropy field $2K_1/M$ should decrease sharply.

Table 5.8 Parameter values for single-ion anisotropy in yttrium–iron garnet

Parameter	Geschwind [67]	Rodrigue, Meyer, and Jones [66]
	$cm^{-1} \times 10^{-3}$	$cm^{-1} \times 10^{-3}$
a_a	19.4	—
F_a	2.7	—
a_d	6.5	—
F_d	−4.2	—
K_{1a}	4.1	−0.4
K_{1d}	3.4	4.2

Other investigations involving hysteresis properties that showed a decrease in anisotropy field $2K_1/M$ suggested that small radii ($\approx$0.51 Å) Al^{3+} ions in d sites are more effective at reducing K_1 than more average radii ($\approx$0.62 A) Ga^{3+}, which unexpectedly produced a rise in $2K_1/M$ [71]. However, this story becomes more confusing when it is recognized that a lower initial fraction of Al^{3+} enters d sites and therefore is less effective than Ga^{3+} at lowering the net $4\pi M_d$, but more effective in lowering K_{1a} according to the site distribution studies of [47,48] of Chap 4. If $|K_{1a}| \gg |K_{1d}|$ in YIG as concluded from experiment, M can decline less than K_1 with Al^{3+}, but the reverse is true for than Ga^{3+} dilution, based on the simple relation

$$\frac{2K_1}{M} = \frac{2\,(K_{1a} - K_{1d})}{M_a - M_d}. \tag{5.50}$$

Studies of single-ion contributions to anisotropy in spinels have confirmed the conclusion that octahedral sites produce a K_1 contribution that is not only negative, but five times greater than a small positive contribution from tetrahedral sites [46, 64, 72].

It is clear that more is to be learned about the nature of the source of anisotropy, including the role of ionic dilutant sizes on the local crystal fields. In all of this speculation, the possible influence of the spin canting caused by the dilution has been ignored. Recalling our earlier discussion of spin canting, we recognize that an In^{3+} or Sc^{3+} ion that replaces a smaller Fe^{3+} in an octahedral site may reduce K_1 by lowering the anisotropy contributions of the neighboring d-site Fe^{3+} ions in addition to removing the a-site contribution of the Fe^{3+} ion that it replaces. A similar argument could be made for the effects of the small Al^{3+} ions entering d sites, or even more radically, the dilution of ions of different valence altogether, such as a V^{5+} or Zr^{4+} ion in a d or a site, respectively, whereby the actual crystal field charges are altered. Furthermore, the ratio $2K_1/M_s$ which influences the magnetization process and ferrimagnetic resonance, is sharply reduced by a-site dilution because of the increase in M_s. As noted above, the result of d-site dilution by Ga^{3+} in particular appears to have the opposite effect. Some of these issues are reflected in the qualitative dilution guidelines provided in Tables 4.4 and 4.5.

In the foregoing analysis, emphasis has been placed on crystal fields of cubic symmetry. In hexagonal ferrites such as M-type $BaFe_{12}O_{19}$, a minority of trigonal bipyramid sites with large axial symmetry components greatly influence the overall anisotropy once the iron spins are ordered by magnetic exchange. In terms of (5.18), one could conclude that the ε parameter increases and produces a greater mixing of the relevant excited orbital term with the ground $^6S_{5/2}$ term. More interesting cases, however, are those that involve lower symmetries and cations with strong intrinsic anisotropies that arise from unquenched orbital angular momentum.

5.3.5 Cooperative Single-Ion Effects: Anisotropy

Although Fe^{3+} ions collectively dominate the magnetic moments of many oxides, the major contributions to magnetic anisotropy and magnetostriction are often provided by small concentrations of ions that have unquenched orbital angular momentum. Interactions between the magnetic moments and the lattice can be significant in both the iron group and rare-earth ions. In the rare-earth $4f^n$ series spin–orbit coupling is strong for all of the members (except Gd^{3+} because of its $L = 0$ ground state), but crystal field effects are small. Nonetheless, the spin–orbit–lattice interactions are large enough to produce very short spin–lattice relaxation times in the microwave band (discussed in Chap. 6), and can create giant magnetoelastic effects in certain noncubic intermetallic compounds such as NdFeB alloys for permanent magnets and $Tb_{1-x}Dy_xFe_2$ (terfenol-D) for magnetostrictive transducers. In oxides, however, the large magnetoelastic effects of rare-earth ions can be equaled by selected ions of the $3d^n$ series that are stabilized in an exchange field by spontaneous local lattice distortions.

Recalling the discussions in Sect. 2.4, we first recognize the orbital angular momentum as the key to the coupling between the spin and lattice systems. Spin–orbit stabilization of the $l_z = \pm 1$ doublet, therefore, would be a prerequisite for anisotropy and spin–lattice relaxation rate $\left(\tau^{-1}\right)$, with the attendant lattice distortion contributing to magnetostrictive extension or compression, depending on the sign of the stabilization. Local Jahn–Teller stabilizations of the $l_z = 0$ singlet would be expected to contribute to magnetostriction once they become cooperative, with anisotropy and relaxation effects appearing as lower order phenomena. In Table 5.9, the expected results are compiled for the various iron-group ions in octahedral and tetrahedral situations, based on the discussions in Chap. 2. Although each member of the series has the potential to cause local perturbations, only five (d^4 through d^9) have consistently demonstrated exchange coupling strong enough to influence cooperative magnetoelastic effects in magnetically ordered compounds. In the following text, reference will be made to this summary in the context of specific ions.

For the discussion of anisotropy, we begin with the case of the T_{2g} triplet term, which is contrasted with the E_g case in Fig. 5.20 using the one-electron examples of Co^{2+} and Mn^{3+} in an octahedral site. Here the ground state from the crystal-field distortion can be either a singlet or doublet, depending on the sign of the splitting parameter δ. To understand these effects, consider the case of Co^{2+}, now compared with Fe^{2+} $\left(3d^6\right)$ in Fig. 5.21. Both present a T_{2g} triplet in an octahedral field and

Table 5.9 Single-ion anisotropy and magnetostriction contributions of high-spin $3d^n$ ions as substitutions in Fe^{3+} ferrite sites[a]

N	Ion	K_1	λ_{100}	λ_{111}	ΔH_i
		Oct/Tet	Oct/Tet	Oct/Tet	Oct/Tet
1	Ti^{3+}, V^{4+}	$\Downarrow$ *t*/–	–/ $\Uparrow$ *e*	–/–	$\Uparrow$ *t*/–
2	Ti^{2+}, V^{3+}, Cr^{4+}	$\Uparrow$ *t*/–	$\Downarrow$ *t*/–	–/–	$\Uparrow$ *t*/–
3	Cr^{3+}, Mn^{4+}	–/ $\Uparrow$ *t*	–/ $\Downarrow$ *t*	–/–	–/ $\Uparrow$ *t*
4	Mn^{3+}, Fe^{4+}	–/ $\Uparrow$ *t*	$\Uparrow$ *e*/ $\Uparrow$ *t*	–/–	–/ $\Uparrow$ *t*
5	Mn^{2+}, Fe^{3+}	–/–	–/–	–/–	–/–
6	Fe^{2+}[b], Co^{3+}[c]	$\Downarrow$ *t*/–	$\Downarrow$ *t*[c]/ $\Uparrow$ *e*	$\Uparrow$ *t*[b]/–	$\Uparrow$ *t*/–
7	Co^{2+}, Ni^{3+}	$\Uparrow$ *t*/–	$\Downarrow$ *t*/–	–/–	$\Uparrow$ *t*/–
8	Ni^{2+}, Cu^{3+}	–/ $\Uparrow$ *t*	–/ $\Downarrow$ *t*	–/–	–/ $\Uparrow$ *t*
9	Cu^{2+}	–/ $\Uparrow$ *t*	$\Uparrow$ *e*/ $\Uparrow$ *t*	–/–	–/ $\Uparrow$ *t*

[a]Symbols embedded in table have the following meanings: the up and down arrows indicate positive and negative contributions, respectively; and the *t* and *e* symbols indicate the particular orbital group responsible for the effects
[b]Fe^{2+} features a trigonal ⟨111⟩-axis expansion
[c]Co^{3+} features a tetragonal ⟨100⟩-axis contraction or expansion

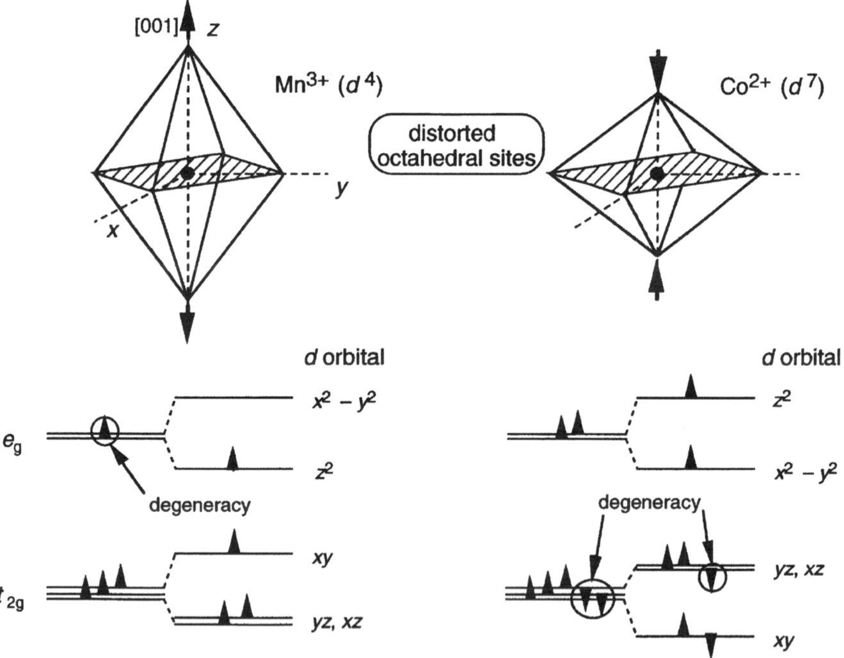

Fig. 5.20 Comparison of a J–T [001]-axis expansion (d^4, Mn^{3+}) and a S–O [001]-axis contraction (d^7, Co^{2+})

both have the option of stabilizing a singlet or doublet by inducing an axial distortion either tetragonally or trigonally. In magnetically ordered systems, the spin–orbit energy can enhance exchange energy through a $(\lambda \boldsymbol{L} + g m_{\mathrm{B}} \boldsymbol{H}_{\mathrm{ex}}) \cdot \boldsymbol{S}$ combination, as demonstrated by [73]. When $\lambda \geq 2\delta/3$ in an exchange field of sufficient magnitude, stabilization of the doublet is usually the result. Based on a discussion from Ballhausen [74], the threshold for doublet stabilization in an exchange field of energy $E_{\mathrm{ex}} = g m_{\mathrm{B}} H_{\mathrm{ex}}$ is computed in Appendix 5B for the d^1 $(S = 1/2)$ case and determined in (5.79) as [16]

$$\frac{\lambda}{\delta} \geq \frac{2}{3}\left(1 + \frac{2}{3} \frac{\delta}{g m_{\mathrm{B}} H_{\mathrm{ex}}}\right). \tag{5.51}$$

In Fig. 5.21, both Co^{2+} and Fe^{2+} are depicted with S–O stabilized ground states. A convenient way to view the Fe^{2+} degeneracy is to consider a single hole occupying the ground doublet. The former case is supported by many measured results that show a z-axis compression and strong magnetoelastic behavior. The Fe^{2+} situation is less obvious, although Goodenough has pointed out that a trigonal extension is the result in FeO [75]. Furthermore, the trigonal extension is consistent with the large λ_{111} magnetostriction constant observed in spinel ferrites containing ferrous

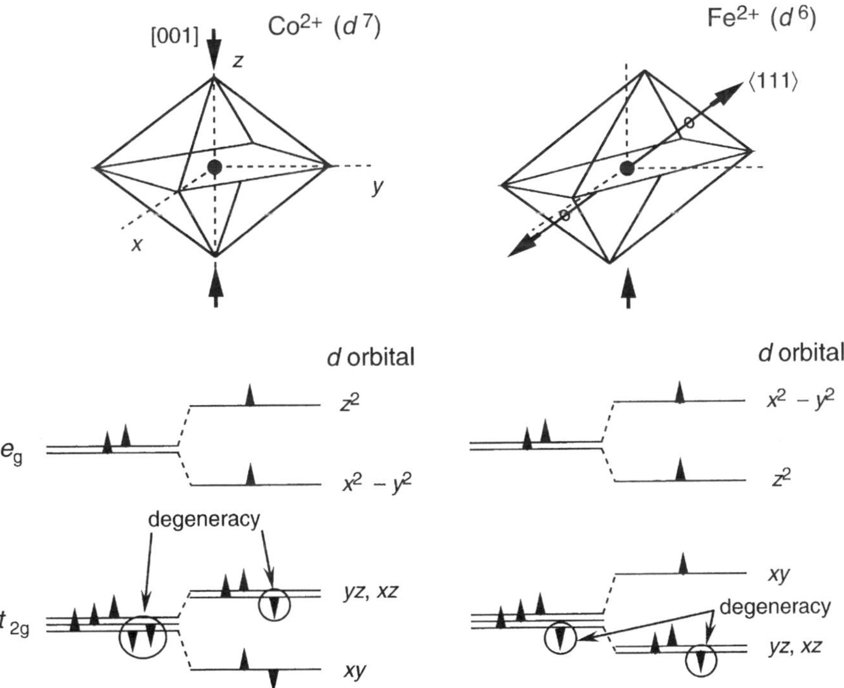

Fig. 5.21 Comparison of a S–O [001]-axis contraction (d^7, Co^{2+}) and a S–O [111]-axis expansion (d^6, Fe^{2+}). For Fe^{2+} in a tetrahedral site, e.g., ZnO, a J–T [001]-axis contraction could be expected (see footnote in Sect. 2.4.3)

iron ions. Figure 2.13 illustrates the energies of the orbital doublets in the T_{2g} and T_2 groups that, respectively, carry the remaining unquenched orbital angular momentum under the influence of a tetragonal compression and a trigonal extension.

The independent behavior of the magnetic moments associated with spontaneous ligand distortions leads one to view them as atomic-scale "domains," with single-domain characteristics typical of small crystallites with large K_1 values, which will be discussed in the next section. As a consequence, single-ion contributions to K_1 can be estimated from the magnitude of the spin–orbit coupling energy. The method is to assign a spin–orbit energy $\lambda \boldsymbol{L} \cdot \boldsymbol{S}$ to each ion when the unquenched $\boldsymbol{L}$ of its partially occupied doublet is directed along its preferred (easy) cubic axis—a compressed $\langle 100 \rangle$ in the case of Co^{2+}, as envisioned in the sketch of the $(1/\sqrt{2})\left(d_{xz} \pm \mathrm{i}d_{yz}\right)$ degenerate hybrid in Fig. 2.24a. Spin–orbit coupling then brings the spin vector $\boldsymbol{S}$ to align with $\boldsymbol{L}$ thereby raising the ground-state energy to $-\lambda \boldsymbol{L} \cdot \boldsymbol{S} \cos\theta_\mathrm{i}$, where θ_i is the angle between an easy $\langle 100 \rangle$ and the direction of the ion magnetic moment. Upon referring to Fig. 5.15b, we see that the maximum value of θ_i is 55° and $\cos\theta_\mathrm{i} = 1/\sqrt{3}$ when the moment is along a hard $\langle 111 \rangle$ direction. If the spins are coupled collinearly, the average anisotropy energy contribution from one Co^{2+} ion will be given by

$$\Delta K_1^{\mathrm{ion}} = -\lambda \boldsymbol{L} \cdot \boldsymbol{S} \left\langle 1 - \cos\theta_\mathrm{i} \right\rangle , \tag{5.52}$$

where $\langle 1 - \cos\theta_\mathrm{i} \rangle$ is the average over all equivalent Co^{2+} sites, and the net magnetization direction in an exchange-ordered system is fixed by the aligned $\boldsymbol{S}$ vectors parallel to a magnetic field.

For the $\boldsymbol{S}$ vectors along a $\langle 111 \rangle$ axis each $\theta_\mathrm{i} = 55°$, and $\Delta K_1^{\mathrm{ion}} \approx -0.43\lambda LS$. If the spontaneous magnetostrictive distortion along the a $\langle 100 \rangle$ axis is ignored, the stabilization of the Co^{2+} orbital doublet can also be interpreted as the resultant effect local trigonal crystal fields along $\langle 111 \rangle$ easy axes at the four equivalent octahedral sites of the host lattice. Slonczewski's calculations for this scenario compared favorably with data for Co^{2+} -substituted Fe_3O_4 (magnetite) and show good agreement over a broad temperature range [73]. A more complete analysis might include crystal-field terms for both tetragonal and trigonal fields with the result that a local orthorhombic ligand symmetry would appear with a conceivably higher λ/δ ratio for greater anisotropy. A similar effect is expected for Fe^{2+} because of its likely spin–orbit stabilization in a trigonal field, favoring an extension of the $\langle 111 \rangle$ axes. Here ΔK_1 has the opposite sign from that of the Co^{2+} case but may be lower in magnitude because of its smaller λ constant.

Another ion that could have an orbital degeneracy in the ground state is Ni^{2+} $\left(d^8\right)$ in a tetrahedral site is included in Table 5.9. According to the occupancies of the one-electron diagram in Fig. 5.22, the doublet would be stabilized with a distortion of the same sign as Co^{2+} in an octahedral site. However, the occurrence of Ni^{2+} in spinels is almost exclusively in octahedral B sites, where its magnetoelastic effects should be negligible. Because evidence of anisotropy, magnetostriction, and spin–lattice relaxation has been associated with the presence of Ni in $NiFe_2O_4$, the presence of a smaller fraction of the Ni^{2+} in the A sublattice must be considered a strong possibility.

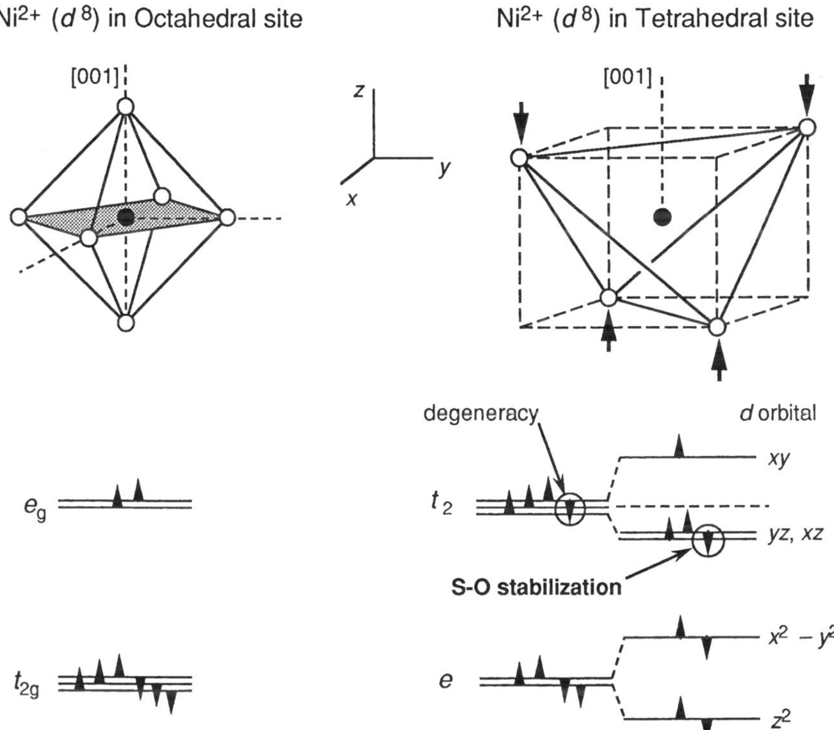

Fig. 5.22 Comparison of spin configurations of d^8 $\left(Ni^{2+}\right)$ in octahedral and tetrahedral sites. In an octahedral site there are no first-order magnetoelastic effects. In the less-common tetrahedral site, S–O stabilization in the t_2 triplet could produce both anisotropy and magnetostriction effect

In the J–T case, a singlet with zero orbital angular momentum is lowest, and the ground state after the spin–orbit coupling operator is applied contains a contribution of orbital interaction from the hybrid with the excited states. For J–T ions, therefore, the magnitude of the anisotropy and spin–lattice interaction of Mn^{3+} $\left(3d^4\right)$ and Cu^{2+} $\left(3d^9\right)$ in octahedral sites is determined by the splitting of the E_g doublet in a tetragonal field that results from an extension of the apical z axis, sketched in Fig. 5.23. Because there is no orbital angular momentum associated directly with the e_g states, anisotropy contributions are not expected to be large. Anisotropy results from second-order contributions from the d_{xz}, d_{yz} orbitals of the T_{2g} shell that enter the ground state hybrid in the fraction amount $\sim\lambda/10Dq$, which is typically only a few percent. Although with less probability, both of these ions can occupy tetrahedral sites and appear to do so in spinel and garnet ferrites. In such cases, the energy-level order is inverted with the T_{2g} term now higher, and small concentrations can undergo spin–orbit stabilization of the $l_z = \pm 1$ doublet, as sketched in Fig. 5.24 for Mn^{3+} $\left(3d^4\right)$. The Cu^{2+} case is analogous, but with an unpaired "hole" in the T_{2g} states. A significant feature is that the sign of the δ splitting is the same as that for the corresponding J–T splitting. As a consequence, the added contribution

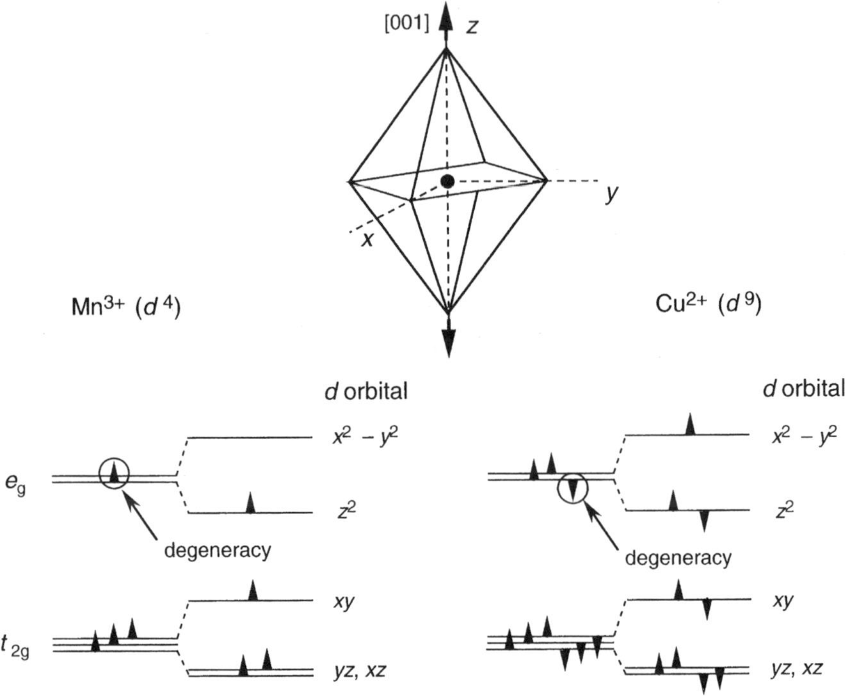

Fig. 5.23 Comparison of spin configurations of classic J–T e_g ions d^4 (Mn^{3+}) and d^9 (Cu^{2+}) [001]-axis expansions in octahedral sites

from the tetrahedral sublattice could explain the negative anisotropy and spin–lattice relaxation enhancements observed in ferrites, as well as the source of magnetic field induced magnetostrictive effects of opposite sign to Co^{2+}, listed in Table 5.9 and discussed in the next section.

5.3.6 Cooperative Single-Ion Effects: Magnetostriction

The single-ion model of anisotropy can also be effective in explaining magnetostriction contributions from isolated sites. For this text, the focus will be on local effects of magnetoelastically active ions. A scholarly overview of this topic, particularly as it applies to the magnetic garnets was reported by [76]. Jahn–Teller and spin–orbit distortions are local, but will influence the overall lattice strain condition in a manner proposed by Dionne [77]. The "local-site distortion" model of magnetostriction provides an useful qualitative guide to the properties of spinel and garnet ferrite compounds where the highly magnetoelastic individual ions can significantly alter the magnetostriction of the iron-dominated sublattices. Local-site distortions are sketched for $\langle 100\rangle$ and $\langle 111\rangle$ axes in Figs. 5.25 and 5.26. For the $\langle 100\rangle$ distortion

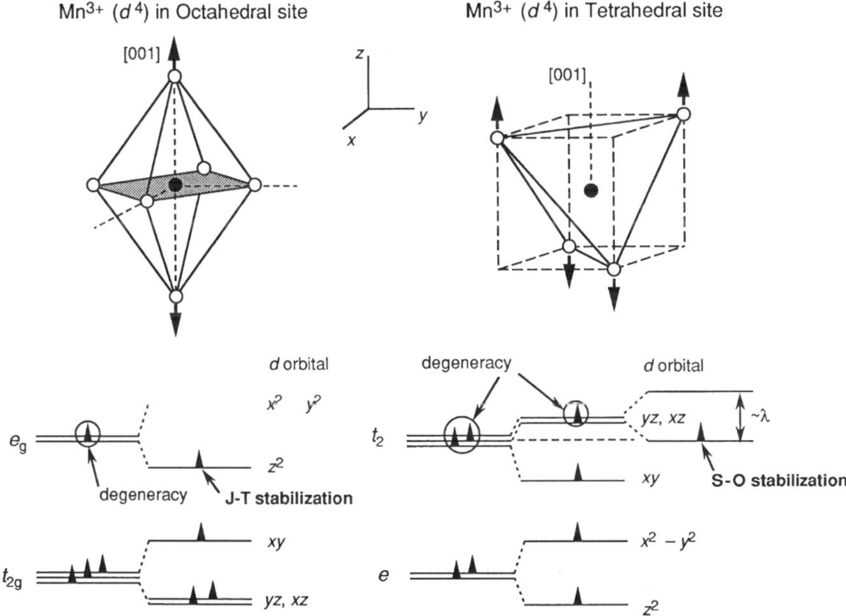

Fig. 5.24 Comparison of spin configurations of d^4 $\left(Mn^{3+}\right)$ [001]-axis expansions in octahedral and tetrahedral sites. Note that the stabilization is J–T in the octahedral case, but S–O in the infrequent tetrahedral case

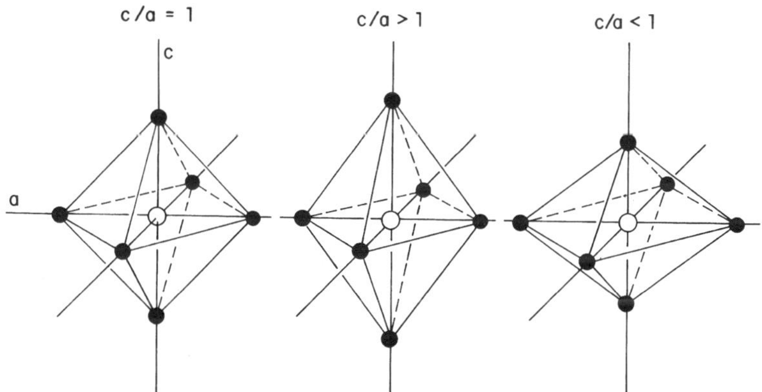

Fig. 5.25 Octahedral sites with tetragonal distortion along [001] axis. Figure reprinted from [77] with permission. © 1979 by the American Institute of Physics

case, in the unmagnetized state the cube edge length parameter l_{100} undergoes local changes of Δl_{100} that are equally distributed among the three equivalent $\langle 100\rangle$ crystallographic axes. As $\boldsymbol{M}$ is aligned with the $\langle 100\rangle$ axis, all three groups of distortions can become cooperative, thereby producing a combined $3\Delta l_{100}$ change along the $\langle 100\rangle$ axis sketched in Fig. 5.27. If $\boldsymbol{M}$ is then rotated to a $\langle 111\rangle$ axis, the individual

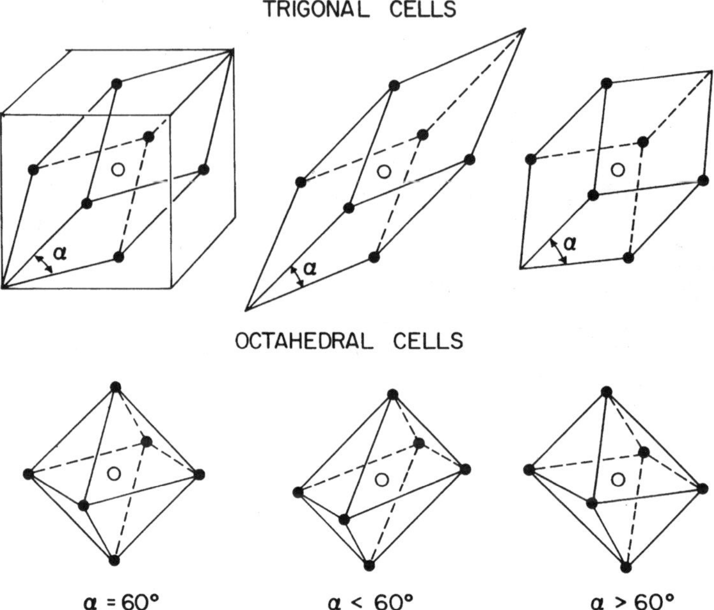

Fig. 5.26 Lattice sites with trigonal distortion along [111] axis. Figure reprinted from [77] with permission. © 1979 by the American Institute of Physics

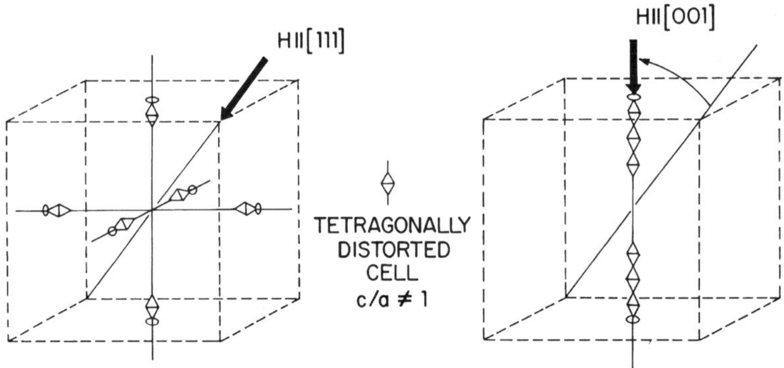

Fig. 5.27 Site former that the local distortion < 100> axis of each site will align with the final ***H*** direction to create a resultant magnetostrictive strain along the [001] axis. Figure reprinted from [77] with permission. © 1979 by the American Institute of Physics

site distortions will relax back to their respective $\langle 100 \rangle$ axes, producing no first-order change in the l_{111} parameter. The single-ion magnetostriction constants can then be expressed as

$$\lambda_{100}^{\text{site}} = \frac{3\Delta l_{100}}{l_{100}}$$

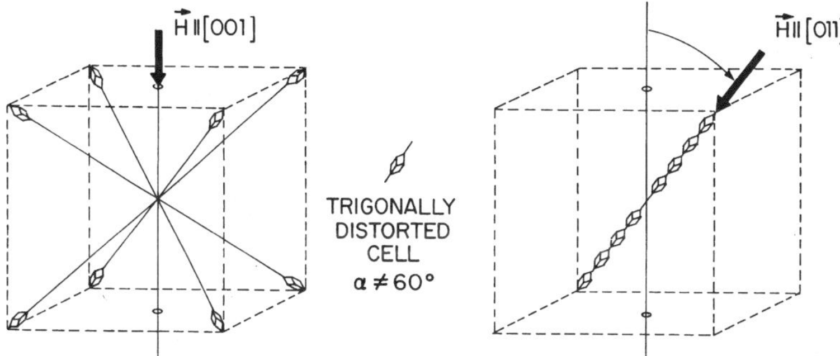

Fig. 5.28 Site distortion switching as $\boldsymbol{H}$ is rotated from the [001] axis to the [111] axis. Diagram is intended to convey that the local distortion <111> axis of each site will align with the final $\boldsymbol{H}$ direction to create a resultant magnetostrictive strain along the [111] axis. Figure reprinted from [77] with permission. © 1979 by the American Institute of Physics

$$\lambda_{111}^{\text{site}} = 0. \tag{5.53a}$$

This result predicts a l_{100} extension for a Mn^{3+} (d^4) ion in either type of cubic site. For the $\langle 100\rangle$ compression typical of Co^{2+} in an octahedral site (and Ni^{2+} in a tetrahedral site) the sign of λ_{100} is reversed. In the case of Fe^{2+} in an octahedral site (Fig. 5.26), four $\langle 111\rangle$ distortions diagonals cooperate to make resultant $4\Delta l_{111}$ extension sketched in Fig. 5.28, and the relations between λ_{100} and λ_{111} are

$$\begin{aligned} \lambda_{100}^{\text{site}} &= \frac{4\Delta l_{100}}{l_{100}} \\ \lambda_{111}^{\text{site}} &= 0. \end{aligned} \tag{5.53b}$$

The impetus for aligning the distorted sites comes from the local spin–orbit generated anisotropy, which, as discussed above, is greater for d^6 and d^7 than for the pure J–T d^4 and d^9 cases. An example of experimental data for these cases in ferrimagnetic oxides is presented in Fig. 5.29. Through the use of controlled substitutions of local-site distortion ions, anisotropy, and magnetostriction can be tailored to produce a range of properties. Qualitative summaries of the more common spinel and garnet systems are given earlier in Tables 4.5 and 4.6. Quantitatively, K_1/M_s values are several times greater for spinels than garnets, typically in the range of 150–300 Oe. Magnetostriction in the spinels is also larger, particularly the λ_{100} constant which is typically -25×10^{-6} in contrast to values of $\sim\!-2 \times 10^{-6}$ for YIG. Nickel ferrite $NiFe_2O_4$ has a $\lambda_{100} \approx -50 \times 10^{-6}$. Ratios of λ_s/K_1, however, are more comparable. If rare-earth ions occupy the garnet c sublattice in sufficient concentrations, both K_1 and λ_s increase, as determined by Iida [78].

The importance of magnetoelastic parameters can be seen in the shape of hysteresis loops, which will be reviewed in the next section, and in the microwave propagation properties to be discussed in Chap. 6.

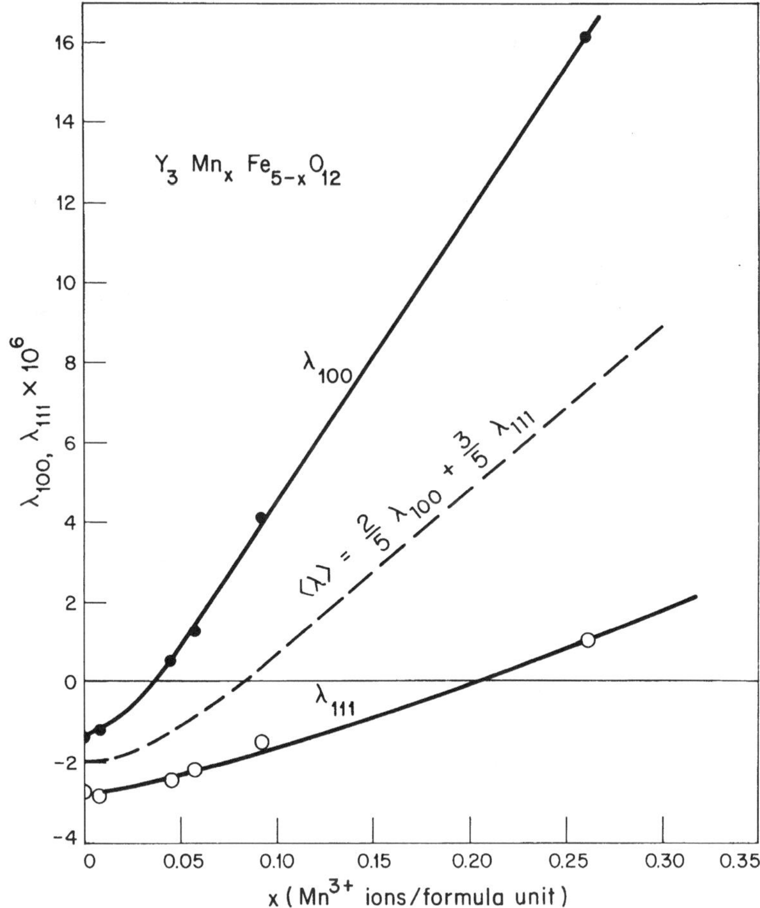

Fig. 5.29 Variation in the room-temperature magnetostriction constants of yttrium–iron garnet with Mn^{3+} concentration. Note that $\langle\lambda\rangle = \lambda_s$. Figure reprinted from [77] with permission. © 1979 by the American Institute of Physics

5.4 Magnetization Process and Hysteresis

Local magnetoelastic effects reveal themselves most commonly in the magnetization process, particularly in relation to hysteresis. These occasions frequently involve ferrites, usually in polycrystalline form. The ability to create a state of magnetization with minimum energy defines a "soft" magnetic material; the ability to retain the magnetized state under conditions of maximum demagnetizing influences determines a "hard" magnetic material. Except for a superficial introduction in Sect. 1.1.4, we have so far ignored the role of magnetic domains largely because domains are the features of the partially magnetized state. Since the emphasis has been on single-ion moments and the establishment of the fully magnetized "single-domain" state, multiple domains have not entered the discussion. Only in the case of hard magnets where the volume of the individual crystallites is too small to support domain walls can the assumption of a single domain particle be made in the unsaturated state. In this section the role of domains and domain-wall energies in the partially magnetized state will be considered as we examine the effects of magnetocrystalline anisotropy and magnetostriction on the magnetization process and the shapes of hysteresis loops that occur because of domains.

5.4.1 *Initial Permeability and Coercivity*

In Fig. 5.30 the essential parameters of a generic $4\pi M$ vs. H hysteresis loop are indicated. From the origin, the magnetization process begins with reversible rotation of individual magnetic domain vectors toward the axis the $\boldsymbol{H}$ field. For a polycrystal with randomly oriented grains of cubic lattice symmetry, the slope (susceptibility) χ_{i} of the curve at $H = 0$ defines the average (isotropic) initial permeability $\mathrm{d}B/\mathrm{d}H$ according to [79]

$$\mu_{\mathrm{i}}'(K_1 < 0) = 1 + 4\pi\chi_{\mathrm{i}} \approx \frac{(4\pi)\,3M_{\mathrm{s}}^2}{4\,|K_1 + \sigma\lambda_{\mathrm{s}}|}\sin^2\theta$$

$$\mu_{\mathrm{i}}'(K_1 > 0) = 1 + 4\pi\chi_{\mathrm{i}} \approx \frac{(4\pi)\,M_{\mathrm{s}}^2}{4\,|K_1 + \sigma\lambda_{\mathrm{s}}|}\sin^2\theta, \tag{5.54a}$$

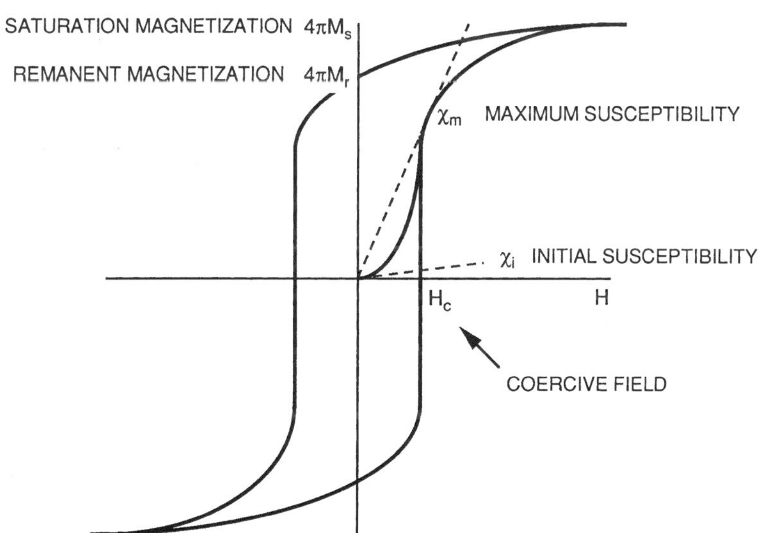

Fig. 5.30 Tutorial diagram of the low-field part of a generic hysteresis loop, with relevant parameters indicated

where θ is the angle between $\boldsymbol{H}$ and the easy axis. When $\sin^2\theta$ is replaced by an average value, e.g., $\langle\sin^2\theta\rangle = 2/3$ if we employ the uniaxial model, (5.54a) can be reduced to an isotropic approximation

$$\mu_i'(K_1<0) \approx \frac{2\pi M_s^2}{|K_1+\sigma\lambda_s|}$$
$$\mu_i'(K_1>0) \approx \frac{4\pi M_s^2}{3\,|K_1+\sigma\lambda_s|}. \tag{5.54b}$$

As H is increased further, domains with $\boldsymbol{M}$ vectors favorable to the $\boldsymbol{H}$ vector grow at the expense of the others through the mechanism of domain wall movement. This beginning of the process is depicted simply in Fig. 1.2b, where two 180° domains are provided with magnetic flux closure by a pair of orthogonal domains bounded by diagonal walls at the long axis ends. In real cases, the domain pattern, e.g., Fig. 1.2c, can be complex and is the basis for the field of micromagnetism where it has been the subject of extensive studies [80, 81]. The resulting pattern is determined by the tradeoff between the increase in energy of the walls that originates from exchange forces and anisotropy, and the reduction in magnetostatic energy that occurs when the external flux is collapsed within the material. The geometrical shape is also an important factor. In terms of the relevant magnetic parameters, a general relation for the domain wall surface energy density is given by

$$\gamma_w \sim [A\,|K_1+\sigma\lambda_s|]^{1/2}, \tag{5.55}$$

where $A = JS^2/a_0$ represents the exchange energy between adjacent spins in the domain wall, and is therefore related to the Curie temperature. Two models of domain walls are generally considered: Bloch walls usually have the lower energy in bulk materials; Néel walls are the preferred approximation for thin films. Our discussion will adopt the Bloch model of (5.55), with typical values for γ_w falling in the range of 1–5 ergs/cm^2. A corresponding expression for the domain wall thickness is

$$\delta_w \sim \left(\frac{A}{|K_1+\sigma\lambda_s|}\right)^{1/2}, \tag{5.56}$$

which yields values on the order of 10^{-5} cm (0.1 μm), or about 200 lattice parameters.

The existence of a domain wall requires a particle dimension exceeding thickness δ_w. Otherwise, the magnetic body behaves as a "single-domain" particle with initial permeability that is determined entirely by magnetic moment rotation from an easy direction to that of the magnetizing field. However, it is the energy of the domain wall balanced against the demagnetizing energy of the magnetized particle that ultimately determines whether a particle can exist as a single domain. For a cubic particle, the critical dimension for single domain is estimated by [82]

$$D_{sd} = \frac{3\sqrt{2}\gamma}{\pi M_s^2} \propto \frac{[A\,|K_1+\sigma\lambda_s|]^{1/2}}{M_s^2}. \tag{5.57}$$

Discussions of this subject can be found in most standard texts, including that by Craik and Tebble [83].

The coercive field (or force) H_c of a hysteresis loop is generally defined as the reverse magnetic field required to reduce the magnetization of a magnetized body to zero. For a single domain in which the only mechanism available to alter the magnetic state is reversible rotation of the $\boldsymbol{M}$ vector, the value of $H_c \sim C_{rotat} |K_1 + \sigma\lambda_s| / M_s$. In most situations of multidomains, demagnetization in an applied field occurs by movement of domain walls irreversibly overcoming impediments such as grain boundaries and nonmagnetic inclusions e.g., air pores. Although irreversible domain rotation will contribute to the coercive field, pinning of domain walls at inclusions or lamellar precipitates caused by grain boundaries is the microstructural source of coercivity [84]. To accommodate the nonmagnetic volume of inclusions or grain boundaries within the wall volume, the surface energy of the wall must increase through expansion of the surface. The resulting increase in wall motion impedance produces two additional contributions to the coercive field which can be combined directly as $H_c{}^{incl} + H_c{}^{grain}$. As a generalization, all three effects could be viewed analogously to a resistor network, with and $H_c{}^{rotat}$ representing an alternative mechanism acting in parallel. Consequently, we approximate

$$H_c = \left(\frac{1}{H_c^{rotat}} + \frac{1}{H_c^{incl} + H_c^{grain}} \right)^{-1}. \tag{5.58}$$

Since the anisotropy and magnetostriction constants influence the coercive field through the wall surface energy density [85], H_c^{incl} and $H_c^{grain\,n}$ can be related to the magnetoelastic parameters according to

$$\begin{aligned} H_c^{incl} &= \frac{|K_1 + \sigma\lambda_s|}{M_s} \frac{\delta_w C_{incl}}{d_{incl}} p^{2/3} \\ \text{and } H_c^{grain} &= \frac{|K_1 + \sigma\lambda_s|}{M_s} \frac{\delta_w C_{grain}}{d_{grain}}. \end{aligned} \tag{5.59}$$

where p is the volume fraction of inclusions (porosity), and d_{incl} and d_{grain} are the average pore and grain dimensions, respectively. Because the rotation coercivity is generally much greater than that allowed by domain wall movement, a more useful relation could be reduced from (5.55), (5.56), and (5.59) as

$$H_c \sim \frac{\gamma_w}{M_s} \left(\frac{C_{incl}}{d_{incl}} p^{2/3} + \frac{C_{grain}}{d_{grain}} \right). \tag{5.60}$$

The magnetization process that begins with magnetic vector rotation (often called domain rotation) will undergo 180°-domain reversal at the coercive field. Beyond this point, rotation ensues until the magnetization reaches alignment with the field (at the anisotropy field). In addition, demagnetizing effects of the porosity continue to be overcome as the body approaches magnetic saturation. These topics will be discussed next.

5.4.2 Anisotropy Field and Remanence Ratio

Other loop parameters of importance are the anisotropy field H_K and remanent magnetization $4\pi M_R$, which is defined as the net moment per unit volume that remains after the material is returned to the $H = 0$ state following saturation. To appreciate the role of K_1 and λ_s in the remanent state, we begin by examining the case of a single crystal magnetized along a hard axis. When the orientation is not along an easy axis, the anisotropy field H_{Kc} defined as $2K_1/M$ delays saturation by imposing an effective demagnetizing effect. The influence of the anisotropy is illustrated in Fig. 5.31, which indicates the magnetization characteristics for materials of cubic and uniaxial crystallographic structure magnetized along a hard axis.

The relations for anisotropy fields with magnetostriction contributions may be found in most of the standard texts cited previously. In a $\{110\}$ plane with hard axis along a $\langle 100 \rangle$ axis (as defined in Fig. 5.15), we restate (5.35)

$$H_{K\sigma} = -\frac{2K_1}{M} + \frac{3}{2}\frac{\sigma\lambda_{100}}{M}. \tag{5.61}$$

In Fig. 5.32 the hard-axis magnetization curve for single-crystal YIG is shown as measured directly from an oscilloscope. Note the near-linear curve between the remanent point and the saturation point at $H = H_{K\sigma}$. Figure 5.16 introduced previously shows the effects of a compressive stress $\boldsymbol{\sigma}$ applied along the hard axis, which reduces $H_{K\sigma}$ and raises $4\pi M$ as the material undergoes a uniaxial strain [59].

The influence of magnetostriction on the remanence ratio of polycrystalline materials can be important in low anisotropy materials where the internal stress is sufficient to compete with K_1, i.e., $\sigma\lambda_{100}/K_1 \geq 1$. Examples of this effect on the remanence region are shown schematically in Fig. 5.33. In practical situations, the issue concerning magnetostriction arises where H_c is small, which often occurs in low switching-energy applications [86]. In these cases, the design strategy is to reduce the value of λ_s by chemical alteration of the material. Figure 5.34

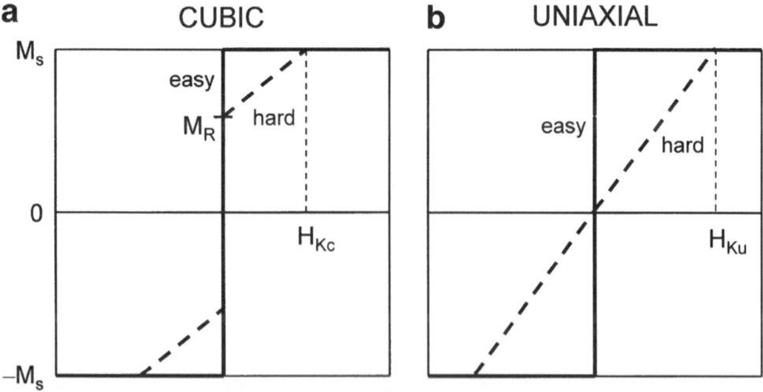

Fig. 5.31 Schematic illustration of the magnetocrystalline anisotropy effects on the magnetization process with zero coercive field H_c is assumed to be zero: **(a)** the cubic case (rotation in a $\{110$ plane$\}$) where the remanence ratio $M_R/M_s = 0.58$ in a *hard* direction, and **(b)**, the uniaxial case, where $M_R = 0$ in a *hard* direction

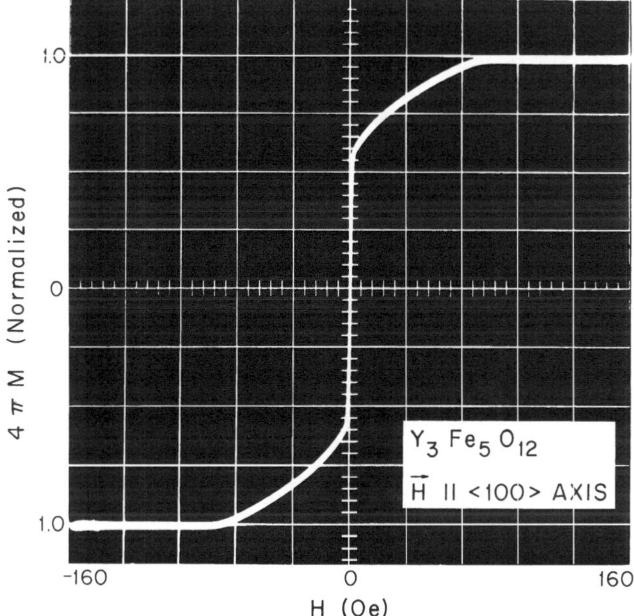

Fig. 5.32 Oscilloscope photograph of hysteresis property of $Y_3Fe_5O_{12}$ single-crystal described in Fig. 5.16. There is effectively no coercive field and both R and H_K agree precisely with theory [59]. Figure reprinted from [59] with permission. © 1969 by the IEEE

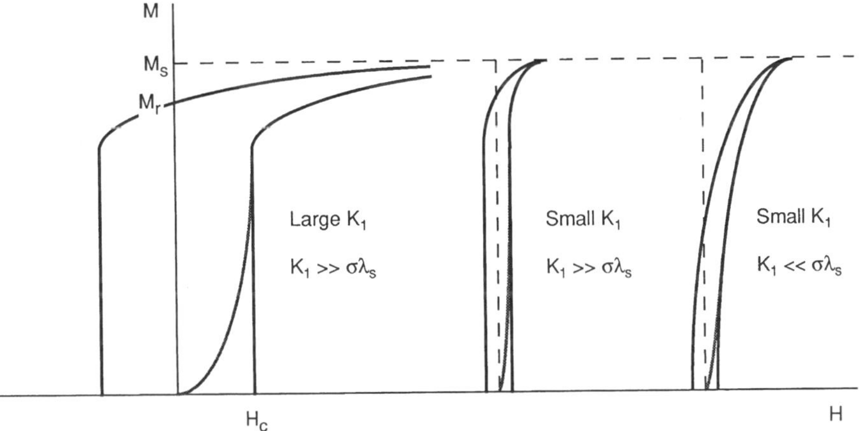

Fig. 5.33 Hysteresis loop models indicating the influence of K_1 and λ

presents measured results of Mn^{3+} substitution mostly in octahedral sites of NiZn spinel ferrite. Similar effects were also reported for YIG-based compounds [87,88]. As described in Sect. 5.3.4, the positive Jahn–Teller extensions of the local ligand group surrounding the Mn^{3+} ion has the effect of compensating the normal negative

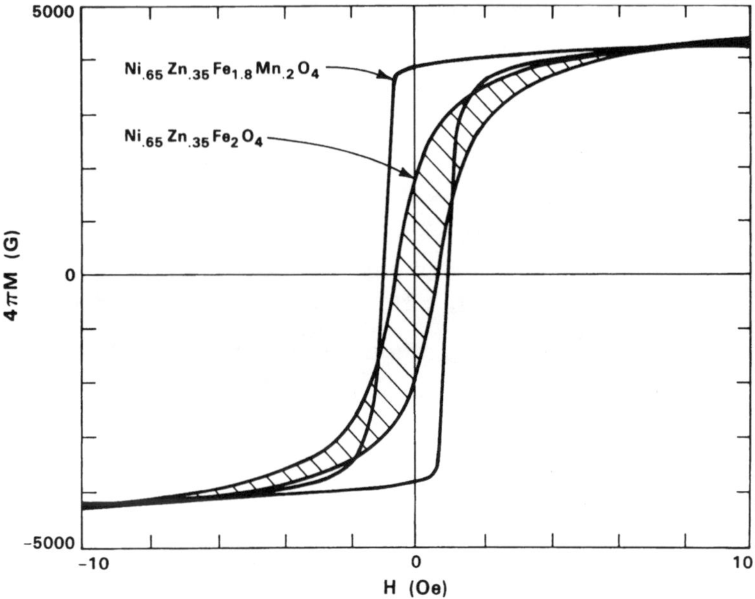

Fig. 5.34 Comparison of low-field hysteresis loops of $Ni_{0.65}Zn_{0.35}Fe_2O_4$ and $Ni_{0.65}Zn_{0.35}Fe_{1.8}Mn^{3+}_{0.2}O_4$ spinel ferrites, showing the cancellation of internal stress effects by local Jahn–Teller effects of Mn^{3+} (d^4) in the octahedral sublattice. Figure reprinted from [86] with permission. © 1986 by the American Institute of Physics

magnetostriction of the Fe^{3+} lattices. A comprehensive discussion of stress effects on hysteresis loops can be found in Bozorth [52].

For the uniaxial case, the relation is similar, $H_{Ku} = -2K_u/M$ and the magnetostriction effect is usually inconsequential because of the extraordinarily large values of K_u. Here the M vs. H relation along a hard axis is expected to be linear for a single crystal, reaching from the origin to the saturation value M_s at H_{Ku}, as sketched in Fig. 5.31b.

In polycrystals with grains of random crystallographic orientation, $H_{K\sigma}$ of each grain varies according to the particular alignment with $\boldsymbol{H}$. The averaging that occurs in the magnetization produces the rounded corners and edges of the loop. Calculated loops that can approximate these effects will be discussed, but first it is important to examine the mechanism of magnetization when H exceeds the coercive field, i.e., in the approach to saturation.

5.4.3 Approach to Saturation

Once the domain walls have moved to complete the reversal of the $\boldsymbol{M}$ direction, the balance of the magnetization with increasing magnetic field beyond H_c is gained by

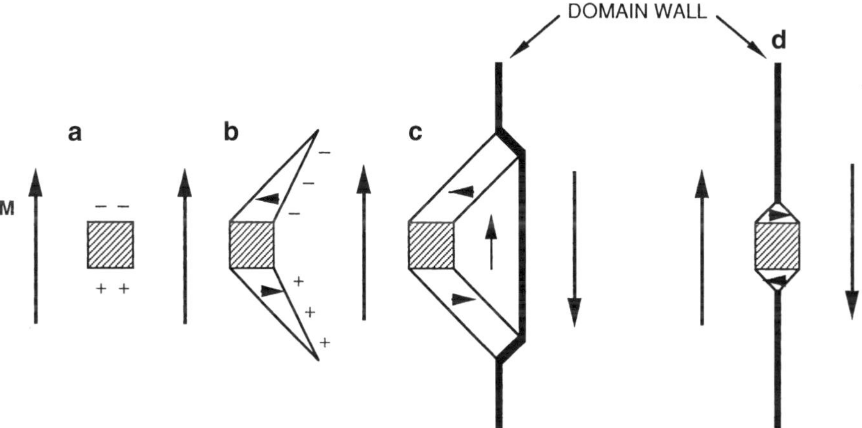

Fig. 5.35 Sequence of diagrams showing how a nonmagnetic inclusion can reduce remanence by nucleating 90° spike domains and also impede the movement of a domain wall by reducing the wall area [82]

M vector rotation up to the maximum $H_{K\sigma}$ and by overcoming the demagnetizing fields centered about pores and other nonmagnetic entities. These latter effects generally involve reverse spike or closure domains on either side of the pores, with the progress of Néel 90° closure domains during switching depicted in Fig. 5.35. The relative magnetization ratio R in the approach to saturation is traditionally written as a semiempirical law

$$R = \frac{M}{M_s} = 1 - \frac{a}{H} - \frac{b}{H^2} - \frac{c}{H^3} - \frac{d}{H^4} \cdots \tag{5.62}$$

where $a \approx (4\pi M_s)\, p_{\text{eff}}$, is the *magnetic hardness* constant that is proportional to the effective porosity p_{eff}, b is related to K_1/M_s, and c, d are the coefficients of higher-order terms that represent magnetostriction contributions from internal stress that show their effects at low magnetic fields [89]. If the field-dependent terms are small and cross terms are ignored, (5.62) can be an approximation to an exponential function

$$R \approx \exp\left(-a/H - b/H^2 - c/H^3 - d/H^4 \cdots\right). \tag{5.63}$$

Careful analysis of R data from the high-field portion of the hysteresis loop yielded accurate values of p_{eff} and K_1/M_s in magnetic garnets and spinels [71]. This model breaks down as $H \to 0$ from the return from saturation, denying the determination of a value for the remanence ratio R_0 or the possibility of completing a hysteresis loop. If the coercive field is treated as a negative bias in the outward leg and as a positive field in the return part, the first term of (5.63) can be expressed as $R \approx \exp\left[-a/\left(H \pm H_c\right)\right]$, thereby allowing the formation of a hysteresis loop and permitting the establishment of a basic remanence ratio $R_0 = \exp\left(-a/H_c\right)$. In a more complete form, a nucleation field H_n for reverse domains is added as a positive bias in the $(H \pm H_c + H_n)$ factor [80, 84].

5.4.4 Demagnetization and Permanent Magnets

For the loops discussed so far in this text, the assumption is made that shape demagnetization effects discussed in Chap. 1 are not present, i.e., the magnetic circuit is fully closed as in the case of a toroid with uniform susceptibility and cross-section. In many practical situations, however, the magnetic circuit is not complete and air gaps that produce demagnetizing fields ranging from insignificant to the full $4\pi M_s$ can influence the resultant magnetization curve. For this reason, "permanent" magnetism requires that the bias field provided by coercivity be maximized to offset the sloping of the loop sides caused by demagnetizing effects. The skewing of the loop can be seen in terms of reduced effective susceptibility by a simple one-dimensional analysis that begins by defining an effective internal field reduced by a demagnetizing field:

$$H_{\mathrm{eff}} = H - H_{\mathrm{D}}, \tag{5.64}$$

where $H_{\mathrm{D}} = N_{\mathrm{D}}(4\pi M_{\mathrm{R}}) = N_{\mathrm{D}}(4\pi M_s)R$. If M is expressed in terms of susceptibility χ and H_{eff},

$$\begin{aligned} M &= \chi\left[H - N_{\mathrm{D}}(4\pi M)\right] \\ &= \frac{\chi}{1 + N_{\mathrm{D}}(4\pi M)\chi}, \end{aligned} \tag{5.65}$$

which leads to the definition of an effective susceptibility

$$\chi_{\mathrm{eff}} = \frac{\chi}{1 + N_{\mathrm{D}} 4\pi\chi}. \tag{5.66}$$

If $\chi \to \infty$ or a single crystal magnetized along an easy direction, $\chi_{\mathrm{eff}} \to 1/4\pi N_{\mathrm{D}}$ and the corresponding magnetization curve is plotted in Fig. 5.36. Since $N_{\mathrm{D}} = 1$ in the case of the ultimate "gap", e.g., a thin disc magnetized normal to its flat surfaces, the field required to saturate the magnetization is exactly equal to $4\pi M_s$.

If a hysteresis loop model following the reasoning of the previous section is constructed for a circuit that contains a demagnetizing section, the demagnetizing field must be included so that

$$R \approx \exp\left(-\frac{a}{H \pm H_c + H_n - H_{\mathrm{D}}} - \cdots\right), \tag{5.67}$$

where $H_{\mathrm{D}} = N_{\mathrm{D}}(4\pi M)$, with $4\pi M$ expressed as $(4\pi M_s)R$ for computational iteration purposes [90]. The issue of shape demagnetization arises mainly in the application of permanent magnets, where $H_c \geq 4\pi M_s$ is the condition to achieve the maximum R value under conditions of $N_{\mathrm{D}} \sim 1$. Figure 5.37 illustrates this extreme condition.

A second concern that enters into the design of a hard magnet, whether oxide or metallic, is the demagnetization "energy product" defined as $(BH)_{\max}$. As indicated

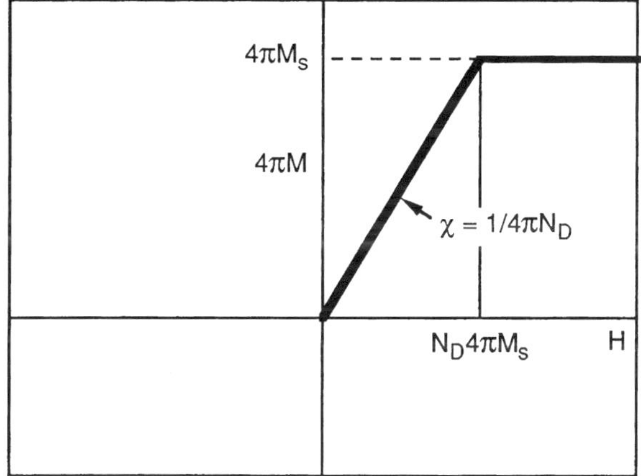

Fig. 5.36 Simple diagram of the shape demagnetizing field along the hard axis of a uniaxial specimen (typical of planar geometry with $\boldsymbol{H}$ normal to the plane)

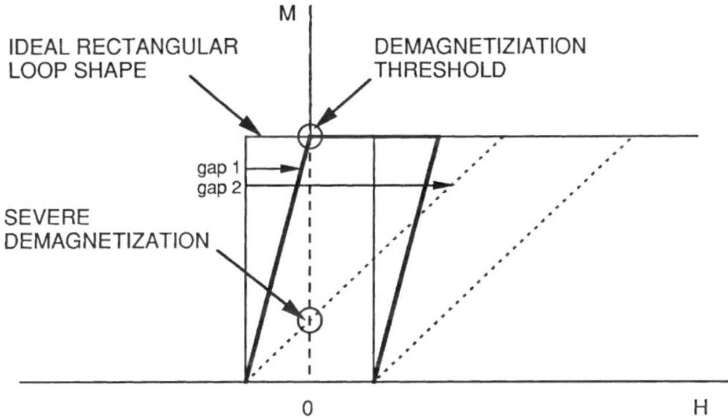

Fig. 5.37 Schematic illustrating the skewing of hysteresis loops caused by demagnetizing gaps in the magnetic circuit

by the rectangular-loop diagram in Fig. 5.38, the field $H_{\max}$ for the maximum BH product is found from the relation $B = H + 4\pi M_{\mathrm{r}}$ that applies in the negative field quadrant, so that

$$H_{\mathrm{D}} = N_{\mathrm{D}}\,(4\pi M)\,, \tag{5.68}$$

resulting in $H_{\max} = -4\pi M_{\mathrm{r}}/2$ and yields

$$(BH)_{\max} = \frac{1}{4}\,(4\pi M_{\mathrm{r}})^2 = (B_{\mathrm{r}}/2)^2\,. \tag{5.69}$$

From a loop design standpoint, we recognize that to achieve the theoretical limit defined by (5.68), $H_{\mathrm{c}} \geq 4\pi M_{\mathrm{r}}/2$. Moreover, it follows that any excess of H_{c} over

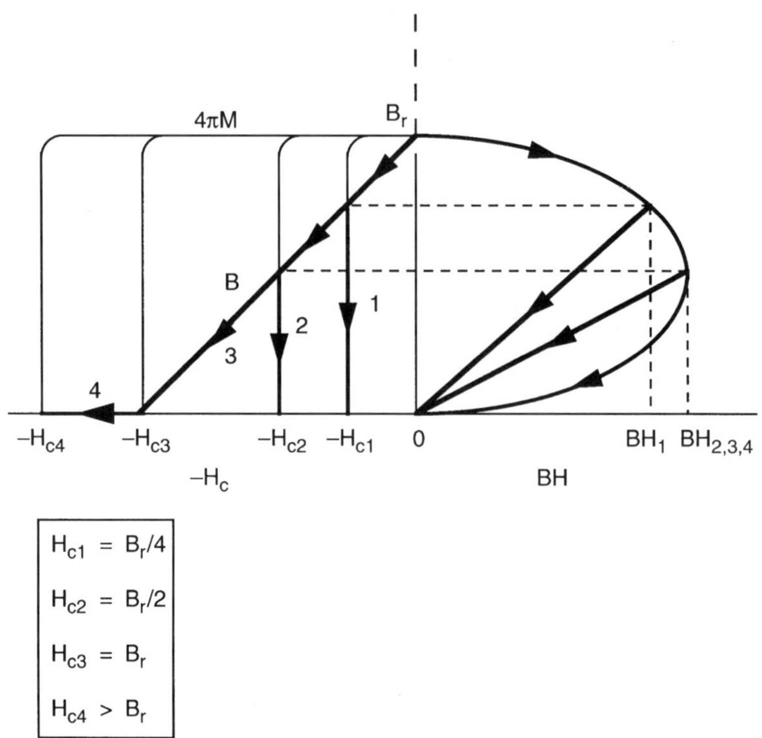

Fig. 5.38 Schematic diagram illustrating the relation between the coercive field H_c and the maximum *BH* energy product, which dictates that $H_c > B_r/2$

$4\pi M_r/2$ does not increase the energy product. However, $H_c \geq 4\pi M_r$ allows the magnet to retain its magnetization up to $H = -H_c$, which is an important feature of any square-knee permanent magnet.

Hexagonal ferrites are vehicles for permanent magnetism. For a comprehensive review of these oxide families, the reader is referred to the chapter by Wijn [91]. Among the various crystallographic structures that comprise the class, uniaxial M-type (magnetoplumbite, with easy c-axis and $K_u < 0$) offers the highest combination of magnetization and anisotropy field. Typically $4\pi M_s \approx 4{,}500$ G and $2K_u/M_s \approx 17{,}000$ Oe in undiluted barium ferrite ($BaFe_{12}O_{19}$). Careful processing that begins with submicron-size single-domain particles and involves cold-pressing in a magnetic field can produce good quality ceramic materials with H_c values that reach 4,000 Oe to satisfy the $H_c \geq 4\pi M_{Rr}$ condition [92]. Easy-plane anisotropy occurs in type-Y structures of formula $Ba_2M_2Fe_{12}O_{22}$, where M is a divalent ion such as Co^{2+} or Zn^{2+}. These compounds provide lower anisotropy fields and are of interest in cases where dilution of the M-type hexaferrites cannot achieve the desired properties.

Appendix 5A Four-Level Degenerate Perturbation Solution for d^1

The following calculation has been excerpted from the analysis of the paramagnetic resonance spectrum of Ti^{3+} (d^1) in the hydrated rubidium alum salt $Rb^{1+}\,Al^{3+}\,(SO_4)^{2-}{}_2 \cdot 12H_2O$ [3]. The energy level diagram for this model is shown in Fig. 5.4 with orbital eigenstates $|0\rangle$ through $|4\rangle$ corresponding to the set listed in (2.12) and (5.3). It begins with the computation of the 10×10 matrix for the elements of the combined crystal field and spin–orbit coupling operators, i.e., $\langle k\,|\mathcal{H}_{\mathrm{cf}} + \lambda \boldsymbol{L} \cdot \boldsymbol{S}|\,n\rangle$, for which no restrictions are placed on the relative strengths of the two contributions. The introduction of the Zeeman perturbation elements can be postponed here because we assume that they are small by comparison.

We begin the process by setting up the full Hermitian matrix:

$$
\begin{array}{c|ccccc|ccccc}
 & |0\alpha\rangle & |1\beta\rangle & |2\beta\rangle & |3\alpha\rangle & |4\beta\rangle & |0\beta\rangle & |1\alpha\rangle & |2\alpha\rangle & |3\beta\rangle & |4\alpha\rangle \\
\hline
\langle 0\alpha| & 0 & -\mathrm{i}\frac{\lambda}{2} & \frac{\lambda}{2} & \mathrm{i}\lambda & 0 & 0 & 0 & 0 & 0 & 0 \\
\langle 1\beta| & \mathrm{i}\frac{\lambda}{2} & \delta_1 & \mathrm{i}\frac{\lambda}{2} & \frac{\lambda}{2} & 0 & 0 & 0 & 0 & 0 & -\frac{\sqrt{3}\lambda}{2} \\
\langle 2\beta| & \frac{\lambda}{2} & -\mathrm{i}\frac{\lambda}{2} & \delta_2 & -\mathrm{i}\frac{\lambda}{2} & 0 & 0 & 0 & 0 & 0 & -\mathrm{i}\frac{\sqrt{3}\lambda}{2} \\
\langle 3\alpha| & -\mathrm{i}\lambda & \frac{\lambda}{2} & \mathrm{i}\frac{\lambda}{2} & \Delta & 0 & 0 & 0 & 0 & 0 & 0 \\
\langle 4\beta| & 0 & 0 & 0 & 0 & \Delta' & 0 & \frac{\sqrt{3}\lambda}{2} & \mathrm{i}\frac{\sqrt{3}\lambda}{2} & 0 & 0 \\
\hline
\langle 0\beta| & 0 & 0 & 0 & 0 & 0 & 0 & -\mathrm{i}\frac{\lambda}{2} & \frac{\lambda}{2} & -\mathrm{i}\lambda & 0 \\
\langle 1\alpha| & 0 & 0 & 0 & 0 & \frac{\sqrt{3}\lambda}{2} & \mathrm{i}\frac{\lambda}{2} & \delta_1 & -\mathrm{i}\frac{\lambda}{2} & -\frac{\lambda}{2} & 0 \\
\langle 2\alpha| & 0 & 0 & 0 & 0 & -\mathrm{i}\frac{\sqrt{3}\lambda}{2} & -\frac{\lambda}{2} & \mathrm{i}\frac{\lambda}{2} & \delta_2 & -\mathrm{i}\frac{\lambda}{2} & 0 \\
\langle 3\beta| & 0 & 0 & 0 & 0 & 0 & \mathrm{i}\lambda & -\frac{\lambda}{2} & \mathrm{i}\frac{\lambda}{2} & \Delta & 0 \\
\langle 4\alpha| & 0 & -\frac{\sqrt{3}\lambda}{2} & \mathrm{i}\frac{\sqrt{3}\lambda}{2} & 0 & 0 & 0 & 0 & 0 & 0 & \Delta'
\end{array}
\qquad (5.70)
$$

where $|\alpha\rangle$ and $|\beta\rangle$ are the $S = \pm 1/2$ eigenfunctions. By inspecting (5.70) we recognize that the full 10×10 matrix can be reduced to two identical 4×4 matrices with only a small sacrifice in accuracy. The equation formed from the secular determinant $\sum_k \left(\langle k|\, H_{\mathrm{cf}} + \lambda L \cdot S\,|n\rangle - E\delta_{\mathrm{kn}}\right) = 0$ of these smaller matrices is expressed as

$$E^4 + \mathrm{A}E^3 + \mathrm{B}E^2 + \mathrm{C}E + \mathrm{D} = 0, \qquad (5.71)$$

where[7] $A = -\left(\delta_1 + \delta_2 + \Delta\right)$,

$$
\begin{aligned}
B &= \delta_1\delta_2 + \delta_1\Delta + \delta_2\Delta - (9/4)\,\lambda^2 \\
C &= (1/4)\,\lambda^2\,(6\delta_1 + 6\delta_2 + 3\Delta) - \delta_1\delta_2\Delta + (5/8)\,\lambda^3 \\
D &= (1/8)\,\lambda^3\,(2\Delta - 4\delta_1 - 4\delta_2) \\
&\quad - (1/4)\,\lambda^2\,(\delta_1\Delta + \delta_2\Delta + 4\delta_1\delta_2) + (3/4)\,\lambda^4.
\end{aligned}
$$

[7] These relations have been corrected for small errors that appeared in the original computation [3].

For the ground doublet (the solution with lowest value of E), the eigenfunctions for the two states can be shown to be

$$\begin{aligned}|e_0+\rangle &= a\,|0\alpha\rangle + b\,|1\beta\rangle + c\,|2\beta\rangle + d\,|3\alpha\rangle \\ |e_0-\rangle &= a\,|0\beta\rangle + b\,|1\alpha\rangle - c\,|2\alpha\rangle - d\,|3\beta\rangle\,,\end{aligned} \tag{5.72}$$

where $a^*a + b^*b + c^*c + d^*d = 1$ for normalization. Similar relations for $|e_1\pm\rangle$, $|e_2\pm\rangle$, and the other eigenfunctions can be derived for the higher energy solutions. The coefficients expressed relative to d that can apply to any of the E solutions and are given by

$$\begin{aligned}a &= \mathrm{i}\frac{\frac{1}{2}\lambda^3 - \lambda\,(\delta_1 - E)\,(\delta_2 - E) - \frac{1}{4}\lambda^2\,[(\delta_1 - E) + (\delta_2 - E)]}{\frac{1}{4}\lambda^2\,[(\lambda - E) + (\delta_1 - E) + (\delta_2 - E)] - E\,(\delta_1 - E)\,(\delta_2 - E)} \times d, \\ b &= \frac{\frac{1}{4}\lambda^2\,(\lambda - E) - \frac{1}{2}\lambda\,(\lambda - E)\,(\delta_2 - E)}{\frac{1}{4}\lambda^2\,[(\lambda - E) + (\delta_1 - E) + (\delta_2 - E)] - E\,(\delta_1 - E)\,(\delta_2 - E)} \times d, \\ c &= \mathrm{i}\frac{\frac{1}{4}\lambda^2\,(\lambda - E) - \frac{1}{2}\lambda\,(\lambda - E)\,(\delta_1 - E)}{\frac{1}{4}\lambda^2\,[(\lambda - E) + (\delta_1 - E) + (\delta_2 - E)] - E\,(\delta_1 - E)\,(\delta_2 - E)} \times d.\end{aligned} \tag{5.73}$$

With the eigenfunctions determined for each solution for E, we can now compute the Zeeman splittings by applying the method used above to each of the Kramers doublets. In cases where the ground state is a degenerate group of Kramers doublets, all of the states must be included in the calculation. Where the ground state $|e_0\rangle$ is a singlet created by lower symmetry crystal fields, the ground doublet alone may be all that is required. If magnetic fields of average strength ($H < 1\,\mathrm{T}$) are applied, the Zeeman energy term is negligible compared with the λ constant and the upper levels $|e_1\rangle$ and $|e_2\rangle$ have little influence except possibly for paramagnetic susceptibilities at higher temperatures. The general procedure is to repeat the above process by forming secular equations from the Zeeman perturbation $\mathcal{H}_\mathrm{h} = m_\mathrm{B}\,(\boldsymbol{L} + 2\boldsymbol{S}) \cdot \boldsymbol{H}$ utilizing the new sets of eigenfunctions obtained from (5.72) and (5.73). If all three doublets are required, we compute the elements to form a 6×6 matrix; if the lowest doublet is needed, the problem is simplified to the 2×2 matrix of the ground doublet $|e_0+\rangle$ and $|e_0-\rangle$, as indicated by

	$\|e_0+\rangle$	$\|e_0-\rangle$	$\|e_1+\rangle$	$\|e_1-\rangle$	$\|e_2+\rangle$	$\|e_2-\rangle$
$\langle e_0+\|$	$E_0 + H_{11}$	H_{12}	H_{13}	H_{14}	H_{15}	H_{16}
$\langle e_0-\|$		$E_0 + H_{22}$	H_{23}	H_{24}	H_{25}	H_{26}
$\langle e_1+\|$			$E_1 + H_{33}$	H_{34}	H_{35}	H_{36}
$\langle e_1-\|$				$E_1 + H_{44}$	H_{45}	H_{46}
$\langle e_2+\|$					$E_2 + H_{55}$	H_{56}
$\langle e_2-\|$						$E_2 + H_{66}$

. (5.74)

For the ground $|e_0\pm\rangle$ doublet submatrix, we can drop the E_0 energy from the diagonal elements and determine that $\mathcal{H}_{22} = -\mathcal{H}_{11}$, and $\mathcal{H}_{12} = \mathcal{H}_{21} = 0$, with the result that we obtain the following relations for the g factors for the three $\boldsymbol{H}$ field directions:

$$
\begin{aligned}
g_z &= 2\left(a^*a - b^*b - c^*c + d^*d\right) + 4\mathrm{i}\left(a^*d - d^*a\right) - 2\mathrm{i}\left(b^*c - c^*b\right) \\
g_x &= 2\left(a^*a + b^*b - c^*c - d^*d\right) + 2\mathrm{i}\left(a^*b - b^*a\right) + 2\mathrm{i}\left(c^*d - d^*c\right) \\
g_y &= 2\left(a^*a - b^*b + c^*c - d^*d\right) + 2\left(a^*c + c^*a\right) - 2\left(b^*d + d^*b\right).
\end{aligned}
\tag{5.75}
$$

Calculations for parameter values $\lambda = 154$, $\delta_1 = 1{,}050$, $\delta_2 = 1{,}320$, and $\Delta = 22{,}000\,\mathrm{cm}^{-1}$ yield $g_z = 1.895$, $g_x = 1.715$, and $g_y = 1.767$, in exact agreement with the measured results [4].

Appendix 5B T_{2g} Solution for d^1 in an Exchange Field

Where spin–orbit coupling is stronger than any likely lower symmetry component of the crystal field, it is advantageous to use a solution that begins with the eigenfunctions of the $\lambda \boldsymbol{L} \cdot \boldsymbol{S}$ operator. The details of the analysis for the d^1 case can be found in Ballhausen [74] or Schläfer and Gliemann (Table 5.10) [93]. This version of the d^1 solution is particularly useful in understanding the behavior of ions with triplet ground states i.e., d^1 and d^6, d^2 and d^7 in octahedral fields, or d^3 and d^4, d^8 and d^9 in tetrahedral fields. For our purposes, we expand this solution to include a tetragonal field component by constructing matrices that include the $\mathcal{V}_{\mathrm{cf}}{}^{\mathrm{T}}$ operator, which produces off-diagonal elements that mix the starting wavefunctions. If we assume that the cubic Dq parameter is large enough to allow the upper E_{g} state to be ignored, we can reduce the problem to the solution of 2×2 matrices generated by applying the $(\lambda \boldsymbol{L} + g m_{\mathrm{B}} \boldsymbol{H}_{\mathrm{ex}})\,\boldsymbol{S}$ operator to the basis vectors of spin–orbit coupling. In order to retain the single-ion nature of the Hamiltonian, the source of the exchange interaction with spins of adjacent ions is treated phenomenologically as a quasimagnetic field $\boldsymbol{H}_{\mathrm{ex}}$ that forms a scalar product with $\boldsymbol{S}$, analogous in certain respects to the point-charge crystal-field concept. Stabilization of only the spin

Table 5.10 Eigenfunction limits and orbital angular momentum for axial crystal field and spin–orbit coupling energy levels of a d^1 electron[a]

δ/λ	φ_0	φ_1	φ_2
$-\infty$	$t_{2g}{}^{-}\alpha$	$t_{2g}{}^{\mathrm{o}}\beta$	$t_{2g}{}^{+}\alpha$
	$m_\ell = -1$	$m_\ell = 0$	$m_\ell = +1$
0	$t_{2g}{}^{-}\alpha$	$-\sqrt{\frac{2}{3}}t_{2g}{}^{\mathrm{o}}\beta + \sqrt{\frac{1}{3}}t_{2g}{}^{+}\alpha$	$-\sqrt{\frac{1}{3}}t_{2g}{}^{\mathrm{o}}\beta + \sqrt{\frac{2}{3}}t_{2g}{}^{+}\alpha$
	$m_\ell = -1$	$m_\ell = +\frac{1}{3}$	$m_\ell = +\frac{2}{3}$
$+\infty$	$t_{2g}{}^{-}\alpha$	$t_{2g}{}^{+}\alpha$	$t_{2g}{}^{\mathrm{o}}\beta$
	$m_\ell = -1$	$m_\ell = +1$	$m_\ell = 0$

[a] For the second set of wavefunctions φ_0', φ_1', and φ_2', the corresponding l_z values will reverse sign

momentum state can be considered in the exchange contribution to the perturbation energy. The magnetic resonance implications of this abbreviated solution in the presence of a "real" magnetic field are discussed further in connection with "exchange" resonance in Chap. 7. As a result, $\boldsymbol{H}_{\mathrm{ex}}$ is assumed to be directed along the z-axis and $g = 2$ for the spin-only value. The complete matrix of the perturbed $T_{2\mathrm{g}}$ term with the crystal field axial splitting parameter δ included is expressed as:

	Γ_8^{c}	Γ_8^{d}	Γ_8^{a}	Γ_7^{a}	Γ_8^{b}	Γ_7^{b}
Γ_8^{c}	$-\left(\frac{1}{2}\lambda+\frac{1}{3}\delta\right)$ $+m_{\mathrm{B}}H_{\mathrm{ex}}$	0	0	0	0	0
Γ_8^{d}	0	$-\left(\frac{1}{2}\lambda+\frac{1}{3}\delta\right)$ $-m_{\mathrm{B}}H_{\mathrm{ex}}$	0	0	0	0
Γ_8^{a}	0	0	$-\left(\frac{1}{2}\lambda-\frac{1}{3}\delta\right)$ $-\frac{1}{3}m_{\mathrm{B}}H_{\mathrm{ex}}$	$\frac{2\sqrt{2}}{3}m_{\mathrm{B}}H_{\mathrm{ex}}$ $-\frac{\sqrt{2}}{3}\delta$	0	0
Γ_7^{a}	0	0	$\frac{2\sqrt{2}}{3}m_{\mathrm{B}}H_{\mathrm{ex}}$ $-\frac{\sqrt{2}}{3}\delta$	λ $+\frac{1}{3}m_{\mathrm{B}}H_{\mathrm{ex}}$	0	0
Γ_8^{b}	0	0	0	0	$-\left(\frac{1}{2}\lambda-\frac{1}{3}\delta\right)$ $+\frac{1}{3}m_{\mathrm{B}}H_{\mathrm{ex}}$	$-\frac{2\sqrt{2}}{3}m_{\mathrm{B}}H_{\mathrm{ex}}$ $-\frac{\sqrt{2}}{3}\delta$
Γ_7^{b}	0	0	0	0	$-\frac{2\sqrt{2}}{3}m_{\mathrm{B}}H_{\mathrm{ex}}$ $-\frac{\sqrt{2}}{3}\delta$	λ $-\frac{1}{3}m_{\mathrm{B}}H_{\mathrm{ex}}$

, (5.76)

where the eigenfunctions are based on the set listed in (2.11) according to:

$$\begin{aligned} \Gamma_8^{\mathrm{c}} &= t_{2\mathrm{g}}^{-}\alpha \\ \Gamma_8^{\mathrm{a}} &= -\sqrt{\frac{2}{3}}t_{2\mathrm{g}}^{\mathrm{o}}\beta + \sqrt{\frac{1}{3}}t_{2\mathrm{g}}^{+}\alpha \\ \Gamma_7^{\mathrm{a}} &= \sqrt{\frac{1}{3}}t_{2\mathrm{g}}^{\mathrm{o}}\beta + \sqrt{\frac{2}{3}}t_{2\mathrm{g}}^{+}\alpha \end{aligned}$$

And

$$\begin{aligned} \Gamma_8^{\mathrm{c}} &= t_{2\mathrm{g}}^{-}\alpha \\ \Gamma_8^{\mathrm{a}} &= -\sqrt{\frac{2}{3}}t_{2\mathrm{g}}^{\mathrm{o}}\beta + \sqrt{\frac{1}{3}}t_{2\mathrm{g}}^{+}\alpha \\ \Gamma_7^{\mathrm{a}} &= \sqrt{\frac{1}{3}}t_{2\mathrm{g}}^{\mathrm{o}}\beta + \sqrt{\frac{2}{3}}t_{2\mathrm{g}}^{+}\alpha. \end{aligned} \tag{5.77}$$

For this exercise, the eigenfunction nomenclature used by Ballhausen is adopted for consistency. The secular equation formed from (5.76) provides the following solutions for the energy states:

$$E\left(\Gamma_8^{\mathrm{c}}\right) = -\left(\frac{1}{2}\lambda + \frac{1}{3}\delta\right) + m_{\mathrm{B}}H_{\mathrm{ex}}$$
$$E\left(\Gamma_8^{\mathrm{d}}\right) = -\left(\frac{1}{2}\lambda + \frac{1}{3}\delta\right) - m_{\mathrm{B}}H_{\mathrm{ex}}$$

$$E\left(\Gamma^{\mathrm{a}}_{8,7}\right)_{\pm} = \frac{1}{2}\left(\frac{1}{2}\lambda + \frac{1}{3}\delta\right) \pm \frac{1}{2}\sqrt{\frac{9}{4}\lambda^2 + \delta^2 - \lambda\delta + 2m_{\mathrm{B}}H_{\mathrm{ex}}\left(\lambda - 2\delta\right) + 4\left(m_{\mathrm{B}}H_{\mathrm{ex}}\right)^2} \quad (5.78)$$

$$E\left(\Gamma^{\mathrm{b}}_{8,7}\right)_{\pm} = \frac{1}{2}\left(\frac{1}{2}\lambda = +\frac{1}{3}\delta\right) \pm \frac{1}{2}\sqrt{\frac{9}{4}\lambda^2 + \delta^2 - \lambda\delta - 2m_{\mathrm{B}}H_{\mathrm{ex}}\left(\lambda - 2\delta\right) + 4\left(m_{\mathrm{B}}H_{\mathrm{ex}}\right)^2}.$$

Computations of E as a function of $m_{\mathrm{B}}H_{\mathrm{ex}}$ for $\lambda < 2\delta/3$, $\lambda = 2\delta/3$, and $\lambda > 2\delta/3$ indicate six distinct levels and the removal of all degeneracy. Of particular interest is the comparison between the lowest energy states of the spin–orbit stabilized case corresponding to $\delta = 3\Delta E_{\mathrm{JT}}/2$ and the Jahn–Teller case with $\delta = -3\Delta E_{\mathrm{JT}}/2$, where ΔE_{JT} is the maximum available spontaneous ligand strain energy. For $H_{\mathrm{ex}} > 0$, the respective ground states are the $E\left(\Gamma^{\mathrm{d}}_{8}\right)$ and $E\left(\Gamma^{\mathrm{a}}_{8,7}\right)_{-}$ solutions of (5.78).[8] Additional discussion of this solution can be found in [16]

By equating these two expressions, the threshold condition for the stabilization of the $E\left(\Gamma^{\mathrm{d}}_{8}\right)$ level (for $\delta > 0$) can be derived by algebraic manipulation, according to

$$\frac{\lambda}{\delta} \geq \frac{2}{3}\left(1 + \frac{2}{3}\frac{\delta}{g m_{\mathrm{B}} H_{\mathrm{ex}}}\right), \quad (5.79)$$

which is introduced as (5.16) in the text.

Orbital angular momentum contributions to magnetoelastic effects are therefore dependent on the strength of the exchange field. Based on this simple model, the threshold ratio λ/δ reaches the value of 2/3 suggested by (2.34) only when $H_{\mathrm{ex}} \to \infty$. Because exchange energy is dependent on the degree of spin-ordering, the collective $g m_{\mathrm{B}} H_{\mathrm{ex}} \cdot S$, and hence the stabilization itself, are temperature dependent. Consequently, a transition from S–O to J–T stabilization as the temperature increases would be theoretically tied to the decline in strength of the molecular field that controls the shape of the Brillouin function dependence of the sublattice magnetization.

Appendix 5C Orbital States of d^5 in a Cubic Field

Two computations were undertaken to determine the energy ladder of orbital states of the various d^n configurations in a cubic crystal field. Orgel began by choosing the free-ion orbital wavefunctions and applying the cubic field potential [94] following the method of Finklestein and Van Vleck [95]. For the d^5 case, the energy

[8] Note that the Hamiltonian for a second ion, with an opposing spin direction needed to comply with the Pauli principle, would be set up with $z \to -z$. The basis wavefunctions would be similar to those of (5.77), but with each orbital and spin component reversed in sign. The corresponding ground states would therefore be $E\left(\Gamma^{\mathrm{c}}_{8}\right)$ and $E\left(\Gamma^{\mathrm{a}}_{8,7}\right)_{+}$ with α and β spin states that complement those of the first ion to provide the spin degeneracy in a two-ion molecular-orbital bonding state. Such a degeneracy would then appear spectrally as a Kramers doublet, removable only by an external magnetic field.

relations of the starting unperturbed states following Racah's terminology defined in Sect. 2.3 are:

$$\begin{aligned} E_{\mathrm{F}}\left({}^4F\right) &= 10A - 13B + 7C \\ E_{\mathrm{D}}\left({}^4D\right) &= 10A - 18B + 5C \\ E_{\mathrm{P}}\left({}^4P\right) &= 10A - 28B + 7C \\ E_{\mathrm{G}}\left({}^4G\right) &= 10A - 25B + 5C \\ E_{\mathrm{S}}\left({}^6S\right) &= 10A - 35B. \end{aligned} \tag{5.80}$$

The calculation proceeds by solution of secular equations obtained from matrices with nonzero elements formed between states of common symmetry group representations. In addition to three orbital singlets, there are six triplets and two doublets that are sorted out according to the following matrices formed with ${}^6S$ as the ground state reference energy:

$$\begin{array}{c|ccc} T_{1g} & {}^4P & {}^4F & {}^4G \\ \hline {}^4P & 7B+7C & 0 & 4\sqrt{5}Dq \\ {}^4F & 0 & 22B+7C & 2\sqrt{5}Dq \\ {}^4G & 4\sqrt{5}Dq & 2\sqrt{5}Dq & 10B+5C \end{array}, \tag{5.81}$$

$$\begin{array}{c|ccc} T_{2g} & {}^4D & {}^4F & {}^4G \\ \hline {}^4D & 17B+5C & 20\sqrt{7}Dq & 0 \\ {}^4F & 20\sqrt{7}Dq & 22B+7C & 10\sqrt{3/7}Dq \\ {}^4G & 0 & 10\sqrt{3/7}Dq & 10B+5C \end{array}, \tag{5.82}$$

$$\begin{array}{c|cc} E & {}^4D & {}^4G \\ \hline {}^4D & 17B+5C & 0 \\ {}^4G & 0 & 10B+5C \end{array}. \tag{5.83}$$

The alternative approach that was carried out earlier in a more comprehensive scope was reported by Tanabe and Sugano [96]. In their work the starting functions were formed from the strong field limit, i.e., the condition obtained when $Dq \to \infty$, the exact reverse of the Orgel method. For details of the computation of the nonzero matrix elements, the reader is directed to the original paper. The matrices for determining the nonsinglet states under the action of a field of strength Dq are as follows:

$$\begin{array}{c|ccc} T_{1g}\left({}^4G, {}^4P, {}^4F\right) & \left(t_{2g}\right)^4 \uparrow\uparrow\uparrow\downarrow\ e_g\uparrow & \left(t_{2g}\right)^3 \uparrow\uparrow\downarrow\ \left(e_g\right)^2 \uparrow\uparrow & \left(t_{2g}\right)^2 \uparrow\uparrow\ \left(e_g\right)^3 \uparrow\uparrow\downarrow \\ \hline \left(t_{2g}\right)^4 e_g & -25B+6C-10Dq & -3\sqrt{2}B & C \\ \left(t_{2g}\right)^3 \left(e_g\right)^2 & -3\sqrt{2}B & -16B+7C & -3\sqrt{2}B \\ \left(t_{2g}\right)^2 \left(e_g\right)^3 & C & -3\sqrt{2}B & -25B+6C+10Dq \end{array}, \tag{5.84}$$

$$\begin{array}{c|ccc} T_{2g}\left({}^4G, {}^4D, {}^4F\right) & \left(t_{2g}\right)^4 \uparrow\uparrow\uparrow\downarrow \left(e_g\right) \uparrow & \left(t_{2g}\right)^3 \uparrow\uparrow\downarrow \left(e_g\right)^2 \uparrow\uparrow & \left(t_{2g}\right)^2 \uparrow\uparrow \left(e_g\right)^3 \uparrow\uparrow\downarrow \\ \hline \left(t_{2g}\right)^2 \left(e_g\right)^3 & -17B + 6C - 10Dq & \sqrt{6}Dq & 4B + C \\ \left(t_{2g}\right)^2 \left(e_g\right)^3 & \sqrt{6}Dq & -22B + 5C & -\sqrt{6}Dq \\ \left(t_{2g}\right)^2 \left(e_g\right)^3 & 4B + C & -\sqrt{6}Dq & -17B + 6C + 10Dq \end{array}, \tag{5.85}$$

$$\begin{array}{c|cc} E\left({}^4G, {}^4D\right) & \left(t_{2g}\right)^3 \uparrow\uparrow\downarrow \left(e_g\right)^2 \uparrow\uparrow & \left(t_{2g}\right)^3 \uparrow\uparrow\uparrow \left(e_g\right)^2 \uparrow\downarrow \\ \hline \left(t_{2g}\right)^3 \left(e_g\right)^2 & -22B + 5C & -2\sqrt{3}Dq \\ \left(t_{2g}\right)^3 \left(e_g\right)^2 & -2\sqrt{3}Dq & -22B + 5C \end{array}. \tag{5.86}$$

For both the Orgel and Tanabe-Sugano analyses the energy states as a function of Dq can be obtained by analytical solution of secular equations that involve nothing more complex than cubic equations. An example of such a calculation for $3d^5$ is shown in Fig. 2.15, which is the identical result for either method using the same values of $B = 1,133\,\text{cm}^{-1}$ and $C = 3,867\,\text{cm}^{-1}$.

Appendix 5D Angular Dependence of Cubic Anisotropy Fields

The effective fields created by the stiffness of the anisotropy torques that affect the rotation of $\boldsymbol{M}$ can be determined analytically for any particular lattice plane by a transformation of coordinates from x, y, z (or r, θ, ϕ) to x', y', z' (or r, θ', ϕ') that places the crystallographic plane of interest in the $x'z'$ plane ($\phi' = 0$), where z' is set to the direction of vector $\boldsymbol{M}$. To establish the location of $\boldsymbol{M}$ in the new plane, the parameter δ is introduced as the angle $\boldsymbol{M}$ makes with a crystallographic reference axis, e.g., [111], [110]. Thus, while δ fixes the position of z' in the crystallographic plane, θ' remains the variable for the determining the stiffness torque $\tau\,(\theta', \delta)$ as $\boldsymbol{M}$ is deflected from z' by the rotation of $\boldsymbol{H}$. With an relation for $E_K\,(\theta', \delta)$ developed from the discussion in Sect. 5.3.1 expressed in terms of these new coordinates, we can then obtain the relation for the in-plane anisotropy field as a function of rotation angle δ. By choosing the $y'z'$ plane ($\phi' = \pi/2$), the out-of-plane stiffness torque and anisotropy field can also be determined. Knowledge of these orthogonal H_K relations is necessary for analyzing ferromagnetic resonance spectra described in Chap. 6.

The relations for H_K were used to obtain the results listed in Table 5.7 and are based on the diagrams of 5.39.

Rotation between axis and orthogonal plane in uniaxial crystal: $\delta = 0$ referenced to the axis.

$$\begin{aligned} H_K{}^{//} &= \frac{2K_1}{M_s}(\cos 2\delta) \qquad \text{(in plane)} \\ H_K{}^{\perp} &= \frac{2K_1}{M_s}(\cos \delta) \qquad \text{(out of plane)}\,. \end{aligned} \tag{5.87}$$

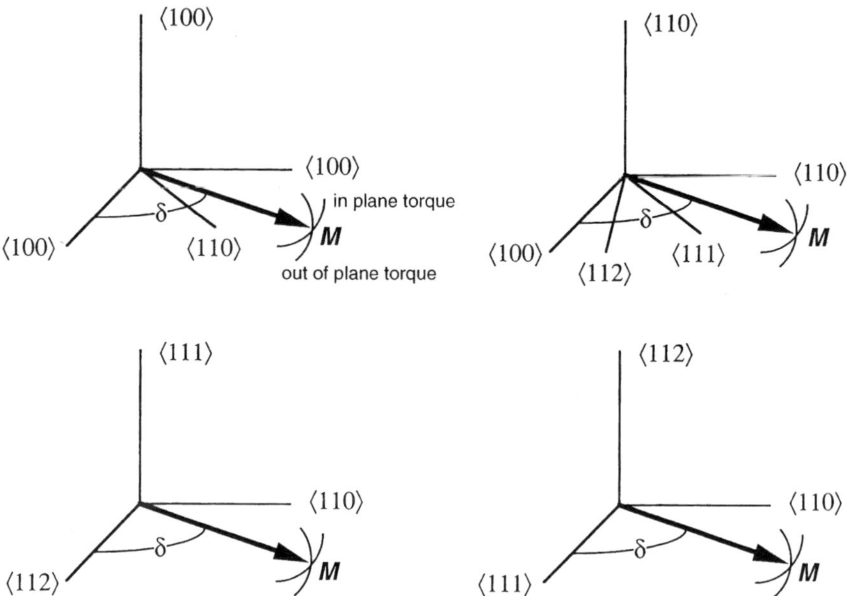

Fig. 5.39 Vector diagrams for $\boldsymbol{M}$ as functions of rotation angle δ in the four planes of interest for a cubic crystal lattice: [100], [110], [111], and [112]

Rotation in {100} planes: $\delta = 0$ referenced to a $\langle 100\rangle$ axis.

$$H_K{}^{//} = \frac{2K_1}{M_s}(\cos 4\delta) \qquad \text{(in plane)}$$

$$H_K{}^{\perp} = \frac{2K_1}{M_s}\left(1 - \frac{1}{2}\sin^2 2\delta\right) + \frac{K_2}{2M_s}\left(\sin^2 2\delta\right) \qquad \text{(out of plane)} \quad (5.88)$$

Rotation in {110} planes: $\delta = 0$ referenced to a $\langle 100\rangle$ axis.

$$H_K{}^{//} = \frac{K_1}{M_s}\left(2 - \sin^2\delta - 3\sin^2 2\delta\right) + \frac{K_2}{2M_s}\left[\sin^2\delta\left(6\cos^4\delta - 11\sin^2\delta\cos^2\delta + \sin^4\delta\right)\right] \qquad \text{(in plane)} \quad (5.89)$$

$$H_K{}^{\perp} = \frac{K_1}{M_s}\left[2 - 4\sin^2\delta - \frac{3}{4}\sin^2 2\delta\right] - \frac{K_2}{2M_s}\left[\sin^2\delta\cos^2\delta\left(3\sin^2\delta + 2\right)\right] \qquad \text{(out of plane)}$$

Rotation in {111} planes: $\delta = 0$ referenced to a $\langle 112 \rangle$ axis.

$$H_K{}^{//} = -\frac{K_2}{3M_s}\left[\cos^3 2\delta - 3\sin^2 2\delta\ \cos 2\delta\right] \qquad \text{(in plane)}$$

$$H_K{}^{\perp} = -\frac{K_1}{M_s} + \frac{K_2}{6M_s}\left[1 - \frac{2}{3}\left(3\sin^2\delta\ \cos\delta - \cos^3\delta\right)^2\right] \qquad \text{(out of plane)} \tag{5.90}$$

Rotation in {112} planes: $\delta = 0$ referenced to a $\langle 111 \rangle$ axis.

$$\begin{aligned} H_K{}^{//} &= -\frac{4K_1}{3M_s}\left(\cos^4\delta + \frac{3}{4}\sin^4\delta - \frac{21}{4}\sin^2\delta\ \cos^2\delta\right) \\ &\quad -\frac{4K_2}{9M_s}\begin{pmatrix} \cos^6\delta - \frac{3}{8}\sin^6\delta + \frac{57}{8}\sin^4\delta\ \cos^2\delta \\ -\frac{41}{4}\sin^2\delta\ \cos^4\delta \end{pmatrix} \qquad \text{(in plane)} \\ H_K{}^{\perp} &= -\frac{4K_1}{3M_s}\left(\cos^4\delta + \frac{3}{4}\sin^4\delta - \frac{3}{4}\sin^2\delta\right) \\ &\quad -\frac{4K_2}{9M_s}\begin{pmatrix} \frac{1}{2}\cos^6\delta + \frac{1}{2}\cos^4\delta - \frac{3}{4}\sin^4\delta \\ +\frac{9}{8}\sin^4\delta\cos^2\delta - \frac{3}{4}\sin^2\delta\ \cos^2\delta \\ -\frac{3}{2}\sin^2\delta\ \cos^4\delta \end{pmatrix} \qquad \text{(out of plane)} \end{aligned} \tag{5.91}$$

References

1. J.H. Van Vleck, *The Theory of Electric and Magnetic Susceptibilities*, (Oxford University Press, London, 1932)
2. J.S. Griffith, *The Theory of Transition-Metal Ions*, (Cambridge University Press, London, 1961)
3. G.F. Dionne, *Phys. Rev.* **137**, A743 (1965)
4. G.F. Dionne, *Can. J. Phys.* **42**, 2419 (1964)
5. J.A. MacKinnon and G.F. Dionne, *Can. J. Phys.* **44**, 2329 (1966)
6. G.F. Dionne and J.A. MacKinnon, *Phys. Rev.* **172**, 325 (1968)
7. B. Bleaney, G.S. Bogle, A.H. Cooke, R.J. Duffus, M.C.M. O'Brien, and K.W.H. Stevens, *Proc. Phys. Soc.* **A68**, 57 (1955)
8. G.A. Woonton and J.A. MacKinnon, *Can. J. Phys.* **46**, 59 (1968)
9. M.H.L. Pryce, *Proc. Phys. Soc.* **A63**, 25 (1950); A. Abragam and M.H.L. Pryce, *Proc. Roy. Soc.* **A205**, 135 (1951)
10. H. Hartmann and H.L. Schlafer, Z. *Physik Chem.* (Leipzig) **197**, 116, (1951)
11. A.J. Freeman and R.E. Watson, *Phys. Rev.* **127**, 2058 (1962)
12. K.W.H. Stevens, *Proc. Phys. Soc.* **A65**, 209 (1952)
13. M.T. Hutchings, *Solid State Phys.* **16**, 227 (1964)
14. W. Low, *Paramagnetic Resonance in Solids*, (Academic, New York, 1960)
15. H.A. Kramers, *Proc. Amsterdam Acad. Sci.* **33**, 959 (1930); W. Low, *Paramagnetic Resonance in Solids*, (Academic, New York, 1960), p. 34
16. G.F. Dionne, *J. Appl. Phys.* **91**, 7367 (2002)
17. L. Dubicki and M.J. Riley, *J. Chem. Phys.* **106**, 1669 (1997)
18. L.W. Tragenna-Pigott, S.P. Best, M.C.M. O'Brien, K.S. Knight, J.B. Forsyth, and J.R. Pillbrow, *J. Am. Chem. Soc.* **119**, 3324 (1997)
19. L.S. Kornienko and A.M. Prokorov, *Sov. Phys. JETP* **11**, 1189 (1960)

20. F.S. Ham, *Phys. Rev.* **138**, 1727 (1965)
21. R.M. MacFarlane, J.Y. Wong, and M.D. Sturge, *Phys. Rev.* **166**, 250 (1968)
22. C.A. Bates and J.P. Bentley, *J. Phys. C: Solid St. Phys.* **2**, 1947 (1969)
23. K.W.H. Stevens, *J. Phys. C: Solid St. Phys.* **2**, 1934 (1969)
24. M. Abou-Ghantous, C.A. Bates, and K.W.H. Stevens, *J. Phys. C: Solid St. Phys.* **7**, 325 (1974)
25. W. Low, *Phys. Rev.* **101**, 1827 (1956)
26. G.A. Slack, *Phys. Rev.* **134**, A1268 (1964)
27. B. Bleaney, K.D. Bowers, and R.J. Trenam, *Proc. Roy. Soc.* **A228**, 157 (1955)
28. M.H.L. Pryce, *Il. Nuovo Cimento* **6** (Suppl.) 817 (1957)
29. C.A. Bates and P.E. Chandler, *J. Phys. C: Solid St. Phys.* **4**, 2713 (1971)
30. G.M. Zverev and A.M. Prokorov, *J. Exptl. Theoret. Phys.* (U.S.S.R.) **34**, 1023 (1958)
31. W. Low, *Paramagnetic Resonance in Solids*, (Academic, New York, 1960), p.91
32. A. Abragam and M.H.L. Pryce, *Proc. Roy. Soc.* **A205**, 135 (1951); **A206**, 173 (1951)
33. C.J. Ballhausen, *Introduction to Ligand Field Theory*, (McGraw-Hill, New York, 1962) p. 124
34. A.L. Kipling, P.W. Smith, J. Vanier, and G.A. Woonton, *Can. J. Phys.* **39**, 1859 (1961)
35. J.S. Thorp, *Masers and Lasers: Physics and Design*, (MacMillan, London, 1967) Chapter 4
36. J.P. Gordon, H.J. Zeiger, and C.H. Townes, *Phys. Rev.* **95**, 282 (1954)
37. W. Low, *Paramagnetic Resonance in Solids*, (Academic, New York, 1960), p.92
38. J. Gielessen, *Ann. Physik.* **22**, 537 (1935)
39. D.S. McClure, *J. Phys. Chem. Solids* **3**, 311 (1957)
40. W.P. Doyle and G.A. Lonergan, *Discuss. Faraday Soc.* **26**, 27 (1958)
41. H. Hartmann and H. Müller, *Discuss. Faraday Soc.* **26**, 49 (1958)
42. H. Watanabe, *Prog. Theoret. Phys. (Kyoto)*, **18**, 405 (1957)
43. W. Low, *Paramagnetic Resonance in Solids*, (Academic, New York, 1960), p.120
44. J.H. Van Vleck and W.G. Penney, *Phil. Mag.* **17**, 961 (1934)
45. B. Bleaney and K.W. H. Stevens, *Rep. Prog. Phys.* **16**, 108 (1953)
46. V.J. Folen, *Paramagnetic Resonance Vol. 1*, (Proceed. First Intl. Conf., ed. W. Low (Academic Press, New York, 1962), p. 68
47. R.J. Elliott and K.W.H. Stevens, *Proc. Roy. Soc.* **A219**, 387 (1953)
48. J.W. Orton, *Rep. Prog. Phys.* **22**, 204 (1959)
49. K.D. Bowers and J. Owen, *Rep. Prog. Phys.* **18**, 304 (1955)
50. S. Chikazumi, *Physics of Magnetism*, (Wiley, New York, 1964)
51. J. Smit and H.P.J. Wijn, *Ferrites*, (Wiley, New York, 1959)
52. R.M. Bozorth, *Ferromagnetism*, (D. Van Nostrand, New York, 1951)
53. A.H. Morrish, *The Physical Principles of Magnetism*, (Wiley, New York, 1965)
54. H.J. Williams, *Phys. Rev.* **52**, 747 (1937); A.H. Morrish, *The Physical Principles of Magnetism*, (Wiley, New York, 1965), p. 310
55. L.R. Bickford, *Phys. Rev.* **78**, 449 (1950)
56. E.W. Lee, *Rep. Prog. Phys.* **18**, 184 (1955)
57. E. du Trémolet de Lacheisserie, *Magnetostriction*, (CRC Press, Boca Raton, FL, 1993)
58. C. Kittel, *Rev. Mod. Phys.* **21**, 541 (1949)
59. G.F. Dionne, *IEEE Trans. Magn.* **5**, 596 (1969)
60. G.F. Dionne, *Mater. Res. Bull.* **6**, 80 (1971)
61. J.H. Van Vleck, *Phys. Rev.* **52**, 1178 (1937)
62. C. Zener, *Phys. Rev.* **96**, 1335 (1954)
63. S. Chikazumi, *Physics of Magnetism*, (Wiley, New York, 1964), Section 7.2
64. K. Yosida and M. Tachiki, *Progr. Theoret. Phys. (Kyoto)* **17**, 331 (1957)
65. W.P. Wolf, *Phys. Rev.* **108**, 1152 (1957)
66. G.P. Rodrigue, H. Meyer, and R.V. Jones, *J. Appl. Phys.* **31**, 376S (1960)
67. S. Geschwind, *Phys. Rev.* **121**, 363 (1961)
68. B. Luthi and T. Henningsen, *Proceed. Intl. Conf. Magn. (Nottingham)* 1965, p. 668
69. P. Hansen, *J. Appl. Phys.* **45**, 3638 (1974)
70. J.R. Cunningham, Jr., *J. Appl. Phys.* **36**, 2491 (1965)
71. G.F. Dionne, *J. Appl. Phys.* **40**, 1839 (1969)
72. A.H. Morrish, *The Physical Principles of Magnetism*, (Wiley, New York, 1965), p. 529

73. J.C. Slonczewski, *Phys. Rev.* **101**, 1341 (1958)
74. C.J. Ballhausen, *Introduction to Ligand Field Theory*, (McGraw-Hill, New York, 1962), p.118
75. J.B. Goodenough, *Magnetism and the Chemical Bond*, (Wiley Interscience, New York, 1963), p. 192
76. P. Hansen, *Physics of Magnetic Garnets*, Proc. Int'l School Phys., Course LXX, (North-Holland, New York, 1978), p. 56
77. G.F. Dionne, *J. Appl. Phys.* **50**, 4263 (1979)
78. S. Iida, *J. Phys. Soc. Jpn.* **22**, 1201 (1967)
79. S. Chikazumi, Physics of Magnetism, (Wiley, New York, 1964), p. 263
80. W.F. Brown, *Micromagnetics*, (Wiley, New York, 1963)
81. H. Kronmüller and M. Fähnle, *Micromagnetism and the Microstructure of Ferromagnetic Solids*, (Cambridge University Press, Cambridge, 2003)
82. C. Kittel, *Phys. Rev.* **70**, 965 (1946)
83. D.J. Craik and R.S. Tebble, *Ferromagnetism and Ferromagnetic Domains*, (Wiley, New York, 1965)
84. J.B. Goodenough, *Phys. Rev.* **95**, 917 (1954)
85. A.H. Morrish, *The Physical Principles of Magnetism*, (Wiley, New York, 1965), p.389
86. G.F. Dionne and R.G. West, *Appl. Phys. Lett.* **48**, 1488 (1986)
87. G.F. Dionne and P.J. Paul, *Mater. Res. Bull.* **4**, 171 (1969)
88. G.F. Dionne, P.J. Paul, and R.G. West, *J. Appl. Phys.* **41**, 1411 (1970)
89. M.A. Stel'mashenko, *Sov. Phys. –.Solid State* **9**, 1137 (1967)
90. G.F. Dionne and D.E. Oates, *J. Appl. Phys.* **85**, 4856 (1999)
91. H.P.J. Wijn, *Landolt-Bornstein III/4b*, (Springer, New York, 1970), p. 547
92. G.F. Dionne and J.F. Fitzgerald, *J. Appl. Phys.* **70**, 6140 (1991)
93. H.L. Schläfer and G. Gliemann, *Basic Principles of Ligand Field Theory*, (Wiley-Interscience, New York, 1969), p. 438
94. L.E. Orgel, *J. Chem. Phys.* **23**, 1004 (1955)
95. R. Finklestein and J.H. Van Vleck, *J. Chem. Phys.* **8**, 790 (1940)
96. Y. Tanabe and S. Sugano, *J. Phys. Soc. Jpn.* **9**, 753 (1954)

Chapter 6
Electromagnetic Properties

In the previous chapters, the emphasis is placed on the electronic origins of static magnetism in electrically insulating compounds, beginning with isolated moments of transition metal ions coupled paramagnetically and progressing to the various spontaneous magnetic systems. Apart from occasional references to their application in magnetic recording by the switching of magnetic domains and their use as permanent magnets in applications requiring electrical insulation, few indications are given that relate to their important applications in time-varying magnetic fields and as electromagnetic transmission media.

The most common usage of ferrimagnetism is for high permeability cores of inductors and transformers that typically operate at audio frequencies, i.e., below 20 kHz, and most commonly 60 Hz for power applications. Despite their generally lower magnetization when compared with ferromagnetic metal alloys, the reduction of eddy current losses makes these oxide materials an attractive alternative to laminated magnetic steel in many cases.

In the partially magnetized state, the frequency dependence of the permeability results from three causes (1) magnetization rotation, (2) magnetic domain wall resonance, and (3) gyromagnetic resonance within individual domains. The former two arise from longitudinal coupling between the alternating magnetic field and the domain magnetization vectors, and usually occur in the frequency range below 1 GHz. The third is the same transverse interaction effect that is exploited in magnetic resonance, where a magnetic field renders the material single domain. It can set in below 1 GHz, usually after the longitudinal permeability has been reduced to small values by damping effects, and can cause large absorption over a band that reaches to several GHz, depending on the anisotropy and geometrical demagnetizing fields.

Because of their excellent dielectric properties at microwave frequencies, ferrites can transmit electromagnetic waves with relatively low absorption losses. Above the absorption band mentioned earlier, significant permeability can remain in the tail of the dispersion curve with relatively small magnetic losses. For this reason, the microwave properties of ferrites have been investigated extensively and their frequency-dependent propagation in bands from 1 to 100 GHz have found widespread application in communications and radar technology.

In all of these frequency regimes, the strength of the coupling between the magnetization and the crystal lattice is crucial in determining the efficiency of

G.F. Dionne, *Magnetic Oxides*, DOI 10.1007/978-1-4419-0054-8_6,

transmission and the frequencies at which the material has useful permeability. The results of theoretical analyses point ultimately to the magnetocrystalline anisotropy, and particularly to the energy transfer between spins and lattice vibrations (phonons), which will be the central theme of this chapter.

6.1 Magnetic Relaxation

To begin the examination of the dynamic properties of magnetic systems, it is appropriate to assume a priori that the switching rate of magnetic moments between equilibrium states is limited by the relaxation time. Because of its importance, magnetic relaxation in relation to gyromagnetic resonance was introduced phenomenologically in Chap. 1. For our present purposes, however, some additional concepts must be reviewed. Specifically, the system of magnetic moments, frequently called the *spin* system because of the crystal field quenching of orbital moments in the $3d^n$ ion transition group, relaxes to an equilibrium state with the phonon system of the lattice. Except for special situations where the ground state retains a significant orbital component, the relaxation rate is determined by a spin–lattice time constant $\tau_1 \sim 10^{-7}$ s at room temperature, which increases to 10^{-2} s in the cryogenic range where the phonon density is much lower.

A much shorter time constant $\tau_2 \sim 10^{-10}$ s exists for the coherence lifetime among the individual spins in a typical paramagnetic system. Within the timescale of the spin–lattice relaxation, disturbances of the spin system can generally be ignored. Moreover, since the interaction between spins is essentially dipolar, with no significant influence from the lattice, there is little temperature dependence. The spin–spin relaxation rate, however, increases in proportion to the volume concentration of spins. Where spin alignments become spontaneous, however, the issue of spin coherence in a magnetically ordered ferromagnetic system is moot because strong exchange coupling tends to lock the spins in collective unison and $\tau_2 \to \infty$.

Magnetic relaxation will be discussed first as a general effect under nonresonant conditions. In sections that follow, we examine more closely the phenomenon of magnetic resonance and introduce the quantum mechanical point of view, which arises from specific conditions dictated by the electronic energy-level splittings in a magnetic field (Zeeman effect) combined with an incident electromagnetic wave of matching quantum energy.

6.1.1 Nonresonant Longitudinal Relaxation

Because of the spin–lattice interaction, a phase lag between a magnetization vector $\boldsymbol{M}_{\text{ac}}$ and an alternating longitudinal drive field $\boldsymbol{H}_{\text{ac}}$ of angular frequency ω will occur. If the time dependent component is parallel to the static component $\boldsymbol{H}$ (longitudinal as opposed to the transverse excitation $\boldsymbol{H}_{\text{rf}}$ of conventional radio-frequency magnetic resonance), the total magnetic field can be represented by

$$H(t) = H + H_{ac} \cos \omega t, \tag{6.1}$$

and the magnetization with a phase lag angle ϕ becomes

$$M(t) = M + M_{ac} \cos(\omega t - \varphi) = M + M_{ac} (\cos \omega t \ \cos \varphi - \sin \omega t \ \sin \varphi). \tag{6.2}$$

Inspection of (6.2) leads to extraction of the real and imaginary parts of the complex susceptibility. From the definitions of Sect. 6.1.2, the static susceptibility is $\chi_m = M/H$. By analogy, we can then separate the in-phase (real) and orthogonal (imaginary) components according to

$$\begin{aligned} \chi'_{ac} &= \frac{M_{ac} \cos \varphi}{H_{ac}}, \\ \chi''_{ac} &= \frac{M_{ac} \sin \varphi}{H_{ac}}, \end{aligned} \tag{6.3}$$

with the phase angle defined by $\tan \phi = \chi''_{ac} / \chi'_{ac}$. If H is expressed in conventional complex notation as $H + H_{ac} \exp(\mathrm{i}\omega t)$, (6.2) can be expressed as

$$M(t) = M + \chi_{ac} H_{ac} \exp(\mathrm{i}\omega t), \tag{6.4}$$

where the longitudinal ac susceptibility $\chi_{ac} = \chi'_{ac} - \mathrm{i}\chi''_{ac}$. One useful relation involving the imaginary χ''_{ac} that can be derived by calculation of the energy integral over one cycle $\int \boldsymbol{M} \cdot \mathrm{d}\boldsymbol{H}$ reveals that the power absorbed per unit volume is given by

$$P = \frac{1}{2} \omega \chi''_{ac} H^2_{ac}. \tag{6.5}$$

Because of the association of χ''_{ac} with power absorption, the parameter $\tan \phi = \chi''_{ac} / \chi'_{ac}$ already defined is often referred to as the *loss tangent*. A nonzero χ''_{ac} occurs because of a damping effect that is introduced as a relaxation time τ similar to a capacitance-resistance time constant that can lead to a dispersion/absorption centered at an angular frequency defined by $\omega_\tau = 1/\tau$. However, the underlying physical situation involving magnetic moments comprising excited spins is more complicated because of the various interactions present. In sketches of Fig. 6.1, this longitudinal dispersion is contrasted with the transverse resonance introduced in Chap. 1.

Among the earliest studies of relaxation phenomena in paramagnetism was the thermodynamic analysis of [1]. In the general context of magnetic interactions of all types, this approach is also referred to as "nonresonant" relaxation. Their initial model comprised separate spin and lattice systems, each in thermal equilibrium at respective temperatures T_{spin} and T_{latt} and subject to spin–lattice relaxation time (here characterized by τ for the macroscopic case), as depicted schematically in

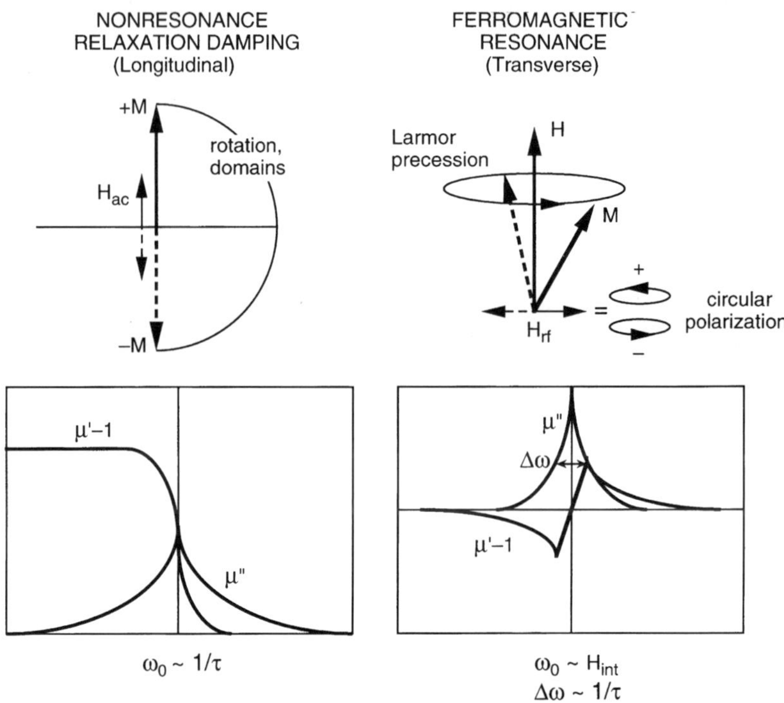

Fig. 6.1 Schematic diagram contrasting the physical origins and permeability effects of longitudinal and transverse interactions between alternating magnetic drive fields and magnetization vectors. Figure reprinted from [80] with permission. © 2003 by the IEEE

Fig. 6.2. In an alternating field of frequency much less than the relaxation rate τ^{-1}, $T_{\text{spin}} \approx T_{\text{latt}}$, because the heat exchange response between the two systems can follow the changes in field and magnetization without a significant phase lag. At higher frequencies, however, the two systems can no longer remain in thermal equilibrium and in the limiting case there is no transfer of heat ($\Delta Q \approx 0$) between them. For these conditions, isothermal and adiabatic susceptibilities are defined, respectively, as

$$\chi_T = \left(\frac{\partial M}{\partial H}\right)_T,$$
$$\chi_Q = \left(\frac{\partial M}{\partial H}\right)_Q. \tag{6.6}$$

The corresponding specific heats at constant M and H are defined as $C_M = (\mathrm{d}Q/\mathrm{d}T)_M$ and $C_H = (\mathrm{d}Q/\mathrm{d}T)_H$, respectively, and are related by

$$\frac{C_H}{C_M} = \frac{\chi_T}{\chi_Q}. \tag{6.7}$$

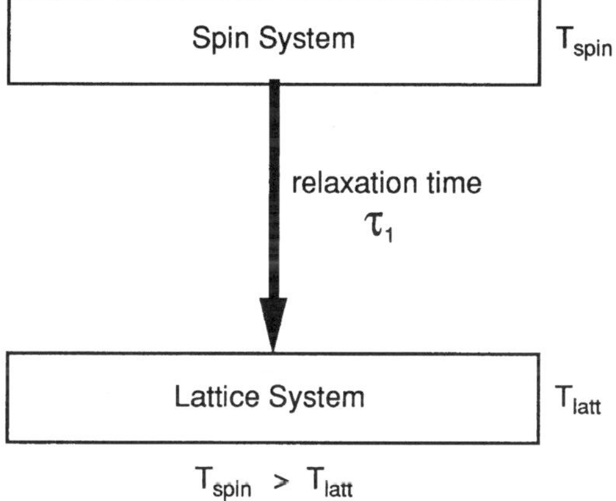

Fig. 6.2 Block diagram of thermodynamic concept of nonresonant relaxation. When the spin and lattice systems have different temperatures, the mechanism that restores thermal equilibrium is characterized by the relaxation time τ_1

The analysis begins with the recognition that the heat transfer rate from spin to lattice baths is proportional to the difference temperature between the systems, according to

$$\frac{\mathrm{d}Q}{\mathrm{d}t} = -\alpha_Q \left(T_{\mathrm{spin}} - T_{\mathrm{latt}}\right) = -\alpha_Q \Delta T. \tag{6.8}$$

where α_Q is the coupling constant between the spin and lattice baths. The expanded solution of this differential equation in thermodynamic formalism with $\mathrm{d}Q$ expressed in terms of entropy may be found in standard references [2, 3]. Here only some of the key steps will be stated. From the basic relation

$$\mathrm{d}Q = C_M \mathrm{d}T - T\left(\frac{\partial H}{\partial T}\right)_M \mathrm{d}M, \tag{6.9}$$

(6.8) can be expressed as

$$-\alpha_Q \Delta T\, \mathrm{d}t = C_M \mathrm{d}T - T\left(\frac{\partial H}{\partial T}\right)_M \mathrm{d}M. \tag{6.10}$$

Time and frequency dependence is introduced with the assumption that ΔT follows the same exponential function as the magnetic field and magnetization, and the susceptibility relations of (6.6) are substituted at the appropriate stage in the derivation. The resulting relations for the frequency-dependent parts of the complex susceptibility are:

$$\chi'_{\mathrm{ac}} = \chi_Q + \frac{\chi_T - \chi_Q}{1 + (\omega\tau)^2},$$

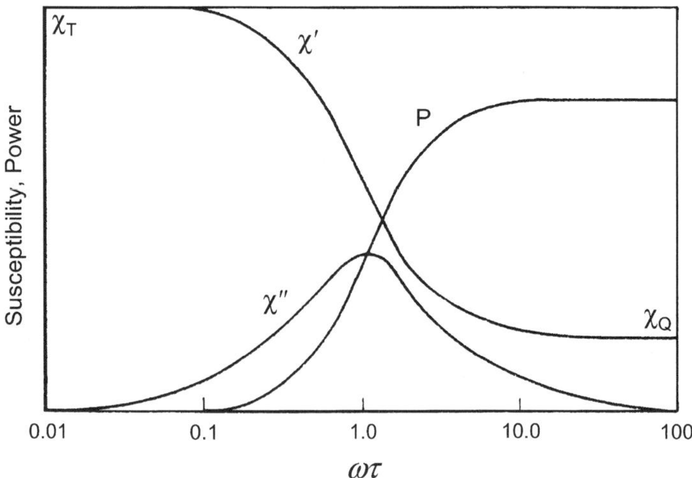

Fig. 6.3 Characteristic susceptibility and power solutions as a function of $\omega\tau$ plotted in arbitrary units

$$\chi''_{\text{ac}} = \frac{\left(\chi_T - \chi_Q\right)\omega\tau}{1 + (\omega\tau)^2}, \tag{6.11}$$

where $\tau = C_H/\alpha_Q$. Since C_H is the specific heat at constant H, there is an implicit temperature dependence which will be examined as we consider the quantum mechanical aspects of this phenomenon. The relaxation time τ in this theory is an average of all contributing mechanisms. In that sense it differs from τ_1 and τ_2 which represent specific paths back to the ground state. It might therefore be expected that τ for the longitudinal case would be shorter than spin–lattice relaxation time τ_1 of a transversely excited ferromagnet.

In Fig. 6.3, χ'_{ac}, χ''_{ac}, and P are plotted from (6.11) and (6.5) as a function of $\omega\tau$ to illustrate the dispersion effect that occurs at $\omega\tau = 1$. Note the appearance of the two susceptibility limits of χ'_{ac}, and the χ''_{ac} peak at the center of the dispersion region, indicating a condition of maximum loss tangent.

6.1.2 Quantum Mechanisms of Spin–Lattice Relaxation

To this point, the concept of relaxation has been used to rationalize the existence of a phase lag that leads to a complex susceptibility in an alternating magnetic field. It has also been seen as a controlling influence in the transfer of energy between spins and the lattice within a thermodynamic system. Ultimately, we seek to describe the origin of τ_1 in quantum mechanical terms. To initiate the analysis of spins transferring between a lower energy E_1 and higher level at E_2, we consider a two-level system comprising N magnetic moments ($m = g m_B S_z$) per unit volume at a

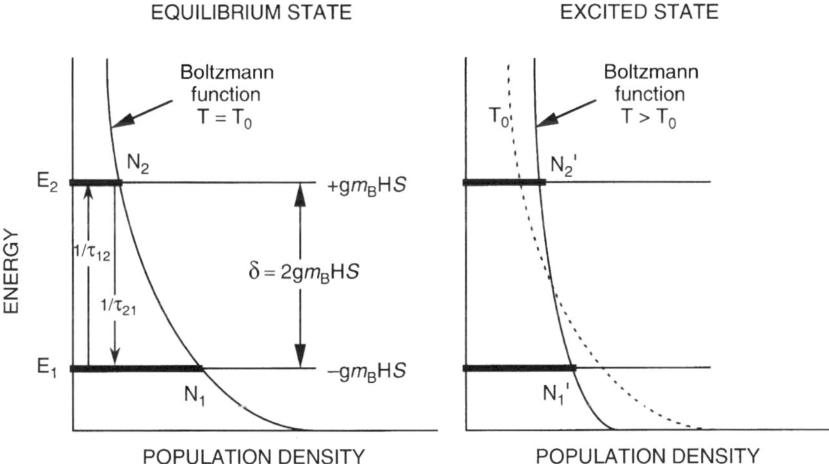

Fig. 6.4 Schematic comparison between equilibrium and excited Boltzmann populations of spins in a two-state gyromagnetic system

temperature T. Referring to Fig. 6.4, we relate the equilibrium populations N_1 and N_2, respectively, to the relaxation rates:

$$\frac{N_1}{\tau_{12}} = \frac{N_2}{\tau_{21}}, \tag{6.12}$$

where τ_{12} and τ_{21} are the corresponding relaxation times, i.e., inverse transition probabilities. If these populations are expressed as a Boltzmann ratio for an energy splitting δ,

$$\frac{N_2}{N_1} = \frac{\tau_{21}}{\tau_{12}} = \exp\left(-\frac{\delta}{kT}\right) \approx 1 - \frac{\delta}{kT} \tag{6.13}$$

and

$$\frac{N_1 - N_2}{N_1 + N_2} = \frac{\tau_{12} - \tau_{21}}{\tau_{12} + \tau_{21}} \approx \tanh\left(\frac{\delta}{2kT}\right) \tag{6.14}$$

for $\delta \ll kT$, thereby expressing a difference between the upward and downward relaxation frequencies. The ground state E_1 population is decreased by the amount of increase of the excited state E_2 population, establishing a transient N_1' and N_2' which can be defined by a Boltzmann function of higher temperature depicted in Fig. 6.4. If we assume that τ_{12} and τ_{21} are independent of the level populations, the rate of change of N_1' can be expressed as

$$\frac{\mathrm{d}N_1'}{\mathrm{d}t} = -\frac{\mathrm{d}N_2'}{\mathrm{d}t} = \frac{N_2'}{\tau_{21}} - \frac{N_1'}{\tau_{12}}. \tag{6.15}$$

From (6.15) it follows that

$$\frac{\mathrm{d}\left(N_1' - N_2'\right)}{\mathrm{d}t} = 2\left(\frac{N_2'}{\tau_{21}} - \frac{N_1'}{\tau_{12}}\right). \tag{6.16}$$

Recalling that $N = N_1 + N_2 = N_1' + N_2'$, we can express (6.16) as

$$\frac{\mathrm{d}\left(N_1' - N_2'\right)}{\mathrm{d}t} = (N_1 + N_2)\left(\frac{1}{\tau_{12}} - \frac{1}{\tau_{12}}\right) - \left(N_1' - N_2'\right)\left(\frac{1}{\tau_{12}} + \frac{1}{\tau_{21}}\right). \quad (6.17)$$

If $N_1 + N_2$ is expressed in terms of $N_1 - N_2$ from (6.14), (6.17) then reduces to

$$\frac{\mathrm{d}\left(N_1' - N_2'\right)}{\mathrm{d}t} = \left[(N_1 - N_2) - \left(N_1' - N_2'\right)\right]\left(\frac{1}{\tau_{12}} + \frac{1}{\tau_{21}}\right). \quad (6.18)$$

When multiplied by m, the factors $(N_1 - N_2)$ and $(N_1' - N_2')$ become magnetizations M and M', respectively. Equation (6.18) reduces to the relaxation relation for precessing magnetization in Fig. 1.14

$$\frac{\mathrm{d}M'}{\mathrm{d}t} = \frac{M - M'}{\tau_1}, \quad (6.19)$$

where the spin–lattice relaxation rate is defined as the combined relaxation rate $1/\tau_1 = 1/\tau_{12} + 1/\tau_{21}$.

The next task is to examine the temperature dependence of τ_1, which will introduce spin–orbit–lattice interactions. To this end, we recall the relation for the creation probability of a phonon of energy δ

$$p\left(\delta\right) = \frac{1}{\exp\left(\delta/kT\right) - 1}. \quad (6.20)$$

Since the spins are coupled to the lattice vibrations by a time-dependent relaxation, perturbation theory described in Sect. 6.1.3 can be used to relate spin and phonon transition probabilities according to

$$\begin{aligned} \frac{1}{\tau_{12}} &= \mathcal{K}[p\left(\delta\right) + 1], \\ \frac{1}{\tau_{21}} &= \mathcal{K}p\left(\delta\right), \end{aligned} \quad (6.21)$$

where $\mathcal{K}$ is a temperature-independent rate parameter that carries units of s^{-1}. If (6.20) and (6.21) are combined, the effective relaxation rate becomes

$$\frac{1}{\tau_1} = \mathcal{K}\coth\left(\frac{\delta}{2kT}\right) \approx \mathcal{K}\frac{2kT}{\delta} = AT \quad \text{for}\, \delta \ll 2kT. \quad (6.22)$$

This linear relation for τ_1 as a function of temperature corresponds to the *direct* process of one spin transition producing one phonon of the same frequency. This relation contains a magnetic field dependence if the splitting is the result of a Zeeman effect where $\delta = 2gm_{\mathrm{B}}HS$, according to the model of Fig. 6.4.

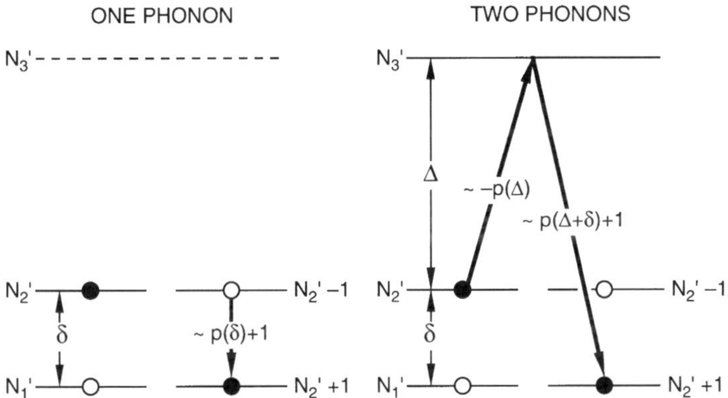

Fig. 6.5 Schematic diagram comparing one- and two-phonon relaxation processes

The above analysis can be extended to an indirect two-phonon process by introducing a third energy level separated from E_2 by energy Δ. Although only a special case of multiphonon relaxation usually referred to as a Raman process, it is particularly relevant to certain fast-relaxing ion situations encountered frequently in ferrimagnetic resonance. Two-phonon relaxation occurs first by an absorption of a phonon to excite a magnetic state from level $|2\rangle$ up to level $|3\rangle$, followed by the emission of a phonon through relaxation of the excited state to the ground state $|1\rangle$, with the net increase δ to the phonon spectrum energy. The transition sequences of one- and two-phonon processes are compared in Fig. 6.5.

In the following discussion of the paramagnetic two-level case the nomenclature employed by [4] is used. Relations similar to (6.21) can be written as

$$\begin{aligned}\frac{1}{\tau_{13}} &= \mathcal{K}_1 p\,(\delta + \Delta)\\ \frac{1}{\tau_{31}} &= \mathcal{K}_1\,[p\,(\delta + \Delta) + 1]\end{aligned} \tag{6.23}$$

and

$$\begin{aligned}\frac{1}{\tau_{23}} &= \mathcal{K}_2 p\,(\Delta)\\ \frac{1}{\tau_{32}} &= \mathcal{K}_2[p\,(\Delta) + 1].\end{aligned} \tag{6.24}$$

For this process, we assume that there is no relaxation between levels $|1\rangle$ and $|2\rangle$, and then express

$$\frac{\mathrm{d}\left(N_1' - N_2'\right)}{\mathrm{d}t} = \mathcal{K}_1\left(\frac{N_3'}{\tau_{31}} - \frac{N_1'}{\tau_{13}}\right) - \mathcal{K}_2\left(\frac{N_3'}{\tau_{32}} - \frac{N_2'}{\tau_{23}}\right). \tag{6.25}$$

After substitutions for the relaxation rates from (6.23) and (6.24),

$$\frac{\mathrm{d}\left(N_1' - N_2'\right)}{\mathrm{d}t} = \mathcal{K}_{\mathrm{eff}}\left[\left(N_1' - N_3'\right) p\,(\delta + \Delta) - \left(N_2' - N_3'\right) p\,(\Delta)\right], \tag{6.26}$$

where $\mathcal{K}_{\mathrm{eff}} = 2\mathcal{K}_1\mathcal{K}_2/\left(\mathcal{K}_1 + \mathcal{K}_2\right)$. In the limit where $\Delta \gg kT$, N_3' is small compared to N_1' and N_2', and (6.26) can be simplified to

$$\frac{\mathrm{d}\left(N_1' - N_2'\right)}{\mathrm{d}t} \approx \mathcal{K}_{\mathrm{eff}}\left[N_1' p\,(\delta + \Delta) - N_2' p\,(\Delta)\right]. \tag{6.27}$$

The corresponding phonon excitation probabilities are determined by applying (6.20) (with the continued assumption that $\Delta \gg kT$) to be

$$p\,(\Delta) \approx \exp\left(-\Delta/kT\right) \quad \text{and} \quad p\,(\Delta + \delta) \approx \exp\left[-\left(\delta + \Delta\right)/kT\right]. \tag{6.28}$$

Substitution of (6.27) into (6.28) followed by proper manipulation of the foregoing relations will result in (6.19), but with a temperature dependence of the relaxation rate different from (6.22):

$$\frac{1}{\tau_1} = K_{\mathrm{eff}}\left(\frac{M}{M + M'}\right) p\,(\Delta) \approx K_{\mathrm{eff}}\left[\exp\left(\Delta/kT\right) - 1\right]^{-1} \tag{6.29}$$

for $\delta \ll kT$ and $M \approx M'$. Since $\Delta \gg kT$, (6.29) simplifies further to become

$$\frac{1}{\tau_1} \approx K_{\mathrm{eff}}\left(\frac{M}{M + M'}\right) \exp\left(-\Delta/kT\right) = C\,\exp\left(-\Delta/kT\right). \tag{6.30}$$

This relation represents the Orbach process [5, 6], which applies where Δ is less than the Debye energy $h\nu_{\mathrm{p}}$ (or $k\Theta_{\mathrm{p}}$), typically less than a 0.1 eV. The reason for this limit is self-evident when one considers that the phonon created by the relaxation from level $|3\rangle$ to level $|1\rangle$ must have a frequency within the allowed phonon (Debye) spectrum which cuts off at ν_{D}, as illustrated in Fig. 6.6. Because $\Delta \gg \delta$ for microwave frequencies, the phonon energy band involved in the Orbach process is very narrow and centered at $h\nu \approx \Delta$. Its importance, however, is particularly significant for the heavy rare-earth ($4f^7$) ions owing to their shallow orbital states that can serve as the third level for a two-phonon process.

In the more general case of two-phonon relaxation, Δ can be greater than $h\nu_{\mathrm{D}}$ if level $|3\rangle$ is treated as virtual. While only low-frequency phonons of energy $h\nu$ are involved in the direct process, in the Raman process the excited spins precessing in level $|2\rangle$ can relax to level $|1\rangle$ by coupling simultaneously with any two lattice vibration modes provided their energy difference equals $h\nu$. Compared with the Orbach process, which we now see as a special case of two-phonon relaxation,

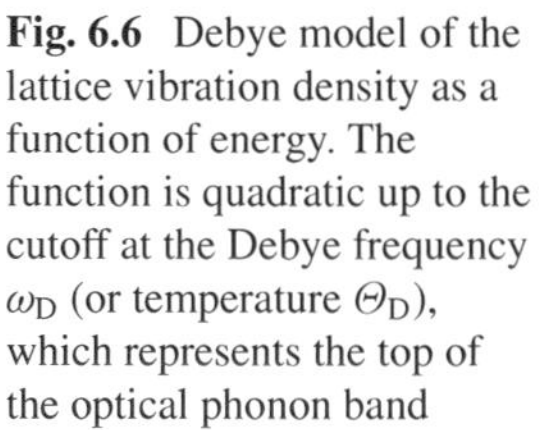

Fig. 6.6 Debye model of the lattice vibration density as a function of energy. The function is quadratic up to the cutoff at the Debye frequency ω_D (or temperature Θ_D), which represents the top of the optical phonon band

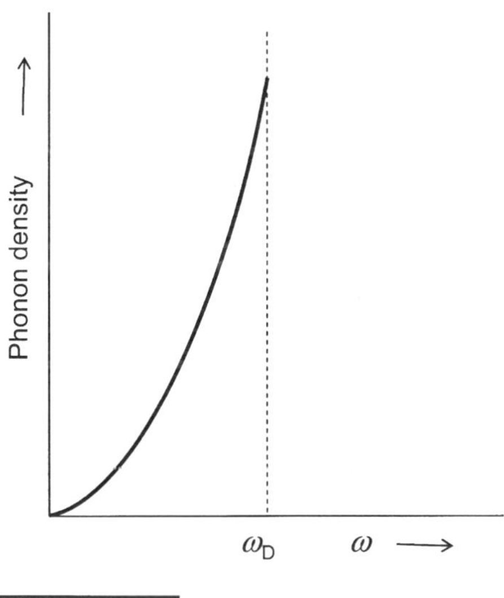

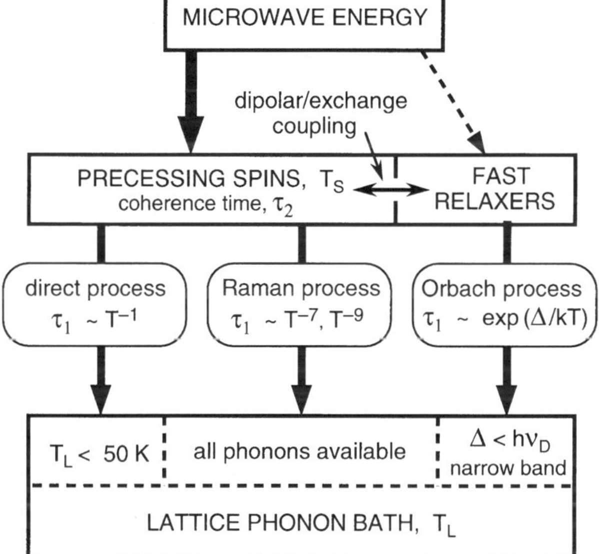

Fig. 6.7 Flow chart diagram showing the paths of microwave energy back to the lattice and their dependence on temperature

Raman processes can occur with any level structure because Δ is a virtual splitting energy. Moreover, they are not restricted to a narrow range of phonon energies around Δ. At this point, the various relaxation processes can be summarized as presented in Fig. 6.7.

The quantum mechanical time-dependent formalism of the spin–phonon interactions used to develop the temperature and magnetic field relations for the τ_1 of the

Table 6.1 Relations for spin-lattice relaxation rates τ_1^{-1}

Process	Non-Kramers even half-integer S	Kramers odd half-integer S
Direct	$AT + A'H^2T$	$A''H^4T$
Raman	BT^7	$B'T^9 + B''H^2T^7$
Orbach	$C_{\text{eff}}[\exp(\Delta/kT) - 1]^{-1}$ $\sim C_{\text{eff}} \exp(-\Delta/kT)$	$C_{\text{eff}}[\exp(\Delta/kT) - 1]^{-1}$ $\sim C_{\text{eff}} \exp(-\Delta/kT)$

Raman process, as well as those for the direct and Orbach processes, are beyond the scope of this text. However, it is necessary to discuss the results of these analyses, which are listed in Table 6.1. One important relation that was not included above is the meaning of the parameter $\mathcal{K}$, which can be expressed as

$$\mathcal{K} = \frac{3}{2\pi}\left(\frac{\delta}{\hbar}\right)^3 \frac{A_{\text{cf}}^2}{\rho\hbar v_{\text{p}}^5}, \tag{6.31}$$

where ρ is the mass density, v_{p} is the phonon velocity, and A_{cf} is a crystal-field parameter in units of potential energy. When this expression is substituted into (6.22), $1/\tau_1$ is found to depend on δ^2. If part of δ is given by a Zeeman splitting energy $gm_{\text{B}}H$, a quadratic dependence on H could be expected. The direct process relaxation rates are given by

$$\begin{aligned} \frac{1}{\tau_1} &= AT + A'H^2T \quad \text{(non-Kramers)}, \\ \frac{1}{\tau_1} &= A''H^4T \quad \text{(Kramers)}. \end{aligned} \tag{6.32}$$

The distinction between the $|1\rangle$ and $|2\rangle$ states as Kramers (half-integer spin number) or non-Kramers (integral spin numbers) arises from the subtleties of the time-conjugate nature of the Kramers doublet, and was explained by [6]. The proportionality parameters for the corresponding Raman processes are defined by convention as B, B', and B'' and are stated without derivation [7]:

$$\begin{aligned} \frac{1}{\tau_1} &= BT^7 \quad \text{(non-Kramers)}, \\ \frac{1}{\tau_1} &= B'T^9 + B''H^2T^7 \quad \text{(Kramers)}. \end{aligned} \tag{6.33}$$

In comparing the direct, Orbach, and Raman processes, we can first dispense with the Orbach case as peculiar to the low-lying excited state that occurs in only rare-earth and certain iron-group situations such as $3d^6$ or $3d^7$. This effect will be seen to play an important role in the low-temperature ferrimagnetic microwave properties of rare-earth iron garnets. The more common tradeoffs occur between the direct and Raman effects as functions of temperature. Therefore, the direct process dominates the low-temperature regime where the Debye spectrum indicated by Figs. 6.6 and 6.7 contains only low-frequency phonons that can match the $h\nu$ of the spin transition. At higher temperatures, multiple combinations of phonons (real or virtual) of greater energy become available for the two-phonon Raman process, providing a more effective means of relaxation than the direct process. Moreover, the Raman process is not restricted by the narrow-band requirement that the difference in energy of the two-phonons equal Δ for the fixed energy of the *real* excited state needed for the Orbach process. Illustration of a transition from Raman to direct processes at low temperatures is shown in the data [8] plotted in Fig. 6.8.

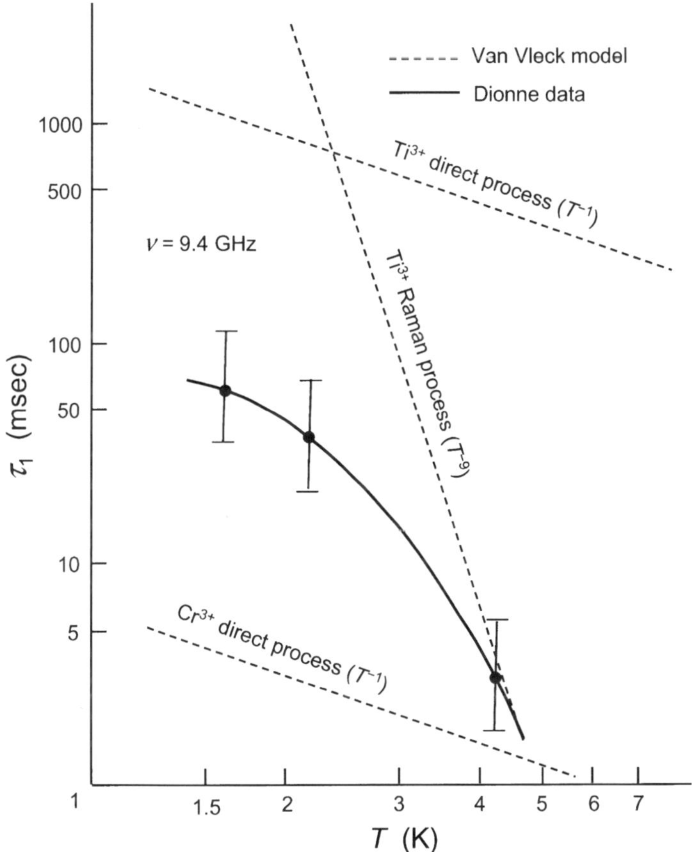

Fig. 6.8 Spin–lattice resonance relaxation of Ti^{3+} in Rb alum at liquid He temperatures [7]. Note the transition from Raman to direct mechanisms as T decreases from 4.2 to 1.2 K. Figure reprinted from [8] with permission. © 1965 by the American Physical Society. http://link.aps.org/doi/10.1103/PhysRev.139.A1648

6.1.3 Perturbation Theories of Spin–Phonon Interaction

To apply the results of relaxation rate mechanisms listed in Table 6.1, it must be realized that all three of them can be in action simultaneously. What distinguish their relative effectiveness are the values of the coefficients A, B, and C, and the excited-state splitting energy Δ for the Orbach process. The underlying cause of the spin–lattice coupling is the vibrations of the ligands bonded to the magnetic cation, which perturb the molecular–orbital states and transfer energy through spin–orbit coupling. To develop a computational formalism, an orbit–lattice Hamiltonian is constructed from a potential energy relation $\mathcal{H}_{\mathrm{ol}}$. When combined with the spin–orbit coupling operator, spin–lattice transition probabilities can be calculated through the standard off-diagonal matrix elements linking the two states involved in the transition. An attempt at formulating an orbit–lattice interaction for the direct process was reported by [9], who defined a crystal-field potential by a lattice strain according to $\mathcal{H}_{\mathrm{ol}} \sim \varepsilon\mathcal{V}_{\mathrm{ol}}$, averaged over all relevant crystal-field orbital states.

For a comprehensive analysis of spin–lattice relaxation, two versions of quantum mechanical perturbation theory can be consulted: first, the *tour de force* by Van Vleck in which Ti^{3+} and Cr^{3+} were examined employing six of the vibronic normal modes of octahedrally coordinated oxygen complexes [10, 11], and a later modification by Mattuck and Strandberg that was more compatible with the spin Hamiltonian perturbation approach [12]. The distinction between these two treatments lies in the order of application of the perturbations. The complete Hamiltonian can be divided into

$$\mathcal{H} = \left(\mathcal{H}_{\mathrm{lattice}} + \mathcal{H}_{\mathrm{spin}}\right) + \mathcal{H}_{\mathrm{spin-lattice}}, \tag{6.34}$$

where the arrangement of Van Vleck is given by

$$\begin{aligned}\mathcal{H}_{\mathrm{lattice}} &= \sum_n h\nu_n\left(a_n^{\dagger}a_n + \frac{1}{2}\right)\\ \mathcal{H}_{\mathrm{spin}} &= \mathcal{H}_0 + \mathcal{H}_{\mathrm{cf}} + g m_{\mathrm{B}}\boldsymbol{S}\cdot\boldsymbol{H}\\ \mathcal{H}_{\mathrm{spin-lattice}} &= \lambda\boldsymbol{L}\cdot\boldsymbol{S} + m_{\mathrm{B}}\boldsymbol{L}\cdot\boldsymbol{H} + \sum_n(\varepsilon\mathcal{V}_{\mathrm{ol}})_n.\end{aligned} \tag{6.35}$$

The subscript n refers to lattice vibration modes, and a_n and $a_n^{\dagger}$ denote phonon creation and annihilation operators. (The reader is cautioned that the above expressions have been simplified.) In the Mattuck–Strandberg version, the $\lambda\boldsymbol{L}\cdot\boldsymbol{S} + m_{\mathrm{B}}\boldsymbol{L}\cdot\boldsymbol{H}$ terms are included as part of $\mathcal{H}_{\mathrm{spin}}$ to render the operators more consistent with the spin-Hamiltonian formalism. For a direct process, the analysis yielded the following relation for the spin–lattice relaxation rate:

$$\frac{1}{\tau_1} \sim \frac{1}{\Delta^4}\left|2\lambda m_{\mathrm{B}} H S + \lambda^2 S_{\mathrm{A}}\right|^2 \approx \frac{\lambda^4}{\Delta^4}\left|S_{\mathrm{A}}\right|^2 \quad (\text{for } H = 0), \tag{6.36}$$

where S_A is the spin anticommutator evaluated between the two spin states. Note once again the dependence of spin–lattice relaxation on the ratio of spin–orbit coupling to crystal-field splitting energies which in this instance is raised to the fourth power.

In Fig. 6.8, results of τ_1 measurements carried out with a single-crystal of Rb alum containing a small concentration of Ti^{3+} ions are presented for the liquid-helium temperature range below 4.2 K [8]. The data are compared with calculations based on the Van Vleck model for the direct process, which is seen to set in as the temperature is lowered toward 1.2 K. At temperatures in the liquid-nitrogen range of 77 K, the Raman process is dominant. At 300 K, spin–lattice relaxation times can be shortened to the order of the spin–spin τ_2 of $\sim 10^{-10}$ s, depending on the values of the A, B, and C coefficients. Only for S-state ions such as Fe^{3+} or Mn^{2+} are the room-temperature τ_1 levels in the microsecond range.

Other topics of interest are cross-relaxation between dissimilar moments through dipolar interactions, phonon bottlenecks at low temperatures when low-energy phonons created by a direct relaxation are slow to disperse their energy to the rest of the bath, and spin-echo phenomena whereby spin–spin coherence (τ_2 relaxation) can be observed directly by absorption of a sequence of microwave pulses. For a comprehensive discussion of these subjects and their historical development, the reader is again directed to [3, 4], and other cited references.

6.2 Gyromagnetic Resonance and Relaxation

As described in the Sect. 6.1, magnetic relaxation is the randomization of magnetic moments upon the removal of an aligning field. In that discussion, energy transfer by spin–phonon interaction accounted for the longitudinal relaxation of moments in an ac field parallel to the moments, i.e., an amplitude-modulated magnetic field. A more exotic phenomenon that manifests relaxation effects occurs where the ac field is *transverse* to the aligning magnetic field – the condition for gyromagnetic resonance. Because these fields are in the radiofrequency and microwave bands, the symbol used to designate their amplitude is $\boldsymbol{H}_{\mathrm{rf}}$.

In Chap. 1, the classical theory of magnetic resonance was introduced by Larmor's theorem of the precessing of a moment vector $\boldsymbol{m}$ about a magnetic field vector $\boldsymbol{H}$ at angular frequency ω_0 (depicted in Fig. 1.16) through the generic vector relation

$$\frac{\mathrm{d}\boldsymbol{m}}{\mathrm{d}t} = \gamma\left(\boldsymbol{H} \times \boldsymbol{m}\right), \tag{6.37}$$

with the Larmor frequency ω_0 and gyromagnetic constant γ defined by (1.70),

$$\omega_0 = \frac{ge}{2m_e c} H = \gamma H. \tag{6.38}$$

and the Larmor spin-flip energy by (1.73),

$$\hbar\omega_0 = g m_B H \Delta\mathcal{M}_S = g m_B H \quad (\text{for } \Delta\mathcal{M}_S = 1). \tag{6.39}$$

Magnetic resonance can occur wherever magnetic moments and fields satisfy the orthogonality conditions that lead to (6.37). The material medium can be a gas, liquid, or solid (crystalline and amorphous), and nuclei as well as electrons can provide the magnetic moment. Although the smaller nuclear magneton (due to the larger proton mass) produces weaker interactions and reduces the Larmor frequency by a factor of 1,836 compared with electron-spin resonance, magnetic resonance from well-shielded nuclei can provide narrow-linewidth signals that are highly frequency selective. Nuclear magnetic resonance (NMR) has become the basis of the important medical diagnostic implement, magnetic resonance imaging (MRI), which operates in the 10–100 MHz range in highly stable and uniform magnetic fields.

Because this text concerns the electronic properties of magnetism, we examine electron paramagnetic resonance (EPR), ferromagnetic or ferrimagnetic resonance (FMR), and to a lesser extent antiferromagnetic resonance (AFMR). There are two main areas of focus (1) EPR as a diagnostic tool for measuring microwave spectra and deducing the parameters of the spin-Hamiltonian and (2) FMR as a means of controlling the propagation of electromagnetic waves in rf and microwave systems through the dependence of the susceptibility on magnetic field and magnetization. Although the phenomenon of magnetic resonance is fundamentally of classical origins, for EPR a quantum mechanical (QM) analog can be used to depict the resonance as an energy transition between two spin states. In effect, it is a magnetic dipole transition, with the selection rule $\Delta\mathcal{M}_S = \Delta S_z = \pm 1$ (for $L \approx 0$) satisfied. In Chap. 7, magnetic-dipole transitions are discussed in relation to electric-dipole transitions in the broader context of magnetooptical phenomena.

6.2.1 Paramagnetic Resonance

For the discussion of paramagnetic ions, we can relate a quantum mechanical model to (6.38) which follows immediately from the realization that the Larmor energy equates to the splitting of individual magnetic moment degeneracies in a magnetic field. As explained in [13], if damping and decoherence effects are ignored (adiabatic fast-passage), a simple two-level system at the resonance condition $\omega = \omega_0$ can be described in terms of population probabilities $p(-1/2)$ and $p(+1/2)$ that vary between 0 and 1 alternatively for the $S = -1/2$ and $+1/2$ states. This model is consistent with the angle θ between spin direction and z-axis $\boldsymbol{H}$ direction resonating between 0 and π at the effective frequency of the polarized rf drive field γH_{rf}, according the relations

$$p\left(-\frac{1}{2}\right) = \sin^2\left(\frac{1}{2}\gamma H_{\mathrm{rf}}t\right) = \frac{1}{2}\left[1 - \cos\left(\gamma H_{\mathrm{rf}}t\right)\right],$$
$$p\left(+\frac{1}{2}\right) = \cos^2\left(\frac{1}{2}\gamma H_{\mathrm{rf}}t\right) = \frac{1}{2}\left[1 + \cos\left(\gamma H_{\mathrm{rf}}t\right)\right]. \tag{6.40}$$

If in the general case, the angular momentum operator is J, and the stationary state component J_z varies from $-J$ to $+J$ in steps of 1. As a result, the classical magnetic energy – $mH\cos\theta$ referred to the polar axis of quantization is accounted for by $-gm_{\mathrm{B}}HJ_z$ averaged over the energy-level ladder and weighted according to the Boltzmann population fraction $\exp\left(-gm_{\mathrm{B}}HJ_z/kT\right)$. The comparison of the two approaches is directly analogous to the reasoning used in the derivation of the Langevin and Brillouin functions for paramagnetism discussed in Chap. 1. For odd-electron systems of the $3d^n$ group, the Kramers theorem would apply to $S = 5/2$, 3/2, as well as 1/2, and the final Zeeman splittings can involve as many as five allowed transitions. In anisotropic crystal fields, these transitions can be resolved in the EPR spectrum, either magnetic field or frequency scanned in accord with (6.39).

Because EPR is intimately tied to the relaxation processes and will be influenced by them in slow-passage situations described in Chap. 1, (6.40) should be accepted only as a basis for discussion in the comparison of the classical and quantum models. In the context of the quantum model of resonance damping, the photons supplied by H_{rf} represent the excitation or pump energy from which the spin–phonon relaxation follows.

Returning to the two standard examples of EPR vehicles, Ti^{3+} $\left(3d^1\right)$ and Cr^{3+} $\left(3d^3\right)$ in octahedral sites discussed in Sect. 5.2, we can examine the crystal-field energy level diagram shown previously in Fig. 5.3 for Ti^{3+} and in Fig. 5.14a for Cr^{3+}. The diagrams indicate orbital state energy splittings that were described in detail in Chap. 2, but now include the Zeeman splittings of the Kramers doublets that are proportional to the magnetic field strength. The diagrams depict the traditional experimental setup for determining the resonance spectra by sweeping the dc magnetic field at a fixed signal frequency (although the reverse approach has also become convenient). To relate this approach to the classical Larmor model, we recognize from (6.39) that $\hbar\omega_0 = gm_{\mathrm{B}}H\Delta S_z = gm_{\mathrm{B}}H$, where $|\Delta S_z| = 1$ for each transition.

In Chap. 5, the Ti^{3+} $\left(3d^1\right)$ ion was discussed as the textbook $S = 1/2$ case for perturbation analysis of electronic structure. A sample spectrum of the 12 equivalent, but differently oriented, complexes of Ti^{3+} in Rb alum is given in Fig. 6.9 [14]. Each magnetic ion of the $3d^n$ series has its own electronic configuration and requires an analytical treatment specific to its peculiarities. In most cases, the spin-Hamiltonian approximation can be used to solve for the eigenstates and their energies. A version of (5.14) defined for an axial crystal field is expressed as

$$\mathcal{H}_{\mathrm{S}} = g_{||}m_{\mathrm{B}}H_zS_z + g_{\perp}m_{\mathrm{B}}\left(H_xS_x + H_yS_y\right) + D\left[S_z^2 - \frac{1}{3}S\left(S+1\right)\right], \quad (6.41)$$

where $g_{||}$ and $g_{\perp}$ are the g-factors parallel and perpendicular to the z-axis and D is an axial symmetry constant that is commonly referred to as the "zero-field" splitting parameter, shown in Fig. 5.14a. For Cr^{3+} with $S = 3/2$ there are four spin levels and therefore three $\Delta S = 1$ transitions. As shown in Fig. 6.9, the absorption spectrum will vary according to the orientation of the magnetic field vector $\boldsymbol{H}$ and the axis of symmetry of each equivalent magnetic complex, in this case,

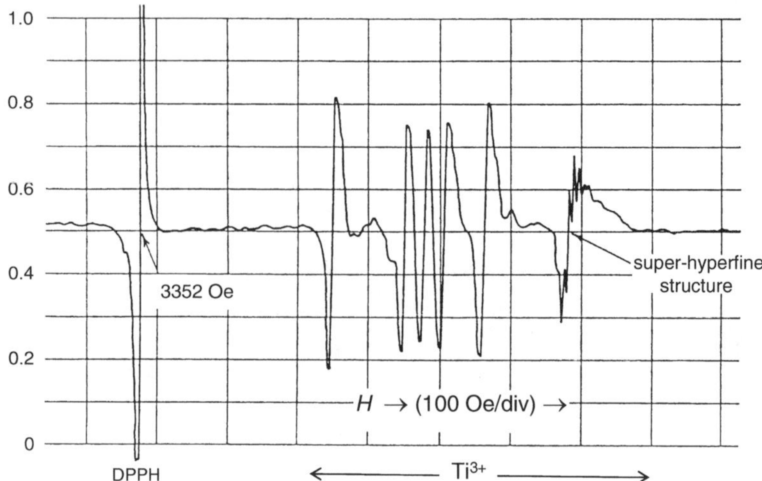

Fig. 6.9 Sample EPR spectrum of the uncommon Ti^{3+} $(3d^1)$ ion in Rb alum with $\nu = 9.4\,\text{GHz}$. Relaxation time measurements were carried out by the saturation method described in [18] and Chap. 6 of [4]. Six of the twelve orientations of the orthorhombic $Ti^{3+} \cdot 6H_2O$ site are resolved. Note also the hyperfine structure that appears at higher fields, possibly due to interactions with neighboring nuclei

from the 12 differently oriented the Ti^{3+} complexes in Rb alum. A straightforward analytical procedure allows the determination of the relevant parameters of the spin-Hamiltonian, $g_{\|}$, $g_{\perp}$, and D (and E, if symmetry is lower than axial). From these values, estimates of the spin–orbit coupling constant λ and the excited-state orbital splitting δ can be made following the reasoning outlined in Sect. 5.1.3.

Table 6.2 summarizes data extracted from EPR studies carried out by early researchers at X-band microwave frequencies ($\sim$10 GHz). The results of this work were compiled in review articles by the British pioneers [15, 16]. Interpretation of spectra can be challenging when higher-order effects cause fine structure in the splitting of energy levels. Interaction terms are added to the spin-Hamiltonian to account for spin couplings between electron and nuclei. In most cases, the energies are great enough to produce discernible lines in a resolved spectrum that allows the measurement of additional coupling constants.

One complication that can arise involves the orientation of local crystal-field site symmetry relative to the main symmetry axes of the crystal lattice. For the standard case of a cubic-lattice hosting complexes of axial crystal-field symmetry aligned with $\langle 111 \rangle$ axes, lines from four different site orientations appear in the spectrum, merging only when $\boldsymbol{H}$ is along one of the $\langle 100 \rangle$ axes where they appear identical. An example of an unusual situation that illustrates this point is shown in Fig. 6.9 for the sample spectrum of Ti^{3+} in Rb alum [14]. From an orbital singlet, Ti^{3+} has only one Kramers doublet transition $(+1/2 \leftrightarrow -1/2)$, and there should be a maximum of only four spectral lines if the site symmetry is axial and $\langle 111 \rangle$-axes aligned, i.e., one line for each $\langle 111 \rangle$ direction. Several spectral lines are observed

Table 6.2 $3d^n$ Transition group data[a]

	$3d^1$	$3d^2$	$3d^3$	$3d^4$	$3d^5$	$3d^6$	$3d^7$	$3d^8$	$3d^9$
Free ion	Ti^{3+}	V^{3+}	Cr^{3+}	Mn^{3+}	Fe^{3+}	Fe^{2+}	Co^{2+}	Ni^{2+}	Cu^{2+}
		Ti^{2+}	V^{2+}	Cr^{2+}	Mn^{2+}	Co^{3+}	Ni^{3+}	Cu^{3+}	
$\lambda\ (\text{cm}^{-1})$	154	104	87	85	–	–100	–180	–335	–852
			55	57					
Hund term	${}^2D_{3/2}$	${}^3F_2$	${}^4F_{3/2}$	${}^5D_0$	${}^6S_{5/2}$	${}^5D_4$	${}^4F_{9/2}$	${}^3F_4$	${}^5D_{3/2}$
High spin	–	–	–	$(e_g)^2$	$(e_g)^2$	$(e_g)^2$	$(e_g)^2$	$(e_g)^2$	$(e_g)^3$
	$(t_{2g})^1$	$(t_{2g})^2$	$(t_{2g})^3$	$(t_{2g})^3$	$(t_{2g})^3$	$(t_{2g})^4$	$(t_{2g})^5$	$(t_{2g})^6$	$(t_{2g})^6$
Low spin	–	–	–	–	–	–	$(e_g)^1$	$(e_g)^{2b}$	$(e_g)^{3b}$
	$(t_{2g})^1$	$(t_{2g})^2$	$(t_{2g})^3$	$(t_{2g})^4$	$(t_{2g})^5$	$(t_{2g})^6$	$(t_{2g})^6$	$(t_{2g})^6$	$(t_{2g})^6$
S(hs)	1/2	1	3/2	2	5/2	2	3 / 2	1	1 / 2
(ls)				1	1 / 2	0	1 / 2	0[b]	
$\langle g \rangle$	0 to ~1.9	~1.9	1.9–2	1.9–2	~2	>3 to ~7	>4	>2.25	2–2.5

[a]Data obtained from W. Low, *Paramagnetic Resonance in Solids*, (Academic, New York, 1960), Table XIX and J.S. Griffith, *The Theory of Transition-Metal Ions*, (Cambridge University Press, Cambridge, 1961), Appendix 6

[b]Low-spin states can occur in lower symmetry crystal fields that split the E_g degeneracy sufficiently to cause a violation of Hund's rule by creating a spin pair in the lower e_g state

because the local crystal field symmetry is of lower symmetry (orthorhombic) than the alum lattice (cubic) and none of the site axes coincide with the lattice $\langle 111 \rangle$ axes. The net result is spectra with a maximum of 12 lines representing 12 equivalent but distinguishable complexes.

Paramagnetic resonance can be observed in any structure where isolated moments are induced into precession about a magnetic field. Most of the early work was carried out with single-crystals of water-soluble salts of the transition metals. Because of their high melting points, bulk single-crystal oxides require exotic crystal-growth facilities and superior artistry. Nonetheless they became a host for vigorous research in quantum electronics that led to the invention of the microwave amplifier known as the *maser* in three-level Cr^{3+} in Al_2O_3 (ruby) [17], which was the precursor of the ruby *laser*.

One of the drawbacks of EPR is the necessity for cryogenics. In magnetically dilute specimens, maximum utilization of the low spin densities is essential, and spin populations of the ground state are largest at low temperatures. Moreover, signal intensities are also enhanced by resonance linewidth narrowing that is encouraged by longer spin–lattice relaxation times τ_1 at low T. This concern is particularly important in compounds containing fast-relaxing ions for which Raman and Orbach processes are dominant at higher temperatures. Conversely, line broadening becomes a limitation on the useful concentrations of magnetic ions due to spin–spin relaxation or decoherence time τ_2 that shortens in proportion to the separation between magnetic ions. This effect has motivated line-shape studies as a function of concentration [18].

Although not of much use for diagnostic purposes, magnetic resonance in exchange-coupled systems can be observed at room temperature, and therefore can be effective in controlling magnetic permeability and rf propagation parameters. In more recent years these magnetic dipole effects have contributed to the enhancement of electrical permittivity discussed in Chap. 7 that forms the basis of magnetooptical phenomena originating from electric-dipole transitions in transition-metal compounds, including the *magnetic* semiconductors, e.g., $\left(\mathrm{GaMn}^{3+}\right)$ As.

6.2.2 Ferromagnetic Resonance

To this point in the discussion, magnetic resonance has been viewed classically as a moment vector precessing about a dc magnetic field vector, stimulated by the magnetic field of an electromagnetic signal directed normal to the dc field. Within the same vector constraints and through use of the spin-Hamiltonian approximation, a quantum electronics model can be applied to a paramagnetic system of Kramers doublets split in the dc magnetic field. However, for magnetically ordered systems where the individual moments are tightly coupled into alignment by strong restoring forces from the exchange fields, the magnetic ions cannot easily be analyzed as individual quantum entities. There is no convenient QM analog to the precessing collective magnetic moment (or magnetization) vector $\boldsymbol{M}$ in the case of a ferromagnet, although it could be argued that the minimum resonance energy transfer in the coupled system is a photon of energy $gm_{\mathrm{B}}H$, representing the 180° reversal of a single electron spin ($\Delta S_z = \pm 1$). The subject of Kramers doublets in an exchange field is examined in Chap. 7.

For ferromagnetic resonance, the system is treated conventionally following the basic classical model of a magnetization vector $\boldsymbol{M}$ comprising collective magnetic moments ($\Sigma \boldsymbol{m}$) with gyromagnetic constant γ as defined previously (and presumed to be positive unless a sign change is required) for paramagnetism. The Larmor precession relation of (6.37) can then be applied to FMR as

$$\frac{\mathrm{d}\boldsymbol{M}}{\mathrm{d}t} = \gamma \left(\boldsymbol{M} \times \boldsymbol{H}_{\mathrm{i}}\right), \tag{6.42}$$

with the appropriate value of γ for the individual magnetic moments in the collective precessing group and the effective internal dc magnetic field H_{i}. This result is based on the assumption that the uniformity of the precession is not affected by the presence of spin waves that will be introduced later. For the present, we shall confine the discussion to the basic uniform precession case and proceed to examine the effects of magnetocrystalline and shape demagnetizing fields on the resonance frequencies of single-crystal specimens of ellipsoidal geometry.

For a ferro- or ferrimagnetic specimen, the effective magnetic field for resonance H_{r} is usually not equivalent to the internal dc field H_{i}, because the demagnetizing factors from the transverse directions influence the value of the rf field. Only in the case of a semi-infinite medium or a thin film (where $N_{Dx} = N_{Dy} \approx 0$, and $N_{Dz} \approx 1$) that is crystallographically isotropic can we write

$$H_r = H_i = H - N_{Dz} 4\pi M \approx H - 4\pi M. \tag{6.43}$$

The general relation for $\omega_r = \gamma H_r$ is found from(6.42) by treating all magnetic fields as vectors, with demagnetization factors introduced as tensors. Because the details have been documented in many publications, beginning with the seminal work of Kittel [19] and expounded in [20] and [21], they will not be repeated here. However, specific solutions for some common situations will be reviewed.

For a fully magnetized specimen with $\boldsymbol{H}$ and $\boldsymbol{M}$ aligned with the z-axis, Kittel determined that

$$\omega_r = \gamma \left\{ \begin{array}{l} [H + (H_{Kx} - H_{Kz}) + (N_{Dx} - N_{Dz})\, 4\pi M] \\ \times [H + (H_{Ky} - H_{Kz}) + (N_{Dy} - N_{Dz})\, 4\pi M] \end{array} \right\}. \tag{6.44}$$

The subscripts x and y refer to the two major axes orthogonal to the z direction of $\boldsymbol{H}$ in the coordinate system selected. Note that effective H_r reduces to (6.43) when all of the demagnetizing factors approach zero, and $N_{Dz} \to 1$. For particular cases, the appropriate relation for H_K from the list in Appendix 5D of Chap. 5 can be inserted directly into (6.44). The appropriate shape demagnetizing factors N_D can be determined from the analysis summarized in Sect. 1.1.3.

For resonance to occur, $\boldsymbol{H}_{rf}$ must have a component in the x–y plane, but values of the N_K and N_D factors will be sensitive to its exact direction within the plane. Applied to the limiting case of a thin flat plate with $N_{Dx} = 1$, and $N_{Dy}, N_{Dz} = 0$, and H_K terms ignored, (6.44) can be expressed as

$$\begin{aligned} \omega_r &= \gamma\, [H\,(H + 4\pi M)]^{\frac{1}{2}} \quad (\boldsymbol{H} \text{ in plane}) \\ \omega_r &= \gamma\, (H - 4\pi M) \quad (\boldsymbol{H} \text{ normal to plane}). \end{aligned} \tag{6.45}$$

For a long slender cylinder aligned with the z-axis, $N_{Dx}, N_{Dy} = 1/2$, and $N_{Dz} = 0$. The resonance frequency is then

$$\begin{aligned} \omega_r &= \gamma\, (H + 2\pi M) \quad (\boldsymbol{H} \text{ parallel to long axis}), \\ \omega_r &= \gamma\, [H\,(H - 2\pi M)]^{\frac{1}{2}} \quad (\boldsymbol{H} \text{ normal to long axis}). \end{aligned} \tag{6.46}$$

For a sphere, $N_{Dx}, N_{Dy}, N_{Dz} = 1/3$, and the shape demagnetizing factors of (6.44) cancel, so that

$$\omega_r = \gamma H. \tag{6.47}$$

Because the effect on the resonance condition from anisotropy fields varies with crystallographic orientation, the relations include the H_K terms only when single crystals are involved. For randomly oriented crystallites in ceramics, H_K serves to produce inhomogeneous broadening of the resonance line. Measurements are usually carried out by rotating the magnetic field $\boldsymbol{H}$ in a particular plane, with the

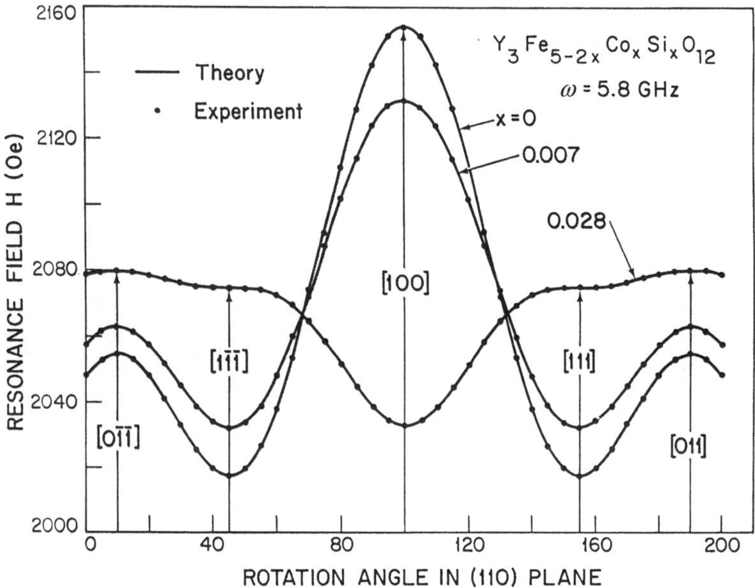

Fig. 6.10 Determination of magnetocrystalline anisotropy fields from measurements of the resonance magnetic field at frequency $\nu = 5.8\,\text{GHz}$. Direction of $\boldsymbol{H}$ is rotated through the [110] of the cubic garnet lattice to expose the influence of Co^{2+} on the anisotropy constants [22]

resonance frequency displayed as a function of angle relative to a major axis of symmetry. Figure 6.10 offers an example of the variation of resonance field $\boldsymbol{H}$ for a fixed $\omega_r = 5.8\,\text{GHz}$ as it is rotated in a $\{110\}$ plane, passing through $\langle 100\rangle$ "hard," $\langle 111\rangle$ "easy," and "intermediate" $\langle 110\rangle$ axes [22]. The effects of small concentrations of Co^{2+} ions on the anisotropy fields are strikingly illustrated. For this case, ω_r as a function of angle δ referenced to the $\langle 100\rangle$ axis is expressed by

$$\omega_r = \gamma \left\{ \begin{bmatrix} H + \frac{K_1}{M}\left(2 - \sin^2\delta - 3\sin^2 2\delta\right) \\ + \frac{K_2}{2M}\left[\sin^2\delta\left(6\cos^4\delta - 11\sin^2\delta\cos^2\delta + \sin^4\delta\right)\right] \end{bmatrix} \times \begin{bmatrix} H + \frac{K_1}{M}\left(2 - 4\sin^2\delta - \frac{3}{4}\sin^2 2\delta\right) \\ - \frac{K_2}{2M}\left[(\sin^2\delta\cos^2\delta\left(3\sin^2\delta + 2\right)\right] \end{bmatrix} \right\}^{\frac{1}{2}}. \quad (6.48)$$

From (6.48), we can extract relations for $\boldsymbol{H}$ along the $\langle 100\rangle$ $(\delta = 0)$, $\langle 112\rangle$ $(\delta = 35°)$, $\langle 111\rangle$ $(\delta = 55°)$, and $\langle 110\rangle$ $(\delta = 90°)$ axes:

$$\omega_r = \gamma\left(H + \frac{2K_1}{M}\right) \quad \boldsymbol{H} \text{ parallel to } \langle 100\rangle$$

$$\omega_r = \gamma\left[\left(H - \frac{K_1}{M} + \frac{K_2}{18M}\right)\left(H - \frac{K_2}{3M}\right)\right]^{\frac{1}{2}} \quad \boldsymbol{H} \text{ parallel to } \langle 112\rangle$$

$$\omega_r = \gamma\left(H - \frac{4}{3}\frac{K_1}{M} - \frac{4}{9}\frac{K_2}{2M}\right) \quad \boldsymbol{H} \text{ parallel to } \langle 111\rangle \tag{6.49}$$

$$\omega_r = \gamma\left[\left(H + \frac{K_1}{M} + \frac{K_2}{2M}\right)\left(H - \frac{2K_1}{M}\right)\right]^{\frac{1}{2}} \quad \boldsymbol{H} \text{ parallel to } \langle 110\rangle\,.$$

The H_K relations for three other common families of planes in a cubic system, i.e., {100}, {111}, and {112}, are listed in Appendix 5D of Chap. 5.

A stress-induced shift in the resonance frequency $\delta\omega_r$ (or field δH) can be used to determine the magnetostriction constants λ_{100} and λ_{111} [23–26]. For these measurements, a spherical specimen geometry remains convenient to cancel shape demagnetizing effects. The relevant relations are expressed as

$$\begin{aligned} \frac{\lambda_{100}}{M} &= -\frac{2}{3}\frac{\delta H_{100}}{\sigma}, \\ \frac{\lambda_{100}}{M} &\cong -\frac{4}{9}\frac{\left(\delta H_{110} + \frac{1}{2}\delta H_{100}\right)}{\sigma}, \\ \frac{\lambda_{100}}{M} &\cong -\frac{2}{3}\frac{\delta H_{111}}{\sigma}, \end{aligned} \tag{6.50}$$

where σ is the uniaxial compressive stress directed along the axis of the magnetic field.

For completeness, we state the expression for the uniaxial case with anisotropy constant K_u (which can apply to a grain-oriented polycrystal as well as a single crystal), according to

$$\omega_r = \gamma\left(H + \frac{2K_u}{M}\cos^2\delta\right), \tag{6.51}$$

and

$$\omega_r = \gamma\left(H + \frac{2K_u}{M}\right) \quad (\boldsymbol{H} \text{ parallel to axis of symmetry})\,, \tag{6.52}$$

which accounts for the very large resonance frequencies of the M-type hexagonal ferrites.

6.2.3 Uniform Precession Damping

For exchange-ordered systems, we must also include the paramagnetic resonance damping mechanism of spin–lattice relaxation. However, the situation can be simplified somewhat because we can neglect spin–spin decoherence in the uniform precession case. From the discussion in Sect. 1.5 and Appendix 1A of Chap. 1, the precessional relations with Bloch–Bloembergen longitudinal and transverse damping ([15, 16] of Chap. 1) can now be written as

$$\frac{\mathrm{d}M_z}{\mathrm{d}t} = \gamma\,(\boldsymbol{M} \times \boldsymbol{H}_i)_z - \frac{M_z - M}{\tau_1}$$
$$\frac{\mathrm{d}M_{x,y}}{\mathrm{d}t} = \gamma\,(\boldsymbol{M} \times \boldsymbol{H}_i)_{x,y} - M_{x,y}\left(\frac{1}{\tau_2} + f\,(\theta)\,\frac{1}{\tau_1}\right) \cong \gamma\,(\boldsymbol{M}\times\boldsymbol{H}_{\mathrm{i}})_{x,y} - \frac{M_{x,y}}{2\tau_1}, \tag{6.53}$$

where $\tau_2 \gg \tau_1$ and $f\,(\theta) \approx 1/2$ when spins are magnetically ordered in a "uniform precession" mode with all moments locked tightly in phase throughout the medium. Longitudinal relaxation back to the z-axis would then be pictured as an inwardly spiraling precession.

In the uniform precession mode, the spin–spin decoherence time τ_2 is assumed to be infinite at temperatures far from the Curie temperature. Under this condition, relaxation and line broadening are determined by spin–lattice coupling, which is characterized by $(2\tau_1)^{-1}$, since we can assume that the canting angle $\theta \approx 0$.

Recalling the discussion of paramagnetic relaxation in Chap. 1, we can examine the effect of $\tau_2 \approx 2\tau_1$ in the susceptibility relations of (1.84) by assuming that $\tau_2^{-1} \approx \Delta\omega = \gamma\Delta H$. In the laboratory frame of reference,

$$\chi'_{\mathrm{rf}} \cong \frac{1}{2}\gamma M \frac{(\gamma H_{\mathrm{i}} - \omega)}{(\Delta\omega)^2 + (\gamma H_{\mathrm{i}} - \omega)^2 + \frac{1}{2}\gamma^2 H_{\mathrm{rf}}^2},$$
$$\chi''_{\mathrm{rf}} \cong \frac{1}{2}\gamma M \frac{\Delta\omega}{(\Delta\omega)^2 + (\gamma H_{\mathrm{i}} - \omega)^2 + \frac{1}{2}\gamma^2 H_{\mathrm{rf}}^2}. \tag{6.54}$$

At resonance, the *saturation* effect resulting from increasing H_{rf} can be seen by inspection after the relations are simplified to

$$\chi'_{\mathrm{rf}} \cong 0$$
$$\chi''_{\mathrm{rf}} = \frac{1}{2}\gamma M \left(\frac{\Delta\omega}{(\Delta\omega)^2 + \frac{1}{2}\gamma^2 H_{\mathrm{rf}}^2}\right) = \frac{1}{2}\gamma M \left(\frac{\Delta H}{(\Delta H)^2 + \frac{1}{2}H_{\mathrm{rf}}^2}\right). \tag{6.55}$$

and

$$\frac{\chi''_{\mathrm{rf}}\,(H_{\mathrm{rf}})}{\chi''_{\mathrm{rf}}\,(0)} = \frac{\Delta H}{\Delta H'} \approx 1 - \frac{1}{2}\left(\frac{H_{\mathrm{rf}}}{\Delta H}\right)^2 \quad (\text{for } H_{\mathrm{rf}} \ll \Delta H)\,, \tag{6.56}$$

where

$$\Delta H' = \Delta H\left[1 + \frac{1}{2}\left(\frac{H_{\mathrm{rf}}}{\Delta H}\right)^2\right]$$

is the broadened half-linewidth at resonance. Note that χ''_{rf} far from resonance is not reduced by H_{rf} and the intrinsic half-linewidth should still be characterized by ΔH (or ΔH_{i} to distinguish it from other broadening mechanisms). In this sense, the departure of the resonance line from a rigorous Lorentzian shape is of consequence only near the line center.

It is important to point out two other models of damping that are used in the derivation of the complex susceptibility relations (see Appendix 6A). Historically, Landau and Lifshitz [27] were the first to propose an FMR damping term, which is expressed as

$$\left(\frac{\mathrm{d}\boldsymbol{M}}{\mathrm{d}t}\right)_{\mathrm{damp}} = -\frac{\lambda}{\boldsymbol{M}^2}\boldsymbol{M} \times (\boldsymbol{M} \times H_{\mathrm{i}}) , \tag{6.57}$$

where λ is a semiempirical damping parameter, a total relaxation time τ can be defined as $\tau = M/\lambda H_{\mathrm{i}}$. It should be noted that in this formalism $\tau \approx \tau_2$ of the Bloch–Bloembergen model, so that τ is mathematically equivalent to $2\tau_1$ in the uniform precession mode of an exchange-ordered system.

A later version of the damping equation was introduced by [28]:

$$\left(\frac{\mathrm{d}\boldsymbol{M}}{\mathrm{d}t}\right)_{\mathrm{damp}} = -\frac{\alpha}{\boldsymbol{M}}\left(\boldsymbol{M} \times \frac{\mathrm{d}M}{\mathrm{d}t}\right), \tag{6.58}$$

where α is the damping parameter. For constant M, it can be shown that (6.57) effectively reduces to (6.58) provided that $\alpha = \lambda/\gamma M$ [29]. Near the resonance frequency, the effective relaxation time can be approximated by (see Appendix 6A)

$$\tau = \frac{1}{\alpha\omega}. \tag{6.59}$$

Although these models are usually treated as equivalent, the Gilbert form in (6.58) is more suitable for the derivation of the rf susceptibility tensor in Sect. 6.4. Relaxation damping of rf signals can be affected greatly in ferrimagnets by even small amounts of fast-relaxing ions and other factors that will be reviewed. These effects, however, must first be seen in the context of the total resonance linewidth within which they are often obscured.

6.2.4 *Inhomogeneous Resonance Line Broadening*

In the foregoing analyses of magnetic resonance damping, the half-linewidth ΔH is proportional to the effective relaxation rate τ^{-1}, which is a combination of τ_1^{-1} and τ_2^{-1}. Before examining the factors that influence the values of τ_1^{-1} and τ_2^{-1} in magnetic lattices, the subject of inhomogeneous line broadening must be introduced. Recalling the discussion in Sect. 1.5, we recognize that an inhomogeneously broadened line does not have the Lorentzian shape assumed in the Bloch–Bloembergen formalism. A Gaussian distribution function is more appropriate. Consequently, for the purposes of using the theory to interpret measurements, only the intrinsic homogeneous component ΔH_{i} of the total ΔH can be applied in (6.56) and (6.57).

Despite their corrupting effect, however, the inhomogeneous broadening can provide useful information for characterizing polycrystalline specimens.

In a polycrystalline ceramic body, inhomogeneities can impact the shape and width of magnetic resonance lines. The first is the random crystallographic orientation of individual crystallites or grains. Where the various symmetry axes are dispersed, the resonance fields of each grain will vary accordingly in proportion to K_1. An actual measurement for cubic YIG that reveals dramatically the nature of these effects is presented in Fig. 6.11 from the work of Van Hook and Euler [30]. Schloemann [31, 32] and Geschwind and Clogston [33] examined the question and produced analytical results that have been helpful in characterizing the anisotropy and grain orientation of ferrites. For a spherical specimen, anisotropy contributions to the half-linewidth were defined according to

$$\Delta H_{\mathrm{K}} \approx \left|\frac{K_1}{M_{\mathrm{s}}}\right| \qquad M_{\mathrm{s}} \ll 2K_1/M_{\mathrm{s}},$$

$$\Delta H_{\mathrm{K}} = \frac{8\pi\sqrt{3}}{21}\frac{(2K_1/M_{\mathrm{s}})^2}{4\pi M_{\mathrm{s}}}\mathrm{G} \quad M_{\mathrm{s}} \gg 2K_1/M_{\mathrm{s}}, \tag{6.60}$$

where G is a shape factor that is dependent on the ratio $\omega/\gamma 4\pi M_{\mathrm{s}}$. For YIG at room temperature, $K_1/M_{\mathrm{s}} \approx 40\,\mathrm{Oe}$ and $4\pi M_{\mathrm{s}} = 1{,}780\,\mathrm{G}$. Here the low K_1/M_{s} case applies and yields $\Delta H_K \approx 8\,\mathrm{Oe}$, in agreement with measured results from dense

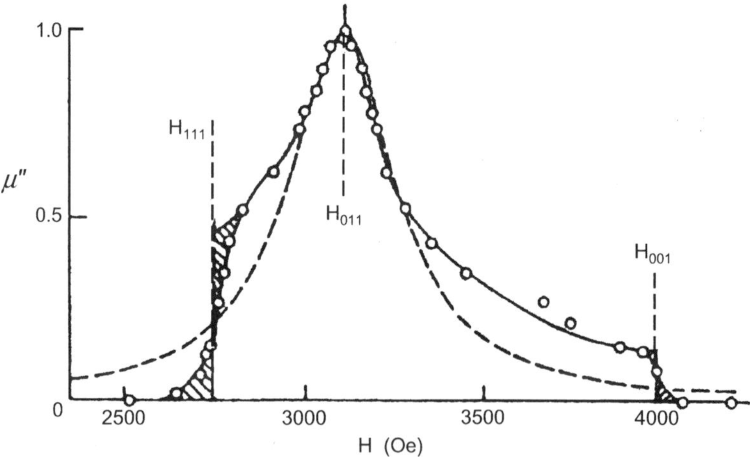

Fig. 6.11 Experimental example of an inhomogeneously broadened FMR permeability of a V-In diluted yttrium–iron garnet ceramic specimen. Individual crystallites (grains) are randomly oriented and have narrow Lorentzian resonance lines along favored axes. The overall width of the combined resonances is determined by the extremes at H_{001} and H_{111}. The peak occurs at H_{011} which is the most numerous symmetry axis of a cube. The dashed line with the extended tails is a homogeneous equivalent of the measured result, indicating that the effective Lorentzian width is considerably smaller than the measurement would suggest. Figure reprinted from [30] with permission. © 1969 y the American Institute of Physics

ceramic specimens. In polycrystalline bodies the saturation magnetization M_s is specified because M can assume various values of partial magnetization. An example of how these relations can be used as design aids for determining variations of K_1 with composition will be given in Sect. 6.2.5.

A second inhomogeneous broadening effect is caused by the local variation of magnetization due to compositional fluctuations, crystallites of different chemical phase, or most commonly, air porosity. Schloemann [31, 34] and Sparks [35] modeled this effect, with the result that a porosity contribution to ΔH was defined as

$$\Delta H_p = \beta\,(4\pi M_s)\left(\frac{p}{1-p}\right), \tag{6.61}$$

where $\beta{\sim}1$ [36]. With(6.1) and (6.2), the total FMR half-linewidth can be expressed as the sum of relaxation rates in linewidth form as

$$\Delta H \approx \Delta H_K + \Delta H_p + \Delta H_i, \tag{6.62}$$

where each individual grain provides an intrinsic Lorentzian ΔH_i contribution to the inhomogeneously broadened Gaussian result that then comprises the frequency spread of smaller intrinsic lines. This is manifested in Fig. 6.11. It should be pointed out here that the relations for the different linewidth contributions can be used effectively in combination with the approach-to-saturation theory for characterization of magnetic polycrystals [37].

To assist in the magnetic loss characterization away from the resonance line center, Patton [38] and Vrehen [39] measured an effective linewidth by a manipulation of (6.54) for χ''_{rf}. With $\omega \gg \gamma H_r$ and $\Delta\omega$, the operating point is far from both the line center and the high-frequency edge of the inhomogeneous loss manifold, and

$$\chi''_{rf} = \frac{\gamma M_s \Delta\omega}{\omega^2} \tag{6.63}$$

for circular polarization. If the effective half-linewidth is now defined as $\Delta H_{eff} = \Delta\omega/\gamma$, (6.63) can be re-expressed as

$$\Delta H_{eff} = \frac{\omega^2 4\pi\chi''_{rf}}{\gamma\,(\gamma 4\pi M_s)} = \frac{\omega^2\mu''_{rf}}{\gamma\omega_M} = \frac{\omega}{\omega_M}\frac{\omega}{\gamma}\mu''_{rf} \approx 2\frac{\omega}{\gamma}\mu''_{rf} \tag{6.64}$$

for a typical microwave device application. As a measure of loss, ΔH_{eff} is proportional to the reduced value of the μ outside of the broadening range and is therefore small enough to resemble the intrinsic ΔH_i, as illustrated by the data for a systematic range of garnet compositions with monotonically changing K_1/M_s values [38] in Fig. 6.12. With ΔH_{eff} seen in the context of spin–lattice relaxation, we can more clearly examine the fundamental factors that influence its true value.

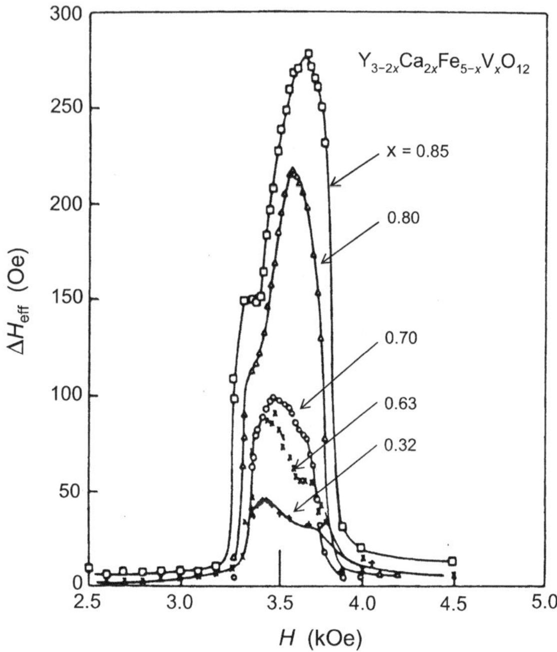

Fig. 6.12 Effective linewidth of polycrystalline garnet specimens of composition $Y^{3+}{}_{3-2x}$ $Ca^{2+}{}_{2x}Fe^{3+}{}_{5-x}V^{5+}{}_xO_{12}$ with a range of anisotropy fields. Figure reprinted from [38] with permission. © 1969 by the American Physical Society. http://link.aps.org/doi/10.1103/PhysRev.179.352

6.2.5 *Fast-Relaxing Ion Effects*

Transmission of microwave energy with minimum relaxation loss is critical to the efficiency of ferrites used as propagation media. Ferrimagnetic resonance (FMR) relaxation is reflected in the intrinsic half-linewidth (ΔH_i), and can be an essential mechanism in the loss of signal intensity. For magnetically ordered spin systems, ΔH_i has been treated as a direct function of the spin–lattice relaxation rate τ_1^{-1} without consideration of its temperature dependence or the disposition of the phonon population. These latter effects were highlighted when measurements of loss in rare-earth (RE) iron garnets revealed a monotonic decrease in ΔH_i with reducing temperatures, but interrupted by a peak in the vicinity of $T \approx 50$ K that is proportional to the concentration of RE ions [40, 41].

As listed in Table 6.3, τ_1^{-1} can be sensitive to temperature through a variety of mechanisms. One effect that was not introduced in the discussions of paramagnetic relaxation was the phonon "bottleneck." If the phonons that are created when the spin relaxes to its ground state are not in thermal equilibrium with the total lattice "bath," an additional relaxation between phonon and lattice must take place. As a result, the effective value of τ_x increases accordingly, and because the nonequilibrium condition between phonon and bath is also temperature dependent, the expressions

Table 6.3 Gyromagnetic relaxation rate–linewidth relations

Precessing system	Theory	τ^{-1} $(=\gamma\,\Delta H)$ (s^{-1})
EPR	Bloch–Bloembergen (B–B)	$(2\tau_1)^{-1}+\tau_2^{-1}$
FMR uniform precession	Bloch–Bloembergen (B–B)	$(2\tau_1)^{-1}$
	Landau–Lifshitz (L–L)	$\lambda\,(H_i/M)$
	Gilbert (G)	αH_i
FMR uniform precession with spin waves	Bloch–Bloembergen (B–B)	$(2\tau_1)^{-1}+\tau_{2k}^{-1}$

τ_1 Spin–lattice relaxation time, τ_2 spin–spin decoherence time (via dipolar interactions), τ_{2k} spin–spin decoherence time (via spin waves)

for τ_{x} must also account for this influence. This subject has been examined by Van Vleck [42], Faughnan and Strandberg [43], and Stoneham [44], and a review can be found in Standley and Vaughan [45]. However, the approach followed here is based on that of de Gennes et al. [40]. In terms of the phonon quantum number n_{p}

$$\tau_1^{\mathrm{eff}} = \tau_1\left[\frac{\left(n_{\mathrm{p}}+1\right)+n_{\mathrm{p}}}{\left(n_{\mathrm{p}}+1\right)-n_{\mathrm{p}}}\right] = \tau_1\left(2n_{\mathrm{p}}+1\right), \tag{6.65}$$

where

$$n_{\mathrm{p}} = [\exp(\hbar\omega/kT)-1]^{-1} = \exp(-\hbar\omega/kT)\,[1-\exp(-\hbar\omega/kT)]^{-1}. \tag{6.66}$$

For the iron sublattices, where $\tau_{\mathrm{Fe}} = \tau^{\mathrm{eff}}$

$$\frac{1}{\tau_{\mathrm{Fe}}} = \left(\frac{1-\exp(-\hbar\omega/kT)}{1+\exp(-\hbar\omega/kT)}\right)\frac{1}{\tau_1} \approx \left(\frac{\hbar\omega}{2kT}\right)\frac{1}{\tau_1}, \tag{6.67}$$

where it is assumed that $\hbar\omega \ll kT$. For an assumed direct process, we can apply a temperature dependence based on (6.67) to the iron intrinsic half-linewidth according to

$$\Delta H_{\mathrm{Fe}} = (\gamma\tau_{\mathrm{Fe}})^{-1} = \left(\frac{\hbar\omega}{2kT}\right)(\gamma\tau_1)^{-1} = \gamma^{-1}\left(\frac{\hbar\omega}{kT}\right)AT^n, \tag{6.68}$$

where n is a data fitting parameter that can exceed the theoretical value of unity.

Where rare-earth or other ions that exchange couple significantly to the Fe sublattices, but weakly enough to each other as to be treated as paramagnets attached to the net iron moments, an additional relaxation rate τ_{RE}^{-1} must be considered. If relaxation rates are transition probabilities as depicted in Fig. 6.13, a straightforward addition of the respective relaxation rates τ_{Fe}^{-1} and τ_{RE}^{-1} for the net iron and RE ions can be used according to [46].

$$\frac{1}{\tau_1} = \frac{1}{\tau_{\mathrm{Fe}}} + \zeta\frac{1}{\tau_{\mathrm{RE}}}, \tag{6.69}$$

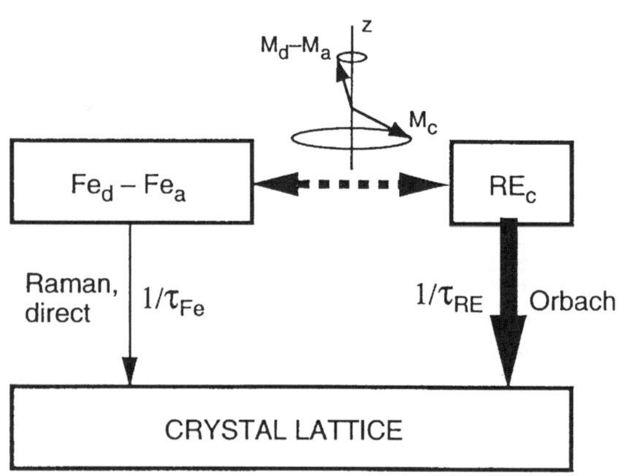

Fig. 6.13 Block diagram of uniform-precession spin energy transfer to the lattice in a magnetic garnet. Note effect of the c-sublattice rare-earth ions and the expected influence the Fe^{3+} spin canting [47]. Figure reprinted from [47] with permission. © 2000 by the American Institute of Physics

where ζ is a factor that is proportional to the concentration of RE ions and the energy of their exchange coupling to the iron sublattices. The RE relaxation rate can be approximated as $\tau_{\mathrm{RE}}^{-1} \approx C_{\mathrm{eff}} \exp(-\Delta/\mathrm{k}T)$, where Δ is the splitting of the lowest excited state of the rare-earth multiplet as described in Sect. 6.1.

To evaluate the parameter ζ, we consider the molecular field surrounding isolated paramagnetic ions of the rare-earth series. From the discussion in Chap. 4, the exchange field can be expressed as

$$H_{\mathrm{ex}}^{c} = \mathcal{N}_{dc}\mathcal{M}_{d} + \mathcal{N}_{ac}\mathcal{M}_{a} + \mathcal{N}_{cc}\mathcal{M}_{c}, \tag{6.70}$$

where $\mathcal{N}_{ij}$ are molecular-field coefficients expressed in mol cm^{-3} and $\mathcal{M}_{\mathrm{i}}$ is the undiluted magnetic moment per mole of the i sublattice ([52, 53] of Chap. 4). The labeling of sublattices is according to crystallographic sites, with d for tetrahedral, a for octahedral, and c for the dodecahedral site of rare-earth ions. Table 6.4 lists the values of $\mathcal{N}_{ij}$ for the RE ions of interest. Note that the intra-sublattice coefficient $\mathcal{N}_{cc}$ is negligible, which allows for the c moments to be treated as coupled only to the iron ions. Figure 6.14 illustrates how the c sublattice moment n_{B} (expressed in Bohr magnetons of Dy^{3+}) decreases sharply with increasing temperature and scales directly with the dilution fraction k_c, while having virtually no effect on the net moment of the opposing iron sublattices. In this sense, the c sublattice comprises relaxation centers with the spin–lattice interaction of isolated paramagnets. The effectiveness of RE ions in transferring microwave energy of spin waves from the Fe^{3+} spin systems to the lattice depends directly on the exchange energy per mole, $E_{\mathrm{ex}}^{c} = M_c H_{\mathrm{ex}}^{c}$.

To compute E_{ex}^{c} as a function of temperature, $M_{\mathrm{c}}(T)$ and $H_{\mathrm{ex}}^{c}(T)$ must be extracted from the complete molecular field solution involving the temperature variations of Brillouin functions for the three-sublattice magnetic garnet. From these

Table 6.4 Rare-earth ion parameters

RE Ion	n_B (RE) Bohr magnetons	$\mathcal{N}_{dc}$ (mol cm^{-3})	$\mathcal{N}_{ac}$ (mol cm^{-3})	C_{eff} (s^{-1} × 10^{13})	Δ (K)
Dy^{3+}	7.07	6.0	−4.0	8.45	50
Ho^{3+}	6.23	4.0	−2.1	24.4	100
Er^{3+}	5.54	2.2	−0.2	5.26	125
Yb^{3+}	1.70	8.0	−4.0	3.47	175[a]

[a]Spectral measurements in paramagnets indicate that $\Delta \sim 500$ K for Yb^{3+}. Because this value is in the order of the Debye temperature, the Orbach process is not fully applicable and could be replaced by a Raman process [4]

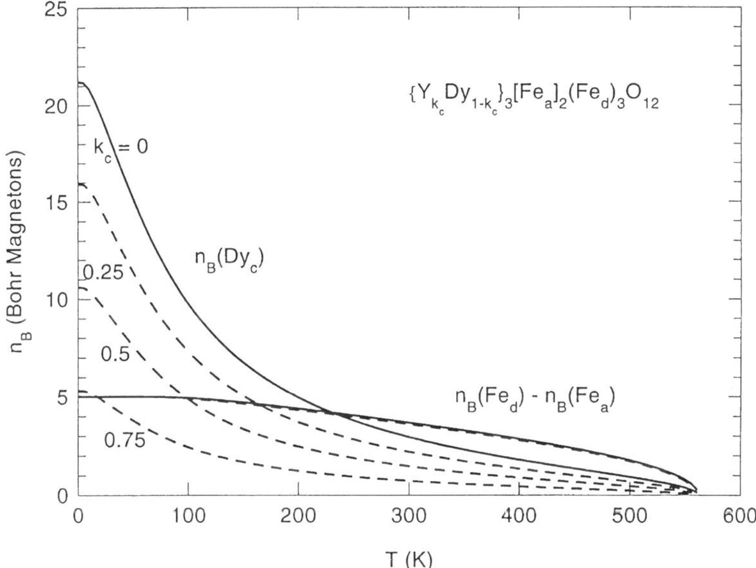

Fig. 6.14 Calculated thermomagnetic curves of $(Y_{1-x}Dy_x)_3Fe_5O_{12}$ with the c sublattice separated from the net of the opposing d and a sublattices. Magnetic moments are expressed in n_B (Bohr magnetons per formula unit) [47]. Figure reprinted from [47] with permission. © 2000 by the American Institute of Physics

concepts, we construct the following models for the relaxation rate and intrinsic linewidth as a function of temperature and RE ion concentration [47]:

$$\tau_1^{-1} = AT^n + \zeta C_{eff} \exp(-\Delta/kT), \tag{6.71}$$

$$\Delta H_i = \gamma^{-1}\left(\frac{\hbar\omega}{kT}\right)\left[AT^n + \zeta_0(1-k_c)\left(\frac{E_{ex}^c(T)}{E_{ex}^c(0)}\right)C_{eff}\exp(-\Delta/kT)\right], \tag{6.72}$$

where A and C_{eff} are proportionality constants in appropriate units, and the exponent $n \geq 1$. For first approximations, $\zeta_0 \sim 1$.

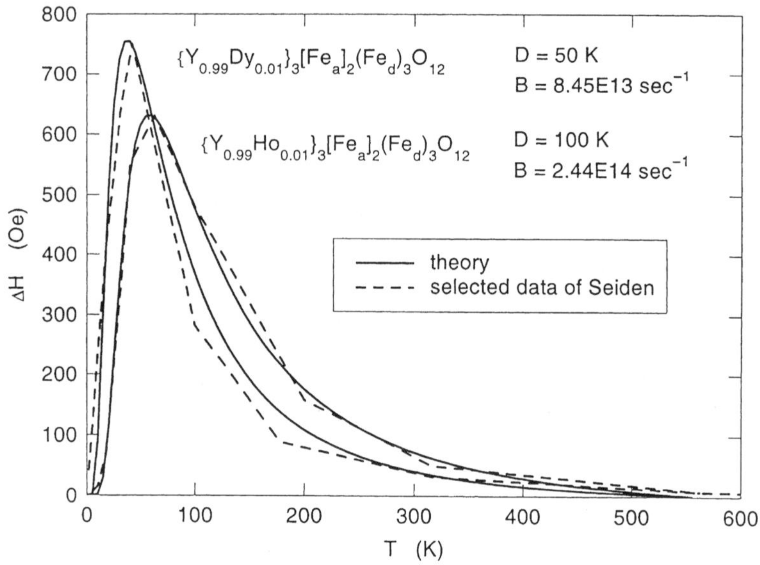

Fig. 6.15 Comparison of theory with selected data of X-band FMR half-linewidth as a function of temperature for $(Y_{0.99}Dy_{0.01})_3\,Fe_5O_{12}$ and $(Y_{0.99}Ho_{0.01})_3\,Fe_5O_{12}$ [47]. Data are from Seiden [41]. Figure reprinted from [47] with permission. © 2000 by the American Institute of Physics

Figure 6.15 presents an example of the fit of theory to Seiden's data scaled to 1% ionic concentration of Dy^{3+}, Ho^{3+} in YIG, i.e., $(Y_{0.99}RE_{0.01})_3\,Fe_5O_{12}$, following his original format [41]. In Table 6.4, the values of C_{eff} and Δ and for a signal frequency of 9.2 GHz as determined by adjustment of computed curves to data are listed for each case. The magnitude of the D splittings are consistent with those expected from the values of the particular RE ion moment [48] and the corresponding exchange fields that are determined by the $\mathcal{N}_{dc}$ and $\mathcal{N}_{ac}$ coefficients ([53] of Chap. 4) listed in Table 6.4. Relaxation rate vs. temperature data from the studies of Spencer and LeCraw on a specimen of high purity single-crystal YIG [49] were analyzed by computation employing (6.72) and plotted as intrinsic ΔH in Fig. 6.16. Because the concentration of RE ions is so small, the major temperature dependence is from the iron sublattices modeled semiempirically with $A = 236\,\text{s}\,\text{K}^{-2.5}$ and $n = 2.5$. Dy^{3+} was selected as the fast-relaxing ion, and the concentration $(1 - k_c) \approx 4 \times 10^{-7}$ necessary to produce the slight peak near $T = 40$ K.

Relaxation of the iron spins in garnets is closer to a T^{-1} direct process, with little evidence of a T^{-9} Raman contribution. As T increases, two factors in (6.13) cause ΔH_i to decrease: $\hbar\omega/kT$ and $E_{\text{ex}}^c(T)$. It should also be pointed out that the Orbach function [6, 7] applies to a paramagnetic system with no exchange field, and is the low-temperature approximation to $\tau_{\text{RE}}^{-1} = C_{\text{eff}}\,[\exp(\Delta/kT) - 1]^{-1}$, from (6.72). If this more rigorous expression were used, the computed ΔH_i would not decrease as quickly above the temperature of the peak. However, if the Orbach formalism included E_{ex}^c as an exchange energy splitting in the two-phonon process, the

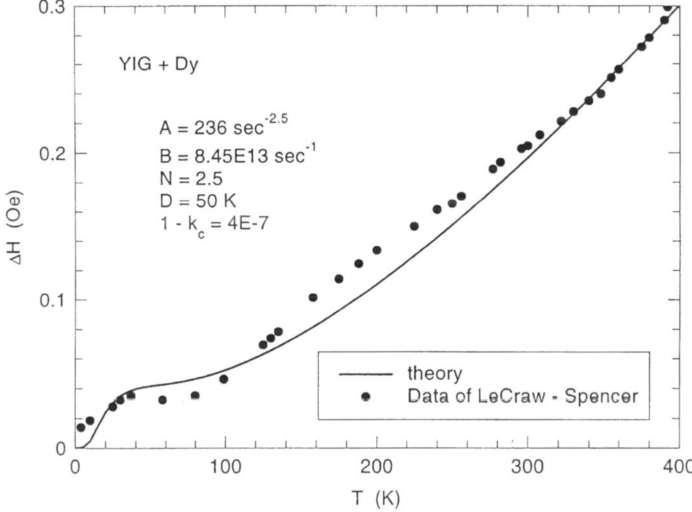

Fig. 6.16 Comparison of theory with X-band FMR half-linewidth data as a function of temperature for high-purity YIG specimen [47]. Data are from LeCraw and Spencer [49]. Figure reprinted from [47] with permission. © 2000 by the American Institute of Physics

coefficient $C_{\rm eff}$ would then be subjected to the strong temperature dependence of $E^c_{\rm ex}$ [50]. $\tau^{-1}_{\rm RE}$ would therefore decrease more sharply at higher T, thereby offsetting the error in $\Delta H_{\rm i}$ introduced by the low-temperature approximation.

Other ions besides rare earths that satisfy the $\lambda/\delta \gg 1$ condition can provide fast relaxation paths to the lattice bath. In the iron sites of both garnets and spinels, Co^{2+} and Fe^{2+} in octahedral sites are common because of their spin–orbit doublet stabilization in an exchange field. Less frequent, but nonetheless significant in select situations, Mn^{3+} in octahedral sites [51], shown in Fig. 6.17, as well as Ni^{2+} in tetrahedral sites can also produce enhanced linewidths at low temperature. Table 5.9 summarizes qualitatively the effects expected for the $3d^n$ series in high-spin states. An instructive example of rapid relaxation from Fe^{3+} $({t_{2g}}^5{e_g}^0)$ in a low-spin state ($S = 1/2$) was reported by Kipling et al. [52] for the host compound $K_3Co^{3+}(CN)_6$.

Another consideration that will be examined in Chap. 7 in connection with magneto-optical Faraday rotation is the influence of Bi^{3+} and Pb^{2+} on the iron sublattices. Because these ions have a wide ranging $6s$ outer orbital that can hybridize with $2p$ orbitals of O^{2-} and the half-filled $3d$ orbitals of Fe^{3+}, all of the relevant overlap integrals could be affected. Evidence of these interactions was found in the Curie temperature of BiIG that exceeded that of pure YIG by 38 K [53]. Later Mossbauer analysis revealed that the presence of Bi^{3+} correlated with a reduced negative exchange interaction in the tetrahedral sublattice, which suggests a possible direct ferromagnetic contribution between d-site Fe^{3+} ions [54]. Further indication of influence on the Fe^{3+} spin system appears in the spin–lattice relaxation rate through hybridization of the $6sp$ excited state with the $6s^2$ excited state. With a very

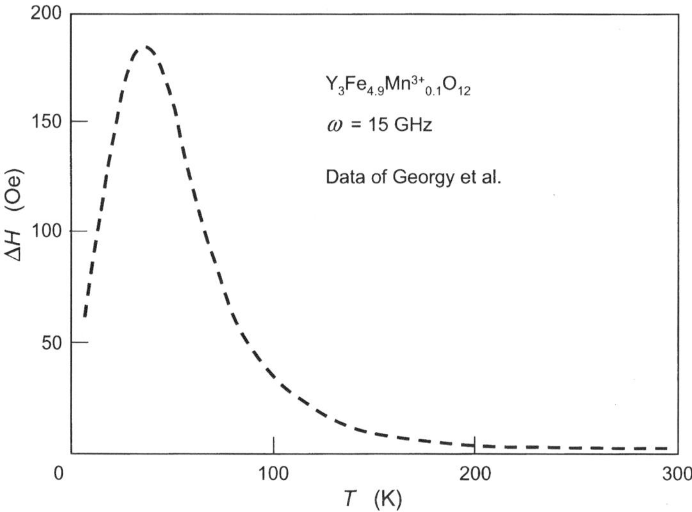

Fig. 6.17 15-GHz FMR linewidth of single crystal $Y_3Fe_{4.99}Mn_{0.01}O_{12}$ at cryogenic temperatures, indicating significant spin–lattice coupling due to the magnetoelastic property of Mn^{3+} $(3d^4)$ in the octahedral a sublattice. Data are from Georgy et al. [51]

large spin–orbit coupling constant ($\sim$17,000 cm^{-1}) similar to RE ions, the formation of hybrid ground states can cause strong microwave damping effects [55].

6.2.6 The Exchange Isolation Effect

Throughout Chaps. 5 and 6 the ratio of spin–orbit coupling to crystal-field energy λ/δ has recurs wherever orbit–lattice interactions are discussed. Its definitive role in determining the anisotropic g factors of Ti alum and other transition-metal ions in paramagnetic systems, its importance in selecting the spontaneous ligand distortions (Jahn–Teller vs. spin–orbit), and its critical contribution to the mechanism of spin–lattice relaxation of electromagnetic spin excitations have all been reviewed. A large value of λ/δ usually means strong magnetoelasticity [56]. Perhaps the best examples are the $4f^n$ rare-earth ions with $\lambda > 10^3$ cm^{-1} and $\delta < 10^2$ cm^{-1}. By contrast, the $3d^n$ series features $\lambda/\delta < 1$, but with notable exceptions, such as Co^{2+} in an octahedral site.

In certain situations, the overall lattice magnetoelastic effects from local distortions are erased due to magnetic dilution. This phenomenon was labeled as an *exchange isolation* effect and was first noted as an absence in sign reversal of the anisotropy constant K_1 with increasing Co^{2+} substitutions in Li spinel ferrite when Ti^{4+} substitutions are made for Fe^{3+} in octahedral sites [57,58]. To preserve charge neutrality, additional Li^{1+} is added to tetrahedral sites. The net result is that small concentrations of Co^{2+} ions in octahedral sites can be surrounded by nonmagnetic

Li^{1+} and Ti^{4+} ions in neighboring sites of either iron sublattices, thereby removing much of the H_{ex} and essentially rendering the Co^{2+} ions paramagnetic. Further evidence that Co^{2+} was removed from the collective spin system but retained its local site spin–lattice coupling was the absence of related spin wave effects and the increase of FMR line broadening from neighboring dipolar interactions. Local effects from isolated Co^{2+} are also apparent at lower frequencies in NiZn spinels where a separate higher μ dispersion frequency peak occurs for small concentrations, only to pull the main Fe^{3+} peak up to it as the Co^{2+} level is raised to the percolation threshold for iron–cobalt magnetic exchange ordering. [59]. In the garnet systems, Llabres et al. reported a similar isolation effect in a Co^{2+}-substituted vanadate iron garnet system [60] and a later observation appears to have been made with tetrahedral-site Ge^{4+} isolating octahedral-site Co^{2+} in yttrium iron garnet [61].

6.3 Exchange-Coupled Modes (Spin Waves)

The subject of propagation losses that can result from the generation of traveling *spin waves* or *magnons* acting as indirect funnels for converting microwave energy into lattice phonons will now be introduced in the context of (1) degenerate spin-wave modes triggered by lattice inhomogeneities and (2) nonlinear growth of spin waves amplitudes from high rf signal power. With specimens of finite dimensions, the uniform precession line can also be divided into a spectrum of gyromagnetic standing waves (*magnetostatic modes*) caused by nonuniformities in internal magnetic fields or variations in the rf field amplitudes within the specimen volume.

6.3.1 *Uniform Precession Decoherence (Degenerate Spin Waves)*

In a system of exchange-coupled spins, decoherence of the uniformly precessing spins can occur analogously to paramagnetic systems, but through a different mechanism. Where individual spins are perturbed thermally by spin–phonon collisions or as a result of local crystal imperfections, magnetic dilution, or high-intensity nonuniform H_{rf} fields, the precession can vary spatially. Periodic phase fluctuations can then propagate in the manner of lattice vibrational modes. Under these conditions, we must take into account the spin-phase variations that are dependent on the restoring force from magnetic exchange field H_{ex} rather than the random dipole–dipole interactions of a paramagnetic system.

Figure 6.18 illustrates the contrast between the uniform precession mode and the spatial variation of $M_{x,y}$ as the wave propagates parallel and transverse to the z-axis of the magnetic field. Because of the wave nature of the phase and the similarity to phonons, these modes are called *spin waves* or *magnons*. The erosion of the uniform precession mode by spin waves is an important concern for microwave propagation in ferrites at high microwave power levels.

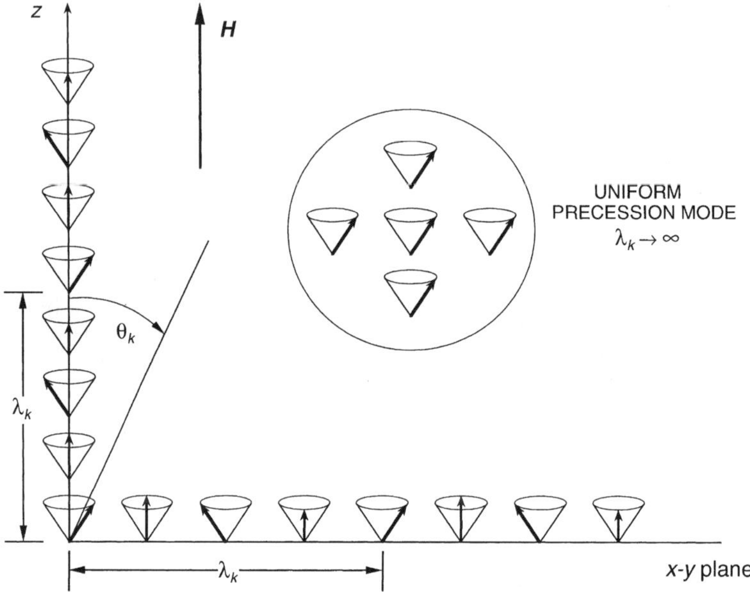

Fig. 6.18 Tutorial diagram contrasting the essential features of gyromagnetic precession in the form of traveling spin waves of wavelength λ_k and the static uniform precession with $\lambda = \infty$

The physical origin of spin waves resides in the exchange fields that align the spin directions. In contrast to dipole–dipole coupling of paramagnets that depends on the spatial proximity of the spins, the restoring force that gives rise to spin waves is derived from the local exchange field. A rate equation for this effect (without spin–lattice damping) can be expressed as [62]

$$\left(\frac{\mathrm{d}\boldsymbol{M}}{\mathrm{d}t}\right) = \gamma\left(\boldsymbol{M}\times\boldsymbol{H}\right) + \omega_{\mathrm{ex}}a^2\frac{\boldsymbol{M}\times\nabla^2\boldsymbol{M}}{\boldsymbol{M}}, \tag{6.73}$$

where a is the lattice distance between spins and $\omega_{\mathrm{ex}} = \gamma H_{\mathrm{ex}} = \gamma N_{ij} M_z$.[1]

The frequency of a spin wave propagating at an angle θ_k to the z-axis in an ellipsoidal specimen of demagnetizing factor N_z can be derived by assuming that the rf magnetization ${M_{\mathrm{rf}}}^{x,y}$ generated by $\boldsymbol{H}_{\mathrm{rf}}$ in the x–y plane varies according to the plane-wave function $\exp\left[\mathrm{i}\left(\omega t - \boldsymbol{k}\cdot\boldsymbol{r}\right)\right]$, to give $\nabla^2\boldsymbol{M} = -k^2\boldsymbol{M}$. With this relation, the component equations in emu can be expressed as

$$\begin{aligned}
\mathrm{i}\omega\left(4\pi M_{\mathrm{rf}}^{x}\right) &= 4\pi M_{\mathrm{rf}}^{y}\left(\omega_0 + \omega_{\mathrm{ex}}a^2k^2\right) - \gamma\left(4\pi M_z\right)H_{\mathrm{rf}}^{y} \quad \text{and}\\
\mathrm{i}\omega\left(4\pi M_{\mathrm{rf}}^{y}\right) &= -4\pi M_{\mathrm{rf}}^{x}\left(\omega_0 + \omega_{\mathrm{ex}}a^2k^2\right) + \gamma\left(4\pi M_z\right)H_{\mathrm{rf}}^{x},
\end{aligned} \tag{6.74}$$

[1] The derivation of the exchange term in (6.73) can be found in Lax and Button [21]

from which the following components of the rf susceptibility analogous to (6.120) in Appendix 6A can be extracted:

$$\chi_{xx} = \chi_{yy} = \frac{M_{\rm rf}^x}{H_{\rm rf}^x} = \frac{1}{4\pi}\frac{\gamma\,(4\pi M_z)\,\omega_k}{\omega_k^2 - \omega^2},$$
$$\chi_{xy} = -\chi_{yx} = \frac{M_{\rm rf}^{x,y}}{H_{\rm rf}^{y,x}} = -{\rm i}\frac{1}{4\pi}\frac{\gamma\,(4\pi M_z)\,\omega}{\omega_k^2 - \omega^2}. \qquad (6.75)$$

For small signals, we can assume that $M_z \approx M$, and express the resonance frequency for the k spin-wave mode as

$$\omega_k = g\left(H_{\rm i} + H_{\rm ex}a^2k^2\right), \qquad (6.76)$$

where the magnitude of the wave vector $k = 0$, $\omega_k = \gamma H_{\rm i}$, the uniform precession frequency. For $k > 0$, ω_k represents the precession frequency of a spin wave of wavelength $2\pi/k$ determined by the sum of the uniform mode frequency and the contribution from the exchange field.

With the inclusion of magnetocrystalline and shape anisotropy demagnetizing fields $H_{\rm K}$ and $N_{\rm D}\,(4\pi M_z)$, respectively, the frequency of spin waves directed at an arbitrary angle θ_k to the z-axis of the applied magnetic field H can be expressed in general terms based on $\omega_{\rm r}$ in (6.44) as

$$\omega_k = \gamma\begin{bmatrix}\left(H - H_{\rm K} - N_{\rm D}4\pi M_z + H_{\rm ex}a^2k^2\right)\\ \times\left(H - H_{\rm K} - N_{\rm D}4\pi M_z + H_{\rm ex}a^2k^2 + 4\pi M_z\sin^2\theta_k\right)\end{bmatrix}^{\frac{1}{2}}. \qquad (6.77)$$

From this relation, the conditions under which the uniform precession and higher modes are degenerate, i.e., share the same ω_k, can be determined. The range of propagation directions from $\theta_k = 0$ to $\pi/2$ gives rise to many possible situations where the waves of different k can couple or scatter without a change in energy as in a direct two-magnon process. Three-magnon scattering in the manner of the Raman processes discussed in Sect. 6.1 in relation to spin–phonon interactions are also possible [63].

By inspection of (6.77), we see that the geometric shape of the specimen and its magnetocrystalline anisotropy are important in determining the degree of degeneracy between the various k waves and the $k = 0$ (uniform precession mode). The limits to the width of the *spin-wave manifold* (assuming $H_{\rm K} = 0$) within which the degeneracy with the $k = 0$ mode can exist are illustrated by the $\theta_k = 0$ and $\theta_k = \pi/2$ curves in Fig. 6.19. As indicated, only an ellipsoidal specimen approximating a thin disk magnetized normal to its axis ($N_{\rm D} = N_{{\rm D}z} = 1$) can have a uniform precession that is not degenerate with spin waves. This is because its dispersion curve touches only the lower edge ($\theta_k = 0$) of the manifold at $k = 0$. A uniform precession frequency within the manifold satisfies the degeneracy requirement for the transfer of energy to spin waves.

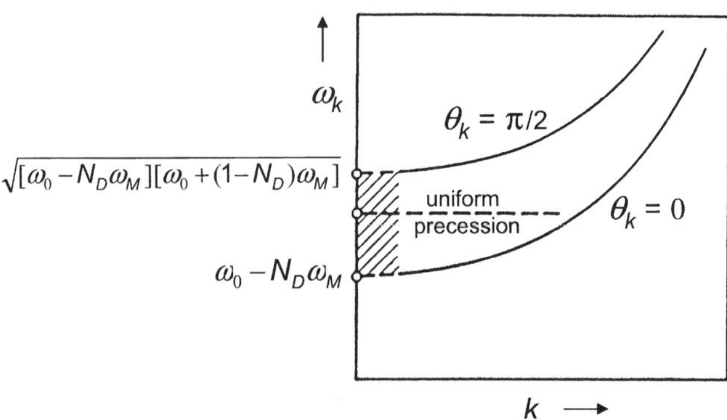

Fig. 6.19 Standard diagram of the spin-wave manifold envelopes within which the uniform precession mode can exist simultaneously with spin waves. Figure reprinted from [68] with permission. © 1956 by the IEEE

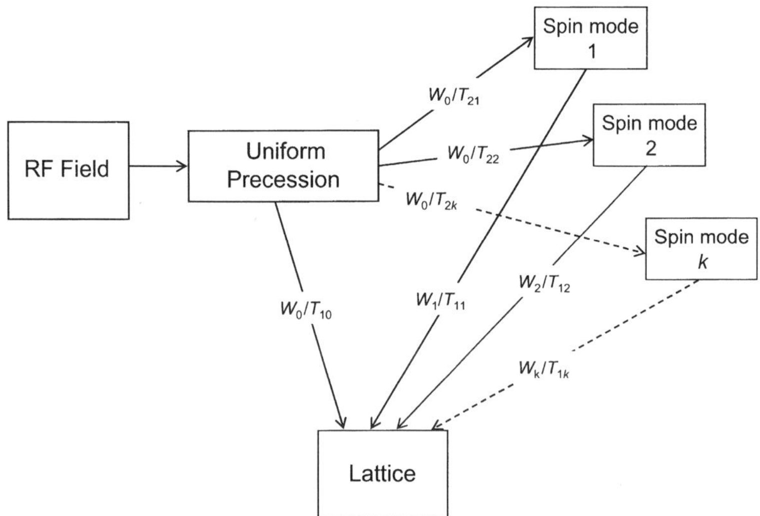

Fig. 6.20 Standard diagram of the distribution of individual spin waves of mode k from the uniform precession ($k = 0$) reservoir. Figure reprinted from R.C. Fletcher, R.C. LeCraw, and E.G. Spencer, *Phys. Rev.* **117**, 1665 (1960) with permission. © 1960 by the American Physical Society. http://link.aps.org/doi/10.1103/PhysRev.117.1665

In terms of magnetic susceptibilities, degenerate spin waves can introduce a line broadening analogous to the dipolar decoherence characterized by the relaxation time τ_2. A spin wave of vector $\boldsymbol{k}$ would be assigned a time τ_{2k}. For the general case, where a spectrum of k values are possible, the rate $\tau_{2k}^{*-1} = \sum_{k=0}^{n} \tau_{2k}^{-1}$ is the sum of n participating modes ranging from $k = 0$ (uniform precession) to n. Figure 6.20

diagrams how this occurs. If we assume that $\tau_{1k} \approx \tau_{10}$ for all modes,[2] the imaginary susceptibility of (6.55) is expressed as the paramagnetic case

$$\chi''_{\mathrm{rf}} \cong \frac{1}{2}\gamma M \frac{\tau_2}{1+\tau_2^2(\gamma H-\omega)^2+\gamma^2 H_{\mathrm{rf}}^2 \tau_{10}\tau_2}, \tag{6.78}$$

where $\tau_{10} = \tau_1$ and $\tau_2^{-1} = (2\tau_1)^{-1} + \tau_{2k}^{*-1}$. Following the model of Schloemann et al. [64], the absorption component at $\omega = \gamma H$ can be expressed as

$$\frac{\chi_{\mathrm{rf}}(H_{\mathrm{rf}})}{\chi_{\mathrm{rf}}(0)} = \left[1 + C\left(\frac{H_{\mathrm{rf}}}{\Delta H}\right)^2\right]^{-1} \approx 1 - C\left(\frac{H_{\mathrm{rf}}}{\Delta H}\right)^2, \tag{6.79}$$

where $H_{\mathrm{rf}}/\Delta H \ll 1$ and $C = \tau_1/\tau_2$. Note that (6.79) reduces to (6.56) when $\tau_2 = 2\tau_1$.

At this point, it should be remarked that the growth of ΔH is not of fundamental origin, but rather the result of random $k > 0$ modes that occur from spatial and structural inhomogeneities within a particular specimen geometry. Surface irregularities probably also contribute to the linewidth of single crystals [65], although inhomogeneous broadening due to direct demagnetization has also been proposed as the cause of ΔH increases in rough-surfaced spheres of polycrystals [66]. The major cause of spin-waves, however, comes from the development of an instability in the uniform precession that occurs above an rf power threshold.

6.3.2 *Instability Threshold (Classical Approximation)*

Although magnetic inhomogeneities can serve as nucleation points for launching spin waves that broaden the uniform precession resonance, a more dramatic important effect occurs when H_{rf} reaches a threshold value where the amplitude of the rf magnetization $M_{\mathrm{rf}}^{x,y}$ is sufficiently great that we can no longer assume that $M_z \approx M$. At this point M_z becomes a meaningful variable, and the system enters a state of accelerated nonlinear spin-wave generation. Since this effect generally arises as the large-signal regime is approached, its occurrence is of concern in higher-power applications, such as long-range radar. To appreciate some of the underlying physics of the uniform precession instabilities that are usually attributed to nonlinear spin-wave generation, we shall introduce three basic classical concepts (1) the origin of the instability when $M_z < M$, (2) the critical rf field $H_{\mathrm{rf}}^{\mathrm{crit}}$, and (3) the relation for the uniform precession resonance peak as H_{rf} increases above $H_{\mathrm{rf}}^{\mathrm{crit}}$.

[2] Because spin–orbit–lattice coupling controls these mechanisms, the propagation of a precessing magnetic moment should not have much influence on its relaxation rate, and the variation among τ_{1k} values can usually be ignored.

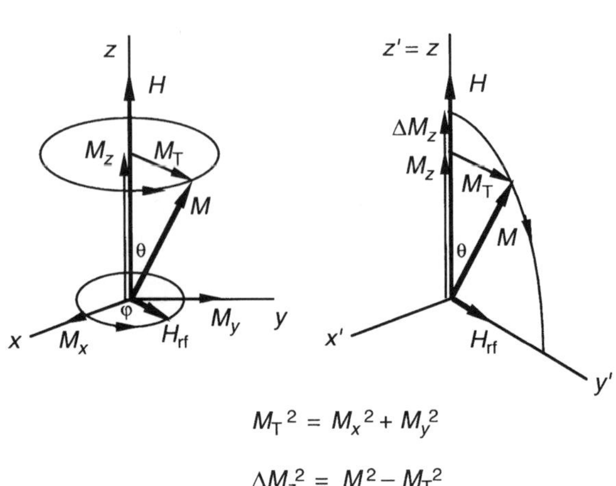

Fig. 6.21 Reduction in M_z under the influence of the torque created by a large H_{rf}, as viewed by stationary and rotating frames of reference

To introduce analytically the nonlinear dependence of spin waves on larger rf magnetic field strengths, we begin by following the reasoning of Anderson and Suhl [67] that the value of M_z decreases when $\boldsymbol{M}$ is rotated from the z-axis by the torque provided by the transverse $\boldsymbol{H}_{rf}$ field, so that

$$M_z = \sqrt{M^2 - M_T^2}, \tag{6.80}$$

where $M_T^2 = M_x^2 + M_y^2$ as seen from the Pythagorean vector triangle in Fig. 6.21. In terms of the rf susceptibility for the circular polarization mode that produces resonance, as defined by (1.84) and (6.54), $M_T^2 = \chi_{rf}^2 + H_{rf}^2$ where

$$\chi_{rf}^2 = \chi_{rf}'^2 + \chi_{rf}''^2 = \frac{\tau_2^2 (\gamma M)^2}{1 + \tau_2^2 (\gamma H - \gamma N_{Dz} 4\pi M_z - \omega)^2}, \tag{6.81}$$

if we ignore the small $\gamma^2 H_{rf}^2 \tau_1 \tau_2$ term in the denominator. Note also that the reduction of the applied field H from the z-axis demagnetizing field $N_{Dz} 4\pi M_z$ in a planar disc magnetized normal to its surface ($N_{Dz} \sim 1$) is now included in the resonance term. This geometry was chosen to emphasize the effect of $\boldsymbol{H}_{rf}$ on the rotation of $\boldsymbol{M}$ from the z-axis.

From (6.138) in Appendix 6B, we repeat the relation that best illustrates the instability:

$$4\pi\, \Delta M_z \cong \frac{1}{2} \frac{4\pi M (\gamma H_{rf})^2}{(\gamma N_{Dz} 4\pi\, \Delta M_z)^2 + \tau_{2k \to 0}^{-2}}. \tag{6.82}$$

The classical instability can be reasoned by envisioning perturbations of the M_z component, defined as ΔM_z. An increase in ΔM_z will reduce the right-hand side, thereby producing a new value of ΔM_z. If the value is reduced, the system will relax to equilibrium; if it increases, the value will diverge and the system will be unstable.

The solution of (6.82) produces the following relations at the critical condition:

$$\gamma N_{Dz} 4\pi \, \Delta M_z^{\text{crit}} = \tau_{2k\to 0}^{-1} = \gamma \, \Delta H_{k\to 0} \tag{6.83}$$

$$\text{and } \gamma H_{\text{rf}}^{\text{crit}} = \tau_{2k\to 0}^{-3/2} \left(\gamma N_{Dz} 4\pi M\right)^{-1/2}, \tag{6.84}$$

$$\text{or } \Delta H_{k\to 0} = \left(H_{\text{rf}}^{\text{crit}}\right)^{2/3} \left(N_{Dz} 4\pi M\right)^{1/3}. \tag{6.85}$$

Note that the subscript $k \to 0$ has been added to τ_2 and ΔH (now without subscript i) in these relations and in Appendix 6B to point out that the threshold for spin-wave onset depends on the initial uniform precession values of the decoherence rate. It is these rates that must increase in order to restore equilibrium to the precessing spin system.

The above exercise reveals interesting effects where $H_{\text{rf}} > H_{\text{rf}}^{\text{crit}}$: from (6.85) ΔH increases in direct proportion to $H_{\text{rf}}^{2/3}$, and from (6.22) and (6.24) the applied magnetic field H at the peak χ''_{rf} decreases by the amount ΔH of the demagnetizing field, as sketched in Fig. 6.22. Because ΔH is typically negligible compared to H or $4\pi M$, the shift in resonance is important only at or very near resonance. However, the broadening of the line as it relates to increased rf loss away from resonance is of great practical significance, as discussed in Sect. 6.2. Because the increase in ΔH is reflected as a decrease in the peak value of χ''_{rf} for a Lorentzian line shape, we can

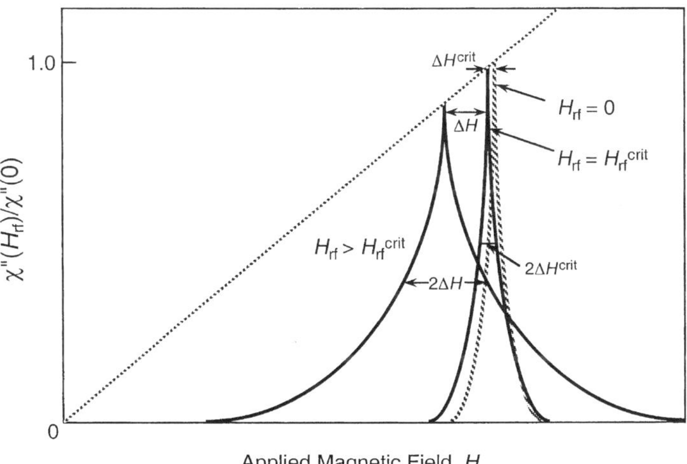

Fig. 6.22 Graphical sketch of the decline in χ'', the growth of the linewidth ΔH and its relation to the downward shift in the resonance field at a fixed frequency, as the rf field amplitude increases above $H_{\text{rf}}^{\text{crit}}$

display the effect of the broadening on the value of $\chi''_{\text{rf}}(H_{\text{rf}})$ normalized to $\chi''_{\text{rf}}(0)$ by merging (6.56) with (6.85), according to

$$\left.\frac{\chi''_{\text{rf}}(H_{rf})}{\chi''_{\text{rf}}(0)}\right|_{\text{peak}} = \left[1+\frac{1}{2}\left(\frac{H_{\text{rf}}}{\Delta H_{k\to 0}}\right)^2\right]^{-1} \cong \left[1+\frac{1}{2}\left(\frac{H_{\text{rf}}}{(H_{\text{rf}}^{\text{crit}})^{2/3}(4\pi M)^{1/3}}\right)^2\right]^{-1}$$

$$\text{for } H_{\text{rf}} \leq H_{\text{rf}}^{\text{crit}} \tag{6.86}$$

$$= \left[1+\frac{1}{2}\left(\frac{H_{\text{rf}}}{4\pi M}\right)^{2/3}\right]^{-1}\left(\frac{H_{\text{rf}}^{\text{crit}}}{H_{\text{rf}}}\right)^{2/3} \cong \left(\frac{H_{\text{rf}}^{\text{crit}}}{H_{\text{rf}}}\right)^{2/3}$$

$$\text{for } H_{\text{rf}} \geq H_{\text{rf}}^{\text{crit}},$$

for $N_{\text{Dz}}4\pi M \gg H_{\text{rf}}$. From the sketch in Fig. 6.23, it can be seen how the generation of spin waves influences the premature decline in the absorption peak and hence, the attendant linewidth broadening.

Although the above model offers rationale for the spin-wave instability, the actual relations used in practical situations must be sensitive to the exchange coupling among specific spin-wave modes. A more exact approach to the derivation of the threshold conditions was developed by Suhl [68], in which the instability was found to occur without a shift in the resonance field. In this case, the restoration of equilibrium is accounted for entirely by an increase in linewidth. As a consequence, the

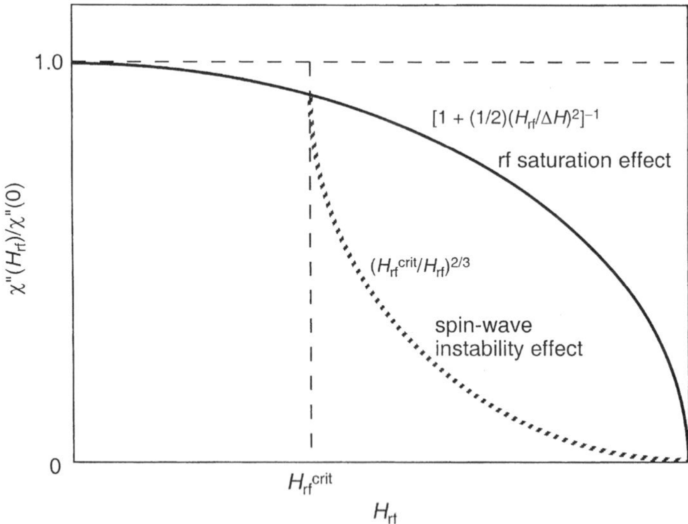

Fig. 6.23 Graphical display of the effects on the peak χ'' below and above the critical field $H_{\text{rf}}^{\text{crit}}$, stated analytically in (6.91). The *solid curve* represents the saturation effect below $H_{\text{rf}}^{\text{crit}}$ that is used as a technique for measuring τ_1 [4, 18]. The sharp decline of the *dashed curve* indicates how the spin waves above $H_{\text{rf}}^{\text{crit}}$ are broadening the linewidth

relation of(6.20) between ΔH and H_{rf} is linear, and the above-threshold part of (6.25) would become

$$\left.\frac{\chi''_{rf}(H_{rf})}{\chi''_{rf}(0)}\right|_{peak} = \frac{H_{rf}^{crit}}{H_{rf}} \quad \text{for } H_{rf} > H_{rf}^{crit}. \tag{6.87}$$

6.3.3 *Instability Threshold (Nonlinear Spin Waves)*

The foregoing classical model does not accurately represent the experimental reality. It is included here to provide an introduction to the instability effect and the role of demagnetizing fields that feedback any growth of the uniform precession angle, thereby leading to a nonlinear increase in absorption. It cannot, for example, be used to interpret the case of spherical specimens where the external demagnetizing term is zero. This situation was addressed by Suhl [68] who reasoned that initial demagnetizing effects of specific $k > 0$ spin waves themselves would destabilize the uniform precession through variations in their exchange fields.

In contrast to the classical model in which the increased rf loss is viewed as a transfer of energy to a collective spin-wave "bath," the rigorous analysis treats the spin waves or magnons as individuals of wave number k (see Fig. 6.20). The computed results are in general accord with experiment. The uniform precession frequency does not shift to lower values above the threshold as in the classical model, but rather a second absorption peak [69] occurs as low as $\omega_0/2$ (or $H_0/2$). This effect is referred to as the *subsidiary* absorption, which is depicted schematically in Fig. 6.24. The changes in χ and M_z as a function of H_{rf} were observed by Bloembergen and Wang [70].

An important consideration is the threshold field away from the resonance peak, where many microwave devices are operated. As expected, the relation for the minimum threshold field is frequency dependent. For a general shape

$$H_{rf}^{crit}(\min) = \frac{2\omega\,(\omega - \gamma H + N_{Dz}\gamma 4\pi M)\,\tau_{2k}^{-1}}{\gamma 4\pi M \sin\theta_k \cos\theta_k\,(\omega/2 + \gamma H - N_{Dz}\gamma 4\pi M)}, \tag{6.88}$$

where the angle θ_k must fall somewhere between 0 and $\pi/2$. From this relation, the celebrated "butterfly" curves in Fig. 6.25 were constructed by Suhl for a spherical geometry. Each point on any curve represents the critical field for the onset of exponential growth of an ω_k spin wave (magnon), with the minimum of the curve occurring for the $k \to 0$ longest available wavelength, i.e., closest to uniform precession. This minimum condition is summarized by the relation

$$H_{rf}^{crit}(\min) = \alpha \Delta H_{k\to 0}\left(\frac{\omega}{\omega_M}\right), \tag{6.89}$$

where $\alpha_{sw}{\sim}1$ is a proportionality parameter that is dependent on the particular conditions chosen for the measurement, and $\Delta H_{k\to 0}$ is the spin-wave half-linewidth,

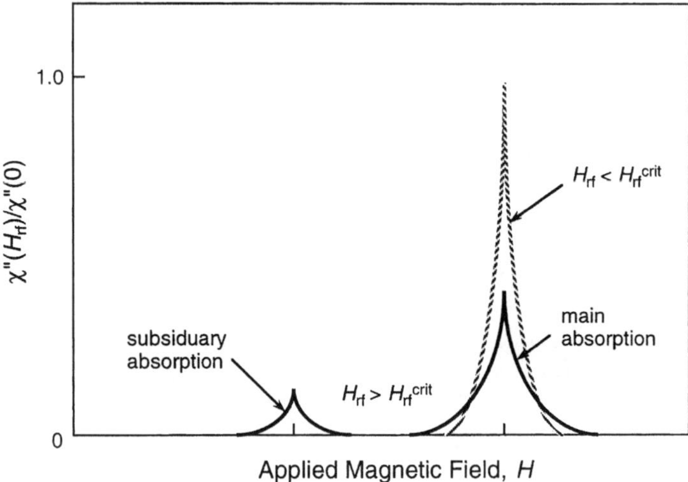

Fig. 6.24 Graphical sketch of the appearance of the subsidiary absorption peak at a magnetic field below the main resonance field, based on the theory of Suhl [69]. The subsidiary absorption is an alternative loss mechanism to the classical broadening of the main resonance line and can appear as low as one-half of the main resonance field

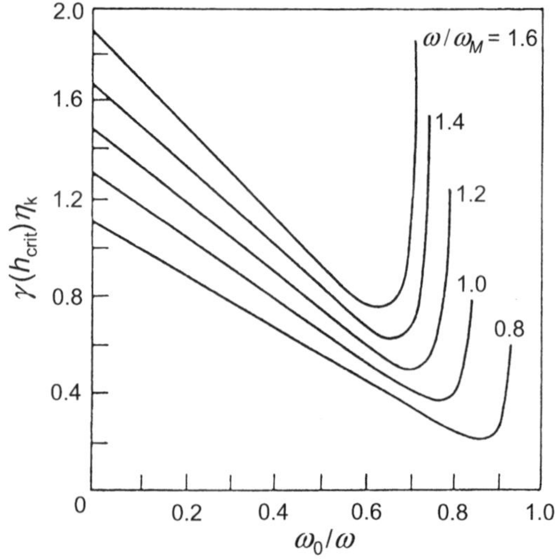

Fig. 6.25 *Butterfly curves* computed from (6.93) for a specimen of spherical geometry. Nomenclature adjustments: $h_{\text{crit}} \equiv H_{\text{rf}}^{\text{crit}}$, and $\eta_k \equiv \tau_{2k}^{-1}$. Minimum values indicate the condition for the onset of the subsidiary absorption depicted in **Fig. 6.25**, which appears in the regime where $\gamma H_0 < \omega$ i.e., the main resonance frequency is less than the signal frequency. Figure reprinted from [68] with permission. © 1956 by the IEEE

which becomes the uniform-precession value in the limit of $k = 0$. A point worth mentioning here is that $H_{\text{rf}}^{\text{crit}}$ can be increased by raising the intrinsic linewidth through the introduction of fast-relaxing impurities, e.g., rare-earth or Co^{2+} ions in particular. In effect, the microwave energy accumulated in the uniform precession reservoir is being depleted, thereby delaying the onset of instabilities (see Figs. 6.13 and 6.20 and the discussion in Sect. 6.2.6). This trade-off between average power loss and the more catastrophic effect of the nonlinear spin-wave generation is a technique used regularly in designing ferrite compounds for high-power applications.

Equation (6.89) is used to determine $\Delta H_{k \to 0}$ as a measure of the intrinsic spin–lattice relaxation rate. With a polycrystalline specimen in which ΔH is masked by the inhomogeneous broadening discussed in Sect. 6.2.4, this technique has proven to be of considerable value in estimating the peak-power capability of ceramic ferrites. Schloemann et al. [64] determined that the nonlinear threshold could also be observed if the rf field was applied parallel to $\boldsymbol{H}$ instead of transverse as Suhl considered. Subsequently, Patton examined the case of the general angle for linear and circular (both Larmor and anti-Larmor) polarizations [71].

6.3.4 Magnetostatic Modes

In the foregoing discussion of spin waves, infinite specimen dimensions were assumed. From (6.76) the k^2 dependence on the spin-wave energy implies that long wavelengths are not likely to be significant. For typical values of ω_{ex} in magnetic oxides which reach toward 10^{12} Hz, λ_k might be expected to be less that 100 lattice parameter lengths. Therefore, the curves defining the limits of the spin-wave manifold in Fig. 6.19, include a cross-hatched region as $k \to 0$ for which spin waves are inconsequential. For specimens of finite dimensions, however, additional "uniform" precession modes can occur where (1) demagnetizing fields produce nonuniform internal fields, particularly near surface boundaries and (2) where the rf signal magnetic field is nonuniform within the specimen. The former case occurs when the geometry is nonellipsoidal and is important in practical situations involving planar geometries. A classic example of this effect was reported by Dillon [72]. The second instance is of concern in waveguides where rf modes provide variable field directions. The most striking example of this condition is the internal magnetostatic (Walker) modes of a YIG sphere [73] that are described the general theory of magnetostatic modes in several texts [74–77].

To remain within the scope of this text, the topics of spin waves and magnetostatic waves will not be explored beyond these basic concepts. The interested reader is encouraged to delve further into the subtleties of instability thresholds and subsidiary absorption peaks in the various textbooks and reviews on the subject of microwave magnetism. Spin waves themselves are also vehicles for studying fundamental physical phenomena of magnetic systems, including the generation and propagation of microwave magnetic envelope (MME) solitons. The interested reader is encouraged to consult the pioneering work of Chen et al. [78] and the results of other investigators in this field.

6.4 Permeability and Propagation

Electromagnetic wave interactions with magnetic materials are important practical as well as fundamental areas of interest. In previous sections, nonresonance (longitudinal $\boldsymbol{H}_{\mathrm{ac}} \cdot \boldsymbol{M}_{\mathrm{s}}$) and gyromagnetic resonance (transverse $\boldsymbol{H}_{\mathrm{rf}} \times \boldsymbol{M}_{\mathrm{s}}$) cases were introduced in classical terms that involved paramagnetic or what amounts to single-domain ferromagnetic systems. In many practical cases, the medium is a polycrystalline ferrite that is operated below the longitudinal dispersion frequency illustrated in Fig. 6.1 for maximum permeability, or above it for rf control applications such as Faraday rotation and phase shift near ferrimagnetic resonance. Despite the fact that these two effects are independent and can occur in widely separated frequency regions, any analytical simplicity that might be assumed disappears quickly once the single-domain assumption is invalidated by the appearance of multiple domains in partially magnetized materials [79, 80].

6.4.1 Low-Frequency Longitudinal Permeability

When the medium is not fully magnetized, the magnetization can be divided into domains in which the individual vectors are aligned according to local crystallographic "easy" axes. In polycrystals, the distribution can occur as individual single-domain grains if they are too small to support the formation of domain walls. As illustrated in Fig. 6.26, the H_{K} fields that fix the directions of the $\boldsymbol{M}_{\mathrm{s}}$ vectors are set by the orientation of the grains themselves, with or without domain walls. In either case, propagating rf magnetic fields will be subject to longitudinal damping effects.

Since there is no resultant of the combined saturation $\boldsymbol{M}_{\mathrm{s}}$ vectors in an unsaturated state, there is no reference direction by which longitudinal and transverse drive fields can be distinguished. Where low anisotropy and ideal microstructure provide negligible impediments to reversible wall movement, H_{ac} can exceed the coercivity field H_{c} thereby allowing domain wall movement to enhance the longitudinal susceptibility. Furthermore, domain walls can resonate as damped forced harmonic oscillators at a characteristic frequency related to the spin–lattice relaxation rate. At higher frequencies approaching the microwave bands, gyromagnetism will occur through transverse coupling about internal demagnetizing fields that originate in part from poles on the surfaces of the walls. A result of this frequency dependence is that absorption occurs beginning at about 1 MHz for longitudinal components of rf drive fields and can reach into the GHz range for the transverse gyromagnetic effects. These coexisting phenomena are contrasted in Fig. 6.1.

The basic susceptibility in the unmagnetized state is influenced by three principal factors. First, the magnitude of the initial susceptibility χ_{i}, which is determined by the ease of magnetization rotation within the randomly oriented domains by the $\boldsymbol{H}_{\mathrm{ac}} \cdot \boldsymbol{M}_{\mathrm{s}}$ interaction in the absence of domain walls. Because the maximum switching or hysteresis-loop coercive field of a single-domain specimen is H_{K}, $\chi_{\mathrm{i}} \sim M_{\mathrm{s}}/H_{\mathrm{K}}$ in simple linear terms. Second, when domain walls are present, an

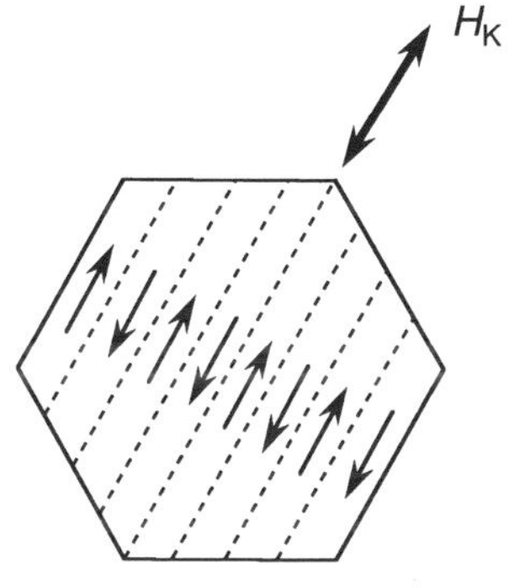

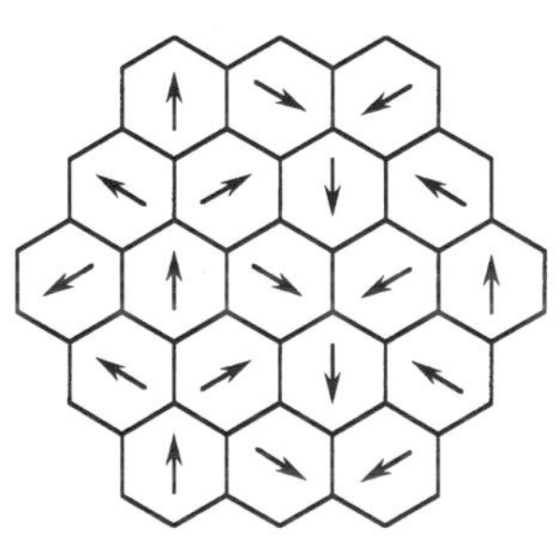

Fig. 6.26 Two-dimensional sketches contrasting saturation magnetization vectors in a large multidomain single-crystal grain and an aggregate of small single-domain grains under the influence of local H_K fields. Figure reprinted from [80] with permission. © 2003 by the IEEE

enhanced susceptibility from reversible wall movement χ_w is obtained as the reduced coercive field $H_c \ll H_K$ allows $\chi_w \sim M_s/H_c$. Third, the spectral character of the resulting permeability is also determined by the lowest frequency ω_r at which transverse $\boldsymbol{H}_{ac} \times \boldsymbol{M}_s$ FMR interactions occur through the gyromagnetic conversion constant γ. A further concern involves the highest frequency above which the gyromagnetic effects influence permeability.

For a specimen of cubic lattice symmetry with $K_1 < 0$ the initial permeability from rotation is expressed in the simplest terms by (5.50) as

$$\mu_i' = 1 + 4\pi\chi_i' \approx \frac{2\pi M_s^2}{|K_1|}. \tag{6.90}$$

For single-domain specimens, a dispersive decline of μ' from its initial value can be viewed as analogous to the paramagnetic system described in Sect. 6.1.1. The dispersion of the complex longitudinal χ is determined by the nonresonant relaxation rate $\omega_\tau = 1/\tau$, where τ is the effective spin–lattice relaxation time.

When unimpeded by microstructural defects and high anisotropy, the more dominant mechanism for producing high χ' in all but the infrequent occurrence of single-domain particles is domain wall movement and resonance. In Appendix 6C, the frequency dependence of susceptibility χ_w' is derived for the harmonic motion of domain walls according to

$$\ddot{\chi}_w' + \left(\frac{1}{\tau_w}\right)\dot{\chi}_w' + \omega_w^2\chi_w' = \frac{(2M_s)^2\varsigma_w}{m_w}\cos(\omega t). \tag{6.91}$$

where the square of the wall resonance angular frequency $\omega_{\mathrm{w}}^2 = \alpha_{\mathrm{w}}/m_{\mathrm{w}} = (8/3)\,\gamma^2\sqrt{\mathrm{A}\,|K_1|}\varsigma_{\mathrm{w}}$, the domain wall mass per unit area $m_{\mathrm{w}} = (1/4\pi)\,\gamma^{-2}\sqrt{|K_1|\,/A}$, τ_{w} is the domain wall damping time constant, and ς_w is the total wall surface area per unit volume for the case of $K_1 < 0$. It should be noted that ω_{w} is dependent jointly on the wall exchange energy A (hence the Curie temperature), the anisotropy, and the disposition and density of the domain configuration.

The susceptibility from reversible wall movement is deduced directly from the solution of (6.86), according to [81]

$$\chi'_{\mathrm{w}} = \frac{(2M_{\mathrm{s}})^2\,\varsigma_{\mathrm{w}}}{m_{\mathrm{w}}\sqrt{\left(\omega_{\mathrm{w}}^2-\omega^2\right)^2+(1/\tau_{\mathrm{w}})^2\,\omega^2}}\cos\left(\omega t-\phi\right), \tag{6.92}$$

where

$$\tan\phi = \frac{\omega}{\tau_{\mathrm{w}}(\omega_{\mathrm{w}}^2-\omega^2)}.$$

It can be easily shown that the maximum frequency ω_{d} to avoid critical damping of the resonance is then given by the minimum of the denominator of (6.92), which becomes

$$\omega_{\mathrm{d}}^2 = \omega_{\mathrm{w}}^2 - \frac{1}{2\tau_{\mathrm{w}}^2}. \tag{6.93}$$

Since ω_{w} already increases with $\sqrt{\mathrm{A}\,|K_1|}$ and ζ_{w}, we now conclude that the threshold for the collapse of χ'_{w} is also dependent on τ_{w}, which requires large values to allow the forced oscillation to remain below the critical damping limit of (6.93).

For the value of χ'_{w} at $\omega = 0$, other useful relations can be deduced from (6.92). Since α_{w} can also be expressed as $4M_{\mathrm{s}}^2\varsigma_{\mathrm{w}}/3\mu_{\mathrm{i}}$ from Appendix 6C, we can introduce the single-domain μ'_{i} to obtain

$$\chi'_{\mathrm{w}}(0) = \frac{4M_{\mathrm{s}}^2\varsigma_{\mathrm{w}}}{m_{\mathrm{w}}\omega_{\mathrm{w}}^2} = \frac{4M_{\mathrm{s}}^2\varsigma_{\mathrm{w}}}{\alpha_{\mathrm{w}}} = 3\mu'_{\mathrm{i}} \approx 12\pi\chi'_{\mathrm{i}}. \tag{6.94}$$

As a consequence, an increase in χ'_{i} of a factor between 30 and 40 over the magnetization rotation value could be expected from reversible domain wall movement. Moreover, at $\omega = \omega_{\mathrm{d}}$ a peak in χ'_{w} occurs, with an increase over the $\omega = 0$ value as estimated from the amplitude of (6.92), according to

$$\frac{\chi'_{\mathrm{w}}(\omega_{\mathrm{d}})}{\chi'_{\mathrm{w}}(0)} = \frac{\omega_{\mathrm{w}}^2\tau_{\mathrm{w}}^2}{\sqrt{\omega_{\mathrm{w}}^2\tau_{\mathrm{w}}^2-1/4}} \approx \omega_{\mathrm{w}}\tau_{\mathrm{w}} \quad (\text{for}\ \omega_{\mathrm{w}}\tau_{\mathrm{w}} \gg 1). \tag{6.95}$$

Note that the denominator of (6.95) is also the condition for the onset of critical damping of the unforced oscillator (at $H_{\mathrm{ac}} = 0$). Above this frequency, the material assumes single-domain character and the advantages to permeability from wall movement are lost. Experimental evidence of this transition is already revealed in

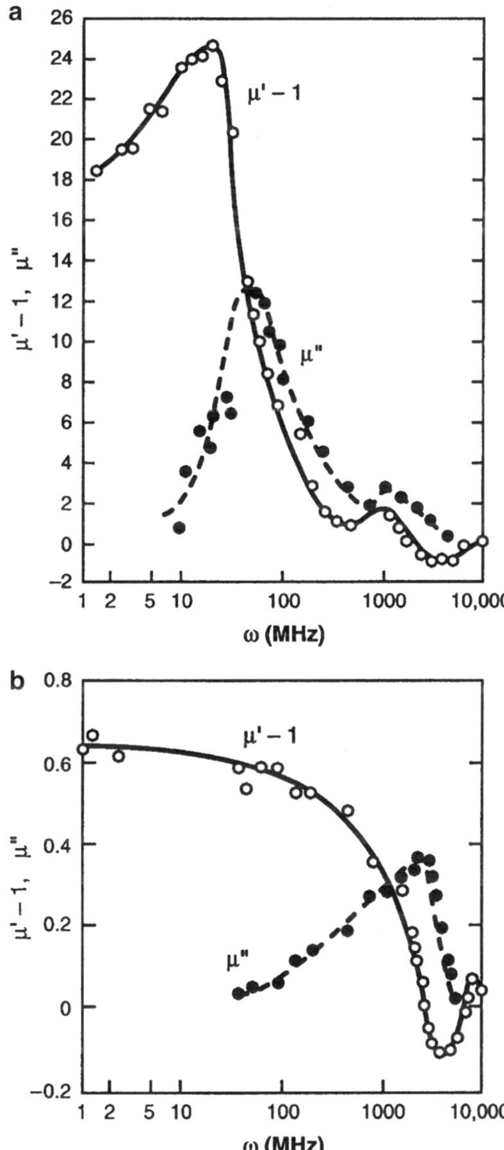

Fig. 6.27 Permeability spectra contrasting (**a**) a solid multidomain magnesium ferrite-based ceramic with (**b**) a 70% (by weight) mixture of single-domain particles of the same compound dispersed in wax. Original data are from Rado et al. [82]. Figure reprinted from [80] with permission. © 2003 by the IEEE

the early data of Rado et al. [82] shown in Fig. 6.27, where single- and multiple-domain behavior are contrasted. In either case, resonance from transverse components is reached at the microwave bands (>1 GHz), which can sometimes overlap the longitudinal dispersion region if ω_τ is large enough.

6.4.2 High-Frequency Transverse Limits

Where gyromagnetic coupling produces the effects described, their appearance in partially magnetized media generally offers an impediment to practical usage due to the absorptive properties. Because the spectral nature of the interaction is unspecific, only the frequency limits of the regimes are of interest. For the cubic systems, this range is defined by

$$\gamma H_K < \omega_r < \gamma \left(H_K + 4\pi M_s\right). \tag{6.96}$$

The resonance frequency from the magnetoelastic fields is the lower limit of the multidomain absorption band, also known as the "low-field" loss region in the microwave lexicon. From Table 5.7 for $K_1 < 0$ with <111> easy axes, the angular resonance frequency is expressed as

$$\omega_r = \gamma H_K = \gamma \frac{4\left|K_1 + \sigma\lambda_s\right|}{3M_s} \approx \gamma \frac{4\left|K_1\right|}{3M_s} \quad \text{for}\ \sigma \sim 0. \tag{6.97}$$

To this point in the discussion dynamic interactions, only materials of cubic lattice symmetry have been included. From the earliest work in this area it was realized that ω_r could be increased by the use of high-anisotropy hexagonal ferrites. The most common of this family is the uniaxial (Ba, Sr) $Fe_{12}O_{19}$ ($\boldsymbol{M}$-type) that features a c-axis polar anisotropy field $H_{K\theta} \sim 20$ kOe. In polycrystalline form, this system is categorized as a "hard" magnet because of its large coercive fields that result from $H_{K\theta}$ $(= gK_{u1}/M_s)$. As a consequence, the longitudinal drive field necessary to cause wall movement is too great to allow any wall movement influence on μ_i'. Beyond permanent magnets, these materials are of interest for gyromagnetism in the millimeter-wave bands.

Above the frequency range of high longitudinal susceptibility, microwave properties come into focus. The gyromagnetic spectral transmission properties of ferrites based on the imaginary part of the susceptibility χ are sketched in Fig. 6.28. To the microwave device engineer, the issues are straightforward: the operating frequency must be well out of the absorption or "low-field loss" regime. This means that $\omega > \gamma\left(H_K + 4\pi M_s\right)$ is a critical design criterion for partially magnetized ferrites. However, if the medium is magnetically saturated, ω can be above or below the resonance frequency without incurring significant absorption losses. A special case of this effect can be realized if uniaxial hexaferrite is utilized with its large $H_{K\theta}$ (or H_{Ku}) as a self-biasing field. As indicated in Fig. 6.28, low-loss transverse susceptibility can be accessed either from below or above the resonance region as was demonstrated in self-biased circulator devices at 31 and 73.5 GHz [83, 84].

Another form of hexagonal ferrite has been utilized to extend the dispersion regime to frequencies above the range of cubic ferrites. In Fig. 6.29, the χ spectra for a cubic NiZn spinel ferrite and the "easy-plane" Co_2Z hexagonal system ($Ba_3Co_2Fe_{24}O_{41}$) are compared. The higher frequency limit of the hexagonal system can be explained as follows: in an easy-plane structure, the large $H_{K\theta}$ forces the $\boldsymbol{M}_s$ vectors away from the c-axis and into the plane where they are subject to

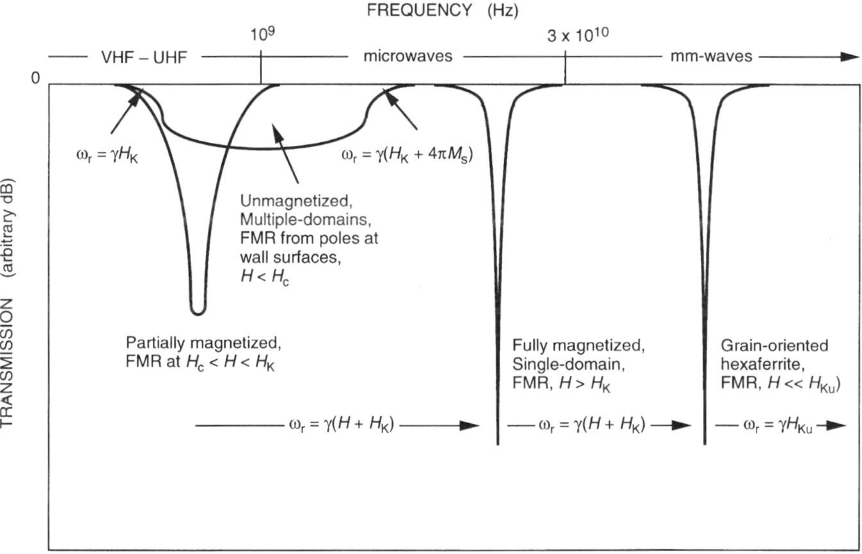

Fig. 6.28 Spectral model of gyromagnetic interaction indicating regions of transmission and absorption. Figure reprinted from [80] with permission. © 2003 by the IEEE

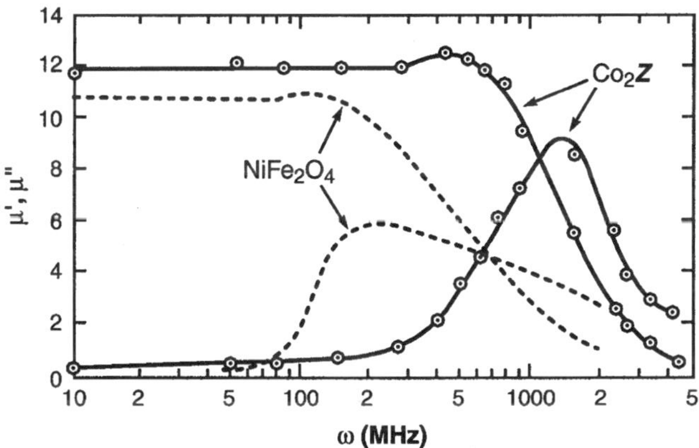

Fig. 6.29 Comparison of permeability dispersion for cubic nickel spinel ferrite and hexagonal cobalt *Z* ferrite [80]. Figure reprinted from [80] with permission © 2003 by the IEEE

the smaller in-plane azimuthal anisotropy field $H_{K\phi}$ ($\ll H_{K\theta}$) field. As a result the resonance frequency for a single-domain platelet following the Kittel theory [19] discussed in Chap. 5 can be written as

$$\gamma\sqrt{H_{K\theta}H_{K\varphi}} < \omega_r < \gamma\sqrt{(H_{K\theta} + 4\pi M_s)\left(H_{K\varphi} + 4\pi M_s\right)}. \qquad (6.98)$$

Table 6.5 Initial susceptibility frequency limits

Mechanism	$\chi_i' = \mu_i' - 1$	$\omega^{\min}$	$\omega^{\max}$
Domain rotation	$2\pi M_s^2/\lvert K_1\rvert$ $(K_1 < 0)$ $4\pi M_s^2/3\lvert K_1\rvert$ $(K_1 > 0)$	–	$\omega_\tau = 1/\tau$
Domain wall movement	$32\pi M_s^2/3\lvert K_1\rvert$ $(K_1 < 0)$ $16\pi M_s^2/\lvert K_1\rvert$ $(\lvert K_1\rvert > 0)$	–	$\left[\omega_w^2 - 1/\left(2\tau_w^2\right)\right]^{1/2}$
Gyromagnetic (cubic)	$4\pi M_s/H_K^a$	γH_K	$\gamma(H_K + 4\pi M_s)$ (multidomain) $\gamma[H_K(H_K + 4\pi M_s)]^{1/2}$ (single-domain, planar)
Gyromagnetic (hexagonal)	$4\pi M_s/H_{K\theta}$ (polar, c-axis) $4\pi M_s/H_{K\phi}$ (azimuthal, in plane)	$\gamma H_{K\theta}$ $\gamma\left(H_{K\theta}H_{K\phi}\right)^{1/2}$	$\gamma(H_{K\theta} + 4\pi M_s)$ $\gamma[(H_{K\theta} + 4\pi M_s) \times \left(H_{K\phi} + 4\pi M_s\right)]^{1/2}$

[a]In the present context, H_K represents the effective maximum anisotropy field for the particular physical situation. As discussed in the foregoing text, it is proportional to the ratio of K_1/M_s

In comparing the magnitude of this range with that of the cubic case, one can deduce that the entire ω_r range can shift to higher frequencies by a ratio on the order of $\left(H_{K\theta}/H_{K\phi}\right)^{1/2}$, which can be at least a factor of ten in most cases, as suggested by the measurement results in Fig. 6.29. A listing of the relevant relations for permeability and the various frequency limits may be found in Table 6.5.

6.4.3 Snoek's Law Considerations

When (6.90) and (6.97) are combined into a product, the relation known as Snoek's law for unmagnetized materials is obtained [85]:

$$\omega_r^{\min}\mu_i' \approx \frac{2}{3}\gamma 4\pi M_s. \tag{6.99}$$

The implications of this relation for ac properties below the dispersion region where most applications of ferrites are found can be seen graphically from Snoek's data for the NiZn spinel ferrite family shown in Fig. 6.30. According to (6.99), Snoek's limit for an isotropic specimen can be characterized simply by the value of $4\pi M_s$. Consequently, the traditional strategy is to seek the highest magnetization. The room temperature $4\pi M_s$ limit of high-resistivity ferrites appears to be about 5,000 Gauss (although reports of Fe^{2+}-laden Fe_3O_4 $(1 - x)$ + $ZnFe_2O_4$ (x) combinations with room temperature $4\pi M_s > 7{,}000$ Gauss have been published [86]). One obvious approach would be to lower the operating temperature, for which the spinel models discussed in Appendix 4B of Chap. 4 could be applied. In addition, the garnet $Gd_3Fe_5O_{12}$ also offers high magnetizations if temperatures were reduced to the liquid helium range, i.e., 4 K.

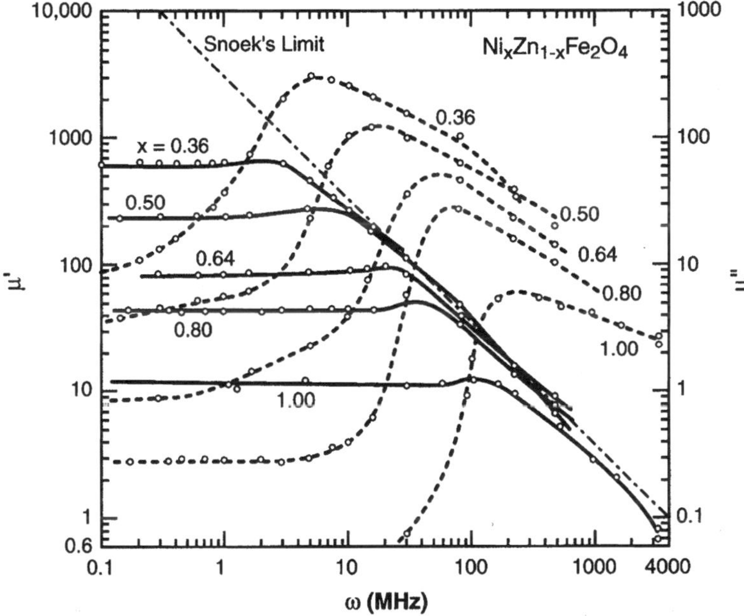

Fig. 6.30 Comparison of permeability dispersion for a range of members of the $Ni_\delta Zn_{1-\delta}Fe_2O_4$ family, illustrating the origin of Snoek's law. Original data are from Gorter [85]. Figure reprinted from [80] with permission. © 2003 by the IEEE

As pointed out above, the single-domain μ_i' can be greatly enhanced by wall movement. However, practical considerations dictate that only domain rotation can be active at the higher frequencies where gyromagnetic effects occur. For spinel ferrites with $4\pi M_s = 5{,}000$ Gauss, $\omega_r^{min}\mu_i'$ translates into 10 GHz for a system of single-domain particles. Therefore, to obtain a permeability in the unmagnetized state of at least 10, for example, the maximum frequency of operation is about 1 GHz, even without considering the losses from absorption.

Caution should be exercised when Snoek's law is used to estimate the upper limit of frequency for acceptable permeability in traditional nonresonant applications. As stated, the frequency limit ω_r^{min} is derived from transverse coupling which is configurationally independent of μ_i'. One must therefore recognize that the relation's validity can be justified only if the dispersion edge at $\omega_\tau \left(= \tau^{-1}\right)$ exceeds the gyromagnetic lower limit of $\omega_r^{min} = \gamma H_K$. This apparent contrivance can be justified if it is recognized that τ^{-1} also varies as H_K through their common dependence on the ratio of spin–orbit coupling to low-symmetry crystal-field splitting energies characterized by λ/Δ that appears in (6.36) from Mattuck and Strandberg [12], and in the analysis of anisotropic g-factors by Dionne [87] in Sect. 5.1.1. As a consequence, Snoek's law should be used with the caveat that the upper frequency limit be ω_τ or ω_r^{min}, whichever is smaller.

Figure 6.31 presents permeability data that illustrate how the variation in K_1 from the substitution of fast-relaxing Co^{2+} ions in $(NiZnFe)_{3-x}Co_xO_4$ influences the frequency limits [59]. At $x = 0.03$, the positive contribution of Co^{2+} to K_1 that off-

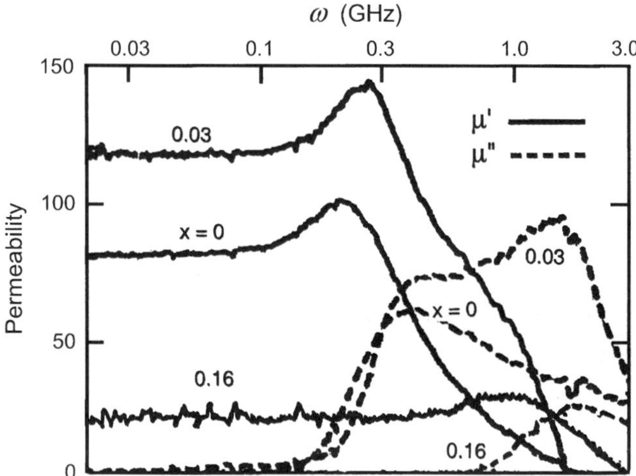

Fig. 6.31 Permeability spectra of Co^{2+}-substituted NiZn spinel ferrite indicating the influence of positive anisotropy contributions to the relaxation and anisotropy field effects on the frequency limits. Original data are from [59]. Figure reprinted from [80] with permission. © 2003 by the IEEE

sets the negative anisotropy field, thereby allowing μ_i to increase, while introducing some local gyromagnetic effects at higher frequencies. When x is raised to 0.16, K_1 is strongly positive, causing μ_i to drop (possibly extinguishing domain-wall movement), while defining a higher frequency range of gryomagnetism because both the high K_1 and coincidentally the faster τ^{-1} rate. Note also that the upper limit remains dominated by the value of $4\pi M_s$.

For hexaferrite specimens of planar geometry, provided that they are single-domain or comprise crystallographically oriented single-domain particles with magnetization vectors and domain rotation is confined to the plane, Snoek's law can be modified to produce a more favorable frequency limit. Based on the expressions for μ_i' and ω_{min} listed in Table 6.5:

$$\omega_r^{min}\mu_i' = \gamma 4\pi M_s \sqrt{\frac{H_{K\theta}}{H_{K\phi}}}. \tag{6.100}$$

For Co_2Z ferrite, $H_{K\theta} = 13{,}000$ Oe, $H_{K\phi} = 112$ Oe, and $4\pi M_s = 3{,}350$ Gauss. Therefore, the Snoek's frequency limit should be increased by a factor of about ten in comparison to a cubic Ni spinel ferrite of equivalent magnetization. The curves in Fig. 6.31 support the reasonable accuracy of this model.

A qualitative summary of this analysis is presented schematically in Fig. 6.32, where it is suggested that the tradeoff between permeability and high-frequency operations is also dependent on the spin–lattice relaxation time. Low-wall damping (large τ_w and ς_w) promises superior low-frequency properties, but with the disadvantage of an earlier onset of dispersion in multiple-domain specimens. Above the

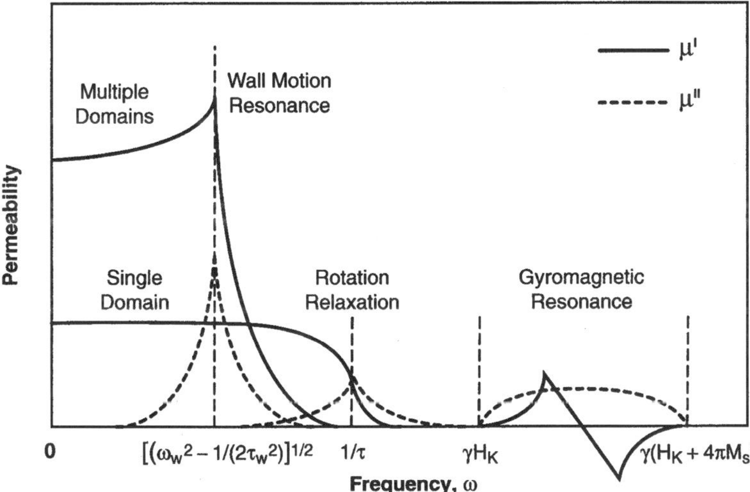

Fig. 6.32 Schematic diagram of permeability variations as a function of frequency caused by various magnetic damping actions. Figure reprinted from [80] with permission. © 2003 by the IEEE

domain wall damping edge, the reduced permeability of single-domain rotation is still present provided that $\omega_\tau > \omega_d$. Amidst these conflicts lies the influence of K_1 which is also dependent on ω_τ. Large τ usually means smaller K_1, higher μ_i, and the earlier appearance of gyromagnetic loss effects because of a reduced value of the anisotropy field H_K.

In conclusion, the issue of domains walls for unmagnetized materials can be resolved into a simple rule: multiple domains, low K_1, high τ for low frequencies; single domain, high K_1, low τ for high frequencies. For microwaves, where ω is above the dispersion region, the upper edge of the gyromagnetic resonance region becomes a concern, such that the low K_1, high τ combination can be preferred, unless very high $H_{K\theta}$ hexaferrites are desired for millimeter waves.

6.4.4 Circular Polarization and Nonreciprocal Properties

To begin the discussion of microwave propagation in magnetically ordered systems (ferrites), we must first point out that only one circular component of the z-axis-directed rf signal has been considered to this point. In Chap. 1, it was recognized that only the mode that is in phase with the Larmor precession can produce the continuous rotation torque necessary for magnetic resonance absorption. However, the counterrotating or anti-Larmor mode (termed negative by convention) is also important in rf propagation, and a relation for its susceptibility needs to be established. The complex susceptibility tensor is derived in Appendix 6A. For our immediate purposes, a shortcut to the relations for the two modes $\chi_{\pm} = \chi'_{\pm} - i\chi''_{\pm}$ can be arrived at if we presume the anti-Larmor solution to be analogous to (6.54), but with the vector direction (sign) of ω reversed. For off-resonance conditions, approximations to the two modes can be expressed as

$$\chi'_{+} \cong \gamma M \left\{ \frac{\gamma H_{\mathrm{r}} - \omega}{(\gamma H_{\mathrm{r}} - \omega)^2 + 4(\Delta\omega)^2} \right\}$$

$$\chi''_{+} \cong \gamma M \left\{ \frac{2\Delta\omega}{\gamma(\gamma H_{\mathrm{r}} - \omega)^2 + 4(\Delta\omega)^2} \right\} \quad \text{(Larmor)}, \tag{6.101}$$

$$\chi'_{-} \cong \gamma M \left\{ \frac{\gamma H_{\mathrm{r}} + \omega}{(\gamma H_{\mathrm{r}} + \omega)^2 + 4(\Delta\omega)^2} \right\}$$

$$\chi''_{-} \cong \gamma M \left\{ \frac{2\Delta\omega}{(\gamma H_{\mathrm{r}} + \omega)^2 + 4(\Delta\omega)^2} \right\} \quad \text{(anti-Larmor)}, \tag{6.102}$$

where the bracketed factors represent the resonance line-shape functions that modify the dc susceptibility M/H_{r} at the resonance field. The rf subscripts have been dropped from χ. The frequency variation of the two modes can be appreciated by the graphical presentation in Fig. 6.33. In effect, the permeability of the magnetic medium can be radically different for the two senses of circular polarization, and the corresponding effects on propagation are the origin of nonreciprocity.

From Appendix 6A, the rf susceptibility tensor is written as

$$[\chi] = \begin{bmatrix} \chi_{xx} & \chi_{xy} & 0 \\ \chi_{yx} & \chi_{yy} & 0 \\ 0 & 0 & 0 \end{bmatrix}, \tag{6.103}$$

from which a permeability tensor can be constructed:

$$[\mu] = 1 + 4\pi[\chi] = \begin{bmatrix} \mu & -i\kappa & 0 \\ +i\kappa & \mu & 0 \\ 0 & 0 & 1 \end{bmatrix}. \tag{6.104}$$

where $\mu = 1 + 4\pi\chi_{xx,yy}$, and $\kappa = i4\pi\chi_{xy} = i4\pi\chi_{yx}$. The respective scalar permeability solutions for the two circular polarization plane-wave modes $(1/2)\left(H^{x}_{\mathrm{rf}} \mp H^{y}_{\mathrm{rf}}\right)$ are the eigenvalues of the determinant, which can be seen by inspection to be $\mu_{\pm} = \mu \pm \kappa$ for the $x - y$ plane after diagonalization is carried out. The desired relations are then obtained from (6.102) and (6.103) as

$$\mu'_{\pm} = 1 + \gamma 4\pi M \left\{ \frac{\gamma H_{\mathrm{r}} \mp \omega}{(\gamma H_{\mathrm{r}} \mp \omega)^2 + 4(\Delta\omega)^2} \right\},$$

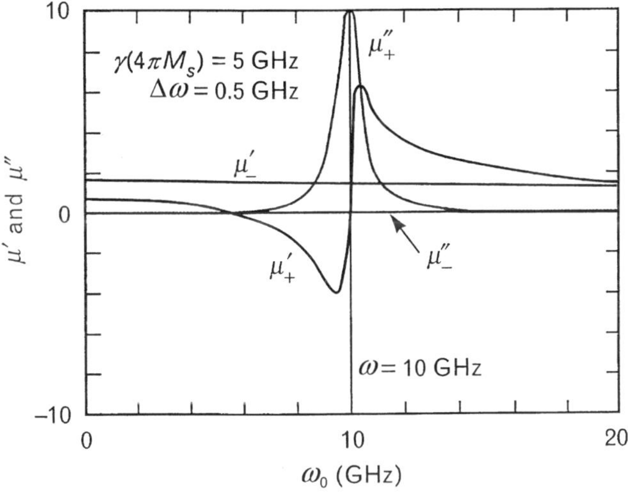

Fig. 6.33 Sketch of the real and imaginary parts of the ferrite complex permeability at an operating frequency $\omega = 10\,\text{GHz}$, illustrating the nonreciprocal features that arise from the different solutions for the two circular modes of a linearly polarized plane wave identified by the $+$ and $-$ subscripts. The resonance frequency is varied by the internal magnetic field according to $\omega_0 = g m_\text{B} H_0$. Note that the differential between μ'_+ and μ'_- is smaller when $\omega_0 > \omega$

$$\mu''_\pm = \gamma 4\pi M \left\{ \frac{2\Delta\omega}{(\gamma H_\text{r} \mp \omega)^2 + 4\,(\Delta\omega)^2} \right\}. \tag{6.105}$$

The Maxwell equations for the respective magnetic and dielectric fields for a fixed value of z are

$$\begin{aligned} \nabla \times \boldsymbol{H} &= -\varepsilon \frac{\partial \boldsymbol{E}}{\partial t} = \text{i}\omega\varepsilon \boldsymbol{E} \\ \nabla \times \boldsymbol{E} &= -\mu \frac{\partial \boldsymbol{H}}{\partial t} = -\text{i}\omega\mu \boldsymbol{H}, \end{aligned} \tag{6.106}$$

from which $\boldsymbol{E}$ can be eliminated to produce

$$\nabla \times \nabla \times \boldsymbol{H} = \omega^2 \varepsilon\,[\mu] \cdot \boldsymbol{H}. \tag{6.107}$$

A general analytical solution for this expression can be found in Lax and Button [88]. For the transverse electric and magnetic plane wave (TEM) in a semi-infinite medium, the permeability enters the propagation constant γ_p. In standard electromagnetic notation $\gamma_\text{p} = \alpha_\text{p} + i\beta_\text{p}$ in terms of the attenuation and phase constants, respectively. In a linearly polarized plane wave, the two equal counterrotating circular modes of initial amplitude $H_\text{rf}/2$ are depicted in Fig. 6.34. As a function of time and distance, the rf field vector (confined in the x–y plane) can be expressed in terms of its components,

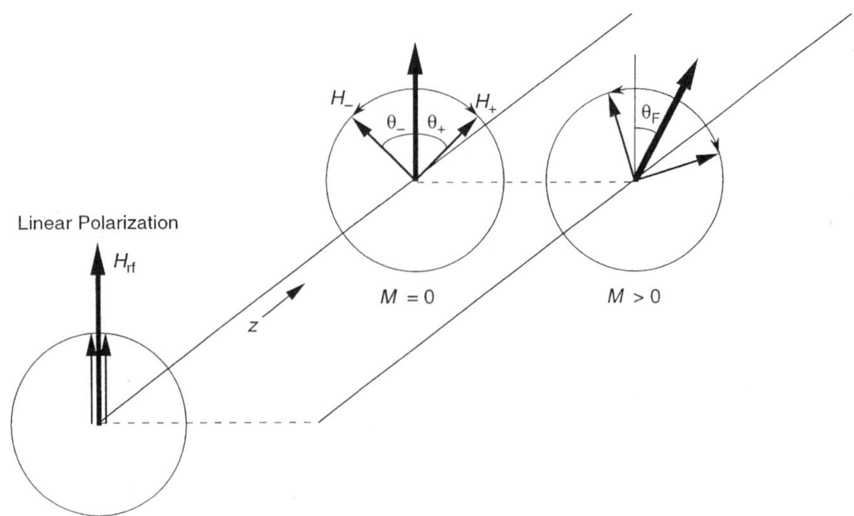

Fig. 6.34 Tutorial diagram illustrating the origin of Faraday rotation of the rf signal axis of a linearly polarized plane wave under the influence of a magnetized medium with nonreciprocal propagation properties

$$\begin{aligned}\boldsymbol{H}_{\mathrm{rf}}(t,z) &= \frac{1}{2}\left(H_{\mathrm{rf}+}\boldsymbol{i} + H_{\mathrm{rf}-}\boldsymbol{j}\right) - \frac{1}{2}H_{\mathrm{rf}}\left(\mathbf{e}^{+\mathrm{i}\omega t-\gamma_{\mathrm{p}\pm}z}\boldsymbol{i} + \mathbf{e}^{-\mathrm{i}\omega t-\gamma_{\mathrm{p}\pm}z}\boldsymbol{j}\right)\\ &= \frac{1}{2}H_{\mathrm{rf}}\left[\mathbf{e}^{-\gamma_{\mathrm{p}+}z}\left(\cos\omega t i + \sin\omega t j\right) + \mathbf{e}^{-\gamma_{\mathrm{p}-}z}\left(\cos\omega t i - \sin\omega t j\right)\right].\end{aligned} \tag{6.108}$$

where $\cos\omega t = H_{\mathrm{rf}}^{x}/H_{\mathrm{rf}}$ and $\sin\omega t = H_{\mathrm{rf}}^{y}/H_{\mathrm{rf}}$. Accordingly, separate propagation constants $\gamma_{p\pm}$ are assigned to each circular polarization mode. By the usual convention, right-hand circular polarization (RHCP) is designated as the clockwise rotation $(+\omega)$ of the vector $(\boldsymbol{H}_{\mathrm{rf}+})$ when viewed along the direction of propagation $(+z)$. It is the RHCP mode that drives the Larmor precession of the magnetic moment into resonance.

The corresponding $\gamma_{\mathrm{p}\pm}$ for the case of both $\boldsymbol{H}$ and the propagation directed longitudinally to the z-axis can be treated as scalars and expressed in terms of the complex permeabilities [88] according to[3]

$$\gamma_{\mathrm{p}\pm}^{2} = \left(\alpha_{\mathrm{p}\pm} + \mathrm{i}\beta_{\mathrm{p}\pm}\right)^{2} = -\left(\frac{\omega}{\mathrm{c}}\right)^{2}\varepsilon\left(\mu \pm \kappa\right), \tag{6.109}$$

[3] Two other propagation situations have been analyzed in a medium of semiinfinite cross section, where $\boldsymbol{H}$ is directed perpendicular instead of parallel to the z-axis. In this case, however, the $\boldsymbol{H}_{\mathrm{rf}}$ of a TEM wave can vary from parallel to perpendicular with respect to $\boldsymbol{H}$. The former produces zero gyromagnetic interaction between $\boldsymbol{H}$ and $\boldsymbol{H}_{\mathrm{rf}}$ $\left(\mu_{\parallel} = 1\right)$, and the latter gives an averaged result $\mu_{\perp} \approx \frac{\mu^2-\kappa^2}{\mu}$.

from which

$$\alpha_{p\pm}^2 - \beta_{p\pm}^2 = -\left(\frac{\omega}{c}\right)^2 \varepsilon\mu'_\pm,$$
$$2\alpha_{p\pm}\beta_{p\pm} = \left(\frac{\omega}{c}\right)^2 \varepsilon\mu''_\pm \tag{6.110}$$

and

$$\alpha_{p\pm} = \left(\frac{\omega}{c}\right)\sqrt{\frac{\varepsilon}{2}}\left(\sqrt{\mu'^2_\pm + \mu''^2_\pm} - \mu'_\pm\right)^{1/2}, \tag{6.111}$$
$$\beta_{p\pm} = \left(\frac{\omega}{c}\right)\sqrt{\frac{\varepsilon}{2}}\left(\sqrt{\mu'^2_\pm + \mu''^2_\pm} + \mu'_\pm\right)^{1/2}.$$

For the important high-frequency case where $\omega \gg \gamma H_r$, $\mu''_\pm/\mu'_\pm \ll 1$, and (6.111) can be simplified to

$$\alpha_{p\pm} \approx \left(\frac{\omega}{c}\right)\frac{\sqrt{\varepsilon}}{2}\frac{\mu''_\pm}{\sqrt{\mu'_\pm}},$$
$$\beta_{p\pm} \approx \left(\frac{\omega}{c}\right)\sqrt{\varepsilon}\frac{\mu'_\pm}{\sqrt{\mu'_\pm}}. \tag{6.112}$$

It is important to examine the physical significance of the $\beta_{p\pm}$ parameter. Equation (6.109) expresses that each circular mode has a different propagation velocity provided that the medium is in a magnetized state. From (6.105) for $\omega \gg \gamma 4\pi M$ and γH_r (the frequency far above resonance) where absorption losses are negligible ($\Delta\omega \ll |\gamma H_r - \omega|$), $\alpha_{p\pm}$, and $\beta_{p\pm}$ can be simplified to

$$\alpha_{p\pm} \approx \frac{\omega\sqrt{\varepsilon}}{2c}\frac{2\omega_M\Delta\omega}{(\gamma H_r \mp \omega)^2} = \frac{\omega\sqrt{\varepsilon}}{c}\frac{(\gamma 4\pi M)\Delta\omega}{(\gamma H_r \mp \omega)^2}, \tag{6.113a}$$
$$\beta_{p\pm} \approx \frac{\omega\sqrt{\varepsilon}}{c}\left(1 + \frac{1}{2}\frac{\gamma 4\pi M}{\gamma H_r \mp \omega}\right) \approx \frac{\omega\sqrt{\varepsilon}}{c}\left(1 \mp \frac{1}{2}\frac{\gamma 4\pi M}{\omega}\right) \tag{6.113b}$$

and the corresponding propagation velocities are then

$$v_{p\pm} \approx \frac{\omega}{\beta_{p\pm}} \approx \frac{c}{\sqrt{\varepsilon}}\left(1 \pm \frac{1}{2}\frac{\gamma 4\pi M}{\omega}\right). \tag{6.114}$$

For a fixed $\boldsymbol{M}$ vector, a circularly polarized wave will produce a phase shift $\phi_\pm = \beta_{p\pm}z$ that is either delayed or advanced relative to the empty space value, depending on the sign of the mode. In further contrast to nonmagnetic media, the phase shift of a TEM wave propagating parallel to the $\boldsymbol{M}$ vector is dependent on its direction, which means that if the signal traverses the medium and is then reflected back to its point of origin along the same path with polarization unaltered, the phase change that occurred initially will not be recovered (unless a reversal of the $\boldsymbol{M}$ direction takes place). This feature of rf propagation is unique in a magnetized medium and is termed "nonreciprocal." Because the velocity of either mode will change to that of the opposite mode if the magnetization direction is reversed, a "differential phase shift" $\Delta\phi$ $(=\phi_- - \phi_+)$ of varying amount can be obtained by switching the magnetization within the full range of $-M_s \leq M \leq M_s$, so that [89]

$$\Delta\varphi = \left(\beta_{p-} - \beta_{p+}\right) z \approx \frac{\sqrt{\varepsilon}}{2c}\gamma 4\pi \left(M_- - M_+\right) z = \frac{\sqrt{\varepsilon}}{c}\gamma 4\pi \left(\Delta M\right) z, \qquad (6.115)$$

since $0 \leq \Delta M \leq 2M_s$ from (6.114), in the context of nonreciprocity.

The unique propagation properties of circular polarization form the basis of microwave devices that can produce adjustable phase shift through choice of frequency and magnetic field by moving the magnetization to selected points on the hysteresis loop. Other applications rely on the presence of both circular modes of a linearly polarized wave that were introduced for the discussion of the classical model of paramagnetic resonance in Chap. 1. Linear polarization in magnetic media has found important applications at optical as well as radio frequencies.

6.4.5 Linear Polarization and Faraday Rotation

Mentioned initially in Chap. 1, linearly polarized waves comprise both circular modes, and are also used in applications where nonreciprocal propagation is required. Because β_{p-} and β_{p+} can have different magnitudes, a rotation of the polarization axis can occur for the same configuration of dc field and propagation axis, as depicted in Fig. 6.34. The resultant $\boldsymbol{H}_{rf}$ of the counterrotating $\boldsymbol{H}_{rf+}$ (RHCP) and $\boldsymbol{H}_{rf-}$ (LHCP) vectors about is rotated through an angle that is half of the differential phase shift of (6.115), given by

$$\theta_F = \frac{\Delta\varphi}{2} = \frac{\left(\beta_{p-} - \beta_{p+}\right)}{2} \approx \frac{\sqrt{\varepsilon}}{2c}\omega_M = \frac{\sqrt{\varepsilon}}{2c}\left(\gamma 4\pi M\right), \qquad (6.116)$$

where θ_F is the Faraday rotation, here expressed as an angle per unit length. Note that θ_F is dependent on the magnetization, but *independent* of the frequency (or wavelength) in this approximation. The corresponding difference in the absorption constant is a measure of the "ellipticity" that results from the unequal amplitudes of $\boldsymbol{H}_{rf+}$ and $\boldsymbol{H}_{rf-}$. It is generally overlooked in rf situations away from resonance, but can be readily derived from (6.113a):

$$\eta_F = \frac{\left(\alpha_{p-} - \alpha_{p+}\right)}{2} = \frac{\sqrt{\varepsilon}}{4c}\frac{4\omega^2\omega_0\omega_M 2\Delta\omega}{\left(\omega_0^2 - \omega^2\right)^2} \approx \frac{2\sqrt{\varepsilon}}{c}\omega_0\left(\frac{\Delta\omega}{\omega}\right)\left(\frac{\omega_M}{\omega}\right). \qquad (6.117)$$

A quasioptical three-port microwave Faraday rotation isolator/circulator was demonstrated at 35 GHz [90]. After exiting a polarizer plate that defines a linearly polarized beam, the axis of polarization is first rotated 45° as it passes through an axially magnetized ferrite medium designed thickness. After reflection, the return beam undergoes an additional 45° rotation when it traverses the ferrite for the second time. Because of the nonreciprocal property of the magnetized ferrite, the polarization axis is then orthogonal to the axis of the polarizer plate and is deflected, thereby accomplishing its isolation function of protecting the signal source and directing the return signal into a receiver channel when desired.

In Chap. 7, this topic is expanded to include quantum mechanical transitions. At wavelengths in the far-infrared region, resonant permeability effects occur from exchange field interactions. In the visible and ultraviolet bands, electrical permittivity frequency spectra arise from selected electric dipole transitions from spin–orbit multiplet structure.

Appendix 6A Transverse Permeability Tensor

The Polder–Smit tensor that relates rf magnetization with transverse rf magnetic field $H_{\mathrm{rf}} = H_{\mathrm{rf}}^{x}\boldsymbol{i} + H_{\mathrm{rf}}^{y}\boldsymbol{j}$ in a dc field $\boldsymbol{H}$ along the z-axis can be expressed in general form as $\boldsymbol{M}_{\mathrm{rf}} = [\chi]\,\boldsymbol{H}_{\mathrm{rf}}$, or

$$\begin{pmatrix} M_{\mathrm{rf}}^{x} \\ M_{\mathrm{rf}}^{y} \\ M_{\mathrm{rf}}^{z} \end{pmatrix} = \begin{bmatrix} \chi_{xx} & \chi_{xy} & 0 \\ \chi_{yx} & \chi_{yy} & 0 \\ 0 & 0 & 0 \end{bmatrix} \begin{pmatrix} H_{\mathrm{rf}}^{x} \\ H_{\mathrm{rf}}^{y} \\ 0 \end{pmatrix}. \tag{6.118}$$

In terms of the corresponding rf permeability, the tensor can be written as

$$[\mu] = \begin{bmatrix} \mu & -\mathrm{i}\kappa & 0 \\ +\mathrm{i}\kappa & \mu & 0 \\ 0 & 0 & 1 \end{bmatrix}. \tag{6.119}$$

For Larmor-related effects of a plane-wave propagating along the z-axis, the components transverse to the z-axis comprise the RHCP and LHCP circular polarization modes $(1/2)\left(H_{\mathrm{rf}}^{x} \mp \mathrm{i}H_{\mathrm{rf}}^{y}\right)$. Standard solutions of the precession equation of motion

$$\left(\frac{\mathrm{d}\boldsymbol{M}}{\mathrm{d}t}\right) = \gamma\left(\boldsymbol{M} \times \boldsymbol{H}_{\mathrm{i}}\right) \tag{6.120}$$

in a semi-infinite medium and without damping were derived as [91–93]

$$\begin{aligned} \chi_{xx} &= \chi_{yy} = \omega_{\mathrm{M}}\left(\frac{\omega_{\mathrm{r}}}{\omega_{\mathrm{r}}^{2} - \omega^{2}}\right), \\ \chi_{xy} &= -\chi_{yx} = -\mathrm{i}\omega_{\mathrm{M}}\left(\frac{\omega}{\omega_{\mathrm{r}}^{2} - \omega^{2}}\right), \end{aligned} \tag{6.121}$$

where $\omega_M = \gamma(4\pi M)$ and $\omega_r = \gamma H_r$, guided where applicable by the various geometric and anisotropic situations described in Sect. 6.2.2. The following solutions are approximations for the case of a semi-infinite medium in which demagnetizing influences on the rf signal are ignored. Effects of demagnetization on M_{rf} are examined in a model outlined in Lax and Button [91].

For the inclusion of damping with (6.120), the two commonly considered versions are the Gilbert (G) form of the Landau–Lifshitz term from (6.58), and the Bloch–Bloembergen (B–B) function listed, respectively, as

$$\left(\frac{\mathrm{d}\boldsymbol{M}}{\mathrm{d}t}\right) \cong \gamma(\boldsymbol{M} \times \boldsymbol{H}) - \frac{\alpha}{M}\left(M \times \frac{\mathrm{d}M}{\mathrm{d}t}\right) \quad [\mathrm{G}] \tag{6.122}$$

and

$$\frac{\mathrm{d}\boldsymbol{M}}{\mathrm{d}t} \approx \gamma(\boldsymbol{M} \times H) - \frac{\boldsymbol{M}_z - \boldsymbol{M}}{\tau_1} \quad [\mathrm{B-B}] \tag{6.123}$$

From (6.122), the Gilbert relations with damping parameter α follow if we replace ω_r with $\omega_r + \mathrm{i}\omega\alpha$ (where $\omega\alpha = 1/\tau_1$) (6.121) to obtain for the complex $\chi_{xx} = \chi'_{xx} - \mathrm{i}\chi''_{xx}$

$$\chi'_{xx} = \omega_M \left\{ \frac{\omega_r\left[\omega_r^2 - \omega^2\left(1-\alpha^2\right)\right]}{\left[\omega_r^2 - \omega^2\left(1+\alpha^2\right)\right]^2 + 4\omega_r^2\omega^2\alpha^2} \right\},$$

$$\chi''_{xx} = \omega_M \left\{ \frac{\omega\alpha\left[\omega_r^2 + \omega^2\left(1+\alpha^2\right)\right]}{\left[\omega_r^2 - \omega^2\left(1+\alpha^2\right)\right]^2 + 4\omega_r^2\omega^2\alpha^2} \right\} \tag{6.124}$$

and for $\chi_{xy} = -\left(\chi''_{xy} + \mathrm{i}\chi'_{xy}\right)$, chosen to comply with the sign convention for $\chi_{xy} = -\mathrm{i}\kappa$ [91],

$$\chi'_{xy} = \omega_M \left\{ \frac{\omega\left[\omega_r^2 - \omega^2\left(1+\alpha^2\right)\right]}{\left[\omega_r^2 - \omega^2\left(1+\alpha^2\right)\right]^2 + 4\omega_r^2\omega^2\alpha^2} \right\},$$

$$\chi'_{xy} = \omega_M \left\{ \frac{2\omega_r\omega^2\alpha}{\left[\omega_r^2 - \omega^2\left(1+\alpha^2\right)\right]^2 + 4\omega_r^2\omega^2\alpha^2} \right\}. \tag{6.125}$$

Conformance to the permeability elements from (6.119) can then be established as $\mu = \mu' - \mathrm{i}\mu''$, and $\kappa = \kappa' - \mathrm{i}\kappa''$,

$$\mu' = 1 + \omega_M \left\{ \frac{\omega_r\left[\omega_r^2 - \omega^2\left(1-\alpha^2\right)\right]}{\left[\omega_r^2 - \omega^2\left(1+\alpha^2\right)\right]^2 + 4\omega_r^2\omega^2\alpha^2} \right\},$$

$$\mu'' = \omega_M \left\{ \frac{\omega\alpha\left[\omega_r^2 + \omega^2\left(1+\alpha^2\right)\right]}{\left[\omega_r^2 - \omega^2\left(1+\alpha^2\right)\right]^2 + 4\omega_r^2\omega^2\alpha^2} \right\} \tag{6.126}$$

and

$$\kappa' = \omega_{\rm M} \left\{ \frac{\omega\left[\omega_{\rm r}^2 - \omega^2\left(1+\alpha^2\right)\right]}{\left[\omega_{\rm r}^2 - \omega^2\left(1+\alpha^2\right)\right]^2 + 4\omega_{\rm r}^2\omega^2\alpha^2} \right\},$$

$$\kappa'' = \omega_{\rm M} \left\{ \frac{2\omega_{\rm r}\omega^2\alpha}{\left[\omega_{\rm r}^2 - \omega^2\left(1+\alpha^2\right)\right]^2 + 4\omega_{\rm r}^2\omega^2\alpha^2} \right\}. \tag{6.127}$$

For the B–B model, damping is introduced into (6.121) by the substitution of $\omega - {\rm i}\,(1/\tau_1)$ for ω instead of $\omega_{\rm r} + {\rm i}\,(1/\tau_1)$. The permeability relations are

$$\mu'' = 1 + \omega_{\rm M} \left\{ \frac{\omega_{\rm r}\left(\omega_{\rm r}^2 - \omega^2 + 1/\tau_1^2\right)}{\left(\omega_{\rm r}^2 - \omega^2 + 1/\tau_1^2\right)^2 + 4\omega^2/\tau_1^2} \right\},$$

$$\mu'' = \omega_{\rm M} \left\{ \frac{2\omega_{\rm r}\omega/\tau_1}{\left(\omega_{\rm r}^2 - \omega^2 + 1/\tau_1^2\right)^2 + 4\omega^2/\tau_1^2} \right\} \tag{6.128}$$

and

$$\kappa' = \omega_{\rm M} \left\{ \frac{\omega\left(\omega_{\rm r}^2 - \omega^2 - 1/\tau_1^2\right)}{\left(\omega_{\rm r}^2 - \omega^2 + 1/\tau_1^2\right)^2 + 4\omega^2/\tau_1^2} \right\},$$

$$\kappa'' = \omega_{\rm M} \left\{ \frac{\left(\omega_{\rm r}^2 + \omega^2 + 1/\tau_1^2\right)/\tau_1}{\left(\omega_{\rm r}^2 - \omega^2 + 1/\tau_1^2\right)^2 + 4\omega^2/\tau_1^2} \right\}. \tag{6.129}$$

For the two counterrotating modes of circular polarization, we define $\mu_\pm = \mu'_\pm - {\rm i}\mu''_\pm = \mu \pm \kappa$, with the Gilbert result

$$\mu'_\pm = 1 + \omega_{\rm M} \left\{ \frac{(\omega_{\rm r} \mp \omega)}{(\omega_{\rm r} \mp \omega)^2 + 4\,(\omega\alpha)^2} \right\},$$

$$\mu''_\pm = \omega_{\rm M} \left\{ \frac{2\omega\alpha}{(\omega_{\rm r} \mp \omega)^2 + 4\,(\omega\alpha)^2} \right\}. \tag{6.130}$$

Solutions equivalent to those of the B–B model in terms of a relaxation time τ_1 can be obtained by substitution for the dimensionless damping parameter $\alpha = (\omega\tau_1)^{-1}$ or $\Delta\omega/\omega$ in terms of half-linewidth $(\Delta\omega)$, according to

$$\mu'_\pm = 1 + \omega_{\rm M} \left\{ \frac{(\omega_{\rm r} \mp \omega)}{(\omega_{\rm r} \mp \omega)^2 + 4\,(\Delta\omega)^2} \right\},$$

$$\mu''_\pm = \omega_{\rm M} \left\{ \frac{2\Delta\omega}{(\omega_{\rm r} \mp \omega)^2 + 4\,(\Delta\omega)^2} \right\}. \tag{6.131}$$

In another investigation, a term was added to the B–B model to make it more compatible with the G version, provided that $\alpha = (\omega_r \tau_1)^{-1} = \Delta\omega/\omega_r$ [94]

$$\frac{\mathrm{d}\boldsymbol{M}}{\mathrm{d}t} \cong \gamma\,(\boldsymbol{M} \times \boldsymbol{H}) - \frac{\boldsymbol{M}_z - \boldsymbol{M}}{\tau_1} + \left|\frac{\boldsymbol{M}}{\boldsymbol{H}}\right| \frac{\boldsymbol{H}}{\tau_1}. \tag{6.132}$$

In this case, the circular polarization permeability components are expressed as

$$\mu'_{\pm} = 1 + \omega_{\mathrm{M}} \left\{ \frac{(\omega_{\mathrm{r}} \mp \omega)}{(\omega_{\mathrm{r}} \mp \omega)^2 + 4\left(\frac{\omega}{\omega_{\mathrm{r}}}\Delta\omega\right)^2} \right\},$$

$$\mu'_{\pm} = \omega_{\mathrm{M}} \left\{ \frac{2\Delta\omega}{(\omega_{\mathrm{r}} \mp \omega)^2 + 4\left(\frac{\omega}{\omega_{\mathrm{r}}}\Delta\omega\right)^2} \right\}. \tag{6.133}$$

Many microwave applications are operated away from resonance, usually where it can be assumed that $\omega \gg \omega_{\mathrm{r}} \gg \Delta\omega$. In these situations, the two damping schemes can be compared directly as

$$\mu'_{\pm} \approx 1 \mp \frac{\omega_{\mathrm{M}}}{\omega},$$

$$\mu''_{\pm} \approx \frac{2\omega_{\mathrm{M}}}{\omega}\frac{\Delta\omega}{\omega} \quad \alpha = \Delta\omega/\omega, \tag{6.134a}$$

$$\mu'_{\pm} \approx 1 \mp \frac{\omega_{\mathrm{M}}}{\omega},$$

$$\mu''_{\pm} \approx \frac{2\omega_{\mathrm{M}}}{\omega_{\mathrm{r}}}\frac{\Delta\omega}{\omega} \quad \alpha = \Delta\omega/\omega_{\mathrm{r}}. \tag{6.134b}$$

Based on inspection of (6A.134), the choice of relation for α will influence mainly the absorption component through the introduction of frequency dependence to the linewidth term.

Appendix 6B Classical Instability Threshold

From (6.80) and (6.81) for the transverse rf susceptibility, an expression for the z-axis magnetization component in a significant H_{rf} field can be expressed as

$$(4\pi M_z)^2 = (4\pi)^2\left(M^2 - M_{\mathrm{T}}^2\right) = (4\pi M)^2 - \frac{(4\pi M)^2\,(\gamma H_{\mathrm{rf}})^2}{(\gamma H - \gamma N_{\mathrm{D}z} 4\pi M_z - \omega)^2 + \tau_{2k\to 0}^{-2}}. \tag{6.135}$$

Since $4\pi M_z = (4\pi)(M - \Delta M_z)$, where ΔM_z is the change in the z component caused by H_{rf}, we can write

$$(4\pi M_z)^2 - (4\pi M)^2 = -\frac{(4\pi M)^2 (\gamma H_{\mathrm{rf}})^2}{[\gamma H - \gamma N_{\mathrm{D}z}\, 4\pi\,(M - \Delta M_z) - \omega]^2 + \tau_{2k\to 0}^{-2}}. \quad (6.136)$$

After factoring the left-hand side of (6.136),

$$-(4\pi\,\Delta M_z)\,[4\pi\,(M_z + M)] = -\frac{(4\pi M)^2 (\gamma H_{\mathrm{rf}})^2}{[\gamma H - \gamma N_{\mathrm{D}z}\, 4\pi\,(M - \Delta M_z) - \omega]^2 + \tau_{2k\to 0}^{-2}}, \quad (6.137)$$

which reduces to

$$4\pi\,\Delta M_z \cong \frac{1}{2}\frac{4\pi M\,(\gamma H_{\mathrm{rf}})^2}{(\gamma N_{\mathrm{D}z}\, 4\pi\,\Delta M_z)^2 + \tau_{2k\to 0}^{-2}}. \quad (6.138)$$

if we assume that $4\pi\,(M_z + M) \cong 2\,(4\pi M)$ and fix the initial resonance condition $(H_{\mathrm{rf}},\, \Delta M_z \to 0)$ as $\gamma H - \gamma N_{\mathrm{D}z}\, 4\pi M - \omega = 0$.

Analytically, (6.138) can be viewed as cubic in ΔM_z, and the possibility of triple solutions was discussed by Anderson and Suhl [67]. The physical reality, however, dictates that there will be only one stable solution, and this can be examined graphically by plotting the left- and right-hand sides (LHS and RHS) separately as functions of $4\pi\Delta M_z$, sketched in Fig. 6.35. A straightforward iteration procedure quickly determines that convergence to a single value of $4\pi\Delta M_z$ will occur only if the slope of the right-hand side is greater than -1. Therefore, we write for the threshold condition

$$-\frac{\gamma N_{\mathrm{D}z}\, 4\pi M\,\left(\gamma H_{\mathrm{rf}}^{\mathrm{crit}}\right)^2 \left(\gamma N_{\mathrm{D}z}\, 4\pi\Delta M_z^{\mathrm{crit}}\right)}{\left[\left(\gamma N_{\mathrm{D}z}\, 4\pi\Delta M_z^{\mathrm{crit}}\right)^2 + \tau_{2k\to 0}^{-2}\right]^2} \geq -1. \quad (6.139)$$

If (6.138) and (6.139) are combined, a simple relation between ΔM_z and ΔH emerges:

$$\gamma N_{\mathrm{D}z}\, 4\pi\Delta M_z^{\mathrm{crit}} = \tau_{2k\to 0}^{-1} = \gamma\Delta H. \quad (6.140)$$

When $\gamma N_{Dz}\, 4\pi\Delta M_z$ is replaced by $\tau_{2k\to 0}^{-1}$ in (6.139), the critical rf field at the new resonance peak (that is now shifted to lower H by an amount $\Delta H = N_{\mathrm{D}z} 4\pi\Delta M_z$) becomes

$$\gamma H_{\mathrm{rf}}^{\mathrm{crit}} = 2\tau_{2k\to 0}^{-3/2}\,(\gamma N_{\mathrm{D}z}\, 4\pi M)^{-1/2}, \quad (6.141)$$

or in terms of uniform precession half-linewidth

$$H_{\mathrm{rf}}^{\mathrm{crit}} = 2\Delta H_{k\to 0}^{3/2}\,(N_{\mathrm{D}z} 4\pi M)^{-1/2}. \quad (6.142)$$

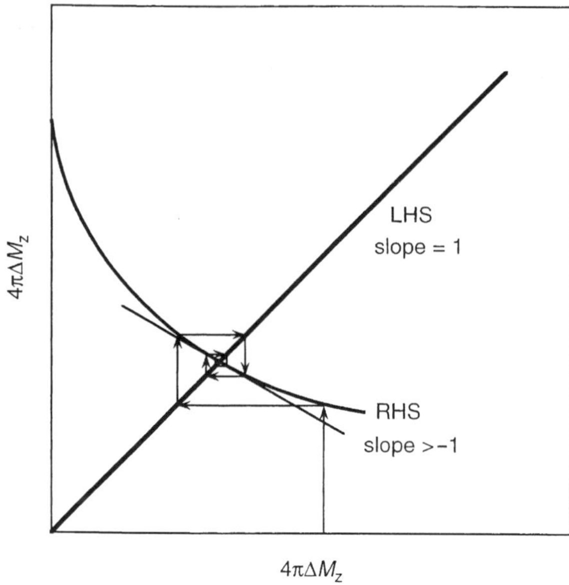

Fig. 6.35 Convergence diagram of the classical iteration solution to determine the magnitude of $4\pi\Delta M_z$ caused by a strong rf drive field

Appendix 6C Domain Wall Susceptibility Equation

Below 1 GHz, the real part of the magnetic susceptibility as a function of frequency is controlled by two mechanisms. The more universal one is magnetization rotation that is a simple extension of the nonresonant relaxation in paramagnets described in Sect. 6.1.1. The more important one, however, is domain-wall resonance that is encountered in multidomain ferromagnets. Domain walls can be shown to have mechanical-like properties of a forced damped harmonic oscillator that is characterized in terms of effective wall mass m_{w}, restoring force $\alpha_{\mathrm{w}}x$ when disturbed from equilibrium in the x direction, and motional damping force $\beta_{\mathrm{w}}(\mathrm{d}x/\mathrm{d}t)$, per unit wall area.

An equation of motion for domain walls under the action of a z-axis alternating magnetic field H_{ac} of angular frequency ω was devised as [80,95,96]

$$m_{\mathrm{w}}\ddot{x} + \beta_{\mathrm{w}}\dot{x} + \alpha_{\mathrm{w}}x = 2M_{\mathrm{s}}H_{\mathrm{ac}}\cos(\omega t)\,, \tag{6.143}$$

where

$$m_{\mathrm{w}} = \frac{1}{4\pi\gamma^2}\sqrt{\frac{|K_1|}{A}} \quad \text{(wall mass area)}$$

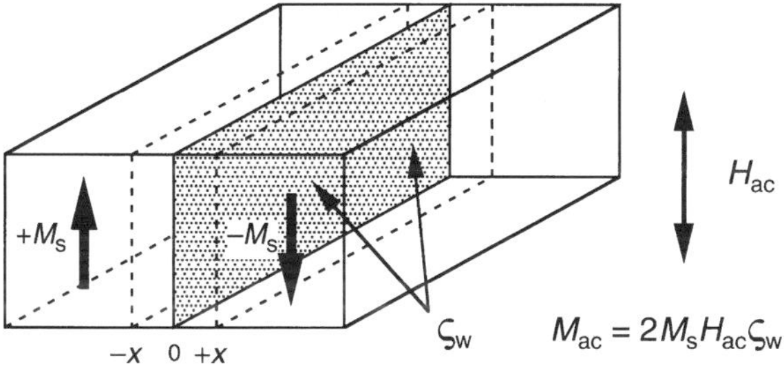

Fig. 6.36 Model of domain-wall displacement under the influence of a longitudinal ac magnetic drive field. Figure reprinted from [80] with permission. © 2003 by the IEEE

$$A = \frac{JS^2}{a_0} \quad \text{(spin exchange energy/lattice parameter)}$$
$$\beta_w = \frac{m_w}{\tau_w} \quad (\tau_w = \text{unforced damping time constant})$$
$$\alpha_w = \frac{4M_s^2}{3\mu_i}\varsigma_w \quad (\varsigma_w = \text{total domain wall surface area/volume})$$
$$\mu_i = \frac{2\pi M_s^2}{|K_1|} \quad (K_1 < 0)\,; \quad \frac{4\pi M_s^2}{3\,|K_1|} \quad (K_1 > 0)$$
$$\delta_w = \pi\sqrt{\frac{A}{|K_1|}} = \pi\sqrt{\frac{JS^2}{a_0\,|K_1|}} \quad \text{(wall thickness)} \tag{6.144}$$

From inspection of Fig. 6.36, we reason that the wall displacement of magnitude x driven by H_{ac} sweeps out a fractional change in volume equal to $\zeta_w x$ and produces a change in net magnetization of

$$M_{ac} = 2M_s\varsigma_w x, \tag{6.145}$$

or in terms of the real part of the initial susceptibility

$$\chi'_{ac} = \frac{M_{ac}}{H_{ac}} = \frac{2M_s\varsigma_w x}{H_{ac}}. \tag{6.146}$$

Therefore, we can now express (6.143) as (after dropping the subscript ac)

$$\ddot{\chi}' + \left(\frac{1}{\tau_w}\right)\dot{\chi}' + \omega_w^2\chi' = \frac{(2M_s)^2\,\varsigma_w}{m_w}\cos(\omega t), \tag{6.147}$$

where $\omega_w^2 = \alpha_w/m_w = (8/3)\gamma^2\sqrt{A\,|K_1|}\varsigma_w$ for the more common case of $K_1 < 0$.

References

1. H.B.G. Casimir, *Magnetism and Very Low Temperatures*, (Dover, New York, 1961); also H.B.G. Casimir and F.K. du Pre, *Physica* **5**, 507 (1938)
2. C.J. Gorter, *Paramagnetic Relaxation*, (Elsevier, Amsterdam, 1947)
3. A.H. Morrish, *The Physical Principles of Magnetism*, (Wiley, New York, 1965), Chapter 3
4. K.J. Standley and R.A. Vaughan, *Electron Spin Relaxation Phenomena in Solids*, (Plenum, New York, 1969), pp. 10–14
5. C.B.P. Finn, R. Orbach, and W.P. Wolf, *Proc. Phys. Soc. (London)* 77, 261 (1961)
6. R. Orbach, *Proc. R. Soc. (London)* **A264**, 458 (1961)
7. K.J. Standley and R.A. Vaughan, *Electron Spin Relaxation Phenomena in Solids*, (Plenum, New York, 1969), p. 37
8. G.F. Dionne, *Phys. Rev.* **139**, A1648 (1965)
9. P.L. Scott and C.D. Jefferies, *Phys. Rev.* **127**, 32 (1962)
10. J.H. Van Vleck, *J. Chem. Phys.* 7, 72 (1939)
11. J.H. Van Vleck, *Phys. Rev.* **57**, 426 (1940)
12. R.D. Mattuck and M.W.P. Strandberg, *Phys. Rev.* **119**, 1204 (1960)
13. G.E. Pake, *Paramagnetic Resonance*, (W.A. Benjamin, New York, 1962), Chapter 2
14. G.F. Dionne, *Can. J. Phys.* **42**, 2419 (1964)
15. B. Bleaney and K.W.H. Stevens, *Rep. Prog. Phys.* **XVI**, 108 (1953)
16. K.D. Bowers and J. Owen, *Rep. Prog. Phys.* **XVIII**, 304 (1955)
17. J.P. Gordon, H.J. Zeiger, and C.H. Townes, *Phys. Rev.* 95, 282 (1954)
18. J. Vanier, *Can. J. Phys.* **42**, 494 (1964)
19. C. Kittel, *Phys. Rev.* **73**, 155 (1948); **76**, 743 (1949)
20. A.H. Morrish, *The Physical Principles of Magnetism*, (Wiley, New York, 1965), Chapter 10
21. B. Lax and K.J. Button, *Microwave Ferrites and Ferrimagnetics*, (McGraw Hill, New York, 1962), Chapters 4–6
22. G.F. Dionne and J.B. Goodenough, *Mater. Res. Bull.* 7, 749 (1972)
23. A.B. Smith and R.V. Jones, *J. Appl. Phys.* **34**, 1283 (1963)
24. A.B. Smith, *Rev. Sci. Instrum.* **39**, 378 (1968)
25. G.F. Dionne, *J. Appl. Phys.* **40**, 4486 (1969); **41**, 831 (1970)
26. G.F. Dionne, *J. Appl. Phys.* **41**, 2264 (1970)
27. L. Landau and E. Lifshitz, *Physik. Z. Sowjetunion* **8**, 153 (1935)
28. T.L. Gilbert, *Phys. Rev.* **100**, 1243 (1955)
29. A.H. Morrish, *The Physical Principles of Magnetism*, (Wiley, New York, 1965), p. 550
30. H.J. Van Hook and F. Euler, *J. Appl. Phys.* **40**, 4001 (1969)
31. E. Schloemann, *Proc. Conf. Magn. Magn. Mater.*, AIEE Spec. Publ. **T-10**, 600 (1956)
32. E. Schloemann, *J. Phys. Chem. Solids* **6**, 242 (1958)
33. S. Geshwind and A.M. Clogston, *Phys. Rev.* **108**, 49 (1957)
34. E. Schloemann, *J. Appl. Phys. 38*, 5027 (1967)
35. M. Sparks, *Ferromagnetic Relaxation Theory,* (McGraw-Hill, New York, 1964); also M. Sparks, Phys. Rev. **36**, 49 (1957)
36. G.F. Dionne, *Mater. Res. Bull.* **5**, 939 (1970)
37. G.F. Dionne, *J. Appl. Phys.* **40**, 1839 (1969)
38. C.E. Patton, *Phys. Rev.* **179**, 352 (1969)
39. Q.H.F. Vrehen, *J. Appl. Phys.* **40**, 1849 (1969)
40. P.-G. de Gennes, C. Kittel, and A.M. Portis, *Phys. Rev.* **116**, 323 (1959)
41. P.E. Seiden, *Phys. Rev.* **133**, A728 (1964)
42. J.H. Van Vleck, *J. Appl. Phys.* **35**, 882 (1964)
43. B.W. Faughan and M.W.P. Strandberg, *J. Phys. Chem, Solids* **19**, 155 (1961)
44. A.M. Stoneham *Proc. Phys. Soc.* **85**, 107 (1965)
45. K.J. Standley and R.A. Vaughan, *Electron Spin Relaxation Phenomena in Solids*, (Plenum, New York, 1969), Chapter 4

46. G.F. Dionne, *IEEE Trans. Magn.* **28**, 3201 (1992); also G.F. Dionne and R.G. West, *AIP Conf. Proc.* **10**, 169 (1972)
47. G.F. Dionne and G.L. Fitch, *J. Appl. Phys.* **87**, 4963 (2000)
48. K.J. Standley and R.A. Vaughan, *Electron Spin Relaxation Phenomena in Solids*, (Plenum, New York, 1969), Appendix 6B
49. R.C. LeCraw and E.G. Spencer, *J. Phys. Soc. Jpn* **17**, Suppl. B-1, 401 (1962)
50. J. H. Van Vleck and R. Orbach, *Phys. Rev. Lett.* **11**, 65 (1963)
51. E.M. Georgy, R.C. Le Craw, and M.D. Sturge, *J. Appl. Phys.* **37**, 1303 (1966)
52. A.L. Kipling, P.W. Smith, J. Vanier, and G.A. Woonton, *Can. J. Phys.* **39**, 1859 (1961)
53. A. Thavendrarajah, M. Pardavi-Horváth, P.E. Wigen, and M. Gomi, *IEEE Trans. Magn.* **25**, 4015 (1989)
54. M. Pardavi-Horváth, L. Bottyán, I.S. Szücs, P.E. Wigen, and M. Gomi, *Hyperfine Interact.* **54**, 639 (1990)
55. F.J. Rachford, M. Levy, R.M. Osgood, Jr., A. Kumar, and H. Bakhru, *J. Appl. Phys.* **85**, 5217 (1999)
56. G.F. Dionne, *J. Appl. Phys.* **91**, 7367 (2002)
57. G.F. Dionne, *J. Appl. Phys.* **57**, 3727 (1985)
58. G.F. Dionne, *J. Appl. Phys.* **64**, 1323 (1988); also G.F. Dionne, *Anisotropy and Relaxation Effects of* Co^{2+} *in LiTi Ferrites*, MIT Lincoln Laboratory Technical Report TR-688, 15 August 1984 AD-A146550
59. N. Matsushita, T. Nakamura, and M. Abe, *IEEE Trans. Magn.* **38**, 3111 (2002)
60. J. Llabres, J. Nicolas, and R. Sproussi, *Appl. Phys.* **12**, 87 (1977)
61. D. Dale, G. Hu, V. Balbarin, and Y. Suzuki, *Mater. Res. Soc. Symp. Proc.*, **603**, 95 (2000)
62. C. Herring and C. Kittel, *Phys. Rev.* **81**, 869 (1951)
63. E.F. Schloemann, *IEEE Trans. Magn.* **34**, 3830 (1998)
64. E.F. Schloemann, J.J. Green, and U. Milano, *J. Appl. Phys.* **31**, 386S (1960)
65. R.C. LeCraw, E.G. Spencer, and C.S. Porter, *Phys. Rev.* **110**, 1311 (1958)
66. G.F. Dionne, *J. Appl. Phys.* **43**, 1221 (1972)
67. P.W. Anderson and H. Suhl, *Phys. Rev.* **100**, 1788 (1955)
68. H. Suhl, *Proc. IRE* **44**, 1270 (1956)
69. H. Suhl, *Phys. Rev.***101**, 1437 (1956)
70. N. Bloembergen and S. Wang, *Phys. Rev.* **93**, 72 (1954); also Fig. 5.9 in B. Lax and K.J. Button, *Microwave Ferrites and Ferrimagnetics*, (McGraw Hill, New York, 1962)
71. C.E. Patton, *J. Appl. Phys.* **40**, 2837 (1969)
72. J.F. Dillon, *J. Appl. Phys.* **31**, 1605 (1960)
73. L.R. Walker, *Phys. Rev.* **105**, 390 (1957)
74. B. Lax and K.J. Button, *Microwave Ferrites and Ferrimagnetics*, (McGraw Hill, New York, 1962), p. 180
75. P.J.B. Clarricoats, *Microwave Ferrites*, (Wiley, New York, 1961), Chapter 4
76. A.H. Morrish, *The Physical Principles of Magnetism*, (Wiley, New York, 1965), p. 563
77. D.D. Stancil, *Theory of Magnetostatic Waves*, (Springer, New York, 1993)
78. M. Chen, M.A. Tsankov, J.O. Nash, and C.E. Patton, *Phys. Rev.* **49**, 12773 (1994)
79. J. Smit and H.P.J. Wijn, *Ferrites*, (Wiley, New York, 1959), Part D
80. G.F. Dionne, *IEEE Trans. Magn.* **39**, 3121 (2003)
81. R.A. Becker, *Introduction to Theoretical Mechanics*, (McGraw-Hill, New York, 1954), p. 144
82. G. Rado, R. Wright, and W. Emmerson, *Phys. Rev.* **80**, 273 (1950)
83. J.A. Weiss, N.G. Watson, and G.F. Dionne, *IEEE MTT-S Int. Microwave Symp. Digest 1989*, p. 145; also J.A. Weiss, N.G. Watson, and G.F. Dionne, *Appl. Microwave*, **Fall**, 74 (1990)
84. Y. Akaiwa and T. Okazaki, *IEE Trans. Magn.* **24**, 863 (1974)
85. E.W. Gorter, *Proc. IRE* **43**, 1945 (1955)
86. D. Stopples, P.G.T. Boonen, U. Enz, and L.A.H. Van Hoof, *J. Magn. Magn. Mater.* **37**, 116 (1983)
87. G.F. Dionne, *Phys. Rev.* **137**, A743 (1965)
88. B. Lax and K.J. Button, *Microwave Ferrites and Ferrimagnetics*, (McGraw Hill, New York, 1962), p. 299

89. G.F. Dionne, D.E. Oates, D.H. Temme, and J.A. Weiss, *Lincoln Lab. J.* **9**, 19 (1996)
90. G.F. Dionne, J.A. Weiss, and G.A. Allen, *IEEE Trans. Magn.* **24**, 2817 (1988)
91. B. Lax and K.J. Button, *Microwave Ferrites and Ferrimagnetics*, (McGraw Hill, New York, 1962), p. 150
92. P.J.B. Clarricoats, *Microwave Ferrites*, (Wiley, New York, 1961), p. 63, with $\alpha = 0$
93. D.M. Pozar, *Microwave Engineering*, (Wiley, New York, 1998), p. 503
94. N. Bloembergen, *Proc. IRE* **44**, 1259 (1956)
95. S. Chikazumi, *Physics of Magnetism*, (Wiley, New York, 1997), p. 348
96. P.J.B. Clarricoats, *Microwave Ferrites*, (Wiley, New York, 1961), p. 92

Chapter 7
Magneto-Optical Properties

In the previous chapters, the emphasis is placed on the electronic origins of local and collective molecular magnetism in transition-metal oxides and their behavior in alternating magnetic fields. Models of magnetic resonance based on precessing magnetic moments provide a classical analog to quantum mechanical transitions provided that the internal magnetic fields are large enough to produce the Zeeman energy splittings for the particular frequency of interest. In the energy range that can be easily reached by fields from laboratory electromagnets, electron paramagnetic resonance (EPR) and ferromagnetic resonance (FMR) occur in the microwave bands. However, resonances can also occur in magnetically ordered systems at the energies of magnetic exchange. Since the exchange effects occur in the submillimeter and far-infrared bands, but have the properties of a magnetic-dipole stabilization, this topic will serve as a transition to the subject of magneto-optics that is based on magnetically polarized electric-dipole interactions with optical waves.

In the visible and ultraviolet bands, electric-dipole transitions can produce magneto-optical phenomena without the need for large applied magnetic fields. In this regime, the dielectric permittivity tensor with off-diagonal terms can produce nonreciprocal propagation at optical wavelengths analogous to those from magnetic interactions with RF waves. Faraday rotation of the linear polarization of plane-wave transmission and its complementary Kerr reflection effect are of major importance for discrete fiber-optical technology. In later developments, optical waveguides that simulate their microwave counterparts have shown promise for integrated photonics technology that can benefit from the nonreciprocal properties of magneto-optical control devices. To remain within the scope of this volume, the discussion of materials systems will be focused on the room temperature properties of the garnet family of magnetic oxides, first on the basic host compound yttrium iron garnet and then on the dramatic effects of Bi^{3+} ion substitutions. The discussion will review the work carried out at Lincoln Laboratory and the Department of Physics of the Massachusetts Institute of Technology where the author was an active participant, but is dawn heavily from the pioneering work of scientists at the Mullard Research Laboratories in England and the Philips Research Laboratories in Eindhoven, the Netherlands and Hamburg, Germany.

G.F. Dionne, *Magnetic Oxides*, DOI 10.1007/978-1-4419-0054-8_7,

7.1 Infrared Exchange Resonance

Before magnetooptical permittivity in transition-metal oxide compounds is addressed, the interionic magnetic resonance that originates from the molecular exchange stabilization of magnetic spins is an appropriate segue topic. Although the exchange energies extend to the infrared bands, approaching those of electric-dipole orbital energy transitions, the actual mechanism involves the reversal of magnetic ion spins between parallel and antiparallel states of a covalently bonded molecule. For convenience, it is treated phenomenologically as the result of a quasimagnetic exchange field. In reality, the exchange resonance effects are the result of neither magnetic nor electric-dipole quantum transitions. Despite the fact that their frequencies are typically several orders of magnitude greater than that of microwave resonance, they greatly enhance the intensity of the microwave resonance.

7.1.1 Classical Precession Model

The phenomenological theory of gyromagnetic resonance has been applied to both electron and nuclear magnetism represented generically as a magnetic moment $\boldsymbol{m}$ precessing about a magnetic field $\boldsymbol{H}$ directed along the z-axis of a Cartesian coordinate system. The angular precession frequency $\omega_0 = \gamma H_0$ is derived from the solution of the customary equation of motion (without damping) [1][1]

$$\frac{\partial \boldsymbol{m}}{\partial t} = \gamma\left(\boldsymbol{m} \times \boldsymbol{H}\right), \tag{7.1}$$

where $\gamma\ (=g m_{\mathrm{B}}/\hbar)$ is the gyromagnetic ratio. Resonance takes place when the frequency and sense of an electromagnetic wave circularly polarized in the x–y plane matches those of the precession. In this basic form, ω_0 is influenced only by the value of H, while m determines the intensity of the observed resonance. In a paramagnet, however, where m represents the combined moment of a collection of independent isolated moments, the intensity is determined by the degree of saturation, which is a function of field and temperature. This situation is affected dramatically when the independence of the separate moments is removed by magnetic exchange coupling through molecular bonding that becomes the dominant aligning agent and creates a second resonance condition.

In a system where the individual spins form a magnetization vector $\boldsymbol{M}$ in a molecular or exchange field $\boldsymbol{H}_{\mathrm{ex}}$, the result is a treated as a Larmor-type precession usually in an infrared band. For the two magnetically opposed sublattices of a

[1] The reader is advised that sign conventions vary in the literature of this subject. Wherever possible, the convention from Chap. 6 will be continued.

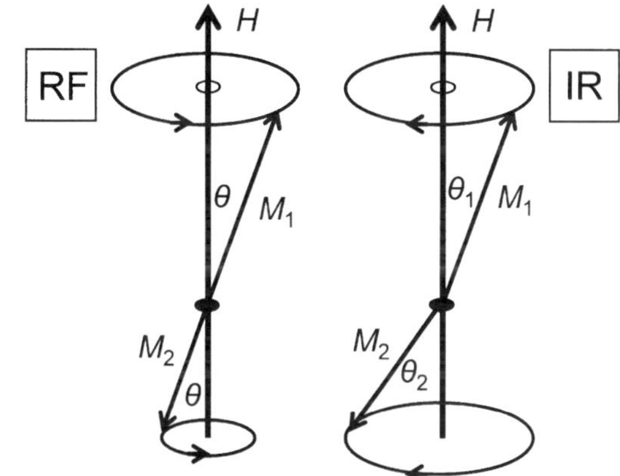

Fig. 7.1 Microwave and exchange resonance modes in a ferrite. Precession directions are opposite for the two modes

ferrite (see Fig. 7.1), (7.1) is elaborated to account for the additional stabilization by [2][2]

$$\frac{\partial \boldsymbol{M}_1}{\partial t} = \gamma_1 \left[\boldsymbol{M}_1 \times (\boldsymbol{H} + \boldsymbol{H}_{\mathrm{K1}} + \boldsymbol{H}_{\mathrm{ex1}})\right],$$
$$\frac{\partial \boldsymbol{M}_2}{\partial t} = \gamma_2 \left[\boldsymbol{M}_2 \times (\boldsymbol{H} + \boldsymbol{H}_{\mathrm{K2}} + \boldsymbol{H}_{\mathrm{ex2}})\right], \tag{7.2}$$

where $H_{\mathrm{K1,2}}$ are the respective sublattice anisotropy fields, and the corresponding exchange fields are $\boldsymbol{H}_{\mathrm{ex1,2}} = N_{12}\boldsymbol{M}_{1,2}$ with N_{12} (<0) as the intersublattice antiferromagnetic molecular-field coefficient. The intrasublattice molecular fields N_{11} and N_{22} are ignored in this approximation.

Simplifying (7.2) for our present purposes, we assume zero-orbital angular momentum and let $\gamma_1 = \gamma_2 = \gamma$ and $H_{\mathrm{K1}} = H_{\mathrm{K2}} = 0$. From (7.2), two solutions, one for microwaves (RF) and another for infrared waves (IR), emerge with precession frequencies:

$$\omega_{\mathrm{RF}} = \gamma H$$
$$\omega_{\mathrm{IR}} = \gamma \left|H + H_{\mathrm{ex}}\right| = \gamma \left[H + N_{12}\left|M_1 - M_2\right|\right], \tag{7.3}$$

which represent the sketches in Fig. 7.1 in their most primitive forms. If the ferrite is now treated as ferromagnetic with magnetization $M = |M_1 - M_2|$, which is a

[2] Note that although the anisotropy and exchange terms are presented in vector notation as magnetic fields, only the anisotropy energy in its role as a demagnetizing field that modifies the applied field H_0 is legitimate in the Maxwell sense. H_{ex} has its origins in the chemical bond of the molecule and draws its value and sense from the implications of the indistinguishability of the electron as manifested by the Pauli Exclusion principle. As a consequence, it should be considered as a scalar, separate from the sequence of perturbations affecting an individual cation, and would not influence the removal of degeneracies in the electronic energy level structure.

conventional approach to treating ferrimagnetism classically, (7.2) simplifies to a single magnetic lattice:

$$\frac{\partial \boldsymbol{M}}{\partial t} = \gamma\,[\boldsymbol{M} \times (\boldsymbol{H} + \boldsymbol{H}_{\text{ex}})] = \gamma\,[\boldsymbol{M} \times (\boldsymbol{H} + N\boldsymbol{M})]\,. \tag{7.4}$$

The dispersive parts of the undamped permeabilities for right- and left-hand circular polarization modes (+ and −) can be expressed as [3]

$$\begin{aligned} \mu'_{\text{RF}\pm} &= 1 + \frac{\omega_{\text{M}}}{\omega \mp \omega_{\text{RF}}} = 1 + \frac{\omega_{\text{M}}}{\omega \mp \gamma H}, \\ \mu'_{\text{IR}\pm} &= 1 + \frac{\omega_{\text{M}}}{\omega \mp \omega_{\text{IR}}} = 1 + \frac{\omega_{\text{M}}}{\omega \mp \gamma\,(H + NM)}, \end{aligned} \tag{7.5}$$

where $\omega_{\text{M}} = \gamma 4\pi M$ and $\gamma = 1.76 \times 10^7$ rad s^{-1} Oe^{-1}.

Inspection of (7.5) indicates that the amplitude of both resonances is proportional to the net magnetization, which is saturated by a large exchange energy that stabilizes the spins into the collective magnetization vector. Confirmation that μ_{IR} follows a Brillouin–Weiss thermomagnetization characteristic was reported by Tinkham [4]. The question that arises is how the applied field H that is responsible for the polarization of the combined moment can produce an independent precession if it adds vectorially to the exchange field. The reasoning applied to the equations of motion [(7.2)] is that both $\boldsymbol{H}$ and $\boldsymbol{H}_{\text{ex}}$ $(=N\boldsymbol{M})$ are vectors that combine in creating and controlling the net $\boldsymbol{M}$, but at the same time are independent when producing separate precessions of the same $\boldsymbol{M}$. If the exchange energy is the result of a real magnetic field, then it follows that there should be no separate RF precession because the two fields would not combine as scalars, leaving only the exchange field H_{ex} with magnitude enhanced slightly by H. On the other hand, if the exchange stabilization is not the result of a magnetic field, there should not be an IR precession. Only the RF precession should occur. In essence, the classical model predicts independent reversals of magnetic moments, i.e., spin "flips" through 180°, from the common ground state into identical excited quantum states by means of plane-wave excitations of radically different wavelengths, one from microwave interactions and the other of lattice vibration origins. To explore further this apparent paradox, it is necessary to consider the quantum mechanical description of the physical situation.

7.1.2 Quantum Spin Transition Model

If the two ferrimagnetic precessions of opposite sense in (7.3) are replaced by an equivalent ferromagnetic model sketched in Fig. 7.2, with both precessions in the same direction, the question of interaction between RF and IR resonance effects can be viewed from a quantum mechanical perspective [5]. A hybrid energy diagram of a single electron spin ($S = 1/2$) in an exchange field of parameter J_S

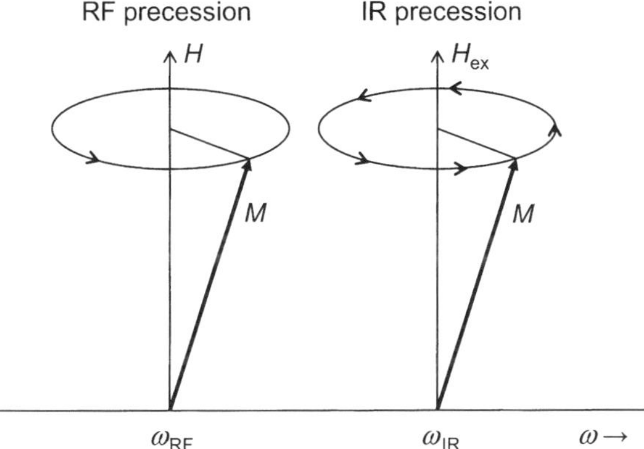

Fig. 7.2 Ferromagnetic model for the two precession modes of a ferrite, shown in Fig. 7.1 *Extra arrows* on the IR precession orbit indicate that $S > 1/2$ and the angular velocity is higher. Figure reprinted from [5] with permission. © 2009 by the American Institute of Physics

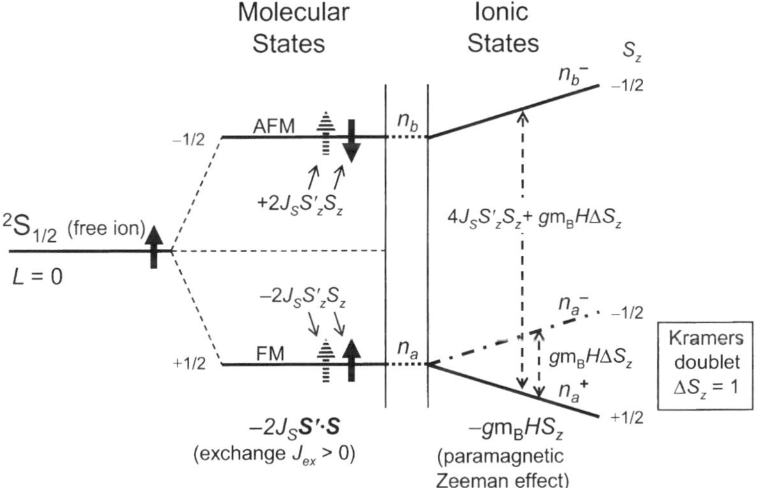

Fig. 7.3 Energy level model for an $S = 1/2$ system coupled ferromagnetically in an exchange "field." Molecular environment stabilizes the spin by a large energy, but leaves Kramers doublets as degenerate levels. Occupation probability of spin ground state is determined by the exchange energy, which has the effect of lowering the temperature of the spin system and permit FMR to be strong up to the Curie temperature. Figure reprinted from [5] with permission. © 2009 by the American Institute of Physics

and real magnetic field $\boldsymbol{H}$ is shown in Fig. 7.3. It must first be recognized that the exchange energy that stabilizes the ^2S spin ground state is not the result of perturbation in a spin-Hamiltonian equation, but rather the result of interionic molecular bonding. Relative to the isolated ion states, the molecular state is stabilized by $2J_S\boldsymbol{S'}\cdot\boldsymbol{S}\left(=\pm 2J_S S'_z S_z\right)$ in the Ising approximation, with $\boldsymbol{S'}$ representing the source of the exchange acting on $\boldsymbol{S}$. Because the exchange coupling of the spin vectors is not the result of a true magnetic field interaction, both $S = \pm 1/2$ states remain as degenerate Kramers doublets.

The two magnetic transitions indicated in Fig. 7.3 correspond to the IR and RF precessions of Fig. 7.2. The paradox is immediately apparent from the energy level diagram. In the quantum model, the distinction between H and H_{ex} becomes a meaningful issue. First, the inability of H_{ex} to split the $S_z = \pm 1/2$ Kramers degeneracy allows the Zeeman effects of energy $gm_{\text{B}}\ HS_z$ (where $g \approx 2$ for a spin and m_{B} is the Bohr magneton) to become the source of the RF precession and ferromagnetic resonance of the individual ions even when the real magnetic field $\boldsymbol{H}$ saturates the already exchange-coupled spins. The point here is that the paramagnetic energy level structure and resulting microwave spectral frequencies are unaffected by the exchange stabilization. What is affected, however, is the Boltzmann population probabilities of the spin ground state that is now determined by the combined splitting energy $4J_S S'_z S_z + gm_{\text{B}} H \Delta S_z$ of the system, and not just by the paramagnetic term that is orders of magnetic smaller than the exchange effect.

If the Boltzmann population ratio is applied to the molecular-orbital states separated in energy by $4J_S S'_z S_z$ in Fig. 7.3,

$$\frac{n_{\text{b}}}{n_{\text{a}}} = \exp\left(-\frac{4J_S S'_z S_z}{kT}\right), \tag{7.6}$$

where $n_a + n_b = n$, the population density of like magnetic ions participating in the distribution. For the energy levels of the Zeeman splitting from the applied field H in the absence of exchange, i.e., $J_S = 0$, the population ratio is

$$\frac{n_a{}^{-}}{n_a{}^{+}} = \exp\left(-\frac{gm_{\text{B}} H \Delta S_z}{kT}\right) = \exp\left(-\frac{gm_{\text{B}} H}{kT}\right), \tag{7.7}$$

where $\Delta S_z = 1$ in accord with the magnetic-dipole selection rule. When the exchange and Zeeman energies are combined as scalars,

$$\frac{n_b{}^{-}}{n_a{}^{+}} = \exp\left(-\frac{4J_S S'_z S_z + gm_{\text{B}} H}{kT}\right). \tag{7.8}$$

An underlying assumption of the Boltzmann theory of population distribution among the various energy states, e.g., vibrational modes of a CO_2 molecule, is the noninteraction between individual members. Multiple energy levels of isolated magnetic ions induced by a common magnetic field (paramagnetism) can also qualify unless their individuality is lost by the spin ordering stabilization imposed by chemical bonding. As a consequence, (7.7) cannot be applied as stated unless $J_S = 0$. Only (7.8) is a valid application of the Boltzmann law because it involves energy changes within the constraints of the collective magnetization that are then responsible to Fermi statistics dictated by the Pauli principle. Because the ground-state population $n_a{}^{+}$ is common to both (7.7) and (7.8), it follows that the population

$n_a{}^- = n_b{}^-$ in order to avoid the impossibility of the same excited state $S_z = -1/2$ with different Boltzmann populations at two widely different energies. However, the question remains as to the actual expressions for the exchange-stabilized Kramers doublet when the physical reality of $n_a{}^- = n_b{}^-$ is imposed. Equating (7.7) and (7.8), we define an effective "spin" temperature T_S in terms of the lattice temperature T_L:

$$T_S = \left(\frac{g m_B H}{4 J_S S_z' S_z + g m_B H}\right) T_L = \left(\frac{H}{2 H_{ex} S_z + H}\right) T_L$$
$$= \left(\frac{H}{H_{ex} + H}\right) T_L \quad \text{for } S_z = 1/2, \tag{7.9}$$

where $H_{ex} = 2 J_S S_z' / g m_B$.

The conclusion to be reached from (7.9) is that a stabilization energy that commits the spin alignment to its neighbors rendering them dependent in a collective system will have the effect of increasing the ground state population, thereby creating a lower effective spin temperature and resulting in a higher intensity of the Zeeman resonance effect. It is for this reason that microwave FMR signals retain their strength in ferrites at room temperature, as indicated by the presence of ω_M in the classical relation for the permeability in (7.5). The approximate fractional population difference derived from (7.7) and (7.8) is expressed as [6]

$$\frac{n_a{}^+ - n_a{}^-}{n} = \frac{M(T)}{M(0)} = \tanh\left(\frac{4 J_S S_z' S_z + g m_B H}{2kT_L}\right) \mathcal{F}$$
$$= \tanh\left(\frac{J_{1/2} + g m_B H}{2kT_L}\right) \mathcal{F} \quad \text{for } S_z',\ S_z = 1/2. \tag{7.10}$$

From (7.6), $\mathcal{F} = n_a/n = \left[1 + \exp\left(-4 J_S S_z' S_z / kT\right)\right]^{-1}$, which approximates to 1 for $J_S S_z' S_z \gg kT$ at low T. These results are also consistent with the Brillouin–Weiss function $\mathcal{B}_S$ for $S = 1/2$ [7]:

$$\mathcal{B}_{1/2}(x) = \tanh\left(\frac{2 J_S S_z' S_z + g m_B H S_z}{kT_L}\right)$$
$$= \tanh\left(\frac{J_{1/2} + g m_B H}{2kT_L}\right), \tag{7.11}$$

where x is the argument of the tanh function. This general concept was used to interpret the M vs. T characteristics of the highly magnetoresistive (La,Sr)MnO_3 manganites [8].

Figure. 7.4 shows a comparison between the d^1–d^1 case (S', $S = 1/2$) of Ti^{3+} or Cu^{2+} and the multispin d^5–d^5 case (S', $S = 5/2$) of Fe^{3+} encountered in ferrites and other magnetic oxides. For the latter case, S_z', $S_z = 5/2$ in (7.10), and the $\Delta S_z = 1$ FMR transition is $S_z = +5/2 \rightarrow +3/2$. If the spin temperature concept of (7.9) is applied

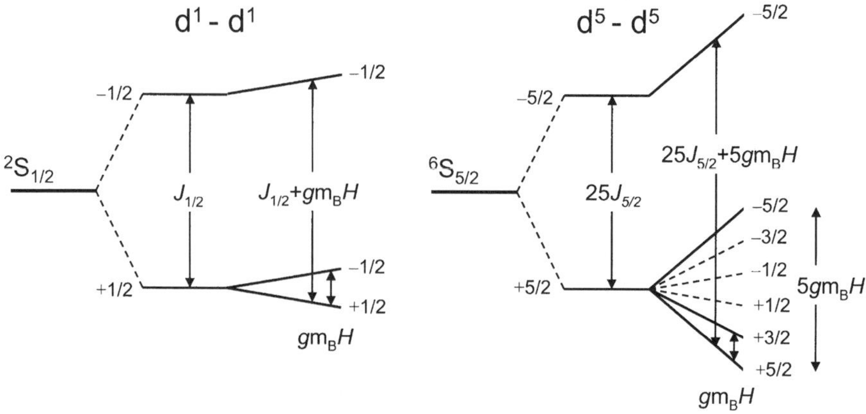

Fig. 7.4 Comparison of $S = 1/2$ with $S = 5/2$ spin systems based on the model of Fig. 7.3 Note that $\Delta S_z = 5$ in the exchange transition. Figure reprinted from [5] with permission. © 2009 by the American Institute of Physics

$$\frac{n_{5/2} - n_{3/2}}{n_{5/2} + n_{3/2}} = \frac{\omega_M(T)}{\omega_M(0)} = \tanh\left(\frac{5J_{5/2} + gm_B H}{2kT_L}\right)\mathcal{F}$$
$$= \tanh\left(\frac{J_{1/2} + gm_B H}{2kT_L}\right)\mathcal{F}. \qquad (7.12)$$

where $J_{5/2} \approx J_{1/2}/5$ from the general relation $J_{q/2} \approx J_{1/2}/q$, and q ($\leq$5) is the number of half-filled orbital states [9]. As $J_S S_z' S_z/kT \to 0$ near the Curie temperature, $\mathcal{F} \to 1/2$ because half of the total population will begin to fill the upper molecular-orbital state when the exchange energy splitting reduces to zero and the system becomes paramagnetic. To obtain additional refinement in the high-temperature regime, both molecular states would have to be included in the formalism. A more general solution would broaden to the case of unlike magnetic ions, e.g., $S \neq S'$, where it must also be remembered that g varies among the different transition metal ions.

Further consideration of (7.12) provides the answer to a long-standing question of why ferromagnetic Zeeman transitions in ferrimagnets are intense well above room temperature. A paramagnetic specimen with a comparable number of magnetic ions usually requires not only cryogenic temperatures but also a spectrometer of much greater sensitivity. As the magnetization dwindles with increasing temperature, the resonance intensity decreases until the Curie temperature (or Neel temperature in some situations) falls below the operating temperature T_L. At this point, the exchange alignment is frustrated by thermal energy and the ferromagnetic system becomes paramagnetic. It is appropriate to comment that without spin ordering by exchange, microwave magnetic devices would not perform at room temperature.

Any stabilizing energy that organizes an otherwise independent population of spins into a collective magnetic moment should be examined in the context of an effective spin temperature when thermal properties are interpreted. Another situation where Boltzmann ground state populations can be increased is by a cooperative magnetostrictive strain (see Sect. 8.3.3).

7.1.3 Experimental Exchange Spectra

Exchange resonance is a unique effect in that it is derived from the cation–cation molecular bonding, mediated by oxygen anions. Hence, there follow the beginnings of a bonding/antibonding band gap included in Fig. 7.3 as a function of J_S. Ex amples of exchange resonance are somewhat sparse in the literature. One reason is the lack of power sources in the submillimeter and far-infrared wavelength bands that confine device applications to passive roles. In many instances, the source is limited to a fixed-frequency low-power laser beam commonly drawn from the rotational or vibrational modes of simple gas molecules such as H_2O, CO_2, or CH_2O_2 (formic acid).

For a comprehensive review of the early work in this area, the reader is directed to Lax and Button [10]. Among the contributions of particular significance is the comprehensive analysis by Geschwind and Walker in which antiferromagnetic resonance is explained in detail [2]. Tinkham and his colleagues have provided experimental results for rare-earth iron garnets and some common antiferromagnetic compounds, FeF_2, MnO, and NiO, listed in Table 7.1. Later work on exchange resonance has been focused on the millimeter and submillimeter bands with a goal set toward device applications.

In subsequent sections, magneto-optical phenomena in transition-metal oxides are described. Before these issues are addressed for specific molecular systems, however, the classical phenomenology of magnetic circular birefringence (MCB) will be reviewed in a context consistent with radiofrequency gyromagnetism.

Table 7.1 Exchange resonance data

Compound	T_N (K)	H_{ex} (Tesla)	ω_{ex} (cm^{-1})	Ref.
MnF_2	67.7	54	–	[57,58]
Cr_2O_3	307	210	–	[58–60]
$MnTiO_3$	65	∼100	–	[61]
FeF_2	78.4	54	52.7	[62]
MnO	120	100	27.5	[63]
NiO	523	400	36.6	[4,64]
YbIG	–	–	14.1	[4]
ErIG	–	–	10.0	[4]
SmIG	–	–	33.5	[4]

7.2 Combined Permeability and Permittivity

As the frontier of signal propagation moved toward shorter wavelengths with the advent of the fiber-optical communications technology, the quest for nonreciprocal devices in the wavelength band near 1 μm became a priority for laser source protection and related applications. Regrettably, the elegance of the microwave band solution that involved simple magnetic-dipole transitions in conventional ferromagnetic media could not be extended to the higher energies because of the magnetic field strengths required are prohibitively large. Moreover, although the classical Larmor precession gyromagnetic effect could still be valid conceptually, its value as a computational tool gives way to a quantum mechanical approach based on atomic spectra. From a phenomenological standpoint, it is important to recognize that the propagation theory introduced in Chap. 6 is only part of the reality that now includes polarization-sensitive electrical permittivity that is variable with frequency. To this end, the classical models of permeability and permittivity are examined jointly [3].

7.2.1 The $[\boldsymbol{\varepsilon}]\cdot[\boldsymbol{\mu}]$ Tensor Solutions

For an insulating medium of tensor permeability $[\boldsymbol{\mu}]$ and permittivity $[\boldsymbol{\varepsilon}]$ transmitting an rf plane-wave $\left(H_{\mathrm{rf}}^{x,y}, E_{\mathrm{rf}}^{x,y}\right)\exp\left(\pm\mathrm{i}\omega t - \gamma_{\mathrm{p}\pm} z\right)$ along the z-axis, the relevant Maxwell equations at a fixed value of z are

$$\begin{aligned} \nabla \times \boldsymbol{H} &= -[\varepsilon]\frac{\partial \boldsymbol{E}}{\partial t} = \mathrm{i}\omega[\varepsilon]\boldsymbol{E}, \\ \nabla \times \boldsymbol{E} &= -[\mu]\frac{\partial \boldsymbol{H}}{\partial t} = -\mathrm{i}\omega[\mu]\boldsymbol{H}, \end{aligned} \tag{7.13}$$

where the rf subscript has been dropped for convenience. From (7.13) a solution for the propagation constant can be obtained by eliminating either $\boldsymbol{H}$ or $\boldsymbol{E}$. As a result, a generalization of the magnetic field case outlined in Appendix 6A of Chap. 6 can be written with permittivity as a tensor, according to

$$\begin{aligned} \nabla \times \nabla \times \boldsymbol{H} &= \omega^2[\varepsilon]\cdot[\mu]\cdot\boldsymbol{H}, \\ \nabla \times \nabla \times \boldsymbol{E} &= \omega^2[\mu]\cdot[\varepsilon]\cdot\boldsymbol{E}, \end{aligned} \tag{7.14}$$

where the combined tensor for a z-axis directed plane wave can be reduced to

$$\begin{aligned} [\varepsilon]\cdot[\mu] &= \begin{bmatrix} \varepsilon_0 & -\mathrm{i}\varepsilon_1 \\ +\mathrm{i}\varepsilon_1 & \varepsilon_0 \end{bmatrix}\cdot\begin{bmatrix} \mu & -\mathrm{i}\kappa \\ +\mathrm{i}\kappa & \mu \end{bmatrix} \\ &= \begin{bmatrix} \varepsilon_0\mu + \varepsilon_1\kappa & -\mathrm{i}(\varepsilon_0\kappa + \varepsilon_1\mu) \\ +\mathrm{i}(\varepsilon_0\kappa + \varepsilon_1\mu) & \varepsilon_0\mu + \varepsilon_1\kappa \end{bmatrix}. \end{aligned} \tag{7.15}$$

Advancing to the solution for the propagation constants of the two modes of circular polarization [11] we obtain

$$\gamma_{\mathrm{p}\pm}^2 = -\left(\frac{\omega}{\mathrm{c}}\right)^2 \varepsilon_\pm \mu_\pm = -\left(\frac{\omega}{\mathrm{c}}\right)^2 \left[(\varepsilon_0\mu + \varepsilon_1\kappa) \pm (\varepsilon_0\kappa + \varepsilon_1\mu)\right]. \tag{7.16}$$

where $\mu_\pm = \mu \pm \kappa$, $\varepsilon_\pm = \varepsilon_0 \pm \varepsilon_1$, and the square-bracketed factor contains the scalar eigenvalue solutions of the secular equation derived from (7.15). The corresponding eigenfunctions are $(1/2)\left(H_{\mathrm{rf}}^x \mp \mathrm{i} H_{\mathrm{rf}}^y\right)$ for the counterrotating right-hand and left-hand circular polarization (RHCP and LHCP) modes. Note that (7.16) reduces to the desired results where dichroism from magnetic only or dielectric only conditions apply, i.e.,

$$\gamma_{\mathrm{p}\pm}^2 = -\left(\frac{\omega}{\mathrm{c}}\right)^2 \varepsilon\mu_\pm = -\left(\frac{\omega}{\mathrm{c}}\right)^2 \varepsilon\,(\mu \pm \kappa) \quad \text{(magnetic interaction)}, \tag{7.17a}$$

$$\gamma_{\mathrm{p}\pm}^2 = -\left(\frac{\omega}{c}\right)^2 \mu\varepsilon_\pm = -\left(\frac{\omega}{c}\right)^2 \mu\,(\varepsilon_0 \pm \varepsilon_1) \quad \text{(dielectric interaction)}, \tag{7.17b}$$

where $\mu = \mu' - \mathrm{i}\mu''$, $\kappa = \kappa' - \mathrm{i}\kappa''$, and correspondingly we define $\varepsilon_0 = \varepsilon_0' - \mathrm{i}\varepsilon_0''$ and $\varepsilon_1 = \varepsilon_1' - \mathrm{i}\varepsilon_1''$.[3]

7.2.2 *Propagation Parameters and Faraday Rotation*

The above relation between propagation constant and the combined permeability and permittivity can be used to generate a more universal expression for the non-reciprocal rotation of a linearly polarized vector introduced for the rf plane-wave case in Sect. 6.4. In addition to restoring the dispersive contributions of ε, the effect of magnetic loss embodied in the imaginary parts of μ and κ can also be included. Equation (7.16) expressed in terms of the propagation constants becomes

$$\left(\alpha_{\mathrm{p}\pm} + \mathrm{i}\beta_{\mathrm{p}\pm}\right)^2 = -\left(\frac{\omega}{c}\right)^2 \varepsilon_\pm \mu_\pm, \tag{7.18}$$

from which

$$\begin{aligned} \alpha_{\mathrm{p}\pm}^2 - \beta_{\mathrm{p}\pm}^2 &= -\left(\frac{\omega}{c}\right)^2 (\varepsilon_\pm \mu_\pm)' \\ 2\alpha_{\mathrm{p}\pm}\beta_{\mathrm{p}\pm} &= \left(\frac{\omega}{c}\right)^2 (\varepsilon_\pm \mu_\pm)'' \end{aligned} \tag{7.19}$$

[3] In many cases, these complex permittivity parameters are defined with a positive sign.

and

$$\alpha_{p\pm}^2 = \left(\frac{\omega}{c}\right)^2 \frac{1}{\sqrt{2}} \left(\sqrt{(\varepsilon_\pm \mu_\pm)'^2 + (\varepsilon_\pm \mu_\pm)''^2} - (\varepsilon_\pm \mu_\pm)' \right)^{1/2},$$

$$\beta_{p\pm}^2 = \left(\frac{\omega}{c}\right)^2 \frac{1}{\sqrt{2}} \left(\sqrt{(\varepsilon_\pm \mu_\pm)'^2 + (\varepsilon_\pm \mu_\pm)''^2} + (\varepsilon_\pm \mu_\pm)' \right)^{1/2}. \quad (7.20)$$

This general expression can be expanded in terms of the μ and κ, ε_0 and ε_1 parameters, each of which contain imaginary terms. For practical situations, the assumption that the loss terms are small is valid, where $\omega \gg$ or $\ll \gamma H_r$, and $(\varepsilon_\pm \mu_\pm)'' / (\varepsilon_\pm \mu_\pm)' \ll 1$. By means of binomial expansions, (7.20) can readily be simplified to

$$\alpha_{p\pm} \approx \left(\frac{\omega}{2c}\right) \frac{(\varepsilon_\pm \mu_\pm)''}{\sqrt{(\varepsilon_\pm \mu_\pm)'}},$$

$$\beta_{p\pm} \approx \left(\frac{\omega}{c}\right) \frac{(\varepsilon_\pm \mu_\pm)'}{\sqrt{(\varepsilon_\pm \mu_\pm)'}}, \quad (7.21)$$

which reduces to (6.118) when $\varepsilon_1 = 0$. As defined previously in (6.122) the Faraday rotation angle per unit path length can now be expressed as

$$\varphi_F = \Delta\varphi/2 = \frac{(\beta_{p-} - \beta_{p+})}{2} z \approx \left(\frac{\omega}{2c}\right) \left(\sqrt{(\varepsilon_- \mu_-)'} - \sqrt{(\varepsilon_+ \mu_+)'} \right) z. \quad (7.22)$$

Where $\alpha \neq 0$, the corresponding amplitudes of $\boldsymbol{H}_{rf-}$ and $\boldsymbol{H}_{rf+}$ experience different amounts of attenuation, which leads to a nonlinear or "elliptical" polarization of the resultant wave in the magnetized medium.

Applications for the combined theory might be found in the frequency regime between microwave and optical bands. In the far-infrared regions beyond where simple microwave Zeeman-type splittings $\sim g m_B H$ are limited by applied magnetic fields of only a few Tesla, exchange fields of 10^2–10^3 T can produce splittings of much higher energy. The corresponding magnetic transitions can overlap the electric-dipole transitions of rotational and vibrational molecular spectra. If both electrical and magnetic contributions overlap in nonabsorptive transmission "windows" at frequencies where the imaginary parts of μ, κ, ε_0, and ε_1 are negligible, i.e., no near-resonance conditions, (7.22) can be written as

$$\varphi_F \approx \left(\frac{\omega}{2c}\right) \left(\sqrt{(\varepsilon_0' - \varepsilon_1')(\mu' - \kappa')} - \sqrt{(\varepsilon_0' + \varepsilon_1')(\mu' + \kappa')} \right) z. \quad (7.23)$$

By setting the conditions to isolate individual magnetic and electrical propagation interactions, the influence of the off-diagonal tensor elements can be readily highlighted as

$$\varphi_{F\mu} \approx -\left(\frac{\omega}{2c}\right)\sqrt{\frac{\varepsilon_0'}{\mu'}}\kappa' z$$

$$\varphi_{F\varepsilon} \approx -\left(\frac{\omega}{2c}\right)\sqrt{\frac{\mu'}{\varepsilon_0'}}\varepsilon_1' z. \tag{7.24}$$

The Faraday rotation angle expressions can be simplified further if the frequency-dependent relations for κ' and ε_1' are substituted into (7.24). For the magnetic case, it can be readily shown that $\phi_{F\mu}$ reduces to (6.122) if $\omega^2 \gg \omega_0^2$. The dielectric case can be treated similarly with the relations given in Sect. 7.3.

7.3 Magneto-Optical Spectra

In the energy bands above where average magnetic fields can create Zeeman splittings that allow magnetic-dipole transitions, or where inter-ion spin flips can occur through magnetoelastic phonon interactions (exchange resonance), the spectra are referred to as optical rather than electromagnetic and are the result of actual excitations within the orbital angular momentum structure set by the Russell–Saunders coupling. In the optical bands, the relevant parameters that describe the interaction between waves and matter are the refractive index n and extinction coefficient k. Propagation properties are then usually characterized in terms of the complex index of refraction $\boldsymbol{N} = n - \mathrm{i}k$. For the two circular polarization modes, $N_\pm$ are related to the propagation constants $\gamma_{\mathrm{p}\pm}$ according to

$$N_\pm^2 = -\left(\frac{\mathrm{c}}{\omega}\right)^2 \gamma_{\mathrm{p}\pm}^2 = \mu\left(\varepsilon_0 \pm \varepsilon_1\right) = \left(n_\pm - \mathrm{i}k_\pm\right)^2 = -\left(\frac{\mathrm{c}}{\omega}\right)^2 \left(\alpha_\pm + \mathrm{i}\beta_\pm\right)^2. \tag{7.25}$$

Therefore,

$$-\mathrm{i}\gamma_{\mathrm{p}\pm} = \left(\frac{\omega}{\mathrm{c}}\right)\mathrm{N}_\pm \quad \text{or} \quad \left(\beta_\pm - \mathrm{i}\alpha\right) = \left(\frac{\omega}{\mathrm{c}}\right)\left(n_\pm - \mathrm{i}k_\pm\right). \tag{7.26}$$

Because the direct magnetic-dipole contributions are negligible in the optical frequency bands, it will be assumed that $\mu = 1$ in the analysis that follows, except for exchange resonance cases where both can apply. In Appendix 7A, the relations for complex off diagonal elements are derived for use in the ensuing sections.

7.3.1 Electric-Dipole Transitions

If the dispersion of $[\mu]$ and $[\varepsilon]$ elements occurs in widely separated frequency regimes, the frequency dependence of the tensors can be analyzed separately, starting with $[\mu]$ and then applying the results to $[\varepsilon]$ by analogy. For low power conditions without damping effects (omitted for simplification), the respective gyromagnetic resonance and nonresonance scalar solutions for the right-hand RHCP(+) and left-hand LHCP(-) circular polarization permeabilities $\mu_{\pm} = \mu \pm \kappa$ at a signal frequency ω are given by [3, 11]

$$\mu_{\pm} = 1 + \omega_{\mathrm{M}}\rho_{\pm} = 1 + \frac{\omega_{\mathrm{M}}}{\omega_0 \mp \omega}, \tag{7.27}$$

where $\omega_0 = \gamma H_0$ and $\omega_{\mathrm{M}} = \gamma 4\pi M$, the frequency equivalent of the magnetization. Note that the (+) mode is dominant at resonance $\omega = \omega_0$, but both functions become comparable away from resonance. To account for damping effects, a relaxation rate parameter Γ is introduced. In spectral terms, Γ represents the half-linewidth (often designated as $\Delta\omega$) of a Lorentzian function. Following the Bloch–Bloembergen model [12], the lineshape functions are expanded to complex relations by substitution of $(\omega - \mathrm{i}\Gamma) \rightarrow \omega$, to yield

$$\rho_{\pm} = \frac{(\omega_0 \mp \omega) \mp \mathrm{i}\Gamma}{(\omega_0 \mp \omega)^2 + \Gamma^2}. \tag{7.28}$$

Transition probabilities for magnetic- and electric-dipole interactions with plane-wave radiation are explained by quantum mechanical time-dependent perturbation theory [13–15]. Rotating-dipole diagrams are presented in Fig. 7.5. The parameter ω_{M} $(= \gamma 4\pi \mathrm{M})$ can be expressed as

$$\omega_{\mathrm{M}} = \left(\frac{g m_{\mathrm{B}}}{\hbar}\right) 4\pi N g m_{\mathrm{B}} S = \frac{4\pi N}{\hbar} \left|\langle \mathrm{g}|\, g m_{\mathrm{B}} S \,|\mathrm{e}\rangle\right|^2, \tag{7.29}$$

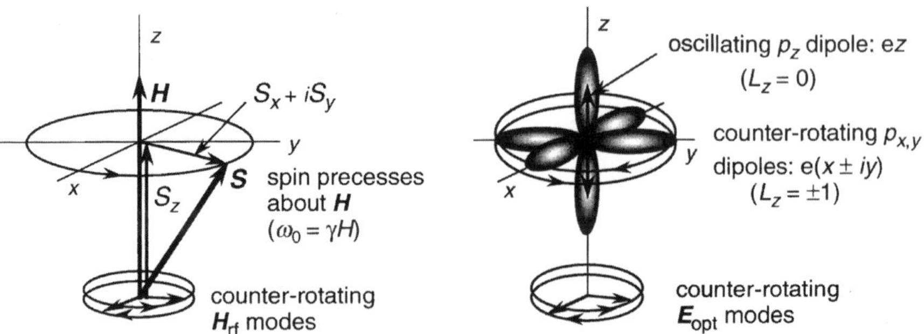

Fig. 7.5 Rotating angular momentum diagrams: (**a**) spin angular momentum $\boldsymbol{S}$ precessing at the Larmor frequency about a z-axis magnetic field $\boldsymbol{H}$, and driven by circularly polarized modes of an rf field $\boldsymbol{H}_{\mathrm{rf}}$ in x–y plane; magnetic resonance occurs when signal frequency equals the Larmor precession frequency of the $\boldsymbol{S}$ vector; and (**b**) split $p_{x,y}$ orbital states of a magnetic molecule with orbital angular momentum $\boldsymbol{L}$ that rotates with the electric field $\boldsymbol{E}_{\mathrm{rf}}$ circular polarization modes in x–y plane; optical transition frequencies correspond to the particular quantum state energies of the electric dipoles. Figure reprinted from [3] with permission. © 2005 by the American Institute of Physics

where N is the net volume density of magnetic ions and S is the spin of one ion.[4] For the basic example of a Zeeman split $S = 1/2$ case (Fig. 7.5a) with an excitation $|+1/2\rangle \rightarrow |-1/2\rangle$,

$$\omega_{\mathrm{M}} = \frac{4\pi N}{\hbar}\left|\left\langle +\frac{1}{2}\right| g m_{\mathrm{B}}\left(S_x + \mathrm{i}S_y\right)\left| -\frac{1}{2}\right\rangle\right|^2 \tag{7.30}$$

thereby observing the basic selection rule $\Delta S_z = +1$.

With $m_{\mathrm{E}}^{\mathrm{a,b}}$ as the electric-dipole moment of the respective a and b quantum orbital states $L_z = \pm 1$ of an orbital triplet P term sketched in Fig. 7.5b, for an orbital singlet ${}^2S$ ground state $L_z = 0$ a parameter $\omega_{\mathrm{E}}^{\mathrm{a,b}}$ is defined analogous to ω_{M} from (7.29) and (7.30) as

$$\omega_{\mathrm{E}}^{\mathrm{a,b}} = \frac{4\pi N}{\hbar}\left|\langle g|\, m_{\mathrm{E}}^{\mathrm{a,b}}\, |e^{\mathrm{a,b}}\rangle\right|^2 = \frac{4\pi N}{\hbar}\left|\left\langle \frac{z}{r}\right| er\left(L_x \pm \mathrm{i}L_y\right)\left|\frac{x \pm \mathrm{i}y}{r}\right\rangle\right|^2 . \tag{7.31}$$

From the definition of Suits [16], the oscillator strength $f^{\mathrm{a,b}} = \left(m\omega_0^{\mathrm{a,b}}/\hbar e^2\right) \left|\langle g|\, m_{\mathrm{E}}^{\mathrm{a,b}}\, |e^{\mathrm{a,b}}\rangle\right|^2$ and $\omega_{\mathrm{E}}^{\mathrm{a,b}} = 4\pi N f^{\mathrm{a,b}}/m\omega_0^{\mathrm{a,b}}$, where e and m is the electron charge and mass, respectively.

In the optical bands, the basic physics involves transitions between atomic states with orbital angular momentum splittings caused by spin–orbit coupling. For this example only one spectral transition pair will be considered. In the magnetic-dipole case, the off-diagonal element κ emerges directly from the solution of the classical Larmor model in the form of a spin-flip transition, as shown in Fig. 7.6a.

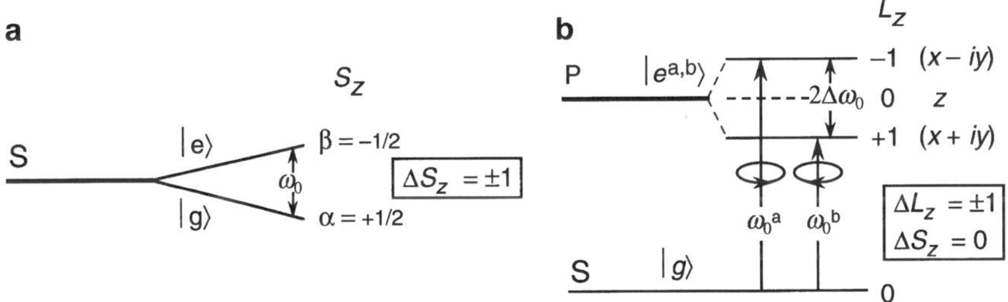

Fig. 7.6 Energy level diagrams corresponding to the physical models of (**a**) and (**b**). Only the split excited state (diamagnetic) case of electric-dipole transitions is shown; for the split ground state (paramagnetic), the levels are inverted. The use of ω to represent quantum energy implies a suppressed Planck's constant multiplier $\hbar$. Note that for the electric-dipole transition the $\Delta S_z = 0$ condition is added as a selection rule. Figure reprinted from [3] with permission. © 2005 by the American Institute of Physics

[4] In this discussion, the symbol N is used instead of n to avoid conflict with the index of refraction n for optical propagation.

A magnetic field creates a frequency splitting ω_0 and determines which circular polarization mode experiences gyromagnetic resonance and which is unaffected, as defined by (7.27). By contrast, there are two transition frequencies ω_0^{a} and ω_0^{b} in the birefringent electric-dipole case, one for each polarization mode. Furthermore, the selection rules for transitions between angular momentum terms are now $\Delta L = \pm 1$ (Laporte's interval rule) and also $\Delta S = 0$.

For separate permittivities from the multiplet splitting, each orbital angular momentum state interacts with the electric vectors of the *both* circular polarization modes, one in resonance and the other in nonresonance, thereby creating four permittivity functions. The models for the counterrotating electric-dipole moments that carry the two opposing states of orbital angular momentum ($L_z = \pm 1$) correspond to the split levels of the ^2P orbital quantum term as illustrated in Fig. 7.6b. Since the lineshape factors are identical for both magnetic- and electric-dipole resonances, the relations of (7.27) and (7.28) can be adopted without loss of generality for each electric-dipole transition to produce the circular polarization permittivities $\varepsilon_{\pm} = \varepsilon_0 \pm \varepsilon_1$.

For the dual transitions between $L_z = \pm 1$ of the ^2P state and $L_z = 0$ of the ^2S state, circular mode permittivities corresponding to states labeled a and b are determined first from a modification of (7.27) and (7.28):

$$\varepsilon_{\pm} = 1 + \omega_{\mathrm{E}}^{\mathrm{a}}\rho_{\pm}^{\mathrm{a}} + \omega_{\mathrm{E}}^{\mathrm{b}}\rho_{\mp}^{\mathrm{b}} = 1 + \omega_{\mathrm{E}}^{\mathrm{a}}\left(\frac{1}{\omega_0^{\mathrm{a}} \mp \omega}\right) + \omega_{\mathrm{E}}^{\mathrm{b}}\left(\frac{1}{\omega_0^{\mathrm{b}} \pm \omega}\right). \tag{7.32}$$

With the aid of (7.32), the tensor elements are obtained from $\varepsilon_0 = (1/2)(\varepsilon_+ + \varepsilon_-)$ and $\varepsilon_1 = (1/2)(\varepsilon_+ - \varepsilon_-)$. For the split excited state of Fig. 7.7a,

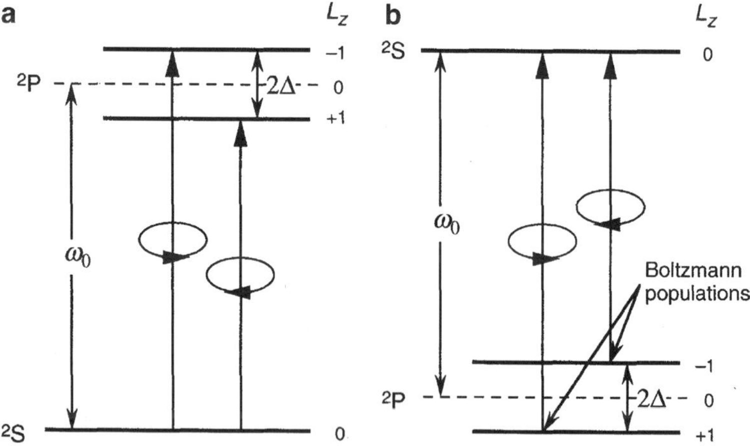

Fig. 7.7 Two basic cases for the model transition $^2\mathrm{S} \leftrightarrow {}^2\mathrm{P}$: (**a**) split excited state (termed diamagnetic) and (**b**) split ground state (termed paramagnetic with temperature-dependent populations). Figure reprinted from [3] with permission. © 2005 by the American Institute of Physics

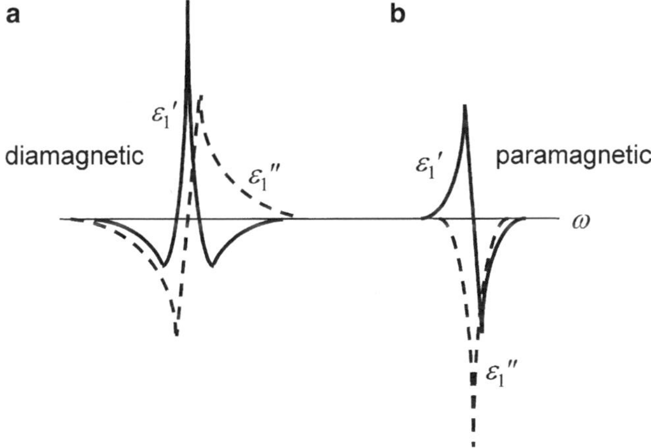

Fig. 7.8 Schematic models of ε_1' and ε_1'' characteristics as a function of frequency: (**a**) diamagnetic transition and (**b**) paramagnetic transition

$$\varepsilon_0 = 1 + \omega_E \left[\frac{(\omega_0 - \omega)}{(\omega_0 - \omega)^2 - \Delta^2} + \frac{(\omega_0 + \omega)}{(\omega_0 + \omega)^2 - \Delta^2} \right],$$
$$\varepsilon_1 = -\omega_E \Delta \left[\frac{1}{(\omega_0 - \omega)^2 - \Delta^2} - \frac{1}{(\omega_0 + \omega)^2 - \Delta^2} \right], \tag{7.33}$$

where $\omega_E^a = \omega_E^b = \omega_E$ and $2\Delta = \omega_0^a - \omega_0^b$. As a complement to the diagrams of Fig. 7.7, the respective curve shapes as a function of ω in the region of resonance ω_0 are modeled in Fig. 7.8.

Because the total population density N is shared among the three orbital components of the ground state, for an absorptive transition the effective population of any one level is one-third of the total. The permittivity relations with both resonance and nonresonance terms for the split ground state (Figs. 7.7b and 7.8b) case can be readily derived as

$$\varepsilon_0 = 1 + \frac{1}{3} \left[\frac{\omega_E (\omega_0 - \omega) - \Delta_E \Delta}{(\omega_0 - \omega)^2 - \Delta^2} + \frac{\omega_E (\omega_0 + \omega) - \Delta_E \Delta}{(\omega_0 + \omega)^2 - \Delta^2} \right],$$
$$\varepsilon_1 = -\frac{1}{3} \left[\frac{\omega_E \Delta - \Delta_E (\omega_0 - \omega)}{(\omega_0 - \omega)^2 - \Delta^2} - \frac{\omega_E \Delta - \Delta_E (\omega_0 + \omega)}{(\omega_0 + \omega)^2 - \Delta^2} \right], \tag{7.34}$$

where $\omega_E \approx (3/2)\left(\omega_E^a + \omega_E^b\right)$ and $\Delta_E = (3/2)\left(\omega_E^a - \omega_E^b\right)$. If the Boltzmann factor is included to account for the temperature dependence of the populations N_a, N_o, and N_b of the split ground state (sketched in Fig. 7.9), the following relation can be written for the condition that $2\Delta \ll kT$,

$$\Delta_E = \omega_E \left(\frac{1 - e^{-2\Delta/kT}}{1 + e^{-2\Delta/kT}} \right) = \omega_E \tanh\left(\frac{\Delta}{kT} \right) \approx \frac{\omega_E \Delta}{kT}, \tag{7.35}$$

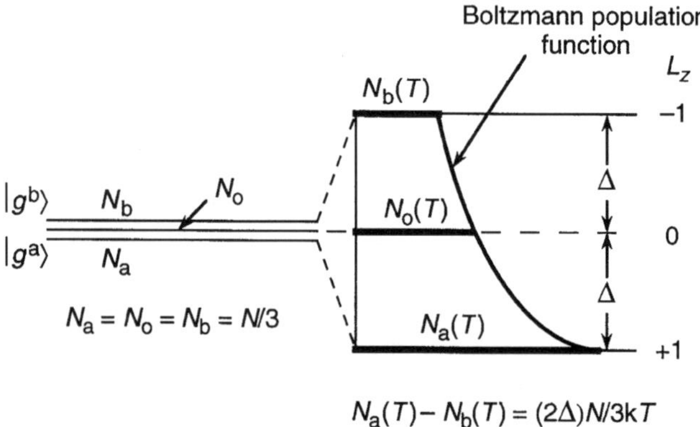

Fig. 7.9 Temperature dependence of the split ground-state (paramagnetic) case as determined by a Boltzmann population distribution function. Figure reprinted from [3] with permission. © 2005 by the American Institute of Physics

and the temperature dependence can be introduced to (7.34) according to

$$\varepsilon_0 = 1 + \frac{\omega_E}{3}\left[\frac{(\omega_0 - \omega) - \Delta^2/kT}{(\omega_0 - \omega)^2 - \Delta^2} + \frac{(\omega_0 + \omega) - \Delta^2/kT}{(\omega_0 + \omega)^2 - \Delta^2}\right]$$
$$\varepsilon_1 = -\frac{\omega_E \Delta}{3}\left[\frac{1 - (\omega_0 - \omega)/kT}{(\omega_0 - \omega)^2 - \Delta^2} - \frac{1 - (\omega_0 + \omega)/kT}{(\omega_0 + \omega)^2 - \Delta^2}\right]. \tag{7.36}$$

Algebraic sums of selected combinations of the four Lorentzian lineshape functions can produce the exact ε_0 and ε_1 tensor elements for any electric-dipole transition with split $L = \pm 1$ states. It is important to recognize that the common practice of omitting the nonresonance term from the these relations can lead to large errors in the frequency regimes where near-infrared isolators are designed to operate, as discussed previously [17]. With this approach, these inaccuracies can be easily avoided. In systems where multiple transition frequencies from different ions occupying different crystal sites, these building blocks can be used with appropriate weighting of the ω_E parameter for each ω_0 of the spectrum. For the degenerate ground-state case, the temperature dependence of the common "paramagnetic" magneto-optical spectra can be readily inserted into the formalism. Damping effects that produce line broadening can be added in a straightforward manner at any stage by substituting a standard relation containing the half linewidth Γ, such as $(\omega - i\Gamma) \to \omega$ [12].

7.3.2 Yttrium Iron Garnet Spectra (Paramagnetic)

The optical and magneto-optical properties of ferrimagnetic yttrium iron garnet $Y_3Fe_5O_{12}$ (YIG) were first investigated in 1959 by Dillon [18], when a large Faraday rotation was discovered with the onset of a strong optical absorption at photon energies near 2.5 eV. Kahn et al. [19] determined that the magneto-optical properties were caused by paramagnetic transitions, i.e., $g^{a,b}$ ground states, in the Fe^{3+} ions. Since the mid-1970s, Scott et al. [20], Wittekoek et al. [21], and Doorman et al. [22] were able to match many of the identified magneto-optical lines with peaks in the various spectra. However, a detailed picture of the quantum structure necessary to produce the observed birefringent spectra originating from orbital angular momentum $L_z = \pm 1$ degeneracies in the ground states has proven to be challenging. The results of a more recent semiempirical fit between permittivity theory and Kerr-effect measurements by Allen [23, 24] will be summarized.

The diagonal tensor elements $\varepsilon_0 = \varepsilon_0' + i\varepsilon_0''$ are deduced from measurements of refractive index n and the extinction coefficient k using the standard relation $\varepsilon_0 = (n + ik)^2$. The off-diagonal elements $\varepsilon_1 = \varepsilon_1' + i\varepsilon_1''$ are calculated from Kerr ellipticity η_K and rotation θ_K measurements from the relations developed in Appendix 7A

$$\begin{aligned} \varepsilon_1' &= \eta_K \left(n^3 - 3nk^2 - n\right) - \theta_K \left(k^3 - 3n^2k + k\right), \\ \varepsilon_1'' &= \eta_K \left(k^3 - 3n^2k + k\right) + \theta_K \left(n^3 - 3nk^2 - n\right). \end{aligned} \tag{7.37a}$$

and

$$\left.\begin{aligned} \varepsilon_1' &\approx \eta_K \left(n^3 - n\right) \\ \varepsilon_1'' &\approx \theta_K \left(n^3 - n\right) \end{aligned}\right\} \text{for small } k \tag{7.37b}$$

Magneto-optical effects occur if the off-diagonal elements ε_1 are *non-zero*. This happens when the initial or final state of the electric-dipole transition includes a splitting of an L_z stationary states that then fix separate spectral energies for plane waves of right- and left-handed circular polarization. Expressions for the tensor elements as a function of angular frequency ω and transition frequency ω_0 are standard in the literature [19]. As illustrated by Suits [16], a $\Delta L_z = \pm 1$ splitting can occur in either ground (*paramagnetic*) or excited (*diamagnetic*) if an effective ^2P-state component is present. The perturbing agent that lifts the L_z degeneracy is spin–orbit coupling $\lambda \boldsymbol{L} \cdot \boldsymbol{S}$. For the paramagnetic case of particular focus here, the split ground states are thermally populated according to a Boltzmann population distribution, as discussed in Sect. 7.3.1. Consequently, a difference in the number of right- and left-handed occupied sites available for selective absorption of circularly polarized light is established. For the present study, the population ratio of the paramagnetic case and its relation to the energy splitting Δ_g can be found from the peak of ε_1'' at resonance frequency ω_0 divided by the corresponding optical absorption ε_0''

$$\left|\frac{\varepsilon_1''(\omega_0)}{\varepsilon_0''(\omega_0)}\right| \approx \frac{N_+ - N_-}{N_+ + N_-} = \frac{\Delta N}{N} \approx \frac{\Delta_g}{2kT}, \tag{7.38}$$

where N_+ and N_- are the populations of the split ground state in the limit of $\Delta_g \ll kT$. It is easily shown that (7.38) represents the difference of oscillator strengths for the separated transitions that interact, respectively, with right- and left-handed circularly polarized waves [16, 19]. A diamagnetic transition results from a $\lambda \boldsymbol{L} \cdot \boldsymbol{S}$ splitting of the excited state, characteristic of BiYIG compounds [25, 26]. The frequency dependence of ε_1' and ε_1'' in the diamagnetic case presents line shapes nearly the reverse of those of the paramagnetic case [16], which is convenient for identification purposes. Note also that because spin–orbit interaction organizes the orbital angular momentum $L_z = \pm 1$ states in either case, the associated spin S_z states must be aligned in a saturating magnetic field to produce an observable cooperative effect.

Figure 7.10 presents the results of the diagonal tensor elements, and Fig. 7.11 shows the Kerr ellipticity and rotation. The identification of the transitions responsible for the spectra of the diagonal elements was accomplished by fitting the imaginary part to the Gaussian peaks shown in Fig. 7.12, with the parameter values listed in Table 7.2. The exercise was begun by assuming energy values corresponding to the features of the measured spectrum. Moreover, to match the data for $\varepsilon_0''(\omega_0)$ at energies above 4 eV found in the literature [21, 22], a strong charge-transfer transition was assumed at 4.35 eV. The Gaussian curve fit to the ε_0'' spectrum compares well with fits achieved by others [27, 28]. Most notable is the agreement with the onset of strong absorption peaks commencing at approximately 2.85 eV.

The ε_1 spectra were fit by assuming a series of paramagnetic transitions [23, 24]. The agreement shown in Fig. 7.13 was accomplished with the parameters listed in Table 7.2. Note the correspondence with the transition energies found from the fit to the spectrum of ε_0''. An exception is a transition at 3.93 eV which has no match in the Gaussian curve fit to ε_0''. To extend the agreement with data in the literature [21, 22] for energies above 4 eV, we added the strong paramagnetic transition in the vicinity of 4.35 eV.

The widths of the magneto-optical peaks listed in Table 7.2 are generally smaller than those of the ε_0'' spectra. The ε_0'' spectrum was modeled with Gaussian functions

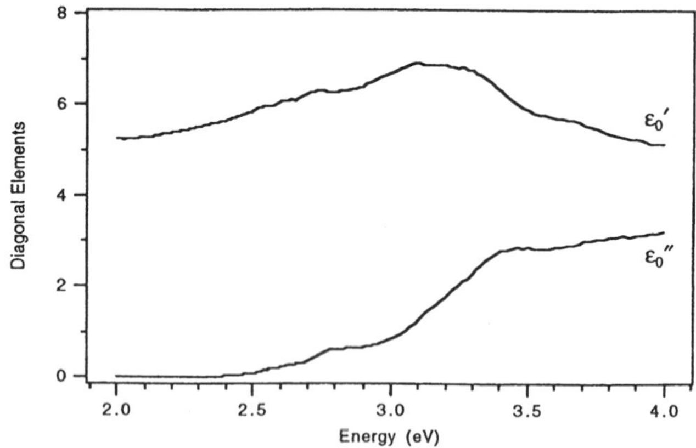

Fig. 7.10 Measured spectra of the real and imaginary parts of ε_0 for YIG. Figure reprinted from [24] with permission. © 2003 by the American Institute of Physics

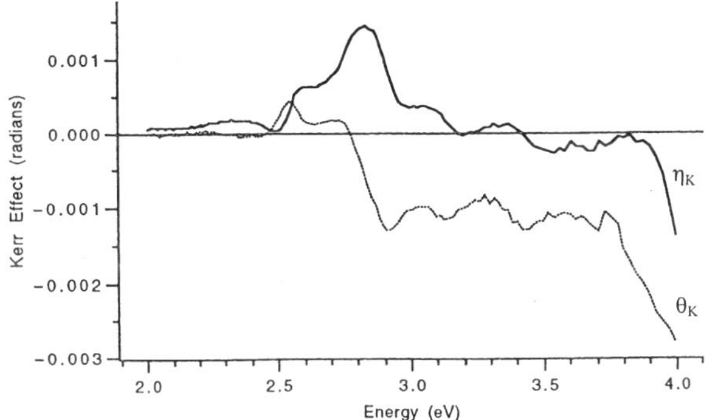

Fig. 7.11 Measured Kerr ellipticity and rotation of YIG. Figure reprinted from [24] with permission. © 2003 by the American Institute of Physics

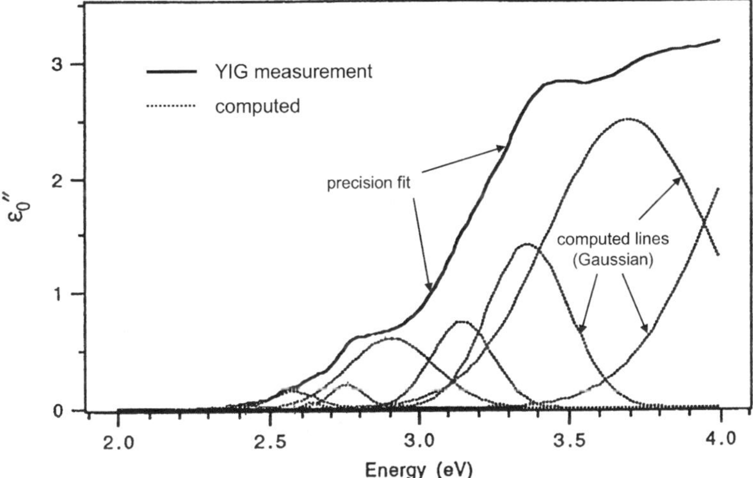

Fig. 7.12 Precision fit of Gaussian curves to data of ε_0''. Computed and measured curves are almost perfectly coincident. Individual computed curves are shown. The parameters used in the fitting are listed in Table. 7.2 Images are reproduced from Allen's doctoral thesis [23]. Figure reprinted from [24] with permission. © 2003 by the American Institute of Physics

of half-width Γ_0 determined at the e^{-1} point relative to the peak value, while the Lorentzian shapes assumed for the ε_1 components have smaller half-widths Γ_1 when characterized by values at half maximum.

From the numerical results obtained from curve fits of ε_0'' and ε_1'', (7.38) was used to determine the quantity $\Delta N/N$ listed in Table 7.2 for each transition. Note that the ε_1'' spectrum consists entirely of paramagnetic transitions. This result is surprising because the ground state of the Fe^{3+} ions is a ^6S term, with no orbital angular momentum. An explanation for the acquisition of orbital angular momentum by the

Table 7.2 Permittivity and ground-state splittings of $Y_3Fe_5O_{12}$

ω_0 (eV)	$\varepsilon_0''(\omega_0) \times 10^{-2}$	Γ_0 $(\text{eV} \times 10^{-2})$	$\varepsilon_1''(\omega_0) \times 10^{-2}$	Γ_1 $(\text{eV} \times 10^{-2})$	$\lvert\Delta N/N\rvert$	$\lvert\Delta_g\rvert$ $(\text{eV} \times 10^{-3})$
2.41	3.0	8.02	–	–	–	–
2.58	14.2	10.6	−1.16	4.2	0.082	4.1
2.64	4.4	3.9	–	–	–	–
2.73	19.4	8.3	−1.23	8.4	0.063	3.15
2.90	58.8	20.7	−3.58	11.0	0.061	3.05
3.15	73.5	14.6	2.40	10.8	0.033	1.65
3.37	141	18.8	2.28	16.5	0.016	0.80
3.70	250	37.1	–	–	–	–
3.93	–	–	5.36	23.8	–	–
4.35	338	43.1	−8.25	35.0	0.024	1.20

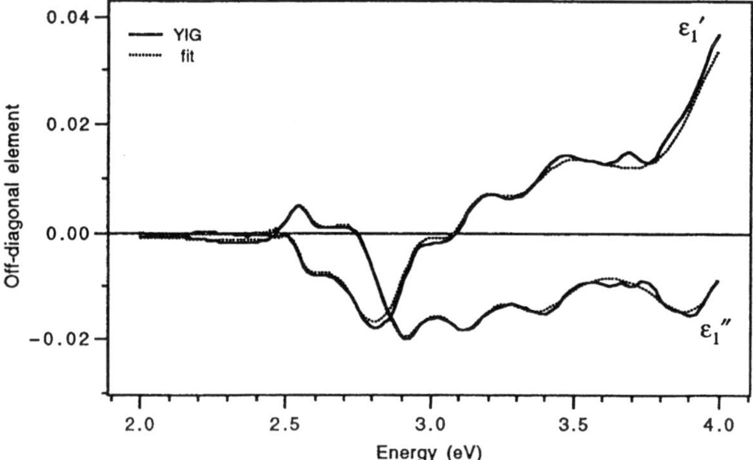

Fig. 7.13 Lorentzian curve fits to ε_1' and ε_1''. The parameters used in the fitting are listed in Table. 7.2 Figure reprinted from [24] with permission. © 2003 by the American Institute of Physics

iron ion ground state from excited terms *within* the ion was proposed by Clogston [29] and later refined by Scott et al. [20]. For the present exercise, a partial solution of an intersublattice covalent model is attempted.

Listed in Table 7.2 are the Δ_g magnitudes of the L_z-degenerate ground-state component split by spin–orbit coupling. In a ferrimagnetic spin system, superexchange stabilizes the maximum stationary value of spin component S_z of the Fe^{3+} ions in both octahedral and tetrahedral sites. The attendant covalent interaction also produces molecular-orbital states that hybridize the orbitals of the octahedral ${}^6S^{\text{o}}$ and tetrahedral ${}^6S^{\text{t}}$ terms in a bonding state that correlates the electrons into an antiferromagnetic spin alignment. More relevant to the present discussion, however, is the hybridization of the octahedral ${}^6S^{\text{o}}$ $(S_z = -5/2)$ term with the lowest excited term ${}^4T_{1g}^{\text{t}}$ $(S_z = +3/2)$ of the neighboring tetrahedral site, and

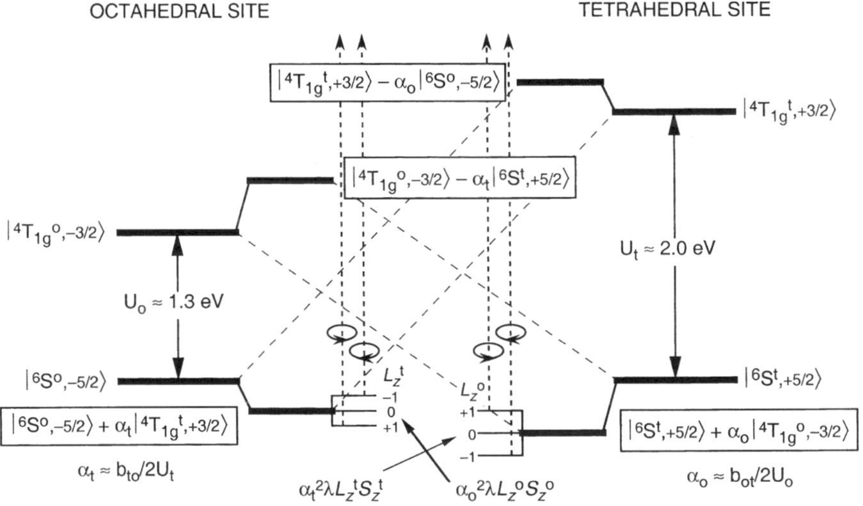

Fig. 7.14 Molecular-orbital model of the $^4T_{1g}$ mixing into the ^6S ground state of the opposing magnetic sublattice. Similar models can be made for the intrasublattice covalent mixing. Figure reprinted from [24] with permission.

vice versa with respect to $^6S^{\rm t}$ and $^4T^{\rm o}_{1g}$. The S_z values are adopted from the exchange energy perturbation used by Clogston [29]. Illustrated schematically in Fig. 7.14 are ground states reflecting a small orbitally degenerate component from the opposing sublattice $^4T_{1g}$ (and higher) terms. The unnormalized hybrid ground terms $|^6S^{\rm o}, -5/2\rangle + \alpha_{\rm t}|\,^4T^{\rm t}_{1g}, +3/2\rangle$ and $|^6S^{\rm t}, +5/2\rangle + \alpha_{\rm o}|\,^4T^{\rm o}_{1g}, -3/2\rangle$, where $\alpha_{\rm t}$ and $\alpha_{\rm o}$ are the respective mixing fractions which are given by the relevant factor $b_{ij}/2U_i$, half of the ratio of the transfer integral to the energy difference between the relevant states of the two cations.

If the T_{1g} triplet is treated as a quasi-P state, the ground-state splitting from $\lambda \boldsymbol{L}\cdot\boldsymbol{S} = \lambda\,[L_zS_z + (1/2)\,(L_+S_- + L_-S_+)]$ arises from the $\Delta L_z = \pm 1$ difference of the i-site diagonal matrix elements, which simplifies to

$$\begin{aligned}\left|\Delta^i_{\rm g}\right| &= 2\left\langle\left\langle ^6S^i, -\frac{5}{2}\right| + \alpha_j\left\langle ^4T^j_{1g}, +\frac{3}{2}\right| \left|\lambda\left(L^i_zS^i_z + L^j_zS^j_z\right)\right| \left|^6S^i, -\frac{5}{2}\right\rangle + \alpha_j\left|^4T^j_{1g}, +\frac{3}{2}\right\rangle\right\rangle \\ &\approx 2\alpha_j^2\left\langle ^4T^j_{1g}, +\frac{3}{2}\right|\lambda L^j_zS^j_z\left|^4T^j_{1g}, +\frac{3}{2}\right\rangle \approx 3\lambda\alpha_j^2. \end{aligned} \tag{7.39}$$

For the $3d^5$, $\lambda{\sim}0.06$ eV and $b_i{\sim}0.5$ eV (in oxides). The $U_{i,j}$ values [24] indicated in Fig. 7.14 result in $\alpha_i^2{\sim}10^{-2}$, which is sufficient to account for the $\Delta_{\rm g}$ splittings that approach 5 meV as deduced from $\Delta N/N$ values.

Although the concept of cross transfers between sublattices is speculative, there is further evidence that these effects can provide indirect avenues to satisfy the $\Delta S = 0$ selection rule. This subject will be revisited after the anomalous diamagnetic contributions of Bi^{3+} in YIG are reviewed.

7.3.3 Iron Garnets with Bismuth Ions (Diamagnetic)

In stark contrast to the theoretical spectra constructed from split ground state paramagnetic transitions that served to provide a high precision fit to the magneto-optical properties of $Y_3Fe_5O_{12}$, the story of the remarkable enhancement of Faraday rotation when Bi^{3+} ions are substituted for Y^{3+}, reported originally by Buhrer in 1969 will be outlined [30]. Rather than a population-sensitive split ground state, it is the splitting of the excited ^3P state by spin–orbit coupling that produces the large value of the off-diagonal tensor element ε_1, but only for select transitions in the Fe^{3+} ions in the magnetically opposed octahedral and tetrahedral sublattices [26]. Similar to the curve-fitting exercise described above, the approach adopted is based directly on the interpretation of reported data. The analysis is based on the overlapping of three diamagnetic-type Fe^{3+} transitions that are influenced by covalent interactions with Bi^{3+} ions.

For ferrimagnetic systems, the orbital singlet ground state ^6S of Fe^{3+} ions is unperturbed by the strong superexchange field that influences only the spin system (Sect. 7.1). For this reason, interpretation of the rotation and ellipticity spectra from $Y_{3-x}Bi_xFe_5O_{12}$ must logically be based on the diamagnetic functions described by the off-diagonal permittivity tensor elements $\varepsilon_1 = \varepsilon_1' + i\varepsilon_1''$, according to [25]

$$\varepsilon_1 = \omega_p^2 \sum_{+}^{-} \frac{f_\pm}{2\omega_0} \frac{\omega\left(\omega_{0\pm}^2 - \omega^2 - \Gamma^2\right) + i\Gamma\left(\omega_{0\pm}^2 + \omega^2 + \Gamma^2\right)}{\left[\left(\omega_{0\pm}^2 - \omega^2 + \Gamma^2\right)^2 + 4\omega^2\Gamma^2\right]}, \tag{7.40}$$

where $\omega_p^2 = 4\pi Ne^2/m$, $\omega_{0\pm} = \omega_0 \pm \Delta$, and N is the density of transition centers.[5] The oscillator strengths for the positive and negative rotations are $f_\pm \approx \pm f/2$ [which denotes a subtraction occurring in (7.40)]. If we introduce the excited state splitting, $f_\pm \approx (\pm f/2)\,(1 \pm \Delta/\omega_0)$, as determined by Allen and Dionne [17].

If this expression is separated into real and imaginary parts without approximations, one obtains separate relations for the ε_1' and ε_1'', which may be used to compute Faraday and Kerr rotations and ellipticities [22]. For the magnetic garnets at energies below 2.5 eV, the diagonal elements of the tensor $\varepsilon_0' \approx n^2$ (n is the index of refraction $\approx$ 2.3 in this regime) and $\varepsilon_0'' \approx 0$. The Faraday rotation constant is given by

$$\theta_F \approx (\omega/cn)\,\varepsilon_1', \tag{7.41}$$

For this system, both ε_0' and ε_0'' are only modestly increased by Bi^{3+} substitutions. These features make (7.41) accurate for use in this energy regime [22].

From inspection of the magnetic garnet magneto-optical data in the literature, one can identify two general groupings (1) Kerr effect ellipsometry measurements

[5] To remain consistent with the literature of this problem, the frequently-used parameter ω_p^2 is introduced. Its relation to the parameter ω_E defined to simplify the analysis in Sect. 7.3.1. is given by $\omega_E\omega_{0\pm} = \omega_p^2 f_\pm$.

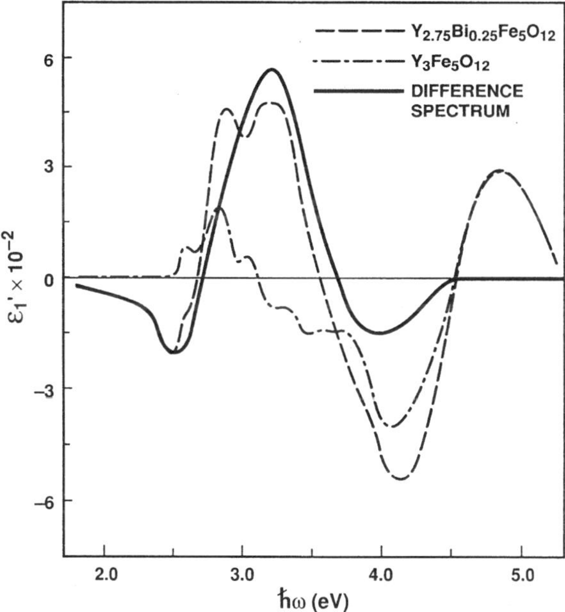

Fig. 7.15 $Y_{3-x}Bi_xFe_5O_{12}$ ε_1' data of Wittekoek et al. [21], showing the difference curve formed from the subtraction of the $x = 0$ and 0.25 curves. Figure reprinted from [25] with permission. © 1993 by the American Institute of Physics

of ε_1' and ε_1'' as a function of spectral energy over the range from 2 to 5 eV and (2) θ_F transmission measurements in the range below 2 eV. Although Faraday rotation effects are of practical importance at lower energies, the major optical events take place at higher energies. It is reflection data from the Kerr effects, therefore, that provide the fundamental clues to the source of the phenomena. According to published data [21,31], the major peak in ε_1' lies between 3 and 3.5 eV.

In Fig. 7.15, ε_1' data of $Y_{3-x}Bi_xFe_5O_{12}$ that have been reduced from Kerr ellipsometry measurements by Wittekoek et al. [22] are reproduced for $x = 0$ and 0.25. If the $x = 0$ (YIG) curve is treated as a base line, the curve for the Bi contribution is found by subtraction of the two curves. Figure 7.16 presents the results of the same procedure applied to the corresponding ε_1'' data. In both cases, the resultant curves reveal smooth Lorentzian-type functions of ω depicting behavior that is strongly suggestive of two or three individual transition bands below 4 eV. Equation (7.40) can be applied directly to the difference curves of Figs. 7.15 and 7.16 by selecting appropriate values of parameters Δ, Γ, and f for two principal diamagnetic transitions of opposite sign (the sign is determined by the direction of the magnetic moment that couples to the electric vector through spin–orbit coupling). As presented in Figs. 7.17 and 7.18, close fits to the experimentally derived curve are made with transitions at 2.6 and 3.15 eV over most of the range of measurement. To refine the interpretation and extend the model beyond 4 eV, a third transition of similar

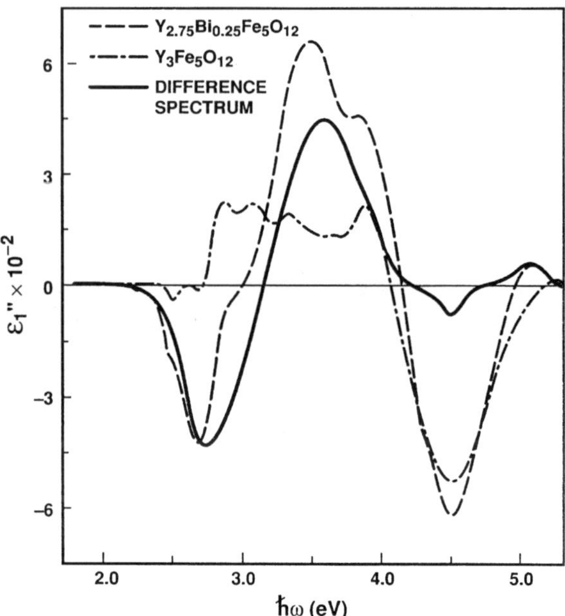

Fig. 7.16 $Y_{3-x}Bi_xFe_5O_{12}$ ε_1'' data of Wittekoek et al. [21], showing the difference curve formed from the subtraction of the $x = 0$ and 0.25 curves. Figure reprinted from [25] with permission. © 1993 by the American Institute of Physics

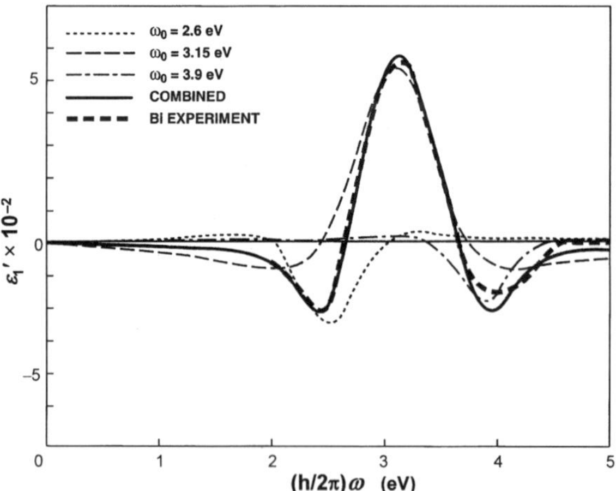

Fig. 7.17 Curves of ε_1' calculated from the parameter values listed in Table. 7.3 showing the combined curve of the Bi effect between energies of 0 and 5 eV. Difference curve from Fig. 7.15 is added for comparison. Figure reprinted from [25] with permission. © 1993 by the American Institute of Physics

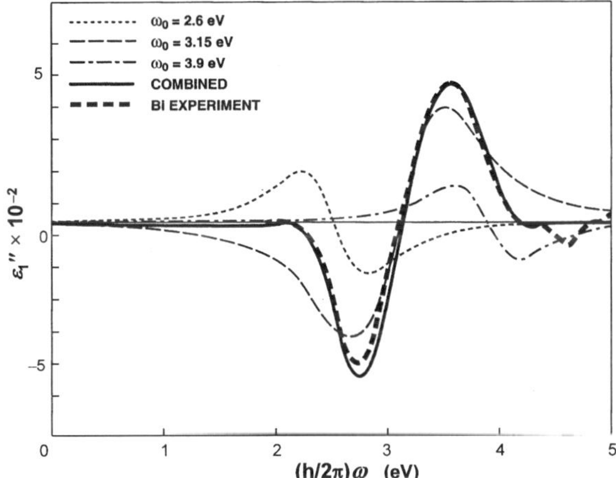

Fig. 7.18 Curves of ε_1'' calculated from the parameter values listed in Table. 7.3 showing the combined curve of the Bi effect between energies of 0 and 5 eV. Difference curve from Fig. 7.16 is added for comparison. Figure reprinted from [25] with permission. © 1993 by the American Institute of Physics

Table 7.3 Spectral parameters of ε_1' enhancement for $x = 0.25$

Lattice site	ω_0 (eV)	$\omega_p^2 f$ (eV)	Γ (eV)	Δ (eV)	Δ/Γ
Tetrahedral	2.6	−2.8	0.44	0.11	0.25
Octahedral	3.15	8	0.54	0.27	0.5
Tetrahedral[a]	3.9	−3	0.44	0.11	0.25

[a]This transition appears to be of tetrahedral Fe origin but may also be influenced by charge transfer excitations (which may not be Lorentzian) or by the strong Bi^{3+} $^1S \rightarrow {}^3P$ transition at 4.5 eV [34]

sign and proportions to the one at 2.6 eV is added at 3.9 eV. It is also important to recognize that the unique spectral shapes of both ε_1' and ε_1'' are fitted with the same set of parameter values listed in Table 7.3.

To highlight further the closeness of the fit between theory and experiment in the lower energy region, the calculated curve for ε_1' from Fig. 7.16 was modified according to (7.41), scaled to $x = 0.44$ (with the assumption that the θ_F dependence on Bi content remains linear at small values of x) [17] and plotted in Fig. 7.19 together with the corresponding θ_F measured curve of Simsa et al. [32] after subtraction of the $Y_3Fe_5O_{12}$ baseline. The close agreement over the range from 1 to 2.5 eV indicates that the principal Bi contributions are not of paramagnetic origin.

The opportunity to separate opposing magnetic sublattice contributions by the signs of the different Faraday peaks is an important aspect of magneto-optical spectra. Scott et al. [27] concluded that the 3.15-eV line originates in the octahedral Fe–O_6 complexes (transitions from ^6S to excited ^4G or ^4D bands [33]). From Fig. 7.16, therefore, one concludes that the weaker 2.6-eV peak is of tetrahedral origin (Fe–O_4). Most significant for practical matters is the dominance of the negative tail of the intense 3.15-eV line in the 1–2-eV region.

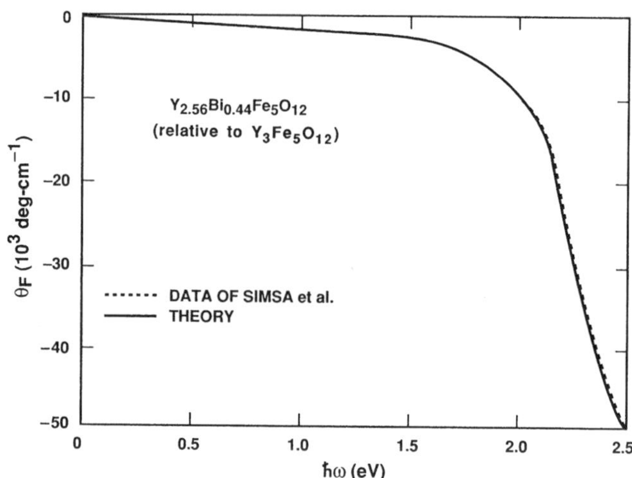

Fig. 7.19 Comparison between theory (scaled to $x = 0.44$) and experiment for energies below 2.5 eV, using parameter values of Table. 7.3 Data curve is from [32] and has also been corrected to remove $Y_3Fe_5O_{12}$ baseline. For fixed N, the largest reasonable Δ/Γ values below saturation levels [17] were chosen to fit the data. Figure reprinted from [25] with permission. © 1993 by the American Institute of Physics

As listed in Table 7.3, the Γ values for each transition are broad (0.25–0.5 eV). If the ε_0'' values reported [21] are not greatly increased by Bi additions, the large enhancement of ε_1' would not be caused by an increased f. For the same reason, the proposition that Bi^{3+} transitions cause the ε_1 anomalies would also have to be ruled out. For a homogeneous distribution, one explanation could be that Bi^{3+} ions perturb and enhance the multiplet splitting of the excited band [29] of the exchange-coupled Fe^{3+} lattices in direct proportion to the density and strength of the Fe^{3+}–O^{2-}–Bi^{3+} bond linkages, thereby producing (to a first approximation) a linear growth in the product $N\Delta$ with x, and a corresponding enhancement of ε_1 for selected Fe^{3+} transitions. This situation would be analogous to the increase of ferrimagnetic resonance linewidths caused by homogeneous distributions of fast-relaxing ions. As a result, the intensities of the calculated lines were fitted by selecting appropriate Δ values, which proved to be larger than reported earlier [21].

Reported saturation of the ε_1' peak at 3.15 eV as $x \to 2$ may be explained by the ratio $\Delta/\Gamma \to 1$. This result further suggests that approximations based on the $\Delta/\Gamma \ll 1$ assumption should be avoided in the interpretation of the Bi effects. It should be pointed out, however, that the ε_1' value at fixed energies in the negative tails may not saturate because the smaller reverse peaks are moved to lower and higher energies, as illustrated in Allen and Dionne [17]. This latter feature would prove beneficial for applications that require materials with the highest θ_F value at lower energies.

The magneto-optical effects at 4 eV and above are less easy to interpret because of the complexity of the excited overlapping bands and the threshold for charge transfer between Fe–O_6 and Fe–O_4 sites, as well as possible contributions from the Bi–O_{12} dodecahedral-site transition at 4.5 eV [34]. Because the key transitions are diamagnetic, meaning that the required orbital splitting occurs in a degenerate excited P state, the construction of a simple molecular-orbital model is a logical next step [26].

For ferrimagnets with exchange coupled Fe^{3+} ^6S ground states, polarization rotation of a wave propagating parallel to the magnetization vector (z direction) is described mathematically as a frequency-dependent off-diagonal element of the permittivity tensor. This requires interaction between the electric vector of the optical wave and the orbital angular momentum $\boldsymbol{L}$ of the excited state. For the magnetic moments to influence nonreciprocal effects, therefore, spin–orbit coupling is a necessary component. Since the excited-state splitting is the result of spin–orbit multiplet structure, contributions from the large Bi^{3+} spin–orbit interaction reflected through covalent molecular-orbital (MO) states become the focus of this discussion.

7.3.4 Fe^{3+}–Bi^{3+} Hybrid Excited States

For ferrimagnets with superexchange coupled Fe^{3+} ^6S ground states, polarization rotation of a wave propagating parallel to the magnetization vector (z direction) is described mathematically as a frequency-dependent off-diagonal element of the permittivity tensor. This requires interaction between the electric vector of the optical wave and the orbital angular momentum $\boldsymbol{L}$ of the excited state. For the magnetic moments to influence nonreciprocal effects, therefore, spin–orbit coupling is a necessary component. Since the excited-state splitting is the result of spin–orbit multiplet structure, contributions from the large Bi^{3+} spin–orbit interaction reflected through covalent molecular-orbital (MO) states is the focus of this discussion. The analysis that follows is based on previous work [26], but has been updated and employs nomenclature and sign conventions in closer conformance with the present text.

For a generic two-level system, the method developed in Chap. 2 can be readily adopted. The secular determinant is given by

$$\begin{vmatrix} E_1 - E & b - E\sigma \\ b - E\sigma & E_2 - E \end{vmatrix} = 0 \tag{7.42}$$

where $b \approx (E_1 + E_2)\,\sigma$ is the transfer integral and σ is the overlap integral. Solution of (7.42) yields the following eigenvalues

$$E_{\pm} = \frac{(E_1 + E_2)\,(1 - 2\sigma^2) \pm \sqrt{(E_1 - E_2)^2\,(1 - \sigma^2) + b^2}}{2\,(1 - \sigma^2)} \tag{7.43}$$

The MO eigenvectors for the two-level case with hybrid coefficients are $\varphi_+ = c_{11}\varphi_1 + c_{12}\varphi_2$ and $\varphi_- = c_{21}\varphi_1 - c_{22}\varphi_2$ for the bonding and antibonding states, respectively. There are two important results here (1) φ_+ is of lower energy than φ_1 by an amount $E_+ - E_1$ and (2) the volume densities of the initial wavefunctions φ_1 and φ_2 hybridize in proportion to the values of c_{11}^2 and c_{12}^2. For this generic example, the fractional participation of φ_2 in the bonding state is c_{12}^2.

As discussed in the foregoing section, the enhanced magneto-optical effects are believed to arise from the cooperative action of Fe^{3+} ions with degenerate excited orbital terms that are split further by covalent interactions with bismuth. The excited state splitting parameter Δ for individual ions is determined principally by the eigenstates of the operator $\lambda \boldsymbol{L} \cdot \boldsymbol{S}$. The z-axis collinearity of the N magnetic moments can occur through an applied field H or an exchange field H_{ex}. In both cases, the degree of alignment follows a Brillouin function $\mathcal{B}$, so that $N_{eff} = N\mathcal{B}(H, T)$. There is, however, an important distinction between these two situations – H influences both L and S by a Zeeman effect, but H_{ex} can affect only the total spin of the ion by flipping the ground S_z state.

The magnetization of the garnet is saturated in the z direction by superexchange fields, causing $\mathcal{B} \to 1$ and $N_{eff} \to N$. Through covalent interactions, the excited states of both Fe^{3+} and Bi^{3+} should also have nondegenerate spin states. If the exchange field perturbation is applied, the lowest state S_z value enters the calculation as a multiplier when the orbital magnetic moment degeneracies of the MO states are lifted by spin–orbit coupling $\lambda \boldsymbol{L} \cdot \boldsymbol{S}$, which now reduces to $\lambda L_z S_z$. Although S_z is captured by H_{ex}, there is no competing crystal field in this compound to prevent $\boldsymbol{L}$ from maintaining its alignment with the z axis, so that the maximum value of S_z for the particular $2S + 1$ manifold simply multiplies each $2L + 1$ state. For a ^4P state, this situation is depicted in Fig. 7.20. Energies of the spin–orbit coupling manifold in the hybrid state can be represented by

$$\begin{aligned} E_{LS} &= \langle \phi_+ | \lambda_1 L_{z1} S_{z1} + \lambda_2 L_{z2} S_{z2} | \phi_+ \rangle \\ &= \langle c_{11}\phi_1 | \lambda_1 L_{z1} S_{z1} | c_{11}\phi_1 \rangle + \langle c_{12}\phi_2 | \lambda_2 L_{z2} S_{z2} | c_{12}\phi_2 \rangle \end{aligned}$$

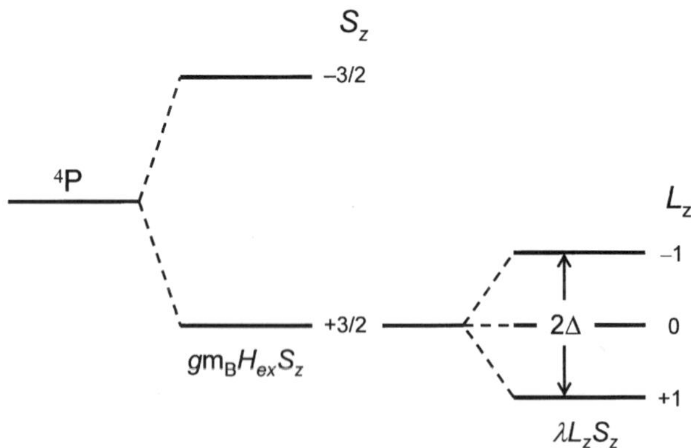

Fig. 7.20 Exchange-field spin quenching and spin–orbit splitting of a ^4P term

$$= c_{11}^2(\lambda_1 S_{z1})L_{z1} + c_{12}^2(\lambda_2 S_{z2})L_{z2}, \tag{7.44}$$

where S_{z1} and S_{z2} are the respective values of the lowest spin state. For the case of ${}^4P\left(Fe^{3+}\right)$, $L_{z1} = 0, \pm 1$ and $S_{z1} = -3/2$; for ${}^3P\left(Bi^{3+}\right)$, $L_{z2} = 0, \pm 1$ and $S_{z2} = -1$. Where $\lambda_2 \gg \lambda_1$ and c_{12}^2 is significant, (7.44) can be simplified by $E_{LS}(L_{z2}) \approx c_{12}^2(\lambda_2 S_{z2})L_{z2}$. The magnitude of the splitting then becomes $2\Delta \approx [E_{LS}(+1) - E_{LS}(-1)] \approx c_{12}^2\lambda_2$.

Before the excited states of the $Bi_3Fe_5O_{12}$ molecule are examined, the energy of the ground state must be determined. This energy may be estimated from the electrostatic interactions between ions and outermost electrons. Here the binding energy of the electrons of Fe^{3+} and Bi^{3+} can be computed from the algebraic sum of the cation ionization potentials (IP) and the energy of the field from the negatively charged anion. For the outermost electron, this repulsive energy (RE) is estimated by dividing the effective lattice energy (LE) by the ionic charge. For Fe^{3+}, IP $= -54.8$ eV and RE (from Fe_2O_3) $\approx +25.5$ eV [35]; for Fe^{3+} and Bi_2O_3, the corresponding values are -45.3 and $+15.8$ eV, so that a common resultant ground state ionic energy $E_g \approx -29$ eV is estimated for both Fe^{3+} and Bi^{3+}. Destabilization energies of the excited states are found by adding $+29$ eV to absorption energy values from spectral data [25].

Interpreting the magneto-optical properties of $Bi_3Fe_5O_{12}$ is begun by analyzing the covalent interactions between the excited ${}^3P$ term of Bi^{3+} in the dodecahedral (c) sites of the garnet lattice, and the excited states of Fe^{3+} in the d and a sites. Since Bi^{3+} excited state is about 4.2 eV above the ${}^1S$ ground state [27], hybridization is possible with several Fe^{3+} excited states. Diagrams of the orbital terms [33, 36] as functions of the crystal-field strength parameter Dq for the d and a sites are shown in Fig. 7.21.

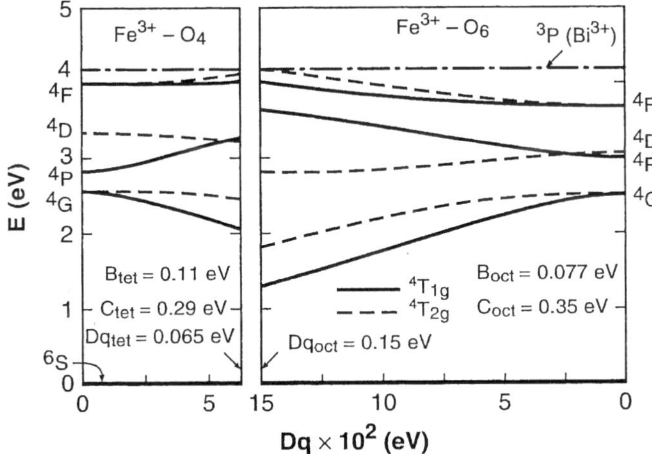

Fig. 7.21 Crystal-field sensitive excited terms for $\left(d^5\right)$ in d and a garnet sites as functions of crystal-field parameter Dq. Energies were computed from Tanabe and Sugano matrices [33], with B and C values based on the spectral analysis of Wood and Remeika [36]. Figure reprinted from [26] with permission. © 1994 by the American Institute of Physics

To construct the hybrid states that represent the bands of half-width Γ between the Bi^{3+} $(6s^1 6p^1)^3 P$ excited state and the Fe^{3+} excited states that would produce electron transition matrix elements with Fe^{3+} ^6S ground states, consider the following rationale: The candidates that satisfy the selection rule $\Delta L_z = \pm 1$, that have symmetry transformation properties similar to the Bi^{3+} ^3P state, and are crystal-field (i.e., Dq) selective are the $^4T_{1g}$ states. Inspection of Fig. 7.21 reveals three $^4T_{1g}$ terms, associated with the ^4G, ^4P, and ^4F terms.

For the principal transitions that appear to cause the magneto-optical anomalies at 2.6 and 3.15 eV [25], the obvious choice is the ^4P term, which has energies of 3.2 and 3.7 eV for the d and a sites, respectively, with corresponding Dq values of 0.065 and 0.15 eV. The bonding state hybrid function for the excited state is then expressed as $\phi_+ = c_{11} |^4 P\rangle + c_{12} |^3 P\rangle$. If these quantities are used in the computations, values that agree with experiment are listed in Table 7.4 for the transition energies $(E_+ - E_g)$ from the ground state to the bonding states for $c_{12}^2 = 0.26$ and 0.33, and $\sigma = 0.03$ and 0.04 for the d and a sites, respectively. The larger value of σ for the d site is consistent with its shorter bond lengths. To determine the Δ parameters listed in Table 7.4, the Fe^{3+} part of E_{LS} in (7.44) is treated as a baseline and the small contribution of λ_1 is ignored. With $\lambda_2 = \lambda_{Bi}$ and $S_{z2} = -1$, the multiplet energies $E_{LS} \approx c_{12}^2 \lambda_{Bi}, 0, -c_{12}^2 \lambda_{Bi}$, thereby yielding $\Delta \approx c_{12}^2 \lambda_{Bi}$. For $\lambda_{Bi} \approx$ 2 eV $(\sim 17,000\,\mathrm{cm}^{-1})$ [37] and $c_{12}^2 < 0.5$, Δ values for d and a sites are 0.52 and 0.66 eV, respectively, in general agreement with the conclusions of experiment. Calculated values of Γ are also in qualitative agreement with experimental results [25].

From the above calculations, Fig. 7.22 may be constructed to illustrate the origin of the Faraday rotation peaks at 2.6 and 3.15 eV. In addition, this work presents an opportunity to suggest that the weaker transition estimated to occur at 3.9 eV, initially labeled as d-site, may result from a partial overlap cancellation of transitions that originate from the MO states formed with the ^4F terms ($E_1 \approx 4.1$ eV) of Fe^{3+} ions in both d and a sublattices. This situation would explain that the location of a second peak from the a sublattice should appear somewhere in this general energy

Table 7.4 Molecular-orbital parameters of $Bi_3Fe_5O_{12}$[a]

Site	$E_1 - E_g$[b] (eV)	$E_+ - E_g$[b] (eV)	$\hbar\omega$ (eV)	σ	Γ[c] (eV)	c_{12}^2	Δ_{theor}[d] (eV)	Δ_{exp}[e] (eV)
tet d	3.2	2.6	2.6	0.04	~0.5	~0.26	0.52	~0.25
oct a	3.7	3.2	3.15	0.03	~0.4	~0.33	0.66	~0.5

[a] $E_2 \approx 4.2$ eV for the Bi^{3+} ^3P state influenced by Bi^{3+}–O^{2-}–Bi^{3+} interactions

[b] $E_g \approx -29$ eV, estimated from ionization potentials and lattice energies. The equivalence of values for both Fe^{3+} and Bi^{3+} in this model is a coincidence

[c] Half-bandwidth of bonding split band Γ $(\sim b/4)$, where $b = (E_1 + E_2)\sigma$, agrees with experimental estimate. The slightly larger value observed for the a site [25] may be due unresolved structure

[d] Based on $\lambda_{Bi} c_{12}^2$, with $\lambda_{Bi} \approx 2$ eV

[e] These values were determined from estimates of $Y_{3-x}Bi_xFe_5O_{12}$ with $x = 0.25$ [25]. Since only part of the full complement of Bi^{3+} interactions was present, the effective σ integrals would be smaller, thereby resulting in smaller c_{12}^2 values and accounting for the above Δ_{exp} values

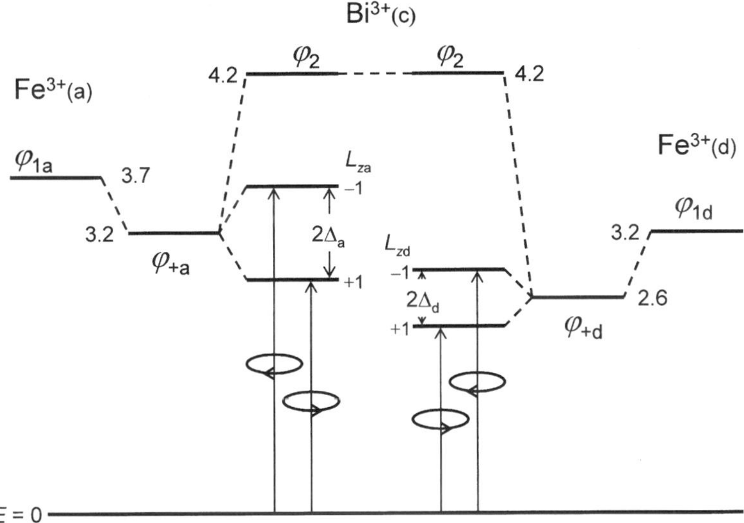

Fig. 7.22 Proposed molecular-orbital energy level diagram of $Bi_3Fe_5O_{12}$ referenced to a ground state of $E = 0$. Oxygen interactions and all antibonding states have been omitted to reduce confusion. Overlap and exchange integrals are treated semiempirically and represent resultants of superexchange (Bi–O–Fe) and direct exchange (Bi–Fe) covalent interactions. Electric-dipole transitions are presented as *intraionic*. Figure reprinted from [26] with permission. © 1994 by the American Institute of Physics

regime. The experimental determination that the peak at 3.9 eV has the same sign as the tetrahedral peak at 2.6 eV is consistent with the ratio (3:2) of d to a sites.

Based on the notion that magneto-optical effects involving excited states are dependent on the product $N_{\text{eff}}\Delta$, we may speculate on the effects of magnetic dilution and temperature variations. If the temperature T remains below the Curie temperature T_C, N_{eff} should track with the magnetizations of the respective sublattices as they apply to the individual transitions. Information about spin canting in both the ground and excited states may be extracted from the peak intensities as functions of Fe^{3+} content. For other ions with a $(6s)^2$ ground-state electron configuration similar to that of Bi^{3+}, i.e., Pb^{2+} and Tl^{1+}, similar magneto-optical effects should be expected.

Hansen et al. [38, 39] reported comprehensive data and analysis of magnetic and magneto-optical properties that revealed thermomagnetic behavior based on the molecular-field approach described in Chap. 4. These results verified the role of superexchange of the spin system in establishing the alignment of electric dipoles necessary for the observed collective magneto-optical effects at room temperature. Their investigations also included anomalous Faraday rotation enhancement at 633 nm wavelength from Pb^{2+} $(6s)^2$ ions substituted in c sites. With smaller λ values, however, the effects of the Δ splitting in these alternative ions might be reduced from those of Bi^{3+}.

7.3.5 *Intersublattice Transitions and the* $\Delta S = 0$ *Rule*

In the foregoing analysis of the hybrid electric-dipole excitations, the electron spin has been largely ignored. In Sect. 7.1, the role of superexchange in freezing the direction on the ionic spins in a collective magnetically ordered system was modeled. It was therefore assumed that magnetic-dipole transitions with selection rule $\Delta S_z = \pm 1$ do not occur in the optical frequency bands. If this assumption is valid, it follows that Laporte's rule for an orbital angular momentum transition $\Delta L_z = 0, \pm 1$ must be augmented by the rule $\Delta S_z = 0$ to deny spin reversals and conserve parity. As concluded in Sect. 7.3.4, however, the transitions that would explain the bismuth Faraday rotation anomaly are ${}^6S \rightarrow {}^4P$ and ${}^3P$ (hybrids), thereby seemingly in violation of this requirement. This issue is examined in the context of ${}^6A_{1a,d} \rightarrow {}^4T_{1d,a}$ intersublattice transfers within the antiferromagnetically paired Fe_a^{3+}–O^{2-}–Fe_d^{3+} molecule depicted in Fig. 7.23 and will be reviewed in this section.

The earliest indication that neighboring Fe^{3+} ions from opposing sublattices participate jointly in magneto-optical transitions was reported by Wood and Remeika, who concluded that the intensities of the spectral lines increased approximately as the square of the combined Fe^{3+} concentration in $Y_3Ga_{5-x}Fe_xO_{12}$ [36]. Similar observations were made by other workers [20,40,41]. In addition to the dramatic enhancement of the off-diagonal permittivity tensor element ε_1 at selected energies by Bi^{3+} substitution for Y^{3+}, moderate increases in the Curie temperature [42] and the diagonal element ε_0 [21,22,43] also occur. These latter effects suggest that Bi^{3+} also augments covalent interactions in the Fe_a^{3+}–O^{2-}–Fe_d^{3+} ground state, which further stabilize antiparallel spin order and boost the intersublattice transition probabilities by increasing the off-diagonal element ε_1 of the permittivity tensor. The magnitude of the element is directly dependent on the orbit and spin selection rules for the

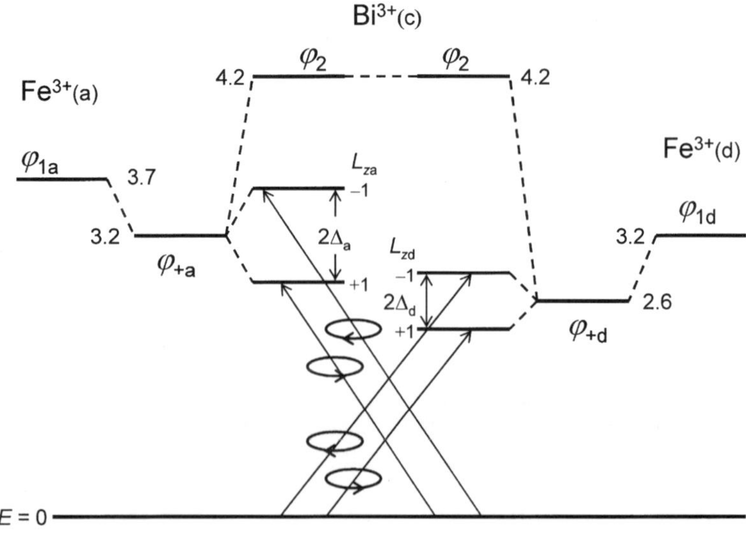

Fig. 7.23 Reproduction of Fig. 7.22 but with electric-dipole transitions presented as *interionic*

transition. In theory, a nonzero off-diagonal matrix element can occur only between even and odd parity orbital wavefunctions where the operator is odd. Therefore, $\Delta L = \pm 1$ for an electric-dipole operator, e.g., $|\langle 0| er |\pm 1\rangle|^2$. If the spin operator and functions are included, that contribution must be unchanged under the transition, thereby dictating that $\Delta S = 0$ and $\Delta S_z = 0$. For Fe^{3+} $(3d^5)$ ions in octahedral (O_6) and tetrahedral (O_4) coordinations, the orbital term energies formed within the $3d$ shell are presented as a function of the crystal-field parameter Dq in Fig. 7.21. Note that none of the spin multiplicities in the spectral energy range of interest will satisfy the $\Delta S = 0$ requirements by an intrasublattice transition. Furthermore, octahedral ${}^6A_{1g} \rightarrow {}^4T_{1g}$ would also violate the parity rule because both ground and excited states have centers of inversion symmetry and would be even functions.

In later work, Allen showed that independent dilution of $[a]$ and (d) sites by In^{3+} and Al^{3+}, respectively, in a $\{Y_{2.53}Bi_{0.47}\}[Fe_2](Fe_3)O_{12}$ host produced almost identical changes in the Bi^{3+}-enhanced complex permittivity ε_1, thereby confirming that both sublattices are involved [23,43]. The spectra of ε_1' and ε_1'' shown in Fig. 7.24 indicate a decrease in magnitude of about 25% across the energy band from 2.0 to 4.0 eV for both compositions. The amounts of In^{3+} and Al^{3+} were selected to reduce the respective sublattice Fe^{3+} concentrations by approximately

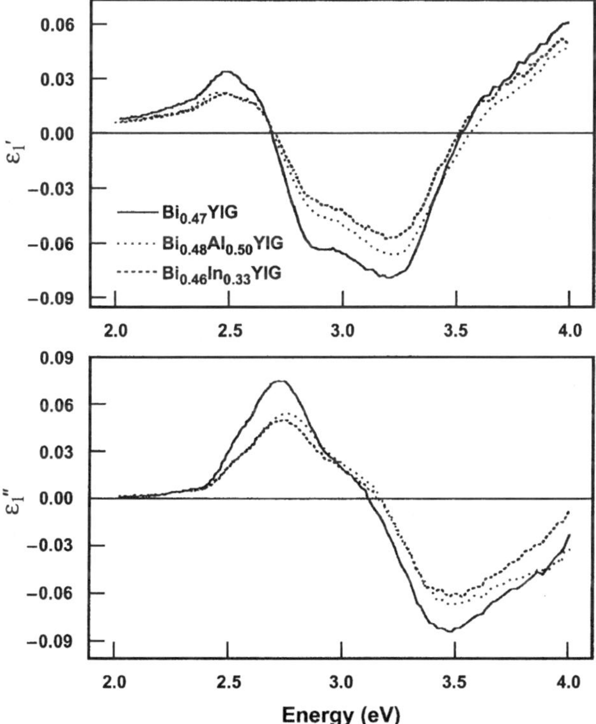

Fig. 7.24 Experimental comparison ε_1 of $\{Y_{2.53}Bi_{0.47}\}[Fe_2](Fe_3)O_{12}$ host with equal fractions of In^{3+} and Al^{3+} substituted into the $[a]$ and (d) sites, respectively. Images are reproduced from Allen's thesis [23]. Figure reprinted from [43] (2004) with permission. © 2004 by the American Institute of Physics

the same fraction of $1/6$ ($\sim$17%). Since the measurements were made at 300 K, spin canting also contributed to the decrease in transition probability. The canting was caused by reductions in the exchange fields that also decline with dilution, and were fitted successfully by a modification of the molecular-field theory for diluted magnetic garnets [44]. Following the suggestion of Wood and Remeika [36], joint participation of [a] and (d) sites proportional to the product of the Fe^{3+} occupation densities $N_{\pm a} \times N_{\pm d} \approx N_a N_d$ was adopted in the model [43].

The intersublattice charge-transfer concept based on the ionic nature of $Fe_a^{3+} - O^{2-} - Fe_d^{3+}$ molecule evolved from a mechanism proposed for MnF_2 [45] that was later applied to the magnetic oxides [28, 46, 47]. The theories used to explain the Bi^{3+} enhancements [25, 26] that considered internal ionic transitions $^6A_{1a,d} \rightarrow {}^4T_{1a,d}$ of the Fe_a^{3+} and Fe_d^{3+} ions are extended to include the interionic charge transfer. The concept is illustrated in Fig. 7.25. Photon absorption probabilities are determined not only by the oscillator strengths of the tensor elements, but also by the efficiency of the two-electron superexchange that now insures $\Delta S = 0$ transitions.

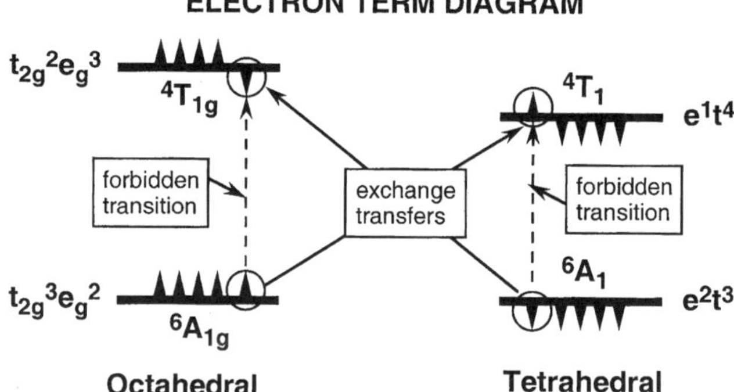

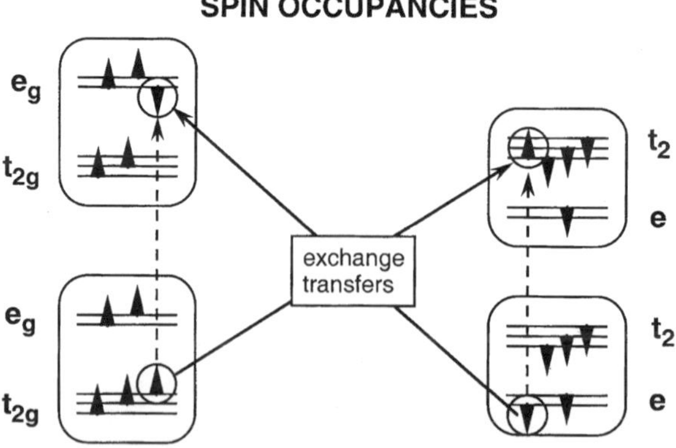

Fig. 7.25 Schematic diagrams of the intersublattice pair transfer between states of magnetically opposed octahedral and tetrahedral sites

Interionic charge transfer by actual ionization is required to create local $Fe^{4+}_{a,d}$ + $Fe^{2+}_{d,a}$ hole–electron pairs (excitons, possibly invoking a O^{1-} peroxide), with coincident magnetic-dipole agents in the form of magnons made possible by intersublattice superexchange. These interrelated events allow electric-dipole transitions without a net change in spin value and permit the mutual ${}^6A_{1a,d} \rightarrow {}^4T_{1d,a}$ intersublattice transfers within the antiferromagnetically coupled $Fe^{3+}_a - O^{2-} - Fe^{3+}_d$ molecule. For this to occur (1) there must be a sufficient concentration of Fe^{3+} ions to create the necessary molecular pairings; (2) the relevant orbital overlap integrals must be large enough for the intersite hybrid wave functions to have a transfer probability that exceeds that of the competing "forbidden" internal transitions; and (3) a coincident electron "back" transfer to an empty orbital state must occur through delocalization exchange to recover the 3+ valence states from the Fe^{4+} and Fe^{2+} ions created as part of the transfer processes

$$Fe^{3+}_a\left(t^3_{2g}e^2_g\right) + Fe^{3+}_d\left(t^3_2e^2\right) \rightarrow Fe^{4+}_a\left(t^2_{2g}e^2_g\right) + Fe^{2+}_d\left(t^4_2e^2\right) - \hbar\omega_a - \Delta(\text{IP}),$$

$$Fe^{4+}_a\left(t^2_{2g}e^2_g\right) + Fe^{2+}_d\left(t^4_2e^2\right) \rightarrow Fe^{3+}_a\left(t^2_{2g}e^3_g\right) + Fe^{3+}_d\left(t^4_2e^1\right) - (\hbar\omega_a + E_{ex}) + \Delta(\text{IP}).$$

These processes result in

$$Fe^{3+}_a\left(A_{1g}\right) + Fe^{3+}_d\left(A_1\right) \rightarrow Fe^{3+}_a\left(T_{1g}\right) + Fe^{3+}_d\left(T_1\right) - (\hbar\omega_a + \hbar\omega_d),$$

where Δ(IP) is the ionization potential energy expended to create the interim 4+ and 2+ states, and E_{ex} ($=\hbar\omega_d$) is the restabilizing energy permitted by the covalent exchange mechanism. These actions are depicted in Fig. 7.25, where the cross transfers are diagrammed. A beam of photons of energy $\hbar\omega_a$ or $\hbar\omega_d$ will cause both ${}^4T_{1g}$ and ${}^4T_1$ to be excited, in the manner of two coupled harmonic oscillators driven at their resonance frequencies by external stimulation of only one of them. As suggested by the spectral data, the orbital bonding mechanisms are based on the unmixed linkages $3t_{2ga}$–$2p$–$3t_{2d}$ and $3e_{ga}$–$2p$–$3e_d$ between the sites of the opposing sublattices. Since both ${}^4T_{1a}$ and ${}^4T_{1d}$ excited terms would occur through rearrangement of spins within the respective $3d^5$ manifolds, the "virtual" spin exchange of the static ferrimagnet would become "real" from the absorption of the photon beam energy.

With the growth of integrated optics and the burgeoning field of photonics by thin-film deposition of optical layers onto semiconductor substrates, the need for Faraday rotation (FR) media for fiber-optic isolators at 1.55 μm has prompted the search for new magneto-optical oxide compounds. In discrete devices, Bi-substituted magnetic garnets have proven to be greatly successful [48]. Due to lattice-mismatch considerations, however, the garnet crystal structure is not compatible with either that of conventional Si or GaAs semiconductors or the available dielectric buffer layers available for such integration purposes. Ferrimagnetic spinel layers grow well as thin films, but were found to be unsatisfactory for low-loss transmission despite high Faraday rotation [49]. Perovskites provide another structure that can be grown with good quality on a range of substrates, and that can show magnetic properties. For example, $BaTiO_3$ substituted with Fe revealed unexpected ferromagnetic properties, but little FR [50]. The FR of orthoferrite perovskites, with formula $\{A\}\,[Fe]\,O_3$, have been explored at shorter wavelengths for bulk materials [19], but there are few data on thin films [51].

The choice of cubic perovskites, however, remains attractive because of their crystallographic compatibility with preferred substrates and buffer layers. In this paper, the underlying physics for the design of two magneto-optically active perovskites are described. Both are charge-ordered transition-metal compounds with mixed alternating cations that satisfy in theory the various requirements for a room-temperature Faraday rotator at 1.55-μm wavelength.

From the development of the $\{Y, Bi\}_3\,Fe_5O_{12}$ class of garnet rotators at infrared (IR) wavelengths, much understanding of the fundamental physics of magnetically aligned electric-dipole transitions is available. For oxide systems of other lattice structures that could be more compatible with integrated photonic applications, some of the main requirements for a successful design of a Faraday rotator can be summarized as follows:

1. To align the orbital angular momentum vectors of the individual ion electric dipoles (through spin–orbit coupling), the material must be spontaneously magnetic (ferro- or ferrimagnetic) with reasonably low anisotropy field while maintaining a Curie temperature $T_C > 300$ K.
2. Electric-dipole transitions of the magnetic ion must satisfy orbital and spin selection rules $\Delta L_z = 0, \pm 1$ (Laporte's rule) and $\Delta S_z = 0$, respectively.
3. Wavelengths of interest must fall in the wings of the Lorentzian-shaped line to avoid the high absorption loss near its center frequency. Because optical spectra are not tunable as in the case of some magnetic-dipole transitions, i.e., by a Zeeman effect, this narrows further the choice of possible candidates.

To satisfy the above conditions in a magnetic insulator, two approaches with possible room-temperature net magnetization have been considered [52]. Both are "double" perovskites of generic formula $\{AA'\}\,[BB']\,O_6$ with octahedral-site B and B′ cations of different magnetic moments and/or ionization states. If the moments differ and antiferromagnetic spin ordering is stabilized, a quasi-ferrimagnet will be the result, thereby raising the possibility of interionic transitions to satisfy the $\Delta S_z = 0$ rule. If, however, the B and B′ ions differ by their valence charge, delocalization exchange can create ferromagnetic ordering [53, 54], which offers the possibility of spin-preserved intraionic transitions if the orbital term structure is suitable.

Although somewhat beyond the scope of this volume, the properties of magneto-optically active $4f^n$ series rare-earth ions in various compounds are of substantive importance to the understanding of the basic physics of these effects. The readers are encouraged to consult the high magnetic field work on these systems, particularly the investigation of paramagnetism that produces Faraday rotation reported by Guillot, Le Gall, Ostorero, and others at magnet laboratories in France [55, 56].

Appendix 7A Magnetic Circular Birefringence and Dichroism

In the near-IR and higher energy bands, the luxury of ignoring attenuation effects can no longer be enjoyed. For that reason, the Faraday rotation parameter is expressed as a complex quantity $\Theta_F = \theta_F + i\eta_F$, where η_F is termed "ellipticity" because it represents the difference in amplitude between the circular polarization modes caused by unequal k_- and k_+. The real and imaginary parts of Θ_F are frequently called magnetic circular birefringence (MCB) and magnetic circular dichroism (MCD). To describe these effects analytically, we adopt a development based on that of Dillon [18], but with the sign conventions already chosen in Chap. 6 for the microwave bands:

$$\begin{aligned}\theta_F + i\eta_F &= \left(\frac{\omega}{c}\right)(N_- - N_+) \\ &= \left(\frac{\omega}{c}\right)[(n_- - n_+) - i(k_- - k_+)], \\ &= \left(\frac{\omega}{c}\right)(\Delta n - i\Delta k)\end{aligned} \tag{7.45}$$

From the earlier definitions $N_\pm^2 = \varepsilon_0 \pm \varepsilon_1$ and $\varepsilon_1 = \varepsilon_1' - i\varepsilon_1''$,

$$\varepsilon_0 \pm \varepsilon_1 = n_\pm^2 - k_\pm^2 - i2n_\pm k_\pm \tag{7.46}$$

and

$$\begin{aligned}\varepsilon_1' &= \frac{1}{2}\left[\left(n_+^2 - n_-^2\right) - \left(k_+^2 - k_-^2\right)\right], \\ \varepsilon_1'' &= (n_+k_+ - n_-k_-).\end{aligned} \tag{7.47}$$

If the average of $n = (n_+ + n_-)/2$ and $k = (k_+ + k_-)/2$ are introduced, (7.47) can be expressed as

$$\begin{aligned}\varepsilon_1' &= -n\Delta n + k\Delta k, \\ \varepsilon_1'' &= -n\Delta k - k\Delta n.\end{aligned} \tag{7.48}$$

When (7.45) and (7.48) are combined, the key relations between permittivity, Faraday rotation, and ellipticity, and refractive index and extinction coefficient are obtained:

$$\begin{aligned}\varepsilon_1' &= -\left(\frac{2c}{\omega}\right)(n\theta_F - k\eta_F), \\ \varepsilon_1'' &= -\left(\frac{2c}{\omega}\right)(k\theta_F + n\eta_F).\end{aligned} \tag{7.49}$$

An alternative to transmission measurements is the Kerr effect. Reflection measurements become necessary if intrinsic properties such as electrical conductivity or microstructure imperfections in ceramic forms of normally transparent compounds render the specimens opaque in the frequency bands of interest. Following the model used by Dillon [18] for the reflected wave, we obtain the complex Kerr rotation from the respective reflection coefficients $r_{\pm}$ according to

$$\Theta_{\mathrm{K}} = \theta_{\mathrm{K}} + \mathrm{i}\eta_{\mathrm{K}} = -\mathrm{i}\left(\frac{r_{-} - r_{+}}{r_{+} + r_{-}}\right) = -\mathrm{i}\left(\frac{N_{-} - N_{+}}{N_{+}N_{-} - 1}\right), \tag{7.50}$$

which can be expressed as

$$\theta_{\mathrm{K}} + \mathrm{i}\eta_{\mathrm{K}} = \frac{-\Delta k - \mathrm{i}\Delta n}{(n_{-}n_{+} - k_{-}k_{+} - 1) - \mathrm{i}\,(n_{-}k_{+} - n_{+}k_{-})}. \tag{7.51}$$

After rationalization of the denominator, (7.51) becomes

$$\theta_{\mathrm{K}} + \mathrm{i}\eta_{\mathrm{K}} = \frac{\left[-\Delta k\left(n^2 - k^2 - 1\right) + \Delta n\,(2nk)\right] + \mathrm{i}\left[-\Delta k\,(2nk) - \Delta n\left(n^2 - k^2 - 1\right)\right]}{\left(n^2 - k^2 - 1\right)^2 + 4n^2k^2}. \tag{7.52}$$

where n and k are the averages defined below (7.47).

The real and imaginary parts of (7.52) are separate equations with two unknowns that allow Δn and Δk to be solved for in terms of θ_{K} and η_{K}, according to

$$\begin{aligned} \Delta n &= \theta_{\mathrm{K}}\,(2nk) - \eta_{\mathrm{K}}\left(n^2 - k^2 - 1\right), \\ \Delta k &= -\theta_{\mathrm{K}}\left(n^2 - k^2 - 1\right) - \eta_{\mathrm{K}}\,(2nk), \end{aligned} \tag{7.53}$$

and with the help of (7.48) the basic relations between the Kerr parameters and the complex permittivity are stated as

$$\begin{aligned} \varepsilon_1' &= \theta_{\mathrm{K}}\left(k^3 - 3n^2k + k\right) + \eta_{\mathrm{K}}\left(n^3 - 3nk^2 - n\right), \\ \varepsilon_1'' &= \theta_{\mathrm{K}}\left(n^3 - 3nk^2 - n\right) - \eta_{\mathrm{K}}\left(k^3 - 3n^2k + k\right). \end{aligned} \tag{7.54}$$

References

1. B. Lax and K.J. Button, *Microwave Ferrites and Ferrimagnetics*, (McGraw-Hill, New York, 1962), Chapter 6
2. S. Geschwind and L.R. Walker, *J. Appl. Phys.* **30**, 163S (1959)
3. G.F. Dionne, *J. Appl. Phys.* **97**, 10F103 (2005)
4. M. Tinkham, *J. Appl. Phys.* **33**, Suppl. 3, 1248 (1962)
5. G.F. Dionne, *J. Appl. Phys.* **105**, 07A525 (2009)
6. K.J. Standley and R.A. Vaughn, *Electron Spin Relaxation Phenomena in Solids*, (Plenum, New York, 1969), Section 1.2
7. A.H. Morrish, *The Physical Principles of Magnetism*, (Wiley, New York, 1965), p. 73
8. G.F. Dionne, *J. Appl. Phys.* **79**, 5172 (1996)
9. G.F. Dionne, *J. Appl. Phys.* **99**, 08M913 (2006)
10. B. Lax and K.J. Button, *Microwave Ferrites and Ferrimagnetics*, (McGraw-Hill, New York, 1962), Section 6-6

11. B. Lax and K.J. Button, *Microwave Ferrites and Ferrimagnetics*, (McGraw-Hill, New York, 1962), Section 7-1
12. N. Bloembergen, *Proc. IRE* **44**, 1259 (1956)
13. Y.R. Shen, *Phys. Rev.* **133**, A511 (1964)
14. Y.R. Shen and N. Bloembergen, *Phys. Rev.* **133**, A515 (1964)
15. N. Bloembergen, *Nonlinear Optics*, (W.A. Benjamin, New York, 1965), p. 27
16. J.C. Suits, *IEEE Trans. Magn.* **8**, 95 (1972)
17. G.A. Allen and G.F. Dionne, *J. Appl. Phys.* **73**, 6130 (1993)
18. J.F. Dillon, *J. Phys. Radium* **20**, 374 (1959)
19. F.J. Kahn, P.S. Pershan, and J.P. Remeika, *Phys. Rev.* **186**, 891 (1969)
20. G.B. Scott, D.E. Lacklison, H.I. Ralph, and J.L. Page, *Phys. Rev.* **B12**, 2562 (1975)
21. S. Wittekoek, T.J.A. Popma, J.M. Robertson, and P.F. Bongers, *Phys. Rev.* **B12**, 2777 (1975)
22. V. Doorman, J.-P. Krumme, and H. Lenz, *J. Appl. Phys.* **68**, 3544 (1990)
23. G.A. Allen, *PhD Thesis*, MIT Department of Physics, 1994
24. G.A. Allen and G.F. Dionne, *J. Appl. Phys.* **93**, 6951 (2003)
25. G.F. Dionne and G.A. Allen, *J. Appl. Phys.* **73**, 6127 (1993)
26. G.F. Dionne and G.A. Allen, *J. Appl. Phys.* **75**, 6372 (1994)
27. G.B. Scott, D.E. Lacklison, and J.L. Page, *Phys. Rev.* **B10**, 971 (1974)
28. G.B. Scott and J.L. Page, *Phys. Stat. Solidi* **b79**, 203 (1977)
29. A.M. Clogston, *J. Phys. Radium* **20**, 151 (1959)
30. C.F. Buhrer, *J. Appl. Phys.* **40**, 4500 (1969)
31. K. Matsumoto, S. Sasaki, K. Haraga, Y. Asahara, K. Yamaguchi, and T. Fujii, *IEEE Trans. Magn.* **28**, 2985 (1992)
32. Z. Simsa, J. Simsova, D. Zemanova, J. Cermak, and M. Nevriva, *Czech. J. Phys. B* **34**, 1102 (1984)
33. Y. Tanabe and S. Sugano, *J. Phys. Soc. (Japan)* **9**, 753 (1954)
34. D.E. Lacklison, G.B. Scott, and J.L. Page, *Solid State Commun.* **14**, 861 (1974)
35. D.R. Lide, Ed., *Handbook of Chemistry and Physics,* 73rd Ed., (CRC Press, Boca Raton, FL, 1992–1993)
36. D.L. Wood and J.P. Remeika, *J. Appl. Phys.* **38**, 1038 (1967)
37. S. Wittekoek and D.E. Lacklison, *Phys. Rev.* Lett. **28**, 740 (1972); also A.B. McLay and M.F. Crawford, *Phys. Rev.* **44**, 986 (1933)
38. P. Hansen, W. Tolksdorf, and K. Witter, *IEEE Trans. Magn.* **17**, 3211 (1981)
39. P. Hansen, K. Witter, and W. Tolksdorf, *Phys. Rev.* B **27**, 6608 (1983)
40. S.H. Wemple, S.L. Blank, J.A. Seman, and W.A. Biolsi, *Phys. Rev. B* **9**, 2134 (1974)
41. S. Wittekoek and T.J.A. Popma, *J. Appl. Phys.* **44**, 5560 (1973)
42. A. Thavendrarajah, M. Pardavi-Horvath, P.E. Wigen, and M. Gomi, *IEEE Trans. Magn.* **25**, 4015 (1989)
43. G.F. Dionne and G.A. Allen, *J. Appl. Phys.* **95**, 7333 (2004)
44. G.F. Dionne, *J. Appl. Phys.* **41**, 4874 (1970)
45. Y. Tanabe, T. Moriya, and S. Sugano, *Phys. Rev.* Letts. **15**, 1023 (1965)
46. J.P. van der Ziel, J.F. Dillon, and J.P. Remeika, *17th Annu. Conf. Magn. Magn. Mater.*, AIP Conf. Proc. No. **5**, 254 (1971)
47. B. Andlauer, J. Schneider, and W. Wettling, *Appl. Phys.* **10**, 189 (1976)
48. G. Winkler, *Magnetic Garnets*, (Vierweg, Braunschweig, 1981), Chapter 4
49. T. Tepper, C.A. Ross, and G.F. Dionne, *IEEE Trans. Magn.* **40**, 1685 (2004)
50. A. Rajamani, G.F. Dionne, D. Bono, and C.A. Ross, *J. Appl. Phys.* **98**, 063907 (2005)
51. D.S. Schmool, N. Keller, M. Guyot, R. Krishnan, and M. Tessier, *J. Appl. Phys.* **86**, 5712 (1999)
52. G.F. Dionne A.R. Taussig, M. Bolduc, L. Bei, and C.A. Ross, *J. Appl. Phys.* **101**, 09C524 (2007)
53. N.S. Rogado, J. Li, A.W. Sleight, and M.A. Subramanian, *Adv. Mater. (Weinhein, Ger.)* **17**, 2225 (2005)
54. H. Guo, J. Burgess, S. Street, A. Gupta, T.G. Calarese, and M.A. Subramanian, *Appl. Phys. Lett.* **89**, 022509 (2006)
55. M. Guillot, H. Le Gall, J.M. Desvignes, and M. Artinian, *J. Appl. Phys.* **70**, 6401 (1991)

56. J. Ostorero and M. Guillot, *J. Appl. Phys.* **83**, 6756 (1998)
57. F.M. Johnson and A.H. Nethercot, Jr., *Phys. Rev.* **114**, 705 (1959)
58. S. Foner, *J. Phys. Radium* **20**, 336 (1959)
59. E.S. Dayhoff, *Phys. Rev.* **107**, 84 (1957)
60. G.S. Heller, J.J. Stickler, and J.B. Thaxter, *J. Appl. Phys.* **32**, 307S (1961)
61. J.J. Stickler and G.S. Heller, *J. Appl. Phys.* **33**, 1302 (1962)
62. R.C. Ohlmann and M. Tinkham, *Phys. Rev.* **123**, 425 (1961)
63. F. Keffer, A.J. Sievers III, and M. Tinkham, *J. Appl. Phys.* **32**, 65S (1961)
64. H. Kondoh, *J. Phys. Soc. Japan*, **15**, 1970 (1960)

Chapter 8
Spin Transport Properties

Electrical conductivity in solids is traditionally associated with collective electron metals, intermetallic compounds, and metal alloys. With outer shell s and p electrons unbound in the sense of a Sommerfeld gas, the analysis of electrical properties usually takes the form of a density of states calculation from a theory that assumes a periodic lattice potential and applies Fermi statistics. The exercise leads to the creation of broadened energy states (bands) and the definition of a Fermi level to serve as the zero energy reference. For transition metals with unfilled "inner" d-shells, complications arise from the hybridization of "free" electron s states with states of unpaired spins in the d shell that forms the collective electron band structure. The first successful attempt to model ferromagnetism in collective electron metals was developed by Stoner employing a phenomenological theory that explained the net magnetization by introducing an exchange field that separated the d-spin populations into up (α) and down (β) bands [1], analogous to the Hund's rule sorting of spins in an individual ion, illustrated by Figs. 2.3 and 2.19. In subsequent years, great strides were made in formalizing these concepts with the help of quantum theory and other aspects of modern physics [2].

Despite the power and elegance of solid-state band theory, the basic assumption of a periodic potential with Bloch-type functions can shroud the role of local interactions that are important in insulator host compounds containing transition-metal cations in varying concentrations. Randomly spaced effects such as lower symmetry crystal-field splittings from electron-lattice vibronic interactions at individual sites, variations in spin–orbit coupling, single-ion magnetoelastic effects, spin–lattice relaxation that controls spin resonance line shapes and spin-wave propagation, and to some extent, the superexchange interactions that determine the type and stability of magnetic ordering, are largely missing from the formalism. In select cases where metallic properties have been observed in homogeneous ferromagnetic oxides, the electrical properties have been interpreted by Hartree–Fock tight-binding approximations based on the spin-density-wave model of Overhauser [3] to split the d shell into the α (majority) and β (minority) spin bands with the Fermi energy set by the highest occupied state, in the manner of Stoner's original model. However, in most cases where electronic conduction in oxides occurs with spontaneous magnetism, the charge transfer is from randomly dispersed polarons.

G.F. Dionne, *Magnetic Oxides*, DOI 10.1007/978-1-4419-0054-8_8,

Charge transfer by electron hopping between mixed-valence cations of the same atomic element is a mechanism of electrical conduction in ferrites, particularly $Fe^{2+} \leftrightarrow Fe^{3+} + e^-$ in the octahedral sublattice. The conductivity that results from these random events generally has the temperature characteristic of an insulator or semiconductor at room temperature. In other systems, however, charge transfer can be metallic in temperature dependence and sometimes sufficiently coherent that superconductivity can exist to temperatures greater than 130 K. These phenomena fall under the general class of polaronic motion, which is strongly influenced by the state of spin ordering in the lattice. The general study of polarons was pioneered by a number of workers, prominent among them being Fröhlich [4] and Mott [5]. For metallic oxides, however, the molecular crystal approach of Holstein [6] and the insightful review by Goodenough [7] are more relevant because of their conformance with molecular-orbital framework on which the present discussion is based [8, 9].

In recent years, discoveries that involve the transport of electron spins have reenergized the field of magnetic oxides. Among these initiatives are high-temperature *colossal* magnetoresistance (CMR) and high-temperature superconductivity (HTS), both dependent on polarized-spin transport in select transition-metal oxide compounds. Promising materials for exploiting spin transport in magnetoresistance and superconductivity are the manganite and cuprate perovskites, respectively. Metallic conduction at room temperature has also been reported in other magnetic oxide structures.

8.1 Polarons and Charge Transfer

A polaron is a charge carrier that resides in an energy trap created by the interaction between the charge and its local crystalline surroundings in the manner of a dipole with one of its charges mobile. The conventional trap is attributed to lattice distortions that are induced by local electric fields and the associated differences in the size of the ion that harbors the charge. Another term that is used to describe a polaron is a "dressed" carrier because it travels together with its lattice accommodation. In ionic compounds, polarons occur as a result of cation chemistry and mixed valence induced by departures from stoichiometry, in contrast to the excitation necessary to create an electron (or hole) carrier in a doped conventional band-gap semiconductor. For electrical conduction, the carrier part of the dipole is not created by excitation, but rather released from its trap by covalent tunneling or by phonons, thermal, or otherwise. The participation of the polaron in an electrical current is usually described as an *activation* of its mobility by overcoming the trap energy created by the stabilizing effects of elastic, electrostatic, as well as magnetic exchange where transition-metal ions with unpaired spins supply the charge carriers. Convenient vehicles for studying these conduction mechanisms are the cubic ABO_3 and tetragonal A_2BO_4 perovskites shown in Fig. 8.1, emphasizing the octahedral B site at which the polaron is centered. An important feature of this crystallographic family in relation to spin exchange and transport are the 180° B–O–B bond angles that form B–O_2 chains along the z (or c) axis and B–O_4 planar layers in the x–y (frequently called the a–b) plane.

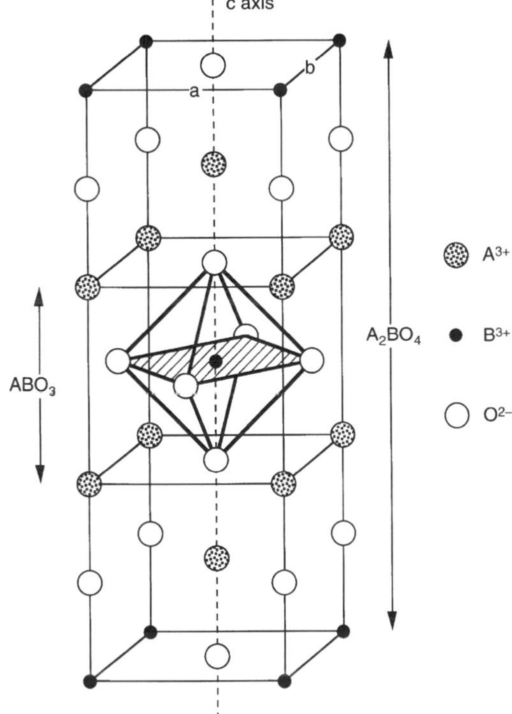

Fig. 8.1 Perovskite unit cells cubic ABO_3 and tetragonal A_2BO_4, highlighting the octahedral B site

Consider the case of a mixed-valence complex transition metal perovskite $A^{3+}B^{3+}O_3^{2-}$ in which different ions of smaller valence charge A'^{2+} are substituted. To restore electrical neutrality (1) either a corresponding number of the O^{2-} ions are converted to O^{1-} to create a partial peroxide $A^{3+}_{1-x}A'^{2+}_{x}B^{3+}O^{1-}_{x}O^{2-}_{3-x}$ or (2) where possible a corresponding number of B^{3+} are converted to B^{4+} to produce the mixed-valence in the B sublattice of $A^{3+}_{1-x}A'^{2+}_{x}B^{3+}_{1-x}B^{4+}_{x}O^{2-}_{3}$. Because larger ionic charges usually produce greater ionic bonding energies and higher stability, once the ionization potential and electron affinity tradeoffs are made between cations and anions, the latter arrangement would likely result in a lower ionic lattice energy. Relative to the neutral background, A'^{2+} is a stationary fixed negative charge and B^{4+} forms the positive half of a dipole that can occupy any one of the several equivalent sites equidistant from A'^{2+}, as depicted in the two-dimensional sketch of Fig. 8.2. Since the charge at the electronic hole labeled as B^{+} is capable of transferring to equivalent sites surrounding the A'^{-} fixed charge (in this ideal case without a net change in energy) it is a polaron, and A'^{-} is the polaron source.

If the model of Fig. 8.2 is examined further, we can identify two parts to the problem of analyzing charge movement in this model: first, the transfer of electrons among sites of equivalent energy, depicted as periodic locations around the rings and second, the transport of charges to sites farther from the source ion, depicted by the transitions to an outer ring of higher energy. In both cases, the gain in binding energy is derived directly from the actual transfer of charge, and favors a

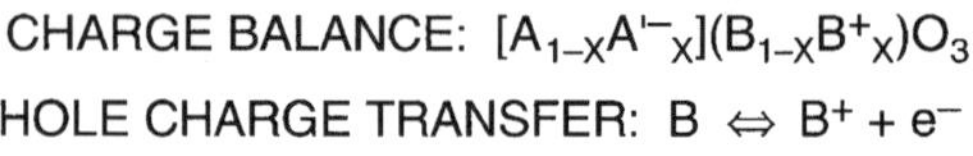

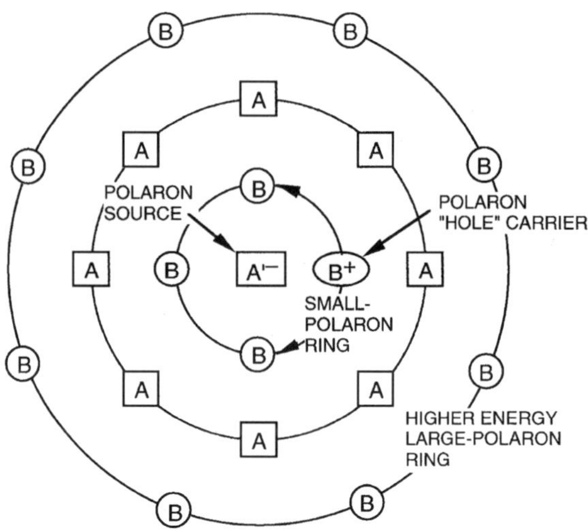

Fig. 8.2 Positive polaron (hole) formation in mixed-valence ABO_3 and the concentric ring model centered about a fixed polaron source. Inner ring is the radius of a small polaron, outer rings of increasingly higher electrostatic potential energy are regions of large polarons

large covalent exchange energy integral to stabilize a kinetic ground state. The stabilization energy associated with the transfer is therefore critically related to the states of spin polarization of the transfer ions because the larger the trap energy, the longer the lifetime of the trapped carrier, and the lower the mobility becomes. For an ideal *small* polaron, the initial and final states are identical and the probability of locating the charge at any site around the inner ring is equal. When the dipole extends outside of the inner ring, the carrier becomes mobile, leading to the formation of a *large* polaron. In this event, the energy of the final state is increased accordingly. The trap is deepened by the increasing Coulomb potential of the dipole attraction and can be examined according to electrostatic theory. Because of the increase in energy of the receptor state, the density of large polaron carriers decays as a function of distance from the source analogous to the tail of an orbital wavefunction away from its nucleus.

8.1.1 Transfer Among Equivalent Energy Sites (Small Polarons)

When an atom is ionized in free space, the removal (or addition) of an electron is accomplished at the cost of an ionization energy of many electron volts. In a crystal lattice, this energy (e.g., the electronic work function) is reduced to a few electron volts because of the lower electrostatic fields due to polarizability of the dielectric medium. In cases where covalent bonding is significant, however, a more efficient mechanism is available by charge transfer between cations through interaction with intermediary anions. Where selection rules for energy-free transfer are satisfied, e.g., spin conservation by $\Delta S = 0$, the transfer can take place spontaneously, and when there is no net loss of energy to the lattice, the action can be termed "adiabatic." The activation energy from this double exchange transfer is therefore dependent on the angle θ_{ij} between adjacent ionic spins S_i and S_j. Where $\theta_{ij} > 0$, an electron spin flip is required to satisfy any intraorbital exchange (Hund's rule) requirements on the receptor ion, [1]and the spontaneous excitation-free sharing of the transfer spin is allowed if sufficient energy is available to maintain the stabilization energy of mobile exchange given by the relation examined previously [8]:

$$E_{\text{ex}}\left(\theta_{ij}\right) = -2z_{ij}J_{ij}S_iS_j\cos^2\left(\frac{\theta_{ij}}{2}\right) = -z_{ij}J_{ij}S_iS_j\left(1+\cos\theta_{ij}\right), \tag{8.1}$$

where $J_{ij} > 0$ is the ferromagnetic exchange constant and z_{ij} is the number of equivalent neighboring ionic spins.

When $\theta_{ij} = 0$, E_{ex} has a minimum value for spins that are parallel to satisfy the $\Delta S = 0$ requirement for the charge transfer. If $\theta_{ij} > 0$, E_{ex} increases by an amount U_{ex} which becomes the energy needed to restore the spin alignment. In the present context, U_{ex} represents the loss in kinetic stabilization energy of spin transfer between sites i and j. In the absence of quantum tunneling that would reduce the effect of the trap on the transfer probability, U_{ex} can also be equated a classical thermal hopping energy according to

$$E_{\text{hop}}^{\text{ex}} \approx U_{\text{ex}} = E_{\text{ex}}\left(\theta_{ij}\right) - E_{\text{ex}}(0) = z_{ij}J_{ij}S_iS_j\left(1-\cos\theta_{ij}\right). \tag{8.2}$$

It should be noted that θ_{ij} values in the range 0 to π are allowed by (8.2), and that the theoretical maximum hopping activation energy from exchange is actually $E_{\text{hop}}^{\text{ex}} = 2z_{ij}J_{ij}S_iS_j$ when the spins are antiparallel $\left(\theta_{ij} = \pi\right)$. For a ferromagnetic to paramagnetic transition that occurs at the Curie temperature, the average angle between spins becomes $\pi/2$. The probability of mobility activation of small polarons among equivalent sites can then be written in the classical Boltzmann sense as $\exp\left[\left(E_{\text{hop}}^0 + E_{\text{hop}}^x\right)/kT\right]$, where $E_{\text{hop}}^0 \approx U_0$ is the corresponding activation energy from the residual elastic trap energy when $\theta_{ij} = 0$.

8.1.2 Transfer to Higher Energy Sites (Large Polarons)

In the simplest case there are no spin polarization limitations, i.e., perfect ferromagnetism, the small polaron ring described above would be likened to a giant molecule

[1] The magnetic implications of a real spin transfer between a pair of covalently linked orbital states arise when the accompanying states of the $3d$ shell are populated with unpaired spins aligned according to Hund's rule. These spin groups can favor parallel or antiparallel exchange interactions. The ferromagnetic energy from the dynamic exchange could then be enhanced or reduced depending on the details of the overall electronic spin structure.

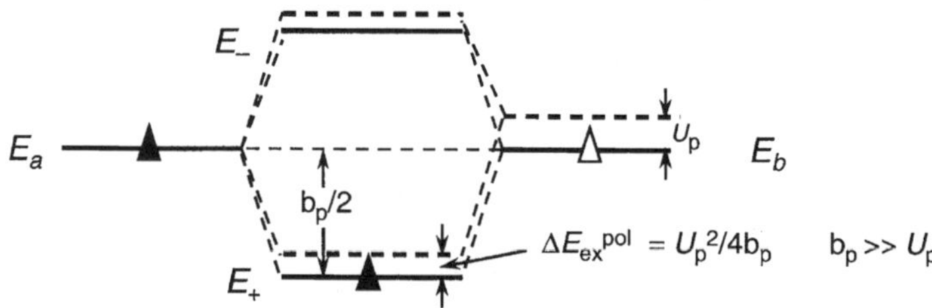

Fig. 8.3 Two-cation molecular-orbital diagram for a polaron in a trap of energy U_{p}

in which the carrier would occupy all sites simultaneously in a quantum mechanical sense. To explore this concept, a giant hybrid wavefunction based on the polaron exchange integral b_{p} must be constructed. In addition, U_{p} is introduced to represent the total trap energy from all sources, including the variable electrostatic force that binds the mobile charge to its source in the lattice. It is therefore assumed that the trap is of multiple origins, with the polaron carrier tethered to its nearby fixed charge of opposite sign.

Estimates of the molecular-orbital functions φ_a and φ_b for a diatomic molecule can be obtained by solving a two-level degenerate perturbation problem in the conventional way based on the one-electron molecular-orbital diagram of Fig. 8.3. The solution is worked out in terms of a polaron activation energy U_{p} between the two lattice cations M_a and M_b with corresponding electron energies E_a and E_b that share an itinerant electron by following the procedures used previously. The electronic stabilization energy and hybrid eigenfunction coefficients may be found by means of a self-consistent approach developed by Wolfsberg and Helmholtz [10]. The degenerate-state perturbation matrix equation for the diatomic molecule with Hamiltonian $\mathcal{H} = \mathcal{H}_a + \mathcal{H}_b$ is expressed as

$$\begin{vmatrix} \mathcal{H}_{aa} - E & \mathcal{H}_{ab} - E\sigma \\ \mathcal{H}_{ab} - E\sigma & \mathcal{H}_{bb} - E \end{vmatrix} = \begin{vmatrix} E_a - E & b_{ab} - E\sigma \\ b_{ab} - E\sigma & E_b - E \end{vmatrix} = 0, \tag{8.3}$$

which is derived from the secular equation $\left|\mathcal{H}_{ij} - E\sigma_{ij}\right| = 0$, where $\mathcal{H}_{ab} = \langle\varphi_a|\,\mathcal{H}\,|\varphi_b\rangle$ and $\sigma_{ab} = \langle\varphi_a \mid \varphi_b\rangle = 1$ for $a = b$, and < 1 for $a \neq b$. If $b_{ab} = \langle\varphi_a|\,\mathcal{H}\,|\varphi_b\rangle$ is approximated by $(E_a + E_b)\,\sigma_{ab}$ and $\sigma_{ab}^2 \ll 1$ are used, the solutions for the bonding (+) and antibonding (−) states can be reduced to

$$E_{\pm} = \frac{1}{2}(E_a + E_b) \pm \frac{1}{2}\sqrt{(E_a - E_b)^2 + b_{ab}^2}, \tag{8.4}$$

with corresponding normalized hybrid eigenfunctions given by

$$\begin{aligned} \varphi_{-} &= \frac{1}{\sqrt{2}}(c_{ba}\varphi_a - c_{bb}\varphi_b) \quad \text{antibonding} \\ \varphi_{+} &= \frac{1}{\sqrt{2}}(c_{aa}\varphi_a + c_{ab}\varphi_b) \quad \text{bonding.} \end{aligned} \tag{8.5}$$

After the standard solutions [11] are applied to (8.5) for the case of interest $b_{ab}^2 \ll (E_a - E_b)^2$,

$$\varphi_- \approx \frac{1}{\sqrt{2}}\left[\left(1 - \frac{E_a - E_b}{b_{ab}}\right)^{\frac{1}{2}}\varphi_a - \left(1 + \frac{E_a - E_b}{b_{ab}}\right)^{\frac{1}{2}}\varphi_b\right],$$
$$\varphi_+ \approx \frac{1}{\sqrt{2}}\left[\left(1 + \frac{E_a - E_b}{b_{ab}}\right)^{\frac{1}{2}}\varphi_a + \left(1 - \frac{E_a - E_b}{b_{ab}}\right)^{\frac{1}{2}}\varphi_b\right]. \tag{8.6}$$

For the limiting cases of small and large polarons, $b_p = b_{ab}$ and $U_p = E_a - E_b$, where both are negative energies, and (8.4) then reduces to

$$E_\pm \approx E_F \pm \frac{1}{2}\sqrt{U_p^2 + b_p^2}. \tag{8.7}$$

where $E_F = (E_a + E_b)/2$ is the mean energy of the free ion states.

In the small-polaron case where $b_p^2 \ll U_p^2$, and the bonding-state solution of (8.4) simplifies to

$$E_+ \approx E_F + \frac{U_p}{2}\sqrt{1 + b_p^2/U_p^2} \approx E_F + \frac{U_p}{2} + \frac{b_p^2}{4U_p},$$
$$E_+ \approx E_a + \frac{b_p^2}{4U_p}. \tag{8.8}$$

The $b_p^2/4U_p$ term represents the increment of stabilization energy that the hybrid state provides for tunneling by the electrons not involved in a thermal hopping process.

For the corresponding large-polaron approximation $b_p^2 \gg U_p^2$, and

$$E_+ \approx E_F + \frac{b_p}{2}\sqrt{1 + U_p^2/\mathrm{b}_p^2} \approx E_F + \frac{b_p}{2} + \frac{U_p^2}{4b_p},$$
$$E_+ \approx E_a + \frac{U_p^2}{4b_p}. \tag{8.9}$$

In this limit, $U_p^2/4b_p$ is the additional stabilization of the single spin in the bonding state. It can be readily recognized that if $U_p = 0$, there is no barrier or trap for the transfer of the spin between φ_a and φ_b because their coefficients in the hybrid orbital state are equal. Therefore, $U_p^2/4b_p$ is the energy that must be gained to eliminate the trap. The effective classical activation energy for general polaron charge transport becomes

$$E_{\mathrm{hop}} = \frac{U_p^2}{4b_p}, \tag{8.10}$$

where $U_p = U_0 + U_0(r) + U_{\mathrm{ex}}$ which now includes a variable electrostatic energy term that is introduced in Sect. 8.1.3.

Incoherent electrical conduction by thermal hopping is significant where E_{hop} is small or temperature is large. Because spin alignment determines U_{ex} in a magnetic material, the term *magnetic* polaron can be added to the lexicon. In simple terms, when neighboring spins of the trapped carrier are aligned parallel, the local environment of the carrier is ferromagnetic and $\cos\theta = 1$, thereby *removing* the U_{ex} part of the polaron trap. The carrier would then be constrained only by its dielectric (elastic) trap. The ferromagnetic polaron therefore has the transport properties of a conventional dielectric polaron.

8.1.3 Transfer by Covalent Tunneling

At low temperatures, a more efficient transfer mechanism can produce remarkable effects when the right conditions of chemical bonding and spin ordering are present. In the context of molecular-orbital states charge transfer can also be viewed as a quantum mechanical probability stemming directly from the covalent bond, i.e., through a "covalent-transfer" mechanism. This concept can be readily appreciated by imagining the concentric ring diagram in Fig. 8.2 as analogous to a hydrogen atom (with reversed polarity in this case), where the rings would represent a series of higher energy states. The wavefunction of the mobile polaron charge could then be viewed as an s orbital wavefunction and its probability of occupancy at a distance r from the source would then be determined by its contribution to the molecular-orbital state as the carrier moves toward the outer rings. Such an exercise can be approximated from the eigenfunctions of (8.6) after being modified to anticipate the variable polaron trap energy $U_p(r)$ of (8.10) and is sketched in the energy level diagram of Fig. 8.3.

For the small polaron limit $b_p^2 \ll (E_a - E_b)^2 = U_p^2$, application of the procedure used to derive (8.6) yields

$$\varphi_+^{\text{small}} \approx \frac{1}{\sqrt{2}}\left(1-\left(\frac{b_p}{2U_p}\right)^2\right)^{\frac{1}{2}}\varphi_a + \left(\frac{b_p}{2U_p}\right)\varphi_b \tag{8.11}$$

and the transfer probability of the polaron carrier from site a to site b can be defined as the square of the coefficient c_{ab} of φ_b. If we define a transfer efficiency as the ratio of the two probabilities

$$\eta(r)_{\text{small}} \approx \frac{(b_p/2U_p)^2}{1-(b_p/2U_p)^2} \approx \left(\frac{b_p}{2U_p}\right)^2, \tag{8.12}$$

which tends toward zero in this approximation, thereby confirming the ineffectiveness of quantum tunneling to extend the range of the carrier beyond the small-polaron ring.

For the large polaron limit $b_\mathrm{p}^2 \gg U_\mathrm{p}^2$, the φ_+ eigenfunction for a bonding state with the coefficients defined in (8.6), would be

$$\varphi_+^{\text{large}} \approx \frac{1}{\sqrt{2}}\left[\left(1+\frac{U_\mathrm{p}}{b_\mathrm{p}}\right)^{\frac{1}{2}}\varphi_a + \left(1-\frac{U_\mathrm{p}}{b_\mathrm{p}}\right)^{\frac{1}{2}}\varphi_b\right], \tag{8.13}$$

and

$$\eta\left(r\right)_{\text{large}} \approx \frac{c_{ab}^2}{c_{aa}^2} \approx \frac{1-U_\mathrm{p}/b_\mathrm{p}}{1+U_\mathrm{p}/b_\mathrm{p}} \approx 1-\frac{2U_\mathrm{p}}{b_\mathrm{p}}. \tag{8.14}$$

From the electrostatic attraction between two oppositely charged particles in a medium of dielectric constant K and M_a to M_b dimension a, the r-dependent part of U_p becomes

$$U_0\left(r\right) \approx \frac{e^2}{K}\left(\frac{1}{a}-\frac{1}{r}\right) \approx \frac{e^2}{Ka}\left(1-\frac{a}{r}\right). \tag{8.15}$$

Therefore, the charge-transfer probabilities can be related to quantum mechanical tunneling to a distance r from the polaron source by substituting (8.15) (or another appropriate function) into (8.12) and (8.14). Figure 8.4 compares the η efficiency parameter variation with r/a for large polarons. A somewhat more rigorous approach would consider consecutive discrete transfers of the polaron carrier from its source. To this point, the only reference to temperature has been in relation to the Boltzmann

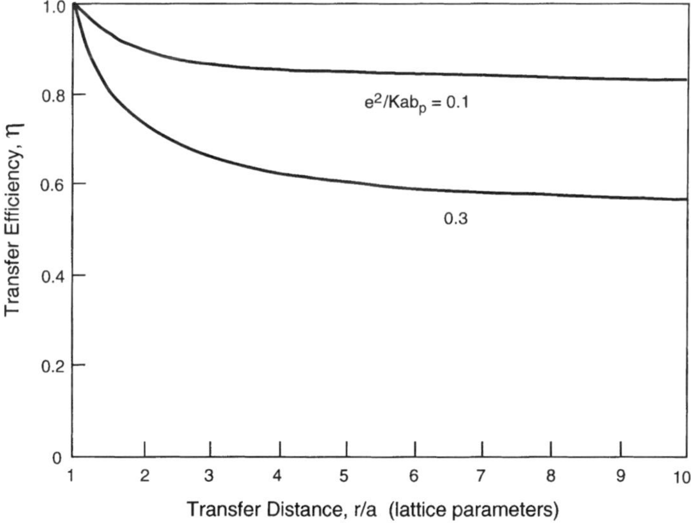

Fig. 8.4 Polaron transfer efficiency as a function of reduced lattice length for two ratios of electrostatic energy to polaron half-bandwidth

activation probability. Although not stated explicitly, there is a T dependence of the spin alignment that is characterized by the Curie temperature through the Brillouin–Weiss theory. The temperature will also affect the value of E_{hop} and the extent to which tunneling is allowed to control the lifetime of the polaron in its trap. However, even if perfect spin alignment were possible, lattice vibrations would still create a temperature dependence of b_{p}, which will now be related to a polaron bandwidth.

8.1.4 The Holstein Polaron Theory

In his quantum mechanical analysis of polaron motion [6]. Holstein defined three stages of conductivity that were determined by the relative magnitudes of two energy parameters (1) the polaronic charge exchange energy b_{p}, which establishes the lifetime τ_p of the polaron carrier in its site according to $\tau_{\mathrm{p}} \sim \hbar/b_{\mathrm{p}}$, and thereby presents b_{p} as the width of the polaron energy band that is the product of the full electronic exchange integral b and a lattice vibronic coupling factor, and (2) the energy U_{p} of the electrostatic potential well in which the carrier resides and which is shaped by the charge of the carrier and a neighboring charge of opposite sign that created it, i.e., the other half of an extendible dipole. The temperature dependence of b_{p} is determined by the number of vibrational modes (phonons) available to interact with the carriers. The vibrational overlap integral is a decreasing function of T and its contribution to b_{p} is maximum at absolute zero, diminishing rapidly with rising temperatures.

Based on these concepts, Holstein defined the criterion for a large polaron (one in which the carrier extends beyond its immediate environment by spontaneous charge transfer) if $b_{\mathrm{p}} \gg E_{\mathrm{hop}}$, and that of a small polaron (which is limited to its immediate neighboring sites) if $b_{\mathrm{p}} < E_{\mathrm{hop}}$. These transfer mechanisms are often referred to as coherent or adiabatic because of their energy conserving nature. In the final stage which sets in at higher temperatures, the metallic conductivity begins to break down and semiconduction by incoherent nonadiabatic electron hopping becomes dominant. With increasing temperatures, the large polaron condition begins to fail because higher-frequency lattice vibrations allow the carriers to stabilize in their traps by permitting elastic adjustments to occur more quickly, thereby lengthening the polaron lifetime of τ_{p}. This action leads to a decrease in b_{p} which begins to set in as the temperature approaches the Debye temperature Θ_{D}.[2]

[2] The relation between the Debye energy $k\Theta_{\mathrm{D}}$ and U_{p} in determining the actual activation energy E_{hop} can be appreciated if one recognizes that the electrostatic potential of the polaron dipole field has an intimate tie to the elastic distortion that dresses the polaron charge site. The distortion of the lattice can be viewed as a reaction to the dipolar field and both energies involve the polarizability of the lattice through the dielectric constant K. As origins of the trap energy, they could be considered to have some equivalence.

The polaron bandwidth b_p is reduced by vibronic (orbit–lattice) coupling that produces a temperature-dependent narrowing given by [12].

$$b_p(T) = b\ \exp\left[-\gamma \coth\left(\frac{\Theta_D}{2T}\right)\right] \tag{8.16}$$

that ranges from $b_p(T) = b \exp(-\gamma)$ at $T = 0$ to the simplification

$$b_p(T) = b\ \exp\left[-\gamma\left(\frac{T}{\Theta_D}\right)\right], \tag{8.17}$$

when $T \gg \Theta_D/2$.

Although not readily discernible in this brief summary, another facet of Holstein's model is that the threshold where thermal activation becomes a significant competitor to tunneling is $T \sim \Theta_D/2$. As seen in the complete function (8.16) plotted in Fig. 8.5, the polaron bandwidth is still a sizable fraction of $b\ \exp(-\gamma)$ at this point. The parameter $\gamma \sim 50/K_{opt}^2$ would be ~0.2 to 2.0 for transition-metal oxides with an optical dielectric constant K_{opt} in the 15 to 5 range. As a consequence, the decrease in b_p will eventually serve to reduce the carrier population according to a probability approximated by the Boltzmann relation $\exp\left(-E_{hop}/kT\right)$ in the usual diffusion relation for electron hopping by thermal activation. Therefore, the transport of a polaronic carrier requires that an increase in energy E_{hop} be supplied to remove it from its trap. A qualitative summary of the polaron stages is presented in Fig. 8.6.

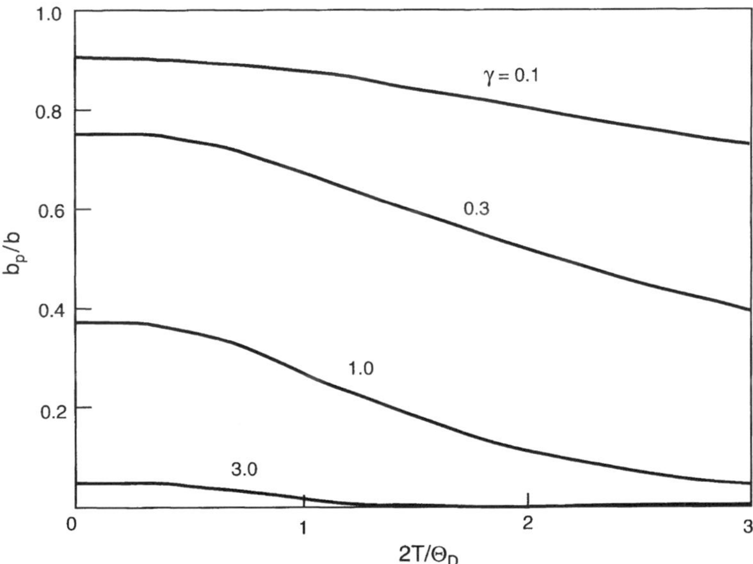

Fig. 8.5 Reduced polaron bandwidth as a function of reduced temperature for different values of γ

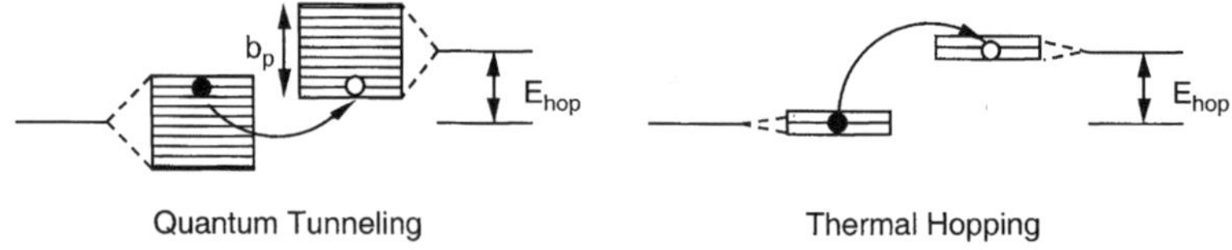

$b_p \sim 1/\tau_p$ (inverse of polaron lifetime in its trap)

$b_p > E_{hop}$	large polarons in broad bands; spontaneous conduction by covalent transfer
$b_p \leq E_{hop}$	small polarons in thermally narrowed bands; local conduction among equivalent neighbors
$b_p << E_{hop}$	hopping electrons from deep in traps; random conduction by thermal activation

Fig. 8.6 Diagram of polaron stages based on the relative magnitudes of b_p and E_{hop}

8.2 Metallic Oxides with Polarized Spins

For electronic conduction in solids, two conditions must be satisfied (1) there must be charge carriers and (2) there must be a mechanism to enable their transport. Carriers can be "free," as in collective-electron metals, excited as in band-model semiconductors, or activated from polaron traps in ionic compounds. The second requirement involves the environment of the destination sites and what lies in between, i.e., the mobility. The temperature dependence of these variables is what generally defines whether the material is a metal (carrier density decreasing with T) or insulator (both carrier density and mobility increasing with T). There is, however, another significant distinction that applies particularly to magnetically ordered materials.

Charge carriers in metals are drawn mainly from s and p states, despite the formation of hybrids with d states in the collective intermingling. As a result, the d-electron magnetic carriers represent only a fraction of the population that produces current. In transition-metal oxides, the s and p electrons of the cations are transferred to the oxygen to establish the anion lattice, and have only negligible hybrid mixing with the d shell. Therefore, in the rare situations where the d electrons become itinerant, the current has the spin polarization of the transfer states, which is usually ferromagnetic in contrast to the more random situation in a ferromagnetic metal. Since the vast majority of the carriers are of only one spin orientation, metallic compounds of the transition-metal series are frequently called "half-metals." The possible origins of polarized spin transport in select metallic oxides can be reviewed in the context of molecular-orbital theory.

In systems where only the t_{2g} orbital states are occupied, i.e., the lighter members of a $3d^n$ transition series, metallic conduction can occur if the t_{2g} levels retain a degeneracy in a lower symmetry crystal field. The most common occurrences are found with d^1 and d^2 configurations in simple monocation-site compounds. Because the cation sites are principally octahedral, the t_{2g} electrons usually form weak π bonds to the O^{2-} anions and therefore feature small exchange stabilization that would normally be expected to provide low antiferromagnetic Néel temperatures T_N. At room temperature, even with dynamic Jahn–Teller (J–T) or spin–orbit (S–O) splittings $(\sim 10^{-2}\,\mathrm{eV})$ of the triplet, the t_{2g} states would then form partially filled t_{2g}–$p\pi$–t_{2g} states resulting from the localized metal–ligand–metal superexchange [8,9].

8.2.1 Simple Oxides

In cases such as TiO and TiO_2, single d electron transfer can occur as a result of hopping electron between 3+ and 4+ mixed-valence states of the Ti ions when the compound is off stoichiometry, e.g., from O^{2-} vacancies. Spontaneous magnetic ordering is usually not involved here because the spin density is too small to produce anything but paramagnetism. A more interesting situation occurs with stoichiometric CrO_2 $(3d^2)$. The crystal structure is rutile (tetragonal cation site symmetry D_{4h} with $c/a < 1$), which splits the t_{2g} triplet into a lower d_{xy} singlet and an upper $d_{xz.yz}$ doublet, as sketched in Fig. 8.7a. The molecular-orbital hybrid states populated according to the Aufbau qualitative guidelines would then have a single electron in each level, but with competing spin-polarization tendencies. In the lower state, the half-filled d_{xy}–$p\pi$–d_{xy} hybrid would be expected to follow the rules of antiferromagnetic superexchange [13]. In the upper $d_{xz.yz}$–$p\pi$–$d_{xz.yz}$ doublet, however, Hund's rule high-spin alignment can exist within the two orthogonal states because the degeneracy allows the two electrons to select different orbitals, thereby avoiding the Pauli principle while providing ferromagnetic exchange stabilization through itinerant charge-transfer characteristic of metallic conduction [8, 9]. The net result would be determined by the spin configuration that produces the lowest energy. Consequently, the antibonding half of d_{xy}–$p\pi$–d_{xy} is stabilized to accommodate the electrons with the lower energy $S=1$ parallel spin alignment. The ferromagnetic order gives the full $2m_B$ magnetic moment per cation (~100% polarization) at low temperatures and coexists with metallic conductivity up to a Curie temperature of 400 K [14, 15]. A more in-depth discussion of the various orbital interactions that lead to the magnetic order and electrical conductivity of these compounds is given by Goodenough [16].

8.2.2 Complex Oxides

Another uncommon situation occurs in mixed (or double) perovskites (Fig. 8.1) in which charge-ordered dissimilar octahedral-site cations can produce metallic conduction and long-range spin order that reach above room temperature. A prime example is $Sr_2{}^{2+}\left[Fe^{3+}Mo^{5+}\right]O_6$ [17]. The charge-transfer and magnetic properties are believed to originate from the half-filled/half-filled correlation

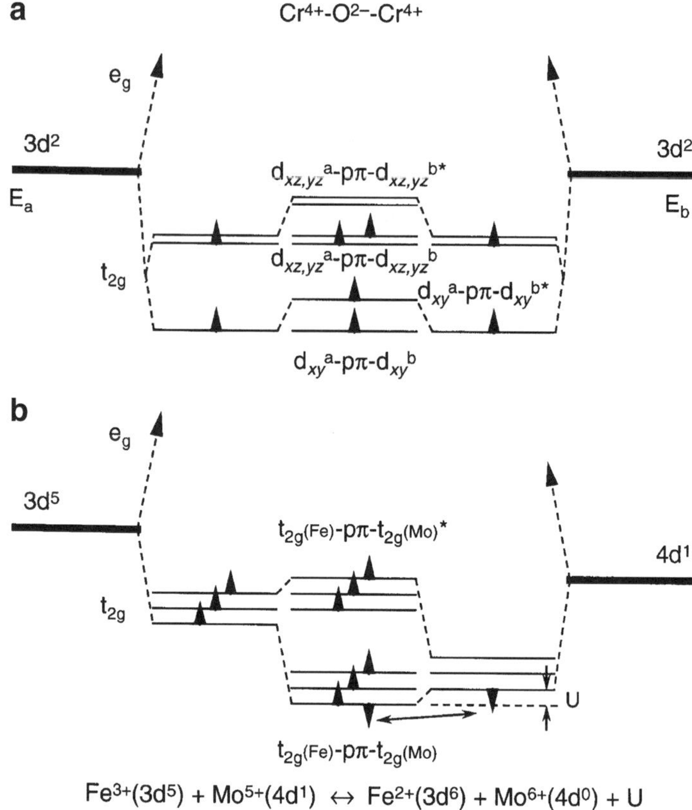

Fig. 8.7 Molecular-orbital approximations for two special cases of spin-polarized metallic charge transfer: (**a**) $Cr^{4+}O_2\left(3d^2\right)$ where a ferromagnetic moment occurs because of the survival of a spin–orbit $d_{xz,yz}$ doublet that forms a "band" with partially filled orthogonal states allowing Hund's rule to apply and (**b**) $Sr_2{}^{2+}\left[Fe^{3+}Mo^{5+}\right]O_6$, which is a paramagnet/unbalanced antiferromagnet (quasi-ferrite) that occurs by delocalization superexchange involving the sharing of the single Mo^{5+} electron. Oxygen ligand states are not shown

superexchange [13] between the two lowest states of $Fe^{3+}\left(3t_{2g}{}^3e_g{}^2\right)$ and $Mo^{5+}\left(5t_{2g}{}^1e_g{}^0\right)$ that form the $3d_{xy}$ (Fe) $-2p\pi-5d_{xy}$ (Mo) hybrid in the B sublattice. Antiferromagnetism is expected in this case, but with a net "ferrimagnetic" moment associated with the remaining four unpaired spins of the Fe^{3+}ion. As a consequence, the compound is a quasi-ferrimagnet with only one crystallographic sublattice and only one molecular-field coefficient. If this material were an ideal uncanted ferrimagnet, the magnetic moment per formula unit would be $4m_B$. Although measurements indicate that the actual value at low temperatures is closer to $3m_B$, the conductivity is still metallic. One possible explanation can be gleaned from the molecular-orbital model in Fig. 8.7b. Beginning with the conductivity issue, we can reason that the single spin resident in the t_{2g} shell of the $Mo^{5+}\left(5d^1\right)$ ion is shared with the $Fe^{3+}\left(3d^5\right)$ ion by delocalization exchange. Since the outermost electron of the Fe^{2+} $(3d^6)$ ion is loosely bound in a filled d_{xy} orbital, i.e., the first one added to the half-filled t_{2g} shell, we can argue that two ionic configurations can exist simultaneously through spin transfer reactions

$$Fe^{3+} + Mo^{5+} \Leftrightarrow Fe^{2+} + Mo^{6+} + U_{eff}$$
$$t_{2g}{}^3 e_g{}^2 + t_{2g}{}^1 e_g{}^0 \Leftrightarrow t_{2g}{}^4 e_g{}^2 + t_{2g}{}^0 e_g{}^0 + U_{eff}$$

The degree of sharing and hence the mobility of the single Mo^{5+} spin will depend on the relative magnitudes of U_{eff} and b, where U_{eff} is the difference in the binding energy between the initial and final states of the transfer and b is the exchange integral between the Fe and Mo ions mediated by the O^{2-} ligand (not shown in Fig. 7b). If the single Mo^{5+} spin becomes itinerant in this sense, spin-polarized transport could take place in the narrow half-filled band constructed from the $3d_{xy}$ (Fe) –$2p\pi$–$5d_{xy}$(Mo) hybrid. Inspection of the above charge-transfer reactions reveals that the left-hand side represents an antiferromagnet (or a quasi-ferrite) and the right-hand side a paramagnet. Therefore, the probability of obtaining the maximum $4m_B$ is reduced by the probability of paramagnetism that occurs when Mo^{5+} is missing its t_{2g} spin. If a paramagnetic contribution is present, experiments at very high magnetic fields might help to clarify the situation.

Another point of interest is that the $3m_B$ value of the net magnetic moment per molecule lends credence to the suggestion that the Goodenough–Kanamori (G–K) rules [9] for predicting the most probable type of spin ordering might not apply in this case. If only the e_g states are considered because of the conventional wisdom that the t_{2g}–$p\pi$ hybrids are effectively nonbonding and can be ignored, the logical conclusion for 180°σ bonding of the perovskites is that ferromagnetic delocalization exchange will result between the empty Mo^{5+} and half-filled Fe^{3+} e_g shells, producing magnetic moment limit of $6m_B$. However, the experimental result of $3m_B$ suggests that the d_{xy}–$p\pi$ bond of the t_{2g} shell is strong enough to stabilize the type of ferrimagnetic ordering proposed.

In a series of compositions with Cr, Mn, and Co used in place of Fe, [18] ferromagnetism was found only with Cr. Since it assumes its 3+ valence with a highly stable electronic configuration $t_{2g}{}^3 e_g{}^0$, $3d^3$-d^1 antiferromagnetic (ferrimagnetic) superexchange is again the likely result. No evidence of metallic conduction was reported, however, which suggests that the value of U_{eff} is greater than that of the Fe case. The other ions also showed insulating properties, but no magnetic ordering. The prospects of antiferromagnetism are reduced because the most stable forms of Mn and Co are divalent, which would force the Mo valence to be raised to the diamagnetic 6+ state, thereby lowering the chance of spontaneous magnetism. Similar analysis can be applied to other combinations of transition metals, such as Ti, W, and Re, but the prospects for strong superexchange with spin-polarized metallic conductivity between the alternating B-site cations probably remains highest with a d^5–d^1 combination.

Ferromagnetism in metallic oxides also occurs with spins in the e_g shell of the mixed-valence perovskite $\left(La^{3+}_{1-x}Ca_x{}^{2+}\right)\left[Mn^{3+}_{1-x}Mn_x{}^{4+}\right]O_3$, for temperatures reaching above 300 K. The origins of this ferromagnetism are complex and are discussed in Sect. 8.3.1. Normally when the e_g shell is occupied, strong σ bonds are formed that produce antiferromagnetic stabilizations that can survive to well above room temperature. Simple oxides of the heavier $3d^n$ elements, MnO through CuO, all feature antiferromagnetism and usually p-type semiconduction when mixed valence occurs, despite the ferromagnetic tendencies that can still exist with the t_{2g} shell partially filled.

8.2.3 Classical Resistivity–Temperature Model

Returning to the discussion that produced (8.10), we can now examine the question of conductivity in polaronic oxides. Most transition-metal oxides are termed mobility-activated semiconductors in contrast to more conventional band model version with the usual description of holes and electrons dictated by Fermi statistics. The electrical resistivity of mixed-valence oxides obeys the standard relation for mobility-activated semiconduction in which the polaron trap energy E_{hop} and the activation energy are equivalent in the temperature regime where random hopping is dominant [19–22],

$$\rho = \left[\frac{x_{\text{eff}}}{V} e \left(\frac{eD}{kT}\right)\right]^{-1} \exp\left(\frac{eD}{kT}\right), \tag{8.18}$$

where the carrier concentration per chemical formula unit, $x_{\text{eff}} = n_{\text{eff}}/V$, the ratio of the volume carrier density to the volume of a formula unit V, the factor $(eD/\text{k}T)$ is the Einstein diffusion mobility, and D the diffusion constant that equals the ratio of the square of the mean-free path to the carrier lifetime $d^2_{\text{hop}}/\tau_{\text{hop}}$. To account for the probability of a transfer being *completed*, an effective carrier concentration is defined by $x_{\text{eff}} = xP$, where P is a polaron dispersal probability equal to $(1-x)$ for a random distribution. Further dependencies of relevance are that $\tau_{\text{hop}} = \nu^{-1}_{\text{hop}}$, which is on the order of the Debye frequency and $d \sim a/(1-x)$ to account for the average hop distance as a function of concentration [23]. Upon substituting appropriately into (8.18), a working expression for the resistivity becomes

$$\rho = \frac{C\,(1-x)\,kT}{x} \exp\left(\frac{E_{\text{hop}}}{kT}\right) = \frac{C\,(1-x)\,kT}{x} \quad \text{for}\, kT \gg E_{\text{hop}}, \tag{8.19}$$

where $C = V/\text{e}^2 a^2 \nu_{\text{hop}}$. As illustrated in Fig. 8.8, ρ plotted as a function of T for $E_{\text{hop}} = 10\,\text{meV}$ from (8.19) will show insulating behavior (decreasing) at low temperatures and pass through a minimum at $T = E_{\text{hop}}/k$. As T increases further, the curve will approach a metallic straight line of (increasing) slope $\text{C}\,(1\text{–}x)\,k/x$. Note that the asymptote of the curve will pass through the origin at $T = 0$, which is a result that is not expected in experiment because residual resistivity ρ_{i} effects would appear. This model is discussed further in relation to superconductivity in Sect. 8.4.

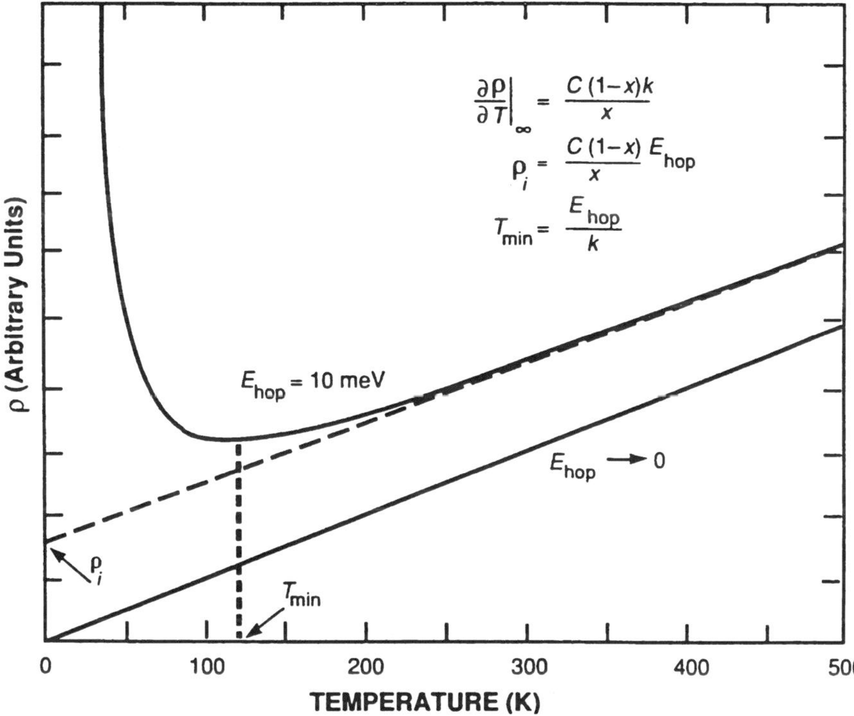

Fig. 8.8 Generic plots of ρ vs. T for $E_{\mathrm{hop}} = 0$ and 10 meV, defining relations for T_{min}, and asymptote slope $\partial\rho/\partial T|_{\infty}$ and intercept ρ_i. Figure reprinted from G.F. Dionne, *IEEE Trans. Magn.* **27**, 1190 (1991) with permission. © 1991 by the IEEE

The temperature of metal–insulator transition therefore resides in the value of E_{hop}, which from (8.10) is controlled greatly by the magnetic state of the spin system. There are two prominent examples where E_{hop} is reduced to the basic elastic polaron limit $E^{0}_{\mathrm{hop}}\left(= U^{0}_{\mathrm{p}}\right)$ by the absence or removal of $E^{\mathrm{ex}}_{\mathrm{hop}}$ $(= U_{\mathrm{ex}})$. If the spins are collinear, $\cos\theta = 1$ and $E^{\mathrm{ex}}_{\mathrm{hop}} \sim 0$. In oxides, this condition is uncommon because superexchange usually dominates to produce antiferromagnetism. An important exception can occur with Mn ions in certain ferromagnetic perovskites compounds that feature metallic properties but also display large magnetoresistance effects at the Curie temperature.

8.3 Magnetoresistance in Oxides (CMR)

In the generic $\left(\mathrm{RE}^{3+}\mathrm{A}^{2+}\right)\mathrm{MnO_3}$ perovskite system, the conditions for metallic conduction in the Mn sublattice can occur in a ferromagnetic phase that appears at lower temperatures. When the ferromagnetism is dissipated at the Curie temperature, a metal–insulator transition takes place. Application of magnetic fields large enough to influence the Curie temperature has been shown to produce dramatic magnetoresistance effects – colossal magnetoresistance (CMR) [24]. To explain the origin of this phenomenon, we first examine the source of the ferromagnetic spin alignment.

8.3.1 Manganese-Ion Exchange Interactions

For the basically cubic octahedral oxygen coordination, crystal-field effects dictate that the t_{2g} orbital states are of lower energy and are half-filled to satisfy the Hund's rule spin polarization requirement for both Mn^{3+} $(3d^4)$ and Mn^{4+} $(3d^3)$. For each combination of exchange linkages, the t_{2g} electrons produce weak antiferromagnetism via covalent π bonding through the O^{2-} anions, e.g., Mn^{4+}–O^{2-}–Mn^{4+}. For the Mn^{3+}–O^{2-}–Mn^{3+}combinations, the larger transfer integrals are the result of the stronger 180°σ-bonding in the e_g states. A single electron in the e_g shell (Mn^{3+} case) can be stabilized by a static Jahn–Teller (J–T) distortion that splits energy levels as shown in Fig. 8.9. Where the distortions are cooperative, a tetragonal/orthorhombic phase will appear (with site axis ratio $c/a, b > 1$) and the half-filled d_{z^2} orbital is stabilized relative to the empty $d_{x^2-y^2}$ state. Superexchange spin ordering possibilities for this system were examined in a seminal paper by Goodenough [25].

From the Curie temperature data of Jonker and Van Santen [26] for $\left(La^{3+}_{1-x}Ca_x{}^{2+}\right)\left[Mn^{3+}_{1-x}Mn_x{}^{4+}\right]O_3$ presented in Fig. 8.10 with the various crystallographic phases as a function of x [25], the variations in exchange field with Mn^{4+} concentration may be analyzed on the basis of changes in the nature of the J–T effect. At $x = 0$, the J–T effect should be mainly static and cooperative, favoring an orthorhombic distortion and antiferromagnetic order in most cases. With increasing Mn^{4+} concentration, ferromagnetism dominates in the regime up to $x = 0.5$, with a peak near $x = 0.3$. As Mn^{4+} ions are introduced, parallel spin alignments

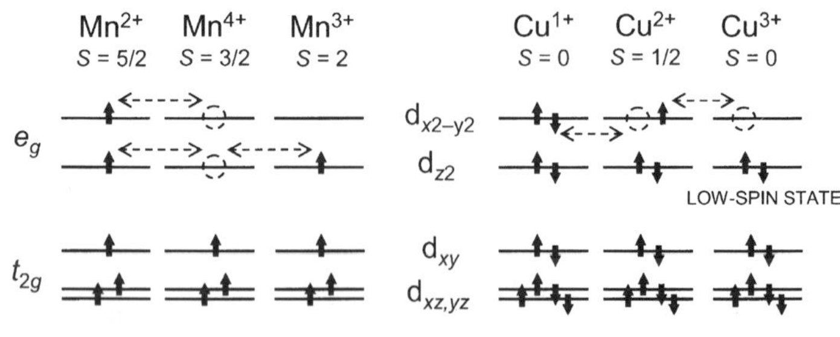

Fig. 8.9 Schematic diagram of delocalization exchange in the e_g shell, comparing the various cases of mixed-valence Mn and Cu

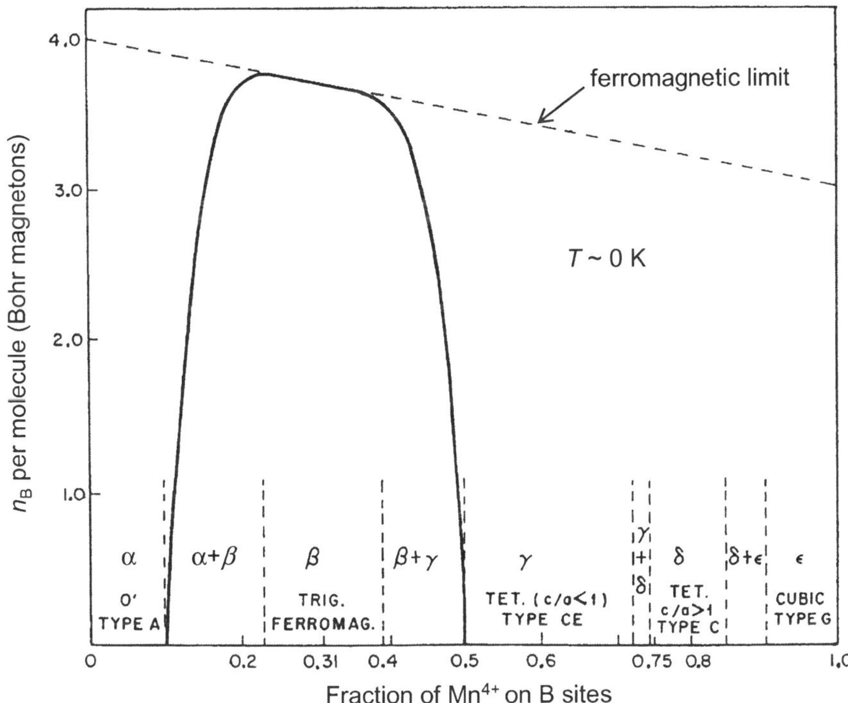

Fig. 8.10 Plot of Curie temperature vs. x for $\left(La^{3+}_{1-x}Ca_x{}^{2+}\right)Mn^{3+}_{1-x}Mn_x{}^{4+}O_3$ with various crystallographic phases indicated. Model is adapted from original data of Jonker and Van Santen [26] presented in **Fig. 3.22**. Figure reprinted from [25] with permission. © 1955 by the American Physical Society. http://link.aps.org/doi/10.1103/PhysRev.127.2058

result from a combination of factors (1) the Mn^{3+}–O^{2-}–Mn^{4+} couplings contribute ferromagnetism by charge transfer among half-filled/empty orbital combinations as studied by Zener [27] and de Gennes [28] and (2) the anticipated antiferromagnetism from the Mn^{3+}–O^{2-}–Mn^{3+} couplings in a static J–T effect is converted to ferromagnetism possibly by vibronic-induced J–T effects proposed by Goodenough [29–32] that alternates the order of the e_g–$2p\sigma$ antibonding levels (bands) and allows the two e_g electrons to be stabilized with parallel spin alignments in separate molecular-orbital states.

Ferromagnetism can therefore occur because of the absence of static tetragonal deformation that leaves the e_g states degenerate and removes the necessity for Pauli spin pairing, similar to the t_{2g} degeneracy in the case of CrO_2 discussed previously. As illustrated in Fig. 8.11, in cases where the Mn^{3+}–O^{2-}–Mn^{3+} couplings dominate and the electronic bandwidth is broad enough for the interactions to be collective, a ligand vibronic mode may cause the d_{z^2} and $d_{x^2-y^2}$ states of adjacent Mn cations to oscillate out of phase and form a quarter-filled e_g shell. This situation allows Hund's rule to apply and gives rise to ferromagnetic order and spin-polarized metallic conductivity. The quasi-static J–T effect is consistent with the absence of

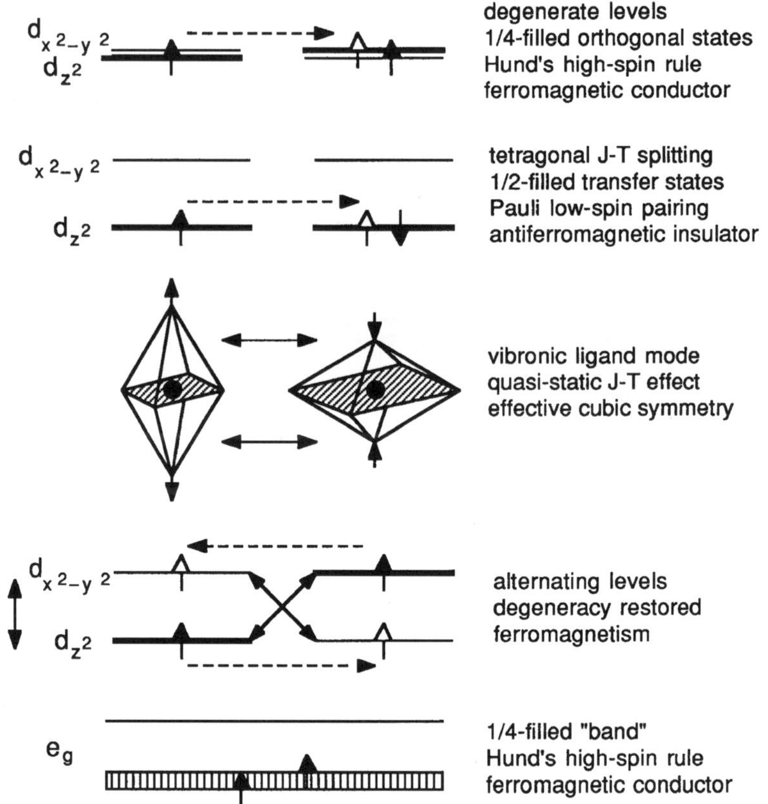

Fig. 8.11 Schematic diagrams of J–T effects, including the quasistatic case in which carrier transfers into empty e_g orbital states that are mixed by vibronic modes may cause ferromagnetism

the static orthorhombic distortion in the regime of the observed ferromagnetism that would normally be expected to stabilize the e_g electron of the Mn^{3+} ions in the lower of the split e_g states. This condition would tend to deny tunneling transfer by increasing the spin-dependent part $E_{\text{hop}}^{\text{ex}}$ of the polaron trap energy. The anticipated couplings for the various combinations were summarized in Chap. 3 (Table 3.7).

Above $x \sim 0.1$, the vibronic actions can influence a change in crystallographic phase from orthorhombic to cubic or rhombohedral (trigonal), thereby restoring the degeneracy of the e_g levels in the crystal field, and remaining such until x approaches 0.5. In some cases the vibronic effect on e_g can occur in the presence of an orthorhombic bias. With half the Mn ions in the 4+ state, the quasi-static behavior breaks down as the lattice symmetry returns to orthorhombic. At this point, the e_g levels are split, and the remaining Mn^{3+}–O^{2-}–Mn^{3+} couplings revert to antiferromagnetism, combining with the existing antiferromagnetic Mn^{4+}–O^{2-}–Mn^{4+} couplings to produce various cation charge order and antiferromagnetic configurations in the range from $0.5 < x < 1.0$. It should also be pointed out that in the ferromagnetic region the maximum available m_B per formula unit is apparent at cryogenic temperatures, confirming the anticipation of complete uncanted spin polarization similar to that of CrO_2.

To account for the magnetic exchange effects that produce the observed ferromagnetism, a single magnetic lattice is assumed. Because both the carrier charges are among the d electrons that provide the magnetic moments, the disposition of spins cannot be static. Valence-charge ordering may occur coincident with spin ordering in spatially variable phases, which could explain the reported observation of a metal–insulator mosaic that probably corresponds to ferro/antiferromagnetic domain patterns [33]. To describe this system by traditional analytical methods is a formidable challenge. Nonetheless, a model based on a random distribution of Mn^{3+} and Mn^{4+} cations carrying spins S_3 and S_4 can be fashioned. For this exercise, three exchange interactions are defined: $J_{33}\boldsymbol{S}_3 \cdot \boldsymbol{S}_3$, $J_{44}\boldsymbol{S}_4 \cdot \boldsymbol{S}_4$, and $J_{34}\boldsymbol{S}_3 \cdot \boldsymbol{S}_4$ (or $J_{43}\boldsymbol{S}_4 \cdot \boldsymbol{S}_3$), from which an effective exchange energy is constructed for use with the Brillouin–Weiss theory.

For an individual charge transfer between Mn^{3+} and Mn^{4+} ions, (8.1) can be applied to express the activation energy as

$$E_{\text{ex}}(\theta_{34}) = zJ_{34}\boldsymbol{S}_3 \cdot \boldsymbol{S}_4 \approx zJ_{34}S_3S_4(1 + \cos\theta_{34}), \tag{8.20}$$

where z is the number of nearest neighbors and θ_{34} is the average angle between the Mn^{3+} and Mn^{4+} spins which will be assumed to be simply θ, the average angle between adjacent spins within the entire system. When $\theta = 0$, E_{34} is a maximum, the spins are collinear, and the $\Delta S = 0$ requirement for the charge transfer is satisfied. If $\theta > 0$, E_{34} decreases and energy must be provided to restore the spin alignment and maintain $\Delta S = 0$. According to (8.1), the additional energy has the effect of a magnetic trap of depth

$$E_{\text{hop}}^{\text{ex}} = E_{34}(\theta_{34}) - E_{34}(0) = zJ_{34}S_3S_4(1 - \cos\theta_{34}), \tag{8.21}$$

and $E_{\text{hop}}^{\text{ex}}$ is also the activation energy necessary to effect the charge transfer between the two cation sites. It should be noted that θ_{34} values in the range 0 to π are allowed by (8.21), and that the theoretical maximum activation energy from exchange is actually $2E_{\text{hop}}^{\text{ex}}$ when the spins are antiferromagnetic. For the present problem, however, the regime of interest is the ferromagnetic to paramagnetic transition that occurs at the Curie temperature where the average angle between spins becomes $\pi/2$.

8.3.2 Magnetoresistivity-Temperature Model

Since the Brillouin–Weiss function $\mathcal{B}_S$ represents the average z-axis projection of spins within a cone of half-angle θ, it also represents the average angle between a spin and the direction of the exchange field in which it resides. Consequently, $\cos\theta$

in (8.21) may be represented by $\mathcal{B}_S$ and the spin canting effect on the binding energy can now be expressed as a function of temperature and magnetic field.

The total activation energy as a function of temperature and magnetic field H may then be expressed in terms of molecular field theory according to

$$E_{\text{hop}} = E^{0}_{\text{hop}} + E^{\text{ex}}_{\text{hop}}\left[1 - \mathcal{B}_S(T, H)\right], \tag{8.22}$$

where E^{0}_{hop} is the polaron trap energy in the absence of spin-polarization constraints (chosen as 0.004 eV), and $E^{\text{ex}}_{\text{hop}}$~0.1 eV) is the magnetic exchange contribution that reaches its full value in this system when the spins become disordered at $T > T_C$ [34, 35].

In Fig. 8.12, the Brillouin–Weiss function is plotted as a function of T for an average molecular-field coefficient $\mathcal{N} = 114\,\text{mol cm}^{-3}$ (derived from the J_{33}, J_{34}, and J_{44} parameters of the randomly dispersed Mn^{3+} and Mn^{4+} ions with $x = 0.23$), resulting in a Curie temperature of 300 K. The effect of an external field $H = 10\,\text{T}$ in extending the ordered ferromagnetic region above T_C is shown together with the corresponding $\Delta\mathcal{B}_S$ that occurs when the field is applied. From (8.22), it is seen that $\Delta\mathcal{B}_S$ reflects the change in $E^{\text{ex}}_{\text{hop}}$ due to the applied field that causes the magnetoresistance effect.

By combining (8.22) and (8.18), magnetoresistance curves can be computed for any set of material parameters or external field values. In Fig. 8.13, ρ vs. T data [36] for a composition estimated as $\left(La^{3+}_{0.77}Ca^{2+}_{0.23}\right)MnO_3$ subjected to H fields of 0, 1, 3, 5, and 14 T are fitted by curves generated from (8.19) in combination with (8.22) using the above values for E^{0}_{hop} and $E^{\text{ex}}_{\text{hop}}$, $x = 0.23$, and $C = 6\,\text{m}\Omega\,\text{cm}\,(\text{eV})^{-1}$. Except for the $H = 0$ curve, which did not reach its full peak probably due to the inhomogeneously broadened tail of the thermomagnetism curve and the possibility that the specimen was not magnetically saturated, and the one for 14 T, which may exceed the range of validity of the approximations, theory and data are in reasonably good agreement. Because of the magnetocrystalline

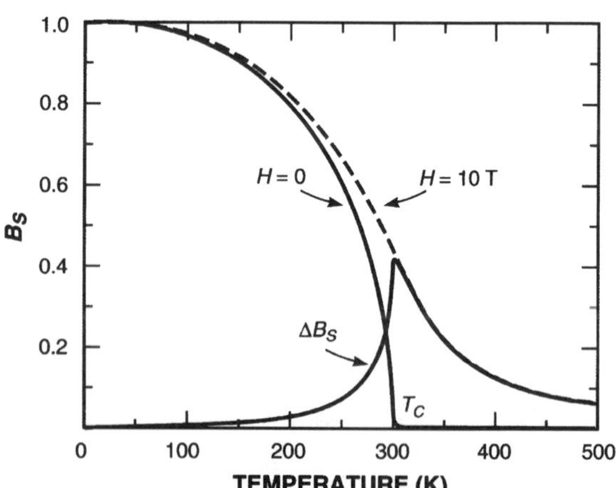

Fig. 8.12 Calculated plots $\mathcal{B}_S$ and $\Delta\mathcal{B}_S$ vs. T for $T_C = 300\,\text{K}$ with $H = 0$ and 10 T. Figure reprinted from [35] with permission. © 1996 by the American Institute of Physics

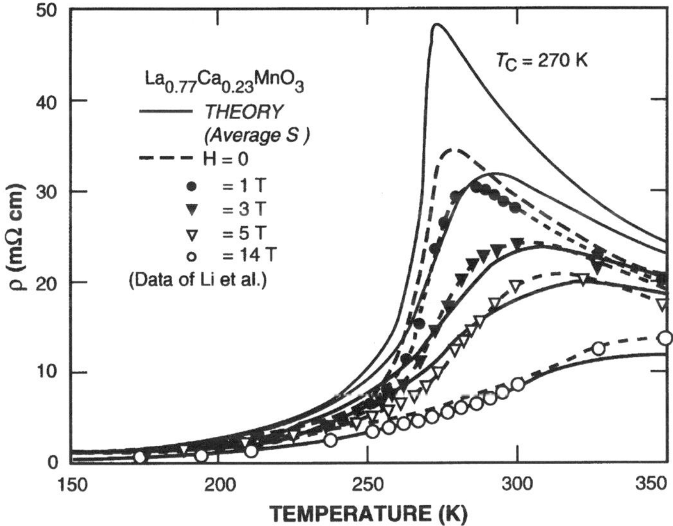

Fig. 8.13 Comparison of $\mathcal{B}_S$ approximation theory with experiment for ρ vs. T with $H = 0, 1, 3,$ 5, and 14 T. Data are from Li et al. [36]. Figure reprinted from [35] with permission. © 1996 by the American Institute of Physics

anisotropy fields, which can probably reach beyond fields of 0.1 T, the presence of domains of varying size and disposition should be expected at low fields and temperatures approaching T_C.

Part of the discrepancy between theory and experiment is the result of the molecular field approximation which represents the z-axis projection of the combined magnetic moment from all of the spins that occupy a cone of average half angle θ. Any difference between θ and the average canting angle between neighboring S_3 and S_4 spins could account for the relatively small disagreement between theory and measurement in the metallic region below T_C. It has also been assumed that the polaron charges are randomly dispersed providing a net molecular field coefficient that is constant with temperature when in fact it probably changes as the various exchange couplings compete for dominance as polaron charges shift about to maintain the lowest lattice energy.

Another consideration is the role of the polaron bandwidth (or inverse lifetime) which narrows with increasing temperature. As indicated by (8.16) and (8.17), carrier transport is likely to be by tunneling at the lowest temperatures, but the coherence could dissipate and give way to random thermal hopping well before the Curie temperature is reached. Since the Debye temperatures of these compounds can be less than 200 K, the onset of nonadiabatic hopping could begin at temperatures below the liquid nitrogen range. This tunneling temperature regime would fall into the range of HTS found in perovskite cuprate lattices. With an appropriate value for b_p, a more refined model that includes the influence of quantum tunneling as a function of magnetization could be constructed to calculate more accurately the resistivity temperature dependence as the material evolves from metal to insulator as $T \to T_C$.

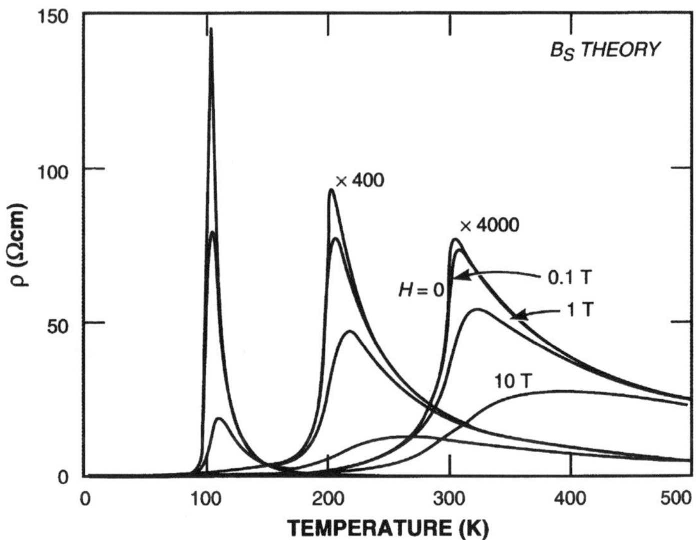

Fig. 8.14 Predicted plots of ρ vs. T by the $\mathcal{B}_S$ approximation for T_C values of 100, 200, and 300 K. Magnetic field strengths are $H = 0$, 0.1, 1, and 10 T. Figure reprinted from [35] with permission. © 1996 by the American Institute of Physics

In Fig. 8.14, example curves of ρ from the $\mathcal{B}_S$ approximation are plotted as functions of T with H values of 0, 0.1, 1, and 10 T for Curie temperatures at 100, 200, and 300 K. Since $E_{\text{hop}}^{\text{ex}}$ is a constant of the transfer ions, the peaks of ρ at $H = 0$ should theoretically touch the insulator-phase envelope (given by a ρ calculation using the full $E_{\text{hop}} \sim 0.1$ eV) at each of the T_C values. From these results, the magnitude of the anomalous increase in ρ at T_C is shown to increase by almost four orders of magnitude between 300 and 100 K.[3]

Some conclusions can be drawn from this analysis (1) the metal–insulator transition occurs at the Curie temperature, which is an intrinsic property of the exchange field and therefore the chemical bonding of the material, (2) the metallic property defined by the positive slope of the ρ vs. T curve is the direct result of the negative slope of the magnetization M vs. T curve, (3) the magnetoresistance is the result of enhancement of the intrinsic exchange field by an external magnetic field, (4) the magnitude of the external field needed to cause significant changes in resistivity must be on the same scale as the exchange field, i.e., greater than 10 T, and (5) the peak resistivity and magnitude of the magnetoresistance decrease with rising temperatures.

[3] Note that if the curves in these figures were extended to $T = 0$, ρ would begin to rise sharply at $T \sim 40$ K because of the elastic trap energy $E_{\text{hop}}^{0} = 4$ meV. In reality, the thermal hopping mechanism may be dominated by polaronic tunneling at these lowest temperatures and the metallic region would theoretically reach $T = 0$, where ρ would also approach some residual value.

The model used above is derived from the notion that spin directions undergo canting as the temperature increases. This is the Boltzmann statistical basis of the Brillouin–Weiss theory that was applied in a direct fashion to the carrier trap energy which is allowed to vary as a continuous function of temperature and magnetic field. A more simplified view of the CMR effect could be constructed by treating the carriers as comprising two groups, each with fixed activation energies: those free to be transported with minimum activation energy (E^0_{hop}) permitted by ferromagnetic ordering and those from a paramagnetic phase with trap energy E_{hop} from (8.10). Partitioning of the carrier populations could be determined by separating the magnetic components according to $\mathcal{B}_S$ (for ferromagnetic spins) and $(1 - \mathcal{B}_S)$ (for paramagnetic spins), each weighted by their respective activation probabilities $\exp\left(-E^0_{\text{hop}}/kT\right)$ and $\exp\left(-E_{\text{hop}}/kT\right)$. Such reasoning would lead to an alternative version of (8.18):

$$\rho = \left[n'_{\text{eff}} e\left(\frac{eD}{kT}\right)\right]^{-1}, \tag{8.23}$$

where $n'_{\text{eff}} = n_{\text{eff}}\left[\mathcal{B}_S \exp\left(-E^0_{\text{hop}}/kT\right) + (1 - \mathcal{B}_S)\left(-E_{\text{hop}}/kT\right)\right]$. By inspecting (8.23), we see that ρ approaches (8.18) in the high temperature limit. In the regime below $T = T_C$, a ρ vs. T curve can be constructed directly from Fig. 8.12. However, this approach could be useful in the immediate vicinity of T_C where the approximation of a two-phase magnetic system might reasonably represent the breaking down of spin ordering. Where magnetic ordering is not a direct issue in determining the degree of carrier availability, a model based on two distinct carrier trap energies can also produce interesting results when applied to the case of HTS in Sect. 8.4.

At higher temperatures, changes in the cation charge distribution could enable the rhombohedral (trigonal) phase to extend beyond $x = 0.5$, giving rise to the peculiar antiferromagnetic/ferromagnetic transition first reported by Jonker and Van Santen [26] for $\left(La^{3+}_{0.3}Sr^{2+}_{0.7}\right)MnO_3$. In the regime where the ferromagnetic stabilization can no longer dominate, a variety of antiferromagnetic ordering configurations can appear labeled as type A, C, and CE, where mixed Mn^{3+} and Mn^{4+} ions compete for spin alignments, and G for the stable antiferromagnetic end member $x = 1$ with only Mn^{4+}ions [37]. The phenomenon of magnetoresistance can still occur when a large enough applied field is able to upset the net exchange field and reverse the sign of the resultant J constant. Structural symmetries react to the magnetic and charge order, as the relative disposition of the d_{z^2} and $d_{x^2-y^2}$ orbitals continue to determine the nature of the spin ordering and the anisotropy of charge transfer, whether along respective z-axis chains or within x–y planes. Detailed low-temperature phase diagrams for the manganite systems that correlate magnetic, crystallographic, and phases can be found in publications by Goodenough [37, 38].

A comment on the conduction properties of the inverted spinel (generic formula $A^{3+}\left[B^{2+}B^{3+}\right]O_4$) magnetite $Fe^{3+}\left[Fe^{2+}Fe^{3+}\right]O_4$ is in order. Although metallic conduction can be attributed to the polaronic charge transfer between octahedral (B-site) Fe^{2+}–Fe^{3+} ions via the incoherent hopping mechanism $Fe^{2+} \leftrightarrow Fe^{3+} + e^-$, above the charge ordering Verwey temperature (120 K) where the trap energy drops from 0.15 eV to 0.04–0.06 [39], there remains the large antiferromagnetic contribution to the trap energy from the A sublattice. Moreover, the ferromagnetic spin transfer is likely between only one of the five orbital states, which leaves the remaining four to oppose it by correlation superexchange [40]. Recent analysis has indicated that the overall exchange interaction between the B-site Fe ions remains antiferromagnetic despite the significant double exchange effect [41]. As a result, the prospects of achieving a high degree of polarized spin transport in a true ferrimagnet seem remote because of the frustration tendencies endemic to systems with opposing sublattices. The relatively small E_{hop} at room temperature, however, supports the suggestion that the value of U_{eff} in the $Sr_2^{2+}\left[Fe^{3+}Mo^{5+}\right]O_6$ compound is also small.

8.3.3 Dilute Magnetic Oxides

As a prelude to a discussion of large polaron superconductivity, some observations concerning polarized spin transport in magnetically dilute compounds are appropriate. Following reports of parallel spin ordering above 300 K, magnetically dilute oxides have been investigated vigorously in the search for room-temperature magnetic semiconductors. Because the spins are isolated as ionic substitutions in crystalline compounds, the apparent ferromagnetic effects that can exist to 1,000 K are not explained by conventional orbital overlap exchange. For magnetically dense ferrites with Curie temperatures (T_C) that range from 500 to 900 K, even modest dilution of the iron will sharply reduce T_C to well below these levels, as explained in Chap. 4. Other features peculiar to these dilute magnetic systems are sketched in Fig. 8.15. Measured thermomagnetism behavior in (a) generally follows a linear

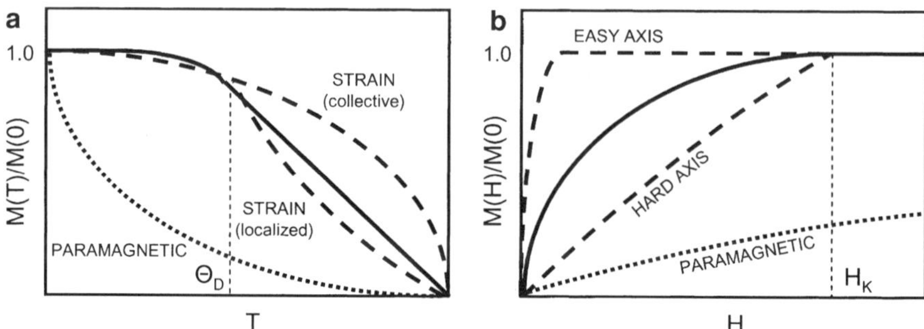

Fig. 8.15 Schematic models of magnetization in dilute magnetic oxides: (**a**) comparison of proposed thermomagnetic concave and convex contours for localized and cooperative magnetoelastic extremes. The more linear intermediate *curve* suggests that magnetoelastic spin ordering (enhanced by ferromagnetic double exchange) percolates at lower concentrations than shorter-range antiferromagnetic exchange. Because the static strain follows the Debye temperature Θ_D, while the important parameter of exchange is the Neel temperature, competition would be expected as temperature and concentration increases. In part (**b**) corresponding magnetization curve models showing the effect of collective magnetocrystalline anisotropy. The *dashed curves* represent the condition where external uniaxial stress defines the limits of collective easy and hard magnetic directions of the cooperative magnetostrictive strain

slope (solid curve) rather than the familiar a Brillouin–Weiss convex contour, suggesting the absence of a magnetic bias field proportional to the magnetization, i.e., a magnetic exchange field $H_{ex} = NM$. Moreover, T_C values vary little among the different magnetic ion/oxide combinations, and are generally insensitive to the magnetic ion concentration at low levels [42]. Magnetic saturation in (b) usually requires several kOe of field, indicating a significant anisotropic demagnetizing field. Vanishingly small remanent moments suggest significant stress demagnetization, thereby indicating further the presence of magnetoelastic ions [43].

Two properties of these compounds have attracted the attention of researchers: (1) the magnetic impurity concentration should be low enough ($<10\%$) to avoid the occurrence of local antiferromagnetic pairs and (2) for good electrical conduction, the impurity concentration must include mixed-valence states and be high enough to produce charge transfer via polarons with aligned spins. This latter issue remains a subject of concern, however. Because of the low densities of magnetic ions, arguments that the reported ferromagnetic spin ordering is the result of itinerant spins stabilized in antibonding states by conventional exchange are difficult to support on theoretical grounds. Without an aligning field of many Tesla, the dilution of the magnetic system will have the magnetization of a simple paramagnet that will not survive to temperatures much above the cryogenic range.

From the discussion of exchange field effects on the intensity of microwave magnetic resonance in Sect. 7.1, it was concluded that when a population of independent spins become stabilized into a collective magnetic moment m, the distribution of the population among available states as a function of temperature will not follow the Boltzmann partition theory unless the effective temperature of the spin system is reduced according to the strength of the alignment interaction [44]. Instead of the scalar addition of a Weiss molecular (exchange) field exponent $m(H_{ex} + H)/kT_L$ that serves to explain the coexistence of microwave (Zeeman effect) and infrared (exchange) resonances discussed, the higher energy stabilization of the magnetic system might also be created by a cooperative magnetoelastic effect induced by a polarizing magnetic field or uniaxial (or planar) stress "field" sufficient to percolate local site distortions into a net strain, e.g., with Mn^{3+} J–T ions in $Y_3Fe_5O_{12}$ [59 of Chap. 5, 45]. Such an energy condensation would be the result of an incipient magnetostrictive strain that overcomes local constraints imposed by the undistorted host sites. The lattice would then respond to an aligning magnetic or strain bias field that encourages the formation of a cooperative magnetoelastic ground state of energy approaching that of the combined unrestrained J–T stabilizations [46].

If a magnetoelastic energy term E_{me} is included, (7.9) can be expressed as

$$T_S = \left(\frac{E_h}{E_h + E_{ex} + E_{me}} \right) T_L \tag{8.24}$$

and the Brillouin function becomes $\mathcal{B}(T) = \tanh(E_h/2kT_S)$. The internal polarizing field term E_h $\left(\sim 1\,\text{cm}^{-1}\right)$ is linearly dependent on the spin alignment, while the spontaneous exchange term E_{ex} $\left(\sim 500\,\text{cm}^{-1}\right)$ has a quadratic dependence on spin ordering. Note that E_h can also represent an anisotropy field arising from crystal-field distortion caused by a local site deformation or an elastic strain induced by external stress. In theory, the magnetoelastic (Jahn–Teller) term E_{me} can be as large as $10^4\,cm^{-1}$ if all the distortion energy were coupled into the spin system. However, the dependence on spin alignment is more complicated. In general, paramagnetic and exchange terms account for the respective concave and convex curves of Fig. 15a. A trace concentration of isolated magnetoelastic ions in sites with spins stabilized by a spontaneous distortion of the ligands will initially produce a characteristic paramagnetic tail. In a prestressed condition that occurs from film/substrate mismatches, the strain bias can organize the local site strains into an energy stabilization that is linearly dependent on spin alignment. With increasing concentrations, the initial concave shape of the thermomagnetism curve can be offset when a cooperative strain begins to form as the individual sites percolate toward a full magnetostrictive system. The growth of these effects was observed in studies if Fe was substituted into stoichiometric In_2O_3 specimens in bulk ceramic form [47].

For significant spin alignment at 300 K, the numerator of the exponent must approach or exceed $kT_L \approx 200\,cm^{-1}$, which occurs in magnetically dense oxides, e.g., ferrites, where the transfer integrals b can stabilize spin alignment enough to produce Curie or Neel temperatures of several hundred Kelvin degrees. When overlap integrals decrease sharply with magnetic dilution, however, breakdown in long-range spin order begins immediately by spin canting that is discussed in Chap. 4, and quickly leads to paramagnetism. This is particularly true of compounds based on d^5 S-state ions Fe^{3+} and Mn^{2+} for which spin–orbit–lattice interactions are insignificant in the nonhybridized ground states.

The contribution of magnetoelastic spin stabilization in dense magnetic oxide systems is reflected in the magnetostriction property, either directly as a cooperative strain induced by polarization of spins through spin–orbit coupling into a net M by an applied field, or inversely by the application of external stress to polarize M. In highly dilute systems, the local site distortions are either dynamic Jahn–Teller effects or are largely constrained by the host lattice. Isolated magnetoelastic ions have Boltzmann spin populations determined by small local E_h/kT_L exponents at room temperature. As the concentration of magnetic ions increases, local dynamic J–T type effects begin to percolate into a net static strain through intersite lattice interactions that are in turn subject to the Debye temperature ($\Theta_D \sim 500$ K). Because E_{me} is temperature dependent according to the approximation $E_{me}(T) = E_{me}(0)\tanh(\Theta_D/2T_L)$, which is based on the growth of optical vibronic phonons from (8.16) for $\gamma = 1$ [42], (7.11) can be refined as

$$\mathcal{B}(T) = \tanh\left(\frac{E_h + E_{ex} + E_{me}(0)\tanh(\Theta_D/2T_L)}{2kT_L}\right). \tag{8.25}$$

As concentrations increase, antiferromagnetic spin pairing will frustrate the net magnetization. If polarized spin transfer between mixed-valence cations contributes ferromagnetic double exchange, the parallel spin alignment can survive to higher concentrations. These magnetoelastic phenomena were modeled semiquantitatively for Cr^{2+} and $Fe^{4+} + Cu^{2+}$ in In_2O_3 [42]. In another study, the conflict between magnetoelastic ordering and antiferromagnetic exchange was illustrated for Fe in $SrTiO_3$ where ferromagnetism and strain ordering dominate until antiferromagnetism appears to take over at Fe concentrations above 40% [48]. In this sense, ions of the $3d^n$ series such as high-spin $3d^4$ and $3d^6$ could provide both the alignment and mobility for spin-polarized conduction at temperatures determined by the stabilization energy of the overall magnetoelastic system.[4]

8.4 Superconductivity in Oxides

Superconductivity remains one of the most intriguing phenomena in the physical sciences. In contrast to collective electron metals that have been analyzed successfully in momentum space [49], the discovery of HTS in ionic compounds [50, 51] presents a formidable challenge to the conventional wisdom. With only a few percent of the molecules providing carriers, a "real" space interpretation based on local effects in the spirit of the London macroscopic molecule seems a logical alternative approach. The model developed for the transition-metal oxides in the ensuing discussion is based on the coherent transport of large polarons in lattices with mixed-valence cations. For purposes of identification, it will be referred to as the covalent electron transfer theory (CET), or simply the large polaron model [8,9]. The first requirement is that it conform to the London conditions for superconductivity, which will be set forth in Sect. 8.4.1.

8.4.1 Classical Foundations

To relate the polaronic transfer mechanism with traditional superconductor phenomenology, it is first necessary to review the macroscopic concepts on which any microscopic theory must be based.

8.4.1.1 The London Equations

Since the electric field $\boldsymbol{E} = 0$ in a hypothetical perfect conductor, it follows that the magnetic flux density $\boldsymbol{B}$ must be constant to satisfy Faraday's law $\nabla \times \boldsymbol{E} = -(1/c)\,\partial \boldsymbol{B}/\partial t$. For this reason, such a material would also be described as a perfect

---

[4] Note that X. Liu, X. Sasaki, and J.K. Furdyna, *Phys. Rev. B* **67**, 205204 (2003) reported strong inverse magnetostriction effects from dilute concentrations of Mn in the host semiconductor GaAs at $T = 5\,\mathrm{K}$. In a tetrahedral coordination, Mn^{3+} $(3d^4)$ would offer either J-T or spin-orbit spin stabilization (Fig. 2.27) sufficient to contribute to the magnetization.

magnetic shield. An unchanging value of $\boldsymbol{B}$, however, is not a sufficient condition for superconduction, because the Meissner effect requires the expulsion of flux from the interior of the specimen as it becomes superconducting, i.e., $\boldsymbol{B} \to 0$. When an external field $\boldsymbol{H}$ is removed from a normal conductor that might have attained a zero resistance state, the existing $\boldsymbol{B}$ would be sustained (flux trapping) by induced surface eddy currents. It follows, therefore, that a superconductor differs from a normal "perfect" conductor by the manner in which currents induced by changes in $\boldsymbol{H}$ are somehow constrained to insure the $\boldsymbol{B} = 0$ condition.

Since superconducting materials are never spontaneously magnetic, $\boldsymbol{B} \sim \boldsymbol{H} = 0$ (i.e., permeability $\mu \sim 1$), and it follows from the Maxwell equation of magnetic induction $\nabla \times \boldsymbol{H} = -(4\pi/c)\,\boldsymbol{i}$ that the supercurrent density $\boldsymbol{i}_s$ is also zero in the interior, where $\boldsymbol{B} = 0$. This means that the current must exist only at the surface, and that the material behaves as a perfect diamagnet, with $\boldsymbol{i}_s$ inducing a field exactly equal and opposite to $\boldsymbol{H}$. It is clear from inspection that Faraday's law alone cannot account for the $\boldsymbol{E} = \boldsymbol{B} = 0$ condition. As a result, the London phenomenological equations [52] were formulated to augment Maxwell's equations for superconductors:

$$\boldsymbol{E} = \left(4\pi\lambda_L^2/c^2\right)\left(\partial \boldsymbol{i}_s/\partial t\right), \tag{8.26a}$$

and

$$\boldsymbol{H} = -\left(4\pi\lambda_L^2/c\right)\nabla \times \boldsymbol{i}_s, \tag{8.26b}$$

where the London penetration depth $\lambda_L = \left(mc^2/4\pi e^2 n_s\right)^{1/2}$ is a constant inversely dependent on the carrier density n_s, with m and e as the carrier mass and charge, respectively. For a stationary state, $\partial \boldsymbol{i}_s/\partial t = 0$, and (8.26a) fulfills the $\boldsymbol{E} = 0$ requirement. If (8.26b) is then combined with the Maxwell induction law $\nabla \times \boldsymbol{H} = -(4\pi/c)\,\boldsymbol{i}_s$, the following differential equations emerge, provided that $\nabla \cdot \boldsymbol{H} = 0$ and $\nabla \cdot \boldsymbol{i}_s = 0$:[5]

$$\nabla^2 \boldsymbol{H} = \boldsymbol{H}/\lambda_L^2,$$

And

$$\nabla^2 \boldsymbol{i}_s = \boldsymbol{i}_s/\lambda_L^2. \tag{8.27}$$

The solutions of (8.27) yield $\boldsymbol{H}$ and $\boldsymbol{i}_s$ as exponential functions of distance x from the specimen surface, e.g., $\boldsymbol{H} = \boldsymbol{H}_0 \exp\left(-x/\lambda_L\right)$. Therefore, both $\boldsymbol{H}$ and $\boldsymbol{i}_s$ are maxima at the surface and decay inward with a profile characterized by London penetration depth λ_L, in compliance with the Meissner effect. If flux and current are expelled from the interior of the material from $\boldsymbol{B} = \boldsymbol{E} = 0$ conditions, they coexist in surface layers of depth λ_L as solenoidal vectors normal to each other, i.e., $\nabla \cdot \boldsymbol{H} = 0$, $\nabla \cdot \boldsymbol{i}_s = 0$, and may be illustrated by the simple geometries of Fig. 8.16.

[5] Here it is necessary to use the vector identity: $\nabla \times (\nabla \times \mathbf{v}) = \nabla(\nabla \cdot \mathbf{v}) - \nabla^2\mathbf{v} = -\nabla^2\mathbf{v}$, with $\nabla \cdot \boldsymbol{H} = 0$ and $\nabla \cdot \boldsymbol{i}_s = 0$.

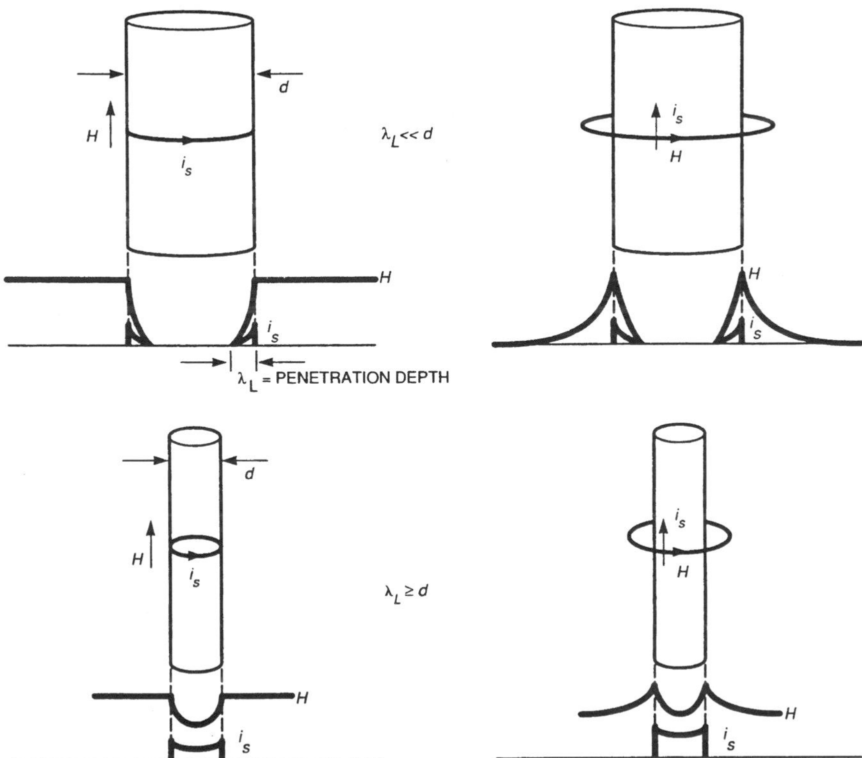

Fig. 8.16 Perpendicular relations between current and magnetic field for superconducting cylinders of large and small diameters. In the *upper cases*, the current responds to an applied external field within a depth λ_L, and in the lower ones, the field is generated to a depth λ_L by a current passed through the superconductor. For the thin cylinders where $\lambda_L \ll d$, the penetration can be almost complete [9]

A more useful relation is the implied interdependence of $\boldsymbol{i}_s$ and $\boldsymbol{H}$ in a superconducting environment, where $\boldsymbol{E} = 0$. Since $\nabla \cdot \boldsymbol{H} = 0$, $\boldsymbol{H}$ may also be defined in terms of the magnetic vector potential according to $\boldsymbol{H} = \nabla \times \boldsymbol{A}$, and it follows from (8.26b) that

$$\boldsymbol{i}_s = -c/\left(4\pi\lambda_L^2\right)\boldsymbol{A} = -\left(e^2/mc\right)n_s\boldsymbol{A}, \tag{8.28}$$

a relation frequently referred to as the "Ohm's law" of superconduction. It should be noted that (8.28) is not gauge invariant, and that a further constraint must be placed on $\boldsymbol{A}$ for application to specific phenomena. In the most general case of simply connected superconductors, the "London" gauge $\nabla \cdot \boldsymbol{A} = 0$ is chosen to conform to the $\nabla \cdot \boldsymbol{i}_s = 0$ condition that defines the observed supercurrent rigidity, i.e., no current components normal to surface.

Although superconductors feature zero resistance, to conform to the phenomenological dictates of the London theory, which concludes that supercurrents are controlled by a magnetic field and not an electric field, supercurrents are constrained in paths near the specimen surface in a manner that prevents eddy currents and leads to internal flux expulsion in external magnetic fields.

8.4.1.2 The Macroscopic Molecule

A more fundamental derivation of (8.28) may be obtained from classical electrodynamics, where the mean local canonical momentum of individual carriers $\langle \boldsymbol{p} \rangle = m\langle \mathbf{v} \rangle + (e/c)\,\boldsymbol{A}$, with $\langle \mathbf{v} \rangle$ as the local mean carrier velocity. Since statistical mechanics dictates that $\langle \mathbf{v} \rangle = \mathbf{0}$ in a normal conductor, it follows that $\langle \boldsymbol{p} \rangle = (e/c)\,\boldsymbol{A}$ in the normal state. For a superconducting ground state, however, a Bloch theorem [53] concluded that $\langle \boldsymbol{p} \rangle = 0$, implying a certain rigidity or inability of the momentum to respond to $\boldsymbol{H}$, and therefore that $\langle \mathbf{v} \rangle = -(e/mc)\,\boldsymbol{A}$. For a chain of ordered carriers of number density n_{s}, it follows that the supercurrent density $\boldsymbol{i}_{\mathrm{s}} = n_{\mathrm{s}} e\langle \mathbf{v} \rangle = -\left(n_{\mathrm{s}} e^2/mc\right)\boldsymbol{A}$, thereby producing an alternate derivation of (8.28). For this electrodynamic approach to comply totally with the constraints of the phenomenological result that $\nabla \cdot \boldsymbol{i}_{\mathrm{s}} = 0$, it is not only necessary that the gauge condition $\nabla \cdot \boldsymbol{A} = 0$ apply, but also that $\nabla n_{\mathrm{s}} = 0$.[6] Therefore, *the basis for describing the supercurrent as "spatially rigid", i.e., the $\langle \boldsymbol{p} \rangle = 0$ condition, must include the condition that the distribution of carriers be uniform (ordered) along the current path.*

Since the earliest superconductors were metals, the notion of spatially ordered carriers was not readily applicable to a free-electron gas. A quantum mechanical extension to this classical concept evolved from the nonlocal ideas of Pippard [54], i.e., the analogy of supercarrier coherence length ξ_0 to normal carrier mean-free-path, giving rise to the Ginsburg and Landau [55] ensemble-average wavefunction ψ_{s}, which in turn is related to the supercurrent electron density by the standard expectation-value relation

$$|\psi_{\mathrm{s}}|^2 = n_{\mathrm{s}}, \tag{8.29}$$

where the number density n_{s} now represents the instantaneous probability of a supercarrier existing at a position vector $\boldsymbol{r}$. As a result, (8.26) may be written as

$$i_{\mathrm{s}} = -\left(e^2/mc\right)|\psi_{\mathrm{s}}|^2 A, \tag{8.30}$$

with the attendant implication that $\nabla|\psi_{\mathrm{s}}|^2 = 0$ to satisfy the condition that $\nabla n_{\mathrm{s}} = 0$.

Four important conclusions may be deduced from this wavefunction rigidity concept (1) the current density vector $\boldsymbol{i}_{\mathrm{s}}$ is directly and exclusively controlled by the magnetic field through the vector potential $\boldsymbol{A}$; (2) the eigenstate of the supercurrent

[6] Recall that $\nabla \cdot \boldsymbol{i}_{\mathrm{s}} = \left(\mathrm{e}^2/\mathrm{mc}\right)(n_{\mathrm{s}}\nabla \cdot \boldsymbol{A} + \boldsymbol{A} \cdot \nabla n_{\mathrm{s}})$, and for $\nabla \cdot \boldsymbol{i}_{\mathrm{s}} = 0$, both $\nabla \cdot \boldsymbol{A}$ and $\nabla n_{\mathrm{s}} = 0$. This latter condition not only is necessary for supercurrent rigidity in the classical argument, but also is sufficient, since a carrier distribution dynamically ordered in *real* space is a rigid current by definition. In effect, it should be considered the fundamental physical requirement for the applicability of (8.28) to superconductivity.

has the properties of a space-invariant wavefunction with zero average momentum ($\langle \boldsymbol{p} \rangle = \nabla\psi_s = 0$); (3) the resulting spatial invariance of the carrier density n_s implies ordered or equispaced supercarriers; and (4) superelectrons cannot be part of the normal free-electron gas. If the carriers have similar quantum states that may be described in terms of a single giant molecular-orbital wavefunction, the current rigidity imposed by the fixed wavefunction provides an immediate explanation for the absence of eddy currents and the presence of flux trapping in the superconducting state.

8.4.1.3 Nonlocal Considerations

Similar to the nonlocal range of normal electrons characterized by a mean-free-path of length ℓ, Pippard [54] proposed that a sphere of radius ξ_0 be considered as a nonlocal region in which each superelectron would exist within the correlation scheme.[7] The coherence length ξ_0 may be estimated from the uncertainty principal once a value for momentum p_s is determined.

Since wavepackets have a spatial profile, Ginsburg and Landau [55] reasoned that the coherence length could be readily introduced through a generalized form of ψ_s with an exponential decay, and proposed a solution of a Schrodinger-type equation with

$$\psi_s(r) = \psi_s(0)\exp(-r/\xi), \tag{8.31}$$

where ξ is a more generalized coherence length. In this context, ξ represents the smallest size of wavepackets that the superconducting charge carriers can form. In a context more appropriate to the discussions that follow, the gradient of the superconduction carrier number density wavefunction may be expressed as

$$\nabla n_s \sim (2/\xi)\,|\psi_s|2 \sim (2/\xi)\,n_s \tag{8.32}$$

As $\xi \to \infty$, $\nabla n_s \to 0$ for spatial ordering of carriers. Figure 8.17 presents a rudimentary contrast between the macroscopic molecule and ensemble wavefunction concepts.

In band-theoretical terms that have been applied to conventional metal superconductors, an intrinsic coherence length $\xi_0 \sim (h/2\pi)\,v_F/kT_c$ is defined in term of the Fermi velocity v_F and superconduction critical temperature T_c. A more general definition was pointed out by Pippard for materials where the coherence length is reduced by impurities that limit the electron mean free path ℓ, according to $1/\xi \approx 1/\xi_0 + 1/\ell$. As a consequence, three classes of superconductors may be defined (a) type-I pure superconductors with large $\xi_0 \gg \lambda_L$ that require a full nonlocal theory treatment (Pippard superconductors), (b) impure superconductors with $\xi_0 \sim \ell$ that are controlled by the mean free path (London limit, where $\xi < \xi_0$) and

[7] For normal conduction, ℓ is the average distance that an electron can be transported without scattering; for superconduction ξ_0 is the distance that a superelectron can remain in coherence with the ensemble as part of the giant molecular state.

THE LONDON EQUATION

$$i_s = -(e^2/mc)\, n_s A$$

where $H = \nabla \times A$.

$$\nabla \bullet i_s = n_s \nabla \bullet A + A \bullet \nabla n_s = 0$$

$$\nabla \bullet A = 0 \quad \text{(London gauge)}$$

CLASSICAL APPROACH (Macroscopic Molecule)	QUANTUM APPROACH (Ensemble Wavefunction)
$\nabla n_s = 0$	$\psi_s = \psi_s^0 \exp(-r/\xi)$
RIGID CURRENT CONCEPT CARRIERS ORDERED IN "REAL" SPACE	$n_s = \lvert\psi_s\rvert^2$ $\nabla n_s = -(2/\xi)\, n_s$
	$\nabla n_s \rightarrow 0$ as $\xi \rightarrow \infty$
	ξ = COHERENCE LENGTH

Fig. 8.17 Contrast between the classical macroscopic molecule concept and the microscopic ensemble wavefunction theory in their approaches to satisfy the requirements of the phenomenological London theory

(c) pure superconductors with $\xi_0 \ll \lambda_L$. For class (b) (8.28) is modified to read $\boldsymbol{i}_s = -\left(n_s e^2/mc\right)(\xi/\xi_0)\,\boldsymbol{A}$, where $\xi^3 \ll \xi_0 \lambda_L^2$; only for class (c) is (8.28) valid as stated. In practice, classes (b) and (c) are type-II superconductors, the former resulting from impurities, and the latter representing the case of small coherence length. The ratio $\kappa = \lambda_L/\xi_0$ is effectively a constant with temperature. In physical terms, the ideal type-I superconductor has features $\kappa \ll 1$, and $\xi_0 \rightarrow \infty$ and $\lambda_L \rightarrow 0$; in the opposite extreme, the magnitudes of these quantities reverse, $\kappa \gg 1$, and $\xi_0 \rightarrow 0$ and $\lambda_L \rightarrow \infty$ in the *natural* type-II case. The essential point is that the coherence length represents a measure of the wavefunction uniformity; it is the quantum mechanical equivalent of the spatially ordered carrier concept of the classical London theory, which will be examined further in the context of coherent electron tunneling in later sections. In magnetic oxides, the conditions necessary to satisfy the London phenomenological requirements can be satisfied in select compositions where large polarons are formed from charge transfer between mixed-valence transition metal ions provided that antiferromagnetic spin ordering is frustrated.

8.4.1.4 Carrier Statistics

The microscopic view of superconductivity in metals is that boson carriers are created from the free electrons with energies near the Fermi level. The resulting supercarrrier ensemble comprises "Cooper pairs" that form the basis for the Bardeen–Cooper–Schrieffer (BCS) microscopic theory of superconductivity [49]. Since the quantum electrodynamic version of the London theory requires that the carrier ensemble form a single wavefunction $\left(n_s = |\psi_s|^2\right)$, the individual carriers cannot be fermions because of the Pauli principle restriction that only one fermion can occupy a state at one time. Supercurrents would therefore occur as bosons that condense to form a superfluid (boson condensation). In the BCS theory, bosons are formed from electron (Cooper) pairs with opposite spins in k-space mediated by lattice vibrational modes (phonons), so that the particles or carriers have a double electron charge and zero spin quantum number ($S = 0$). If a chain of uniformly spaced polaron carriers move in real space as a dynamic ferroelectric condensation, there need not be competition for quantum states. Carriers enter cells to occupy states vacated by simultaneously exiting carriers. This situation is analogous to a vacuum diode *without* space charge, where each electron emitted from the cathode arrives at the anode before the next one is emitted.

Since the large polaron model is based on the classical London theory ($\nabla n_s \sim 0$ instead of $\nabla|\psi_s|^2 \sim 0$), the carriers are not assumed to be free electrons, and there is no requirement for paired electrons mediated by phonons or other "entities" in k-space; in fact, there is no requirement for paired electrons based on purely electrostatic grounds. Instead, the individual polarons may be ordered electrostatically by repulsion within a chain of covalent bonds (the giant molecule concept), and real-space spin pairing in a supercurrent could be required only to maintain any existing dynamic antiferromagnetic order (e.g., spin waves) along the molecular transfer paths. The Pauli principle is satisfied if either the polaron spin $S_P = 0$ and magnetic disorder prevails in a lattice that favors antiferromagnetic coupling in an undiluted state, or the host-lattice ion spins S_L themselves are zero. Where the carriers transfer as real-space pairs with the double electron transfer of an $S_P = 0$ polaron, the single ensemble wavefunction solution of Ginsburg and Landau becomes applicable since the limitations of Fermi statistics are circumvented by $S = 0$ bosons. Since these bosons are local and would condense in real space, the overlapping necessitated by the k-space Cooper pair correlation is no longer of concern, and the question of a Schafroth condensation[8] is moot. Bose–Einstein statistics could also apply as in the quantum boson fluid formalism required by the BCS theory. Whether as individuals or pairs, however, covalent rather than conduction electrons are involved, and the superconduction system proposed is more localized than collective, particularly in the systems of low polaron density.

8.4.2 Zero-Spin Polarons and Magnetic Frustration

Superconductivity occurs in oxides where spin alignment and $\nabla S = 0$ issues are moot if the spin of one of the transfer cations is zero and the lattice environment is without long-range antiferromagnetic order. In this case polaron tunneling is the dominant transfer mechanism because the $b_p > E_{hop}$ condition is assured and incoherent thermal hopping is reduced to a secondary role at low temperatures [7]. Spin transport in transition-metal oxides in which significant local magnetic exchange

[8] See, for example, J.M. Blatt [56].

energy is present can also occur in select situations where the minority transfer ion is in a zero-spin state ($S = 0$). A remarkable feature of this situation is seen in p-type Cu^{2+} (d^9)–O^{2-} – Cu^{3+} $(d^8$, low-spin) and n-type Cu^{2+} (d^9)–O^{2-} – Cu^{1+} (d^{10}) configurations, also in 180° bonds of the perovskite structure. For the Cu^{3+} ion, the $S = 0$ state arises from a low-spin d^8 configuration in the e_g shell that occurs because of a large splitting of the e_g doublet first reported in (LaSr) $Cu^{3+}O_4$ [57], the upper $d_{x^2-y^2}$ orbital empty and available to accept a transferred spin. It should be noted that this arrangement dictates a two-dimensional property and differs from the charge-transfer situation in the manganites, which can occur in either e_g orbital, depending on the sign of the crystal-field distortion. Here the splitting of $d_{x^2-y^2}$ and d_{z^2} exists naturally as a result of the tetragonal/orthorhombic symmetry ($c/a, b > 3$) of the layered-type of perovskites and not necessarily from a J–T effect required to create the e_g splitting in the cubic or rhombohedral manganites [32].[9]

The dependence of this effect on the degree of tetragonal distortion of the crystal field as it evolves from a c-axis extension to the formation of a pyramid and finally to a planar ligand arrangement is diagrammed in Fig. 8.18. Unlike the case where ferromagnetism induced as a byproduct of spin-polarized electron transfer in the $Mn^{3+(4+)}$ combinations eliminates the exchange contribution E^{ex}_{hop} to the activation energy, the involvement of $S = 0$ ions removes the internal polarization exchange energy and therefore renders moot any Hund's rule considerations. In addition, charge transfer from these mobile nonmagnetic ions causes local breakdowns

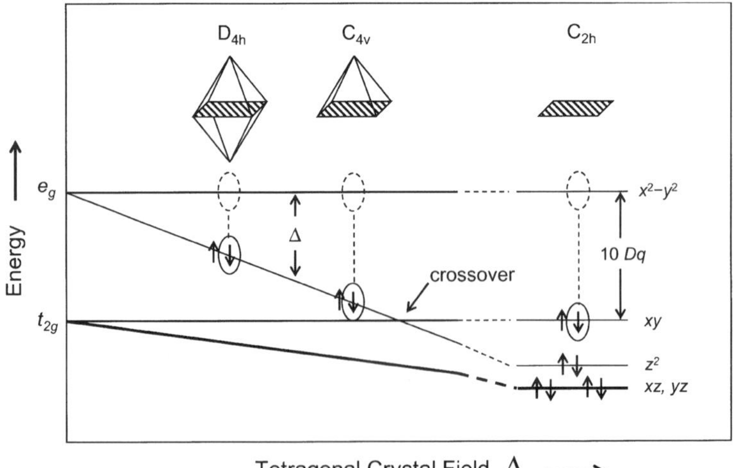

Fig. 8.18 Growth of e_g doublet splitting and stabilization of the d^8 low-spin state as the tetragonal crystal field component increases from z-axis distorted octahedron (D_{4h}) to pyramidal (C_{3v}) to planar (C_{2h}) [9]

[9] A Jahn–Teller condensation in the other end member $La_2Cu^{2+}O_4$ was proposed when an increase in the c/a ratio from 3.30 to 3.46 correlated with antiferromagnetic ordering of the Cu^{2+} J–T ions (J.M. Longo and P.M. Racah, *J. Solid State Chem.* **6**, 526 (1973).

of the antiferromagnetic couplings and can eventually reduce the Néel temperature T_N to zero by causing the spin alignment frustration to spread throughout the entire lattice [23, 58].

The mixed-valence manganites and cuprates are both metallic if the hopping activation energy contribution from antiferromagnetic exchange is compromised. $La_{1-x}Ca_xMnO_3$ is metallic for x <0.5 because of the unusual occurrence of ferromagnetism. However, a net spontaneous magnetic moment represents an internal source of magnetic flux that depresses superconductivity. The existence of a spontaneous magnetic moment, however, eliminates the possibility of superconductivity. $La_{2-x}Sr_xCuO_4$ and other high-T_c cuprates are metallic because the zero-spin Cu ions preclude any type of exchange trap, and thereby frustrate antiferromagnetic order.

As sketched in Fig. 8.19, the polaronic hole carriers created by zero-spin Cu^{3+} ions of concentration $x < 0.08$ in $La_{2-x}Sr_xCuO_4$ would be more than sufficient to disrupt the antiferromagnetic order among the sites immediately surrounding the

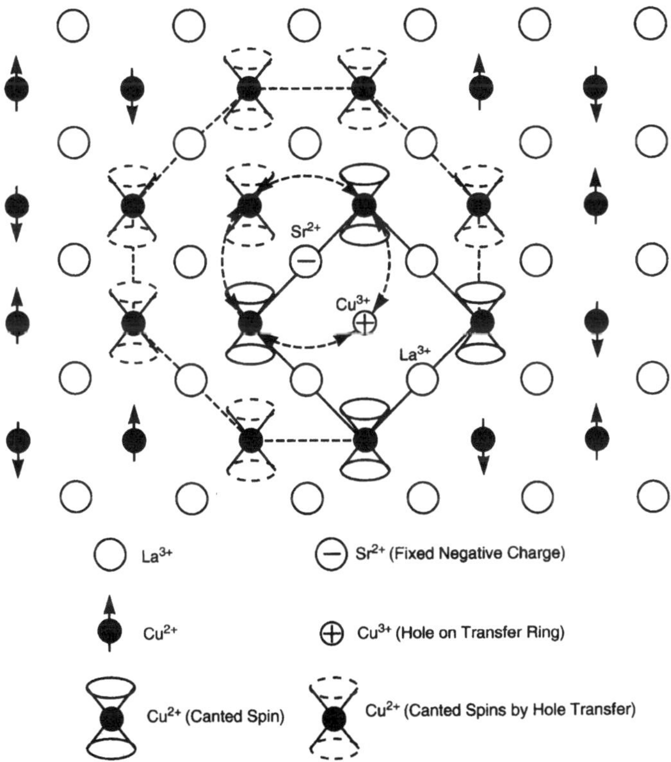

Fig. 8.19 Origin of antiferromagnetic frustration caused by zero-spin Cu^{3+} ions in $La_{2-x}Sr_xCuO_4$. Breakdown in ordering can occur with only one out of 12 ($x \sim 0.08$) $S = 0$ sites in the Cu^{2+} sublattice in the region surrounding an Sr^{2+} "impurity" that acts as a fixed negative charge. Note that the Cu^{3+} hole can transfer around the inner ring in the manner of the model in Fig. 8.2, carrying the canted spins of its four Cu^{2+} neighbors with it [9]

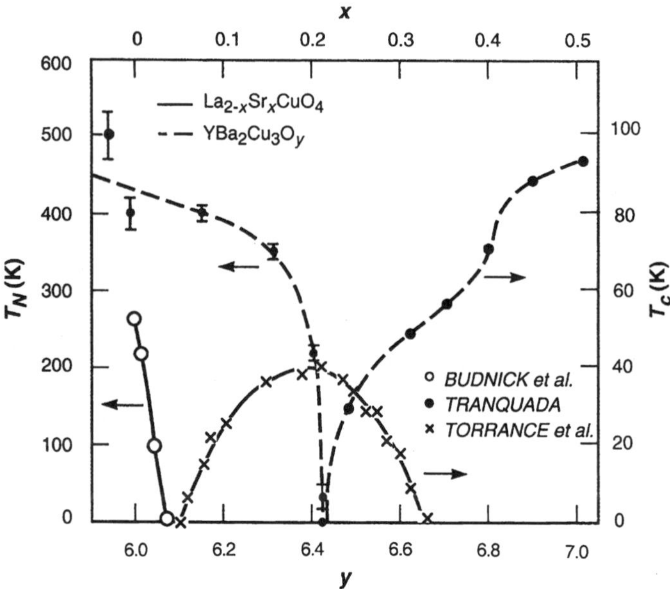

Fig. 8.20 Experimental verification of the magnetic frustration requirement prior to the onset of superconductivity in $La_{2-x}Sr_xCuO_4$ and $YBa_2Cu_3O_y$ systems. Data from Budnick et al. [59], Tranquada [60], and Torrance et al. [61]

fixed negative charge represented by the Sr^{2+} dopant by establishing a ring of mobile carriers (as in Fig. 8.2), and then to frustrate through spin canting the regions beyond these lowest energy polaron sites. Where the charge transfer condenses into a superconducting state at critical temperature T_c, the threshold for frustration could be reduced further. Experimental evidence of this condition is shown in Fig. 8.20 for two common cuprate superconductors [59–61]. Other possible $S = 0$ candidates among the $3d^n$ series with occupied transfer orbitals $d_{x^2-y^2}$ and d_{z^2} are diagrammed schematically in Fig. 8.21.

The influence of magnetic exchange on the Cu^{2+}–Cu^{3+} conductivity in the presence of rare-earth ions occupying the A sites is shown dramatically in Fig. 8.22, where the data of George et al. [62] indicate the absence of activation energy for the nonmagnetic La compound. In all of the others, antiferromagnetic interactions deter the formation of large polarons that are necessary for the coherent activationless charge transfer of superconductivity. As shown in Fig. 8.20, only about 5% zero-spin Cu^{3+} could be sufficient to frustrate the long-range antiferromagnetic ordering in $La_{2-x}Sr_xCuO_4$ and create the metallic phase that allows the onset of superconductivity (the T_N, $T_c = 0$ condition). When one considers that less than one-third of the copper cations supply charge carriers, it would not be surprising to find the superconducting state to consist of spontaneously organized domains of metallic diamagnetism interspersed with regions of insulating or semiconducting antiferromagnetism. Such a result is not unlike the room temperature formation of metallic and insulator regions in the manganites [33].

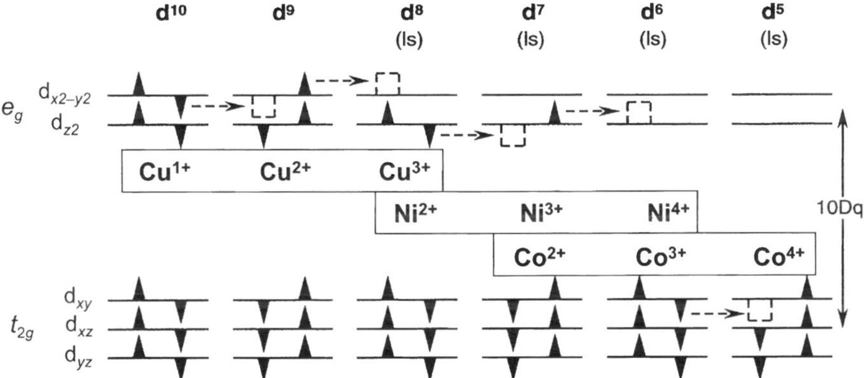

Fig. 8.21 Schematic diagrams of $S = 1/2 \rightarrow S = 0$ transfers among $3d^n$ ions in low-spin states

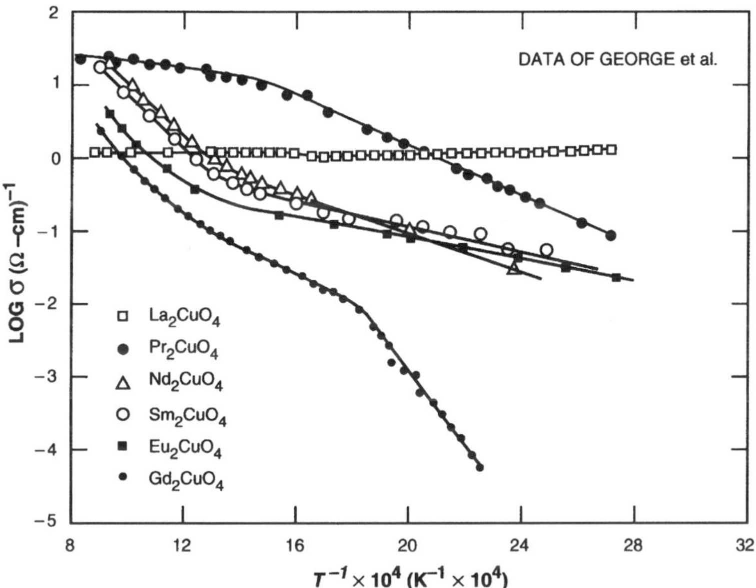

Fig. 8.22 Conductivity σ data as a function of temperature, showing the influence on the activation energies from RE ion exchange interactions with Cu^{2+} ions in a $(RE)_2CuO_4$ series. Note the absence of $E_{\text{hop}}^{\text{ex}}$ in the La_2CuO_4 case. Data are from George et al. [62]

8.4.3 Large-Polaron Superconductivity

Metallic conduction properties in oxides have been demonstrated to condense into superconductivity at temperatures far above those of conventional metal and intermetallic compounds, reaching above 130 K [50, 51, 63]. To understand how covalent tunneling in the form of large polarons can be the agent of HTS, the nature of isolated larger polarons in diamagnetic or magnetically frustrated lattices in the context of superconductivity phenomenology is considered. As introduced in Sect. 8.4.1, the primary condition that evolves from the London equations when applied to a system of supercarriers of volume density n_s in the real-space ideal case, supercarriers must be spatially ordered, thereby rendering their transport coherent, i.e., $\nabla n_s = 0$.

In collective-electron systems this spatial order condition can be treated by the ensemble wave function ψ_{so} with spatial decay (coherence) length ξ according to the proposed $\psi_s = \psi_{so} \exp(-r/\xi)$. Because n_s is represented by the quantum mechanical probability density $|\psi_s|^2$, the gradient expressed from (8.32) can be repeated as

$$\nabla |\psi_s(r)|^2 = -\frac{2}{\xi} |\psi_{so}|^2 \exp\left(-\frac{2r}{\xi}\right) \tag{8.33}$$

or

$$\nabla n_s \approx -\frac{2}{\xi} n_s. \tag{8.34}$$

If $\xi \to \infty$, the London $\nabla n_s = 0$ requirement is satisfied. The magnitude of the coherence length serves as a measure of the purity of the superconducting state [54, 64].

For the polaron case, the equivalent of $|\psi_s|^2$ would be the transfer efficiency $\eta(r)$ of (8.14), which regrettably cannot be expressed as an exponential function in this model because of the central electrostatic field of the polaron trap given by (8.15). To extend the comparison to (8.33), the notion of coherence would be defined by the stabilization energy of the covalent bond according to

$$\nabla \eta(r) \approx \frac{\partial}{\partial r}\left(1 - \frac{2U_p}{b_p}\right) = -\frac{1}{a}\frac{2e^2}{Kab_p}\left(\frac{a}{r}\right)^2. \tag{8.35}$$

Since the bracketed $(a/r)^2$ is dimensionless, a characteristic dimension for an individual polaron can be defined as

$$r_p = a\left(\frac{Kab_p}{e^2}\right), \tag{8.36}$$

which is the same as the result for the large polaron radius derived by Holstein [6], except for a numerical multiplication factor. Consequently, the use of the polaron radius as a measure of coherence of the superconducting state can be adopted with the restriction that polaron dispersal must be achieved by real-space positioning of the polaron sources. This limitation does not exist with collective electron superconductivity formalized in momentum–space.

Within an individual polaron cell, the carrier ranges spontaneously and coherently through correlated transfers to a boundary defined by the covalent strength of the orbital bonds in which it resides and the polarizability of the lattice. In terms of (8.36), the "quality" of the superconductor, i.e., r_p instead of ξ in this model, is determined by the product of the dielectric constant K and the exchange integral b_p.

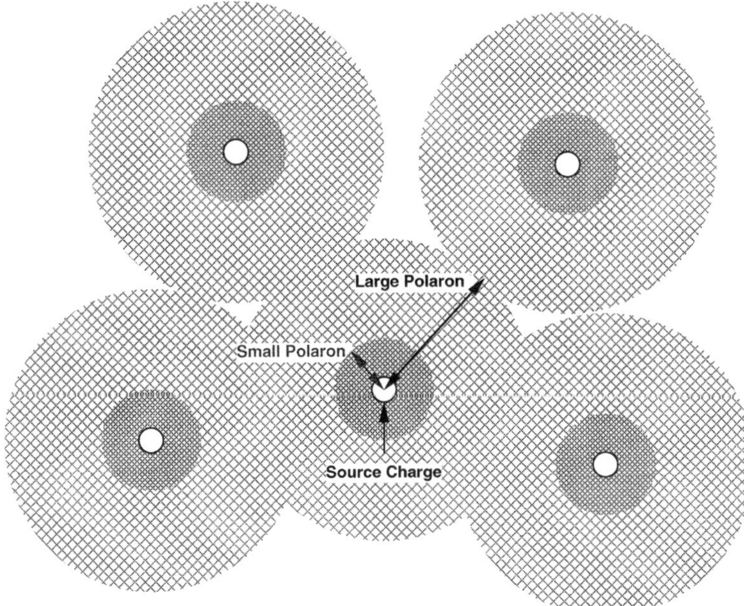

Fig. 8.23 Two-dimensional picture of large polaron cells amidst fixed polaron ionic sources

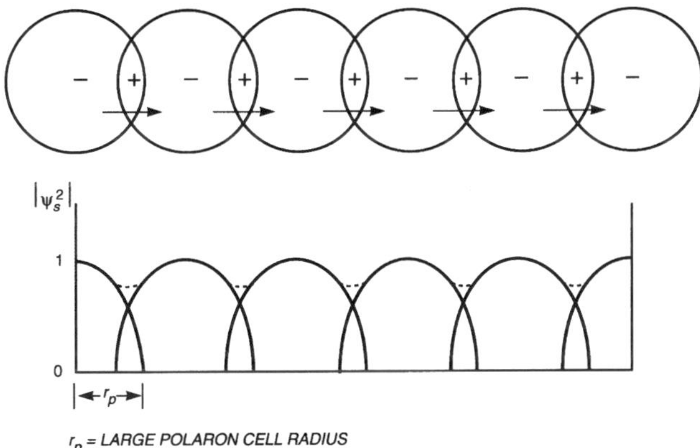

Fig. 8.24 One-dimensional model of a large polaron chain, indicating the merger of carrier density functions $|\psi_S|^2$ to establish a continuous molecular-orbital state

When the concentration reaches a percolation threshold where cells begin to overlap, the dispersed polaron carriers and their fixed source ions sketched in Fig. 8.23 condense into a string or chain of dipoles and a dynamic ferroelectric state of coherently tunneling carriers can form as in Fig. 8.24, still with one carrier per polaron cell. The smallest concentration threshold would then be determined by r_p, subject to the attendant requirement that antiferromagnetic frustration must also occur.

Following the concepts of local vs. collective charge transfer, coherent tunneling would survive only under the conditions (1) that polaron cell overlap be small enough that Pauli scattering would be moot and Fermi statistics not be a concern and (2) that thermal energy would not produce random hopping events to disrupt the correlated charge flow in the superconducting chains. The first condition would be concentration dependent, whereby the orbital levels would merge into a partially filled band at high concentrations, producing normal metallic properties typical of the t_{2g}-band metals such as CrO_2 mentioned in Sect. 8.2. A second limitation, which is reflected in Holstein's conclusion that hopping would dominate over tunneling where $T > \Theta_D/2$, could show its effect at even lower temperatures because of the added requirement that the coherent chains of polaron carriers be continuous.

8.4.4 *Normal Resistivity and Critical Temperature*

From the above qualitative concepts, a "two-fluid" model can be constructed to account for the resistivity characteristics of the HTS cuprate superconductors. This can be accomplished in a straightforward manner; the total carrier concentration x is divided into normal and superconducting fractions by means of the thermal activation probability function. The basic premise of this theory based on large polarons is that carriers that are not activated by random lattice vibrations can be transported through the bonding, and the formation of a coherent tunneling state among this portion of the polaron population can then be established.

To begin the analysis, we define the normal carrier concentration as

$$x_n = x \exp\left(-\frac{E_{\text{hop}}}{kT}\right), \tag{8.37}$$

where E_{hop} is the basic polaron trap energy that was defined previously as E^0_{hop} in our discussion of the manganites. As an example of the normal resistivity behavior of polaronic oxide compounds, (8.34) can be introduced to (8.19) to compute ρ vs. T curves similar to the model of Fig. 8.8. Above the critical temperature T_c, the resistivity behavior of $La_{2-x}Sr_xCuO_4$ superconductors is metallic, as plotted in Fig. 8.25. For the fitting of these data [65] $C = 16\,\text{m}\Omega\,\text{cm}\,\text{eV}^{-1}$, larger than that for $La_{1-x}Ca_xMnO_3$ partly because of 50% higher lattice volume fraction occupied by octahedral B sites and the large angle grain boundaries in these bulk ceramic specimens. In these calculations $E_{\text{hop}} = 4\,\text{meV}$ for each value of $x = 0.10, 0.15,$ and 0.225. Note that the onset of superconductivity occurs at the resistivity minimum, which suggests that the polaron trap could be the key to determine the value of the critical temperature. A second comparison between theory and experiment [66] is presented in Fig. 8.26 for data from the most commonly studied compound $YBa_2Cu_3O_7$ (YBCO) in bulk ceramic [67] and epitaxial film [68] forms. In this latter case, the extrapolated straight-line asymptote appears to reach the origin at $T = 0$.

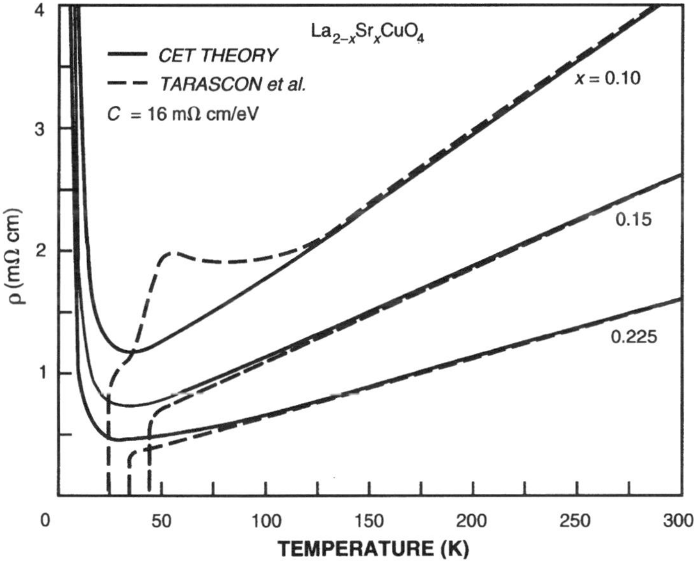

Fig. 8.25 Comparison of theory with measured ρ vs. T for the $La_{2-x}Sr_xCuO_4$ system case. Data are from Tarascon et al. [65]. Figure reprinted from G.F. Dionne, *IEEE Trans. Magn.* **27**, 1190 (1991) with permission. © 1991 by the IEEE

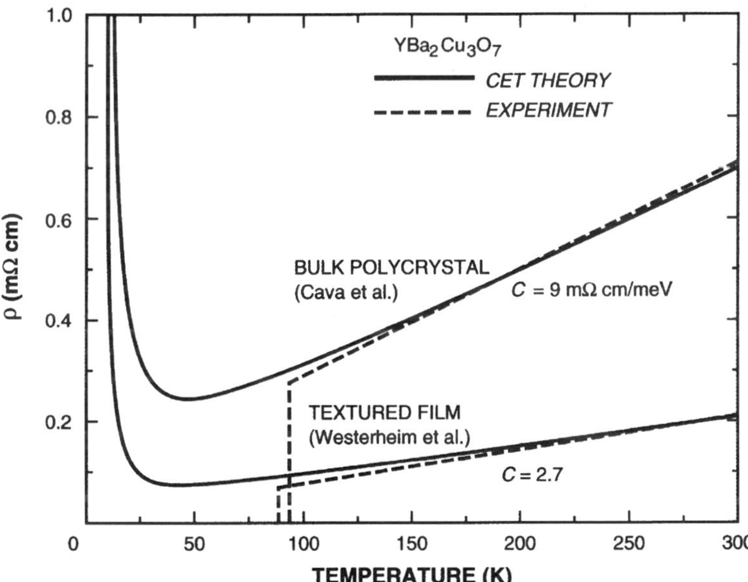

Fig. 8.26 Comparison of theory with measured ρ vs. T for bulk polycrystalline and oriented film $YBa_2Cu_3O_7$. Respective data are from Cava et al. [67] and Westerheim et al. [68]. Figure reprinted from G.F. Dionne, *IEEE Trans. Magn.* **27**, 1190 (1991) with permission. © 1991 by the IEEE

If x_n is now subtracted from x, the concentration fraction that is available for tunneling is the result. From this group, however, the limitations of transfer efficiency or probability and polaron dispersal must be taken into account. To this end, the supercarrier density is expressed as

$$x_s = \eta P\,(x - x_n) = \eta P x \left[1 - \exp\left(-\frac{E_{\text{hop}}}{kT}\right)\right], \tag{8.38}$$

where x_n is substituted by its definition from (8.37). A more general form of the transfer probability $P = (1 - 2\beta x)$ represents the probability that a receptor site is adjacent to a carrier site. For these purposes, a dispersal parameter $0 \le \beta \le 1$ is defined whereby $\beta = 0$ for perfect ordering and 0.5 for random ordering; the higher values represent various degrees of clustering. Because the transfer involves adjacent pairs of mixed-valence ions, the maximum value of x is 0.5 for single transfers. Arguments can be made to support a double transfer as the minimum event (real-space pair transfer). These concepts were discussed previously [8,23] and would be consistent with the suggested presence of spin waves [69]. If the coherence of the condensed state requires a two-carrier transfer as the minimum event, the factor P would be applied twice as $P^2 \sim (1 - 4\beta x)$ and the limit of x would become 0.33.

From a concentration x_s of dispersed large polarons depicted in Figs. 8.23 and 8.24, a percolation threshold, defined as $x_t = a/r_p = e^2/Kab_p$, would be reached at the maximum temperature for which coherent tunneling can exist, i.e., the critical temperature T_c. Above this temperature there would not be enough tunneling carriers to sustain the condensed state; below it, there would be excess carriers to provide supercurrent necessary for the "perfect" diamagnetism and other properties associated with the superconducting state. Accordingly, from (8.38) T_c can be related to x_t according to [8,9][10]

$$T_c = \frac{E_{\text{hop}}}{kW}, \tag{8.39}$$

where $W = \ln\left(1 - (x_t/\eta P x)\right)^{-1}$.

Since the polaron dimension also influences the spatial extent of local magnetic frustration discussed in Sect. 8.4.2, it follows directly that the Neel temperature would decrease monotonically with polaron density, and reach zero where the polaron cells merge or percolate. As a consequence, a minimum concentration for superconduction at $T_c = 0$ (and a maximum for magnetic order at $T_N = 0$) will be defined as $x_0 = x_t/\eta P$, with η and P evaluated at $x = x_t$. Since $\eta P \le 1, x_0$ will be greater than x_t, particularly in the oxides if η is small because of a larger polaron radius. In the data to be examined next, x_t was determined to be ~0.04 and $x_0 \sim 0.075$, consistent with the range of reported minimum polaron concentrations ($0.02 \le x_0 \le 0.09$, from various publications) that also represent the point of total breakdown in long-range antiferromagnetic order ($T_N = 0$) as confirmed by the data in Fig. 8.20.

[10] Equation (8.39) is a correction to (1.8.25) in [8], where the right-hand side was erroneously inverted.

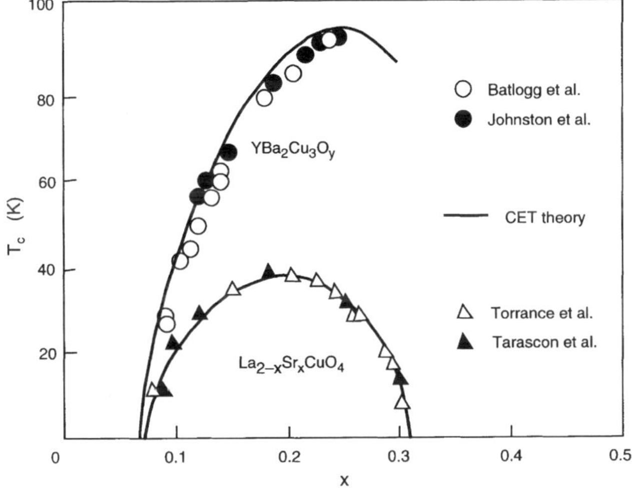

Fig. 8.27 Critical temperature T_c vs. x for $La_{2-x}Sr_xCuO_4$, and $YBa_2Cu_3O_y$ (for which y has been converted to x using the linear relation $y = 0.25x - 1.5$). Data are from Torrance et al. [61], Tarascon et al. [65], Johnson et al. [70], and Batlogg et al. [71]

In Fig. 8.27, the T_c vs. x results of the interpretation of the experiment by means of (8.39) are presented for hole-carrier (p-type) $\left(La^{3+}_{2-x}Sr^{2+}\right)Cu^{2+}_{1-x}Cu^{3+}_xO_4$ and $YBa_2Cu_3O_y$ compounds [61, 65, 70, 71]. To standardize the T_c results in terms of Cu^{3+} polaron ion concentration in $YBa_2Cu_3O_y$ (YBCO), x has been extracted from the charge–balance relation $y = 0.25x - 1.5$ for each value of y. The parabolic character of the curves arises from the introduction of the dispersal factor P to the W parameter. There are two readily discernible differences between the results for the two compounds: (1) the peak in T_c is greater for YBCO, suggesting that the value of β is reduced (from 0.7 to 0.57) because of better spatial ordering of the polaron sources[11] and (2) the entire curve is higher, suggesting that E^0_{hop} is increased (from 2.5 to 4 meV). The implications of this latter possibility are complicated because an increase in the polaron trap energy would result from a decrease in b_p [72], which in turn would mean that U_p could have been increased by the D_{4h} field as suggested by the pyramidal-to-planar descent depicted in Fig. 8.18. If the transfer integral is affected, it would also mean that the transfer probability η and the polaron radius that determines x_t would also change. As an example of the effects of the polaron dispersal variation, Fig. 8.28 is offered for the YBCO values of $E^0_{hop} = 4$ meV, $x_t = 0.035$, and $x_0 = 0.075$, where it is shown that ideal ordering could theoretically produce T_c values approaching room temperature.

[11] In these layered structures, the mixed valence in the Cu cation lattice caused by the charges of the substitutional ions or oxygen vacancies occurs in the crystallographic layer of the $Cu–O_4$ planes.

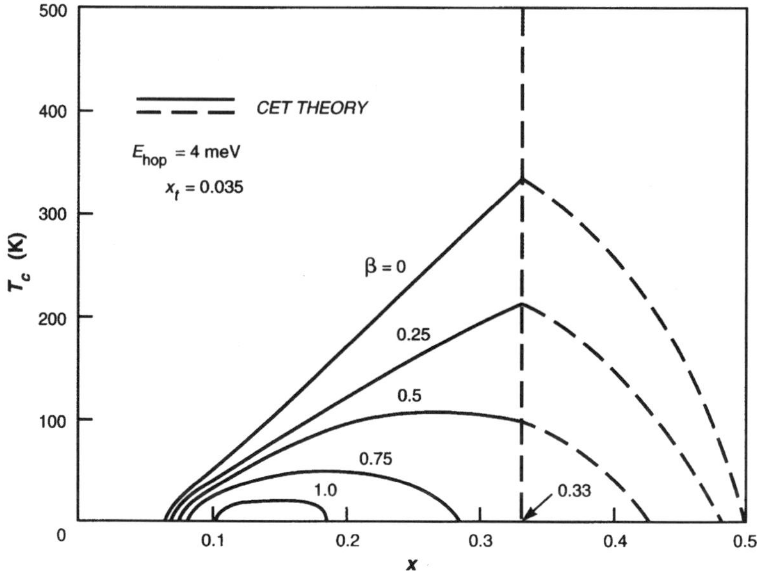

Fig. 8.28 Projected T_c vs. x curves over the range of $0 \leq \beta \leq 2$ for individual carriers and $0 \leq \beta \leq 1$ for pairs. *Dashed curves* indicate that the "real-space" pair model does not apply beyond $x = 0.33$

8.4.5 Layered Cuprate Superconductors

It is appropriate to begin this discussion with a review of the orbital states and occupancies of the Cu^{2+}–O^{2-}–Cu^{3+} superexchange combination, which leads to p-type superconduction that is confined to select Cu–O_4 planes that occur as part of the B-lattice oxygen coordinations in perovskite-type lattices. Although the large-polaron concept implies that the region of mixed-valence condition is local, with carriers tethered to fixed polaron sources, it should be emphasized that the valence state is *not a fixed entity* in cases where itinerant polarons exist through extended covalent delocalization. In accord with the $(CuO)^+$ molecular ion concept,[12] the transfer cations in these partially covalent compounds assume average (noninteger) valences lower than their nominal ionic assignments because the carrier electrons

[12] The $(CuO)^+$ molecule in this model is $Cu^{3+}O^{2-}$, where the "hole" carriers tunnel as part of the antibonding chain formed from the $d_{x^2-y^2} - 2p\sigma$. A second possibility that has received attention is the peroxide option in which the balancing of the electronic charge is not the result of a third ionization of the Cu atom, but rather by the reduction of the negative charge on the O to create a $Cu^{2+}O^{1-}$ molecule. In this case, the Cu^{3+} ion acts as an acceptor in a band model semiconductor sense, leaving the hole carriers in the O^{2-} band. An immediate distinction from the former approach is that the transportable spin could traverse the anion lattice through direct π linkages. The interested reader is encouraged to consult the publications of K. Johnson that describe a novel approach to superconductivity and its relation to dynamic Jahn–Teller effects [74, 75].

become shared among the ions, both Cu and O, within the large-polaron cell.[13] These effects of covalent bonding may be estimated from the *orbital reduction factors* of transition-metal complexes as determined for paramagnetic resonance measurements of *g*-factors and spin–orbit coupling constants. In the case of Cu^{2+} in Tutton salts, for example, the reduction is about 15% [73]; if applied to the oxide, this would mean that the actual ionic charges would be $Cu^{1.7+}O^{1.7-}$. For want of a suitable systematic means for determining these effective valences, however, the integer valence values of the free ion oxidation states will be maintained in the discussions that follow.

To examine covalence involving the unfilled *d* shell of a transition series, it is first necessary to establish the crystal-field (point-charge model of ionic lattice) splittings for the particular system. The order of energy levels for the five 3*d* orbital states shown for the Mn–O_6 octahedral coordination (O_h) in Fig. 2.34 is adjusted to the tetragonal (D_{4h}) case in Fig. 8.29 for the Cu^{2+} cation. For a *c*-axis extension,

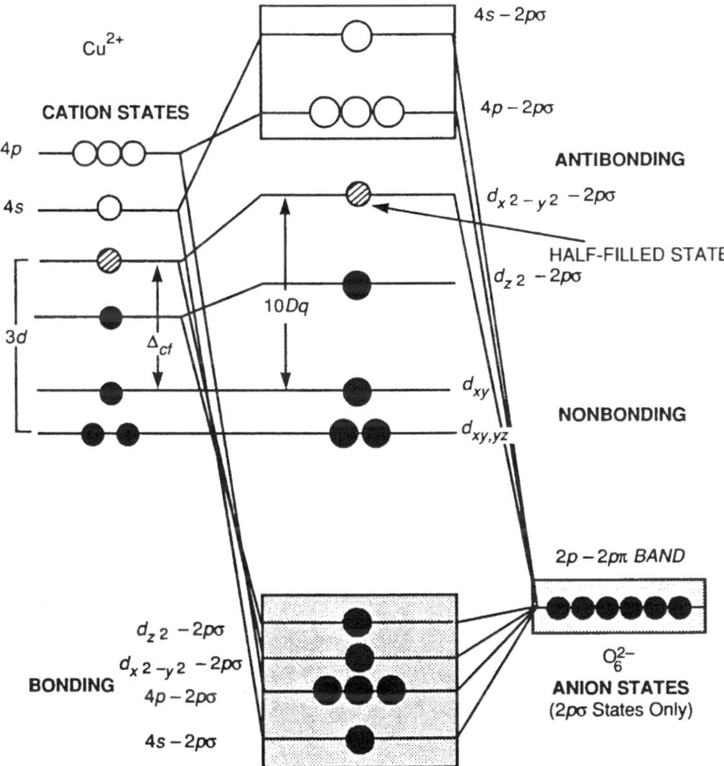

Fig. 8.29 MO diagram for a tetragonally (D_{4h}) distorted CuO_6 complex [9]

---

[13] This traditional view has also been expressed by A.W. Sleight in a review of superconducting oxide chemistry [76].

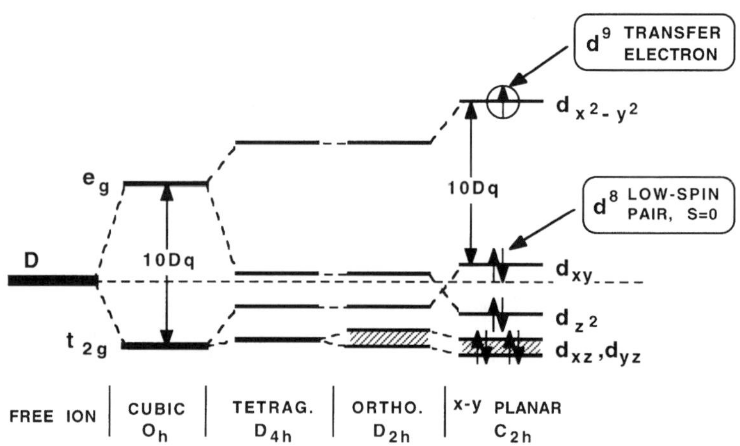

Fig. 8.30 Crystal-field diagram illustrating the d^8 low-spin ($S = 0$) state with the free ion level as zero-energy reference [9]

the antibonding e_g–$p\sigma$ orbital states are split into lower d_{z^2} and upper $d_{x^2-y^2}$, where the unpaired spin is located and from which the polaronic transfer of spins to neighboring ions will take place. Note that the t_{2g} states remain labeled as non-bonding because of their weak π overlaps with the oxygen $2p$ lobes. With reference to Fig. 8.18, the Cu sites of the superconducting perovskites are either tetragonal (with an orthorhombic component in some cases), pyramidal, or square planar. The relevant spin occupancies of the d states are now shown in Fig. 8.30. With $d_{x^2-y^2}$ as the path of transfer, with single occupancy in the Cu^{2+} $\left(d^9\right)$ member and empty for the Cu^{3+} $\left(d^8\right)$ member in a low-spin ($S = 0$) state, as illustrated in Fig. 8.31 and earlier as part of Fig. 8.9.

The source of polarons differs among these compounds. In the simplest case of the $La^{3+}_{2-x}Sr^{2+}_{x}\left[Cu^{2+}_{1-x}Cu^{3+}_{x}\right]O_4$ system with maximum $T_c \approx 40$ K [50], Sr^{2+} ions are fixed negative charges in the A sublattice, and the mixed valence occurs as tetragonally coordinated Cu^{3+} holes that are tethered to the nearest Sr^{2+} ions, thus making the conductivity p-type. A modification of this system that introduces the pyramidal coordinations $La^{3+}_{2-x}Sr^{2+}_{x}Ca\left[Cu^{2+}_{1-x}Cu^{3+}_{x}\right]_2O_4$ increased T_c to 60 K [77]. For the $YBa_2Cu_3O_y$ system with $T_c \approx 95$ K [51], the situation is more complex. The mixed valence occurs here as a result of oxygen vacancies which establish polarons in both the planes of Cu (2)–O_5 pyramids and Cu (1)–O_2 linear chains; chemical formulae highlighting proposed Cu valence distributions that vary linearly with polaron concentration may be written as follows:

$YBa_2\left[Cu^{2+}_{5/2-y/4}Cu^{3+}_{y/4-3/2}\right]_2\left[Cu^{2+}_{11-3y/2}Cu^{3+}_{3y/2-10}\right]O_y$ for $6.67 \leq y \leq 7$

and $YBa_2\left[Cu^{2+}_{5/2-y/4}Cu^{3+}_{y/4-3/2}\right]_2\left[Cu^{1+}_{10-3y/2}Cu^{2+}_{3y/2-9}\right]O_y$, for $6 \leq y \leq 6.67$,

with Cu(1) and Cu(2) site valences $v(1) = 1.5y - 8$ and $v(2) = 0.25y + 0.5$, respectively.

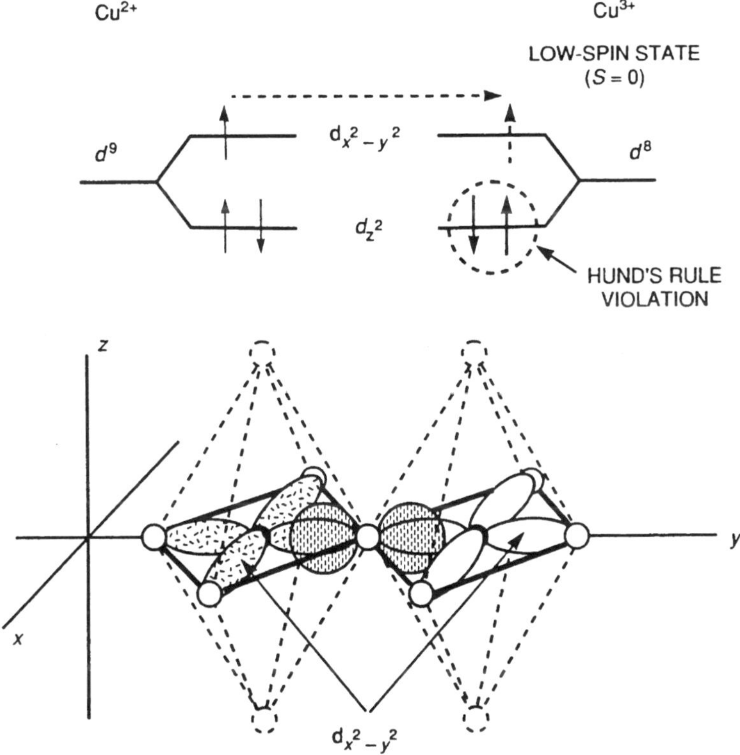

Fig. 8.31 p-Type $3d_{x^2-y^2} - 2p\sigma_y \mathrm{Cu}^{2+} - \mathrm{O} - \mathrm{Cu}^{3+}$ covalent transfer in 180° perovskite bond geometry for $d^9 \rightarrow d^8$ (low-spin) [9]

As suggested by Fig. 8.32, the superconduction is likely to occur in the Cu (2) –O_4 planes of the pyramidal complex, because the $Cu^{2+(3+)}$ content of the Cu (1) –O_4 planes would phase over to $Cu^{1+(2+)}$ at $y = 6.67$ as a result of oxygen vacancies within the plane that create the Cu (1) –O_2 chains. Moreover, these vacancies would break up the continuity of the transfer couplings necessary for superconduction. The origin of positive mobile polarons, therefore, would arise from the fixed negative charges of O^{2-} ions filling the vacancies, as $y \rightarrow 7$. In Sr-free $La_2CuO_{4+\delta}$ [78], the excess oxygen is more correctly described by $La_{2-x}\left[Cu^{2+}_{1-x}Cu^{3+}_{x}\right]O_4$, which is brought about by La cation deficiencies. As determined earlier, a threshold value of $x_0 \approx 0.08$ (or δ=0.04) is all that is necessary for the onset of superconduction.

Partial verification of this valence model was reported by Tranquada et al. [60] who determined experimentally that the average spin of the Cu(2) ions is 0.66 Bohr magnetons (m_B) at $y = 6$, and that the Cu(1) sublattice is diamagnetic. This result indicates that most of the Cu(2) ions are 2+ (with some spin canting likely reducing the effective spin values) and that the Cu(1) ions are 1+, which is consistent with the model in Fig. 8.33. The occurrence of Cu^{1+} ions in the Cu (1) –O_2 chains should be

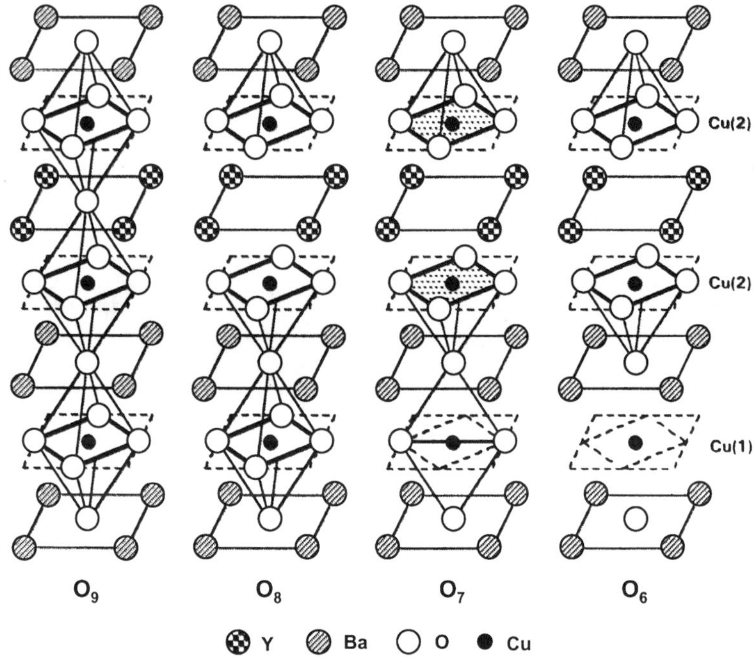

Fig. 8.32 Ordered A-layer structure of $YBa_2Cu_3O_y$, showing breakdown of Cu–O_6 complexes as y decreases from 9 (hypothetical in this case). At $y = 8$, oxygen is removed from Y–O_4 planes and Cu(2) sites are square-pyramids (i.e., Cu–O_5), but retain C_4 symmetry axis. At $y = 7$, Cu(1) ions become linearly coordinated in x–y plane (orthorhombic phase), with uniaxial superconduction expected; Cu(2) ions retain square-planar coordination in x–y plane, with planar superconduction possible. At $y = 6$, Cu(1) planes are fully depleted of oxygen and Cu(2) ions lose mixed-valence with only 2+ species present (see Fig. 8.33) [9]

expected, since its large radius $(\sim 0.96\,\text{Å})$ would preclude its occupancy of the Cu(2) pyramidal sites; furthermore, there is already ample evidence for d^{10} configurations to favor linear coordinations [79].

An even more intriguing confirmation of this originally proposed linear Cu valence distribution has come from the "bond valence sum" analysis of Brown [80]. The results plotted in Fig. 8.34 indicate that the Cu valence distribution is basically linear, but with an oscillation about the relevant portion of the linear curve from Fig. 8.33, added here for comparison.

Together with the compounds discussed above, the parameters for more-complicated "layered" structures are summarized in Table 8.1. In cases where the Cu resides principally in sites with O_4 coordinations, which may provide $E_{\text{hop}} > 4\,\text{meV}$, T_c can reach 120 K. For the $Bi_2^{3+}\left(Sr^{2+}, Ca^{2+}\right)_3 Cu^{2+(3+)}{}_2O_{8+\delta}$ system [63], the optimum Cu^{3+} concentration x=0.33 occurs because of a combination of excess O^{2-} (i.e., $\delta \approx 0.17$) or the occurrence of monovalent calcium [81]. The $Tl_2^{3+}Ba_2^{2+}Ca_{x-1}^{\ 2+}Cu_x^{2+(3+)}O_{4+2x+\delta}$ compounds [82] derive their polaron sources from either fixed-valence cation deficiencies (i.e., excess O^{2-}) or the

Fig. 8.33 Proposed linear valence model of Cu(1) and Cu(2) as a function of the oxygen content variation and distribution depicted in Fig. 8.32 [9]

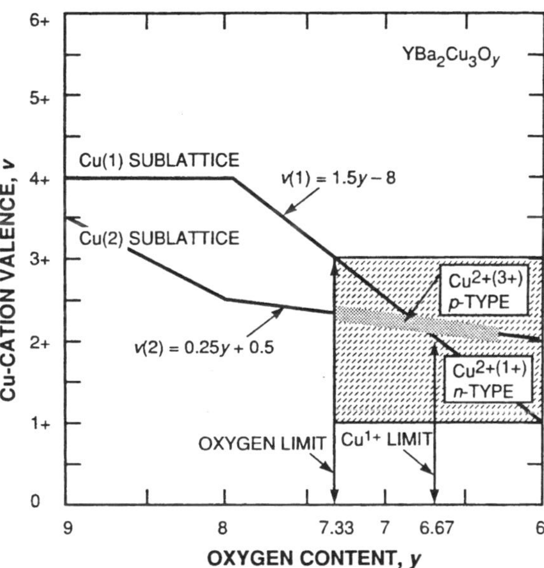

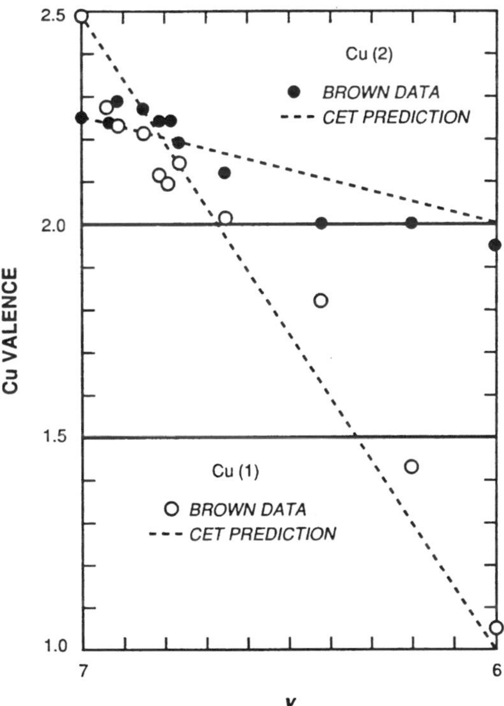

Fig. 8.34 Nominal Cu valence as determined from linear model of Fig. 8.33 compared with valence-bond-sum analysis of Brown [80] for $YBa_2Cu_3O_y$ [9]

Table 8.1 Layered cuprate superconductors[a]

Compound	O^{2-} coord.	T_c (K)	x	E_{hop}[b] (meV)	β[b]	Ref.
p-type $(Cu^{3+})^+$						
$La_{2-x}Sr_xCuO_4$	Cubic O_6	40	0.2	2.5	0.7	[50]
$La_{2-x}Sr_xCaCu_2O_4$	Pyramidal O_5	60	0.2	4	0.7	[77]
$YBa_2Cu_3O_y$	Pyramidal O_5	95	0.25	4	0.57	[51]
$Bi_2(Sr, Ca)_3Cu_2O_{8+\delta}$	Planar O_4	120	0.33	(>4)	(<0.5)	[63]
$Tl_2Ba_2CuO_{6+\delta}$	Cubic O_6	80	–	(2.5)	(<0.5)	[82]
$Tl_2Ba_2CaCu_2O_{8+\delta}$	Pyramidal O_5	110	–	(4)	(<0.5)	[82]
$Tl_2Ba_2CaCu_2O_{10+\delta}$	Planar O_4	120	–	(>4)	(<0.5)	[82]
$Hg_2Sr_2Ca_{n-1}Cu_nO_{2n+4+\delta}$	Pyramidal O_5	135	–	(>4)	(<0.5)	[112][c]
n-type $(Cu^{1+})^-$						
$Nd_{2-x}Ce_xCuO_y$	Cubic O_6	24	0.23	(2.5)	(>0.7)	[83]
$Sr_{1-x}Nd_xCuO_2$	Planar O_4	40	0.14	(4)	(>0.7)	[85]

[a] Some entries of this table are taken from Table 8.1 of reference [86]
[b] Brackets indicate suggested values
[c] This compound was prepared under high pressure which, together with Hg, probably increased the covalent exchange energy b_p between the Cu ions

mixed valence of Tl, which can appear as 1+ or 3+ to suit the ionic size or charge requirements of its locale. With such a dual mixed-valence condition present, the likelihood of higher polaron ordering, i.e., smaller β, is also increased. Since any defects would not be oxygen, these compounds should have chemical stability that is superior to the $YBa_2Cu_3O_y$ system.

Another source of the enhanced T_c values could be larger b values that result from a covalent coupling between the $d_{x^2-y^2}$–$2p\sigma$ antibonding state and the $6s$ orbital of Bi^{3+} or Tl^{1+}. An increased exchange integral would give rise to a smaller x_t. A measurement of the W parameter for the compositions with maximum T_c could help to sort out these possibilities.

Zero-spin polarons can also occur in cuprates with n-type conduction properties. In A_2BO_4 perovskites, Cu^{1+}–O^{2-}–Cu^{2+} combinations can produce superconductivity with Cu^{1+} $(d^{10}, S = 0)$ ions as the minority carrier instead of Cu^{3+} $(d^8, S = 0)$. Through Ce^{4+} substitutions in A sites and the creation of O^{2-} vacancies by a reducing atmosphere anneal, Tokura et al. [83] reported n-type superconductivity with $T_c \approx 24$ K in $Nd^{3+}_{2-x}Ce^{4+}_x\left(Cu^{2+}_{1-x-2y}Cu^{1+}_{x+2y}\right)O_{4-y}$, for $z = 0.15$ and $y = 0.04$. As a consequence, stoichiometry is maintained with a Cu^{1+} concentration of $x = 0.23$.

Although the critical temperatures of this system are not particularly large,[14] these results are very significant for providing insight about the microscopic mechanism of superconductivity. The substitution of tetravalent cations into the

[14] Since Nd^{3+} is a magnetic rare-earth ion with $S = 3/2$, superexchange involving the A sublattice may result in Ce^{4+} clustering that would lead to larger β parameters.

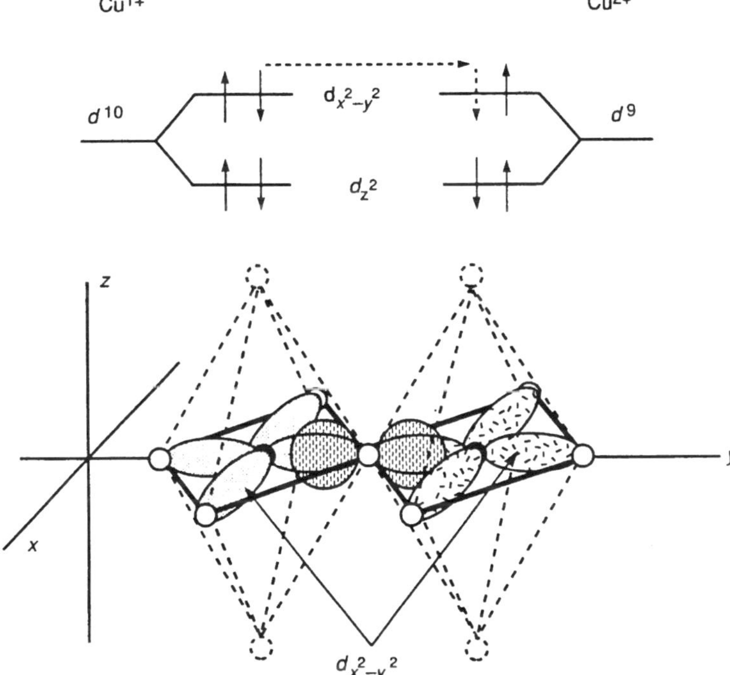

Fig. 8.35 n-Type $3d_{x^2-y^2} - 2p\sigma_y \mathrm{Cu}^{1+} - \mathrm{O} - \mathrm{Cu}^{2+}$ covalent transfer in 180° perovskite bond geometry for $d^{10} \rightarrow d^9$ [9]

A sublattice creates *negative* polarons (electrons), and the appearance of superconduction verifies the prediction of n-type $Cu^{1+} \rightarrow Cu^{2+} + e^-$ orbital transfer predicted in the initial report on the CET model [84]. As shown in the orbital transfer diagram of Fig. 8.35, the $d_{x^2-y^2}$ orbital is again the transfer path. Here the Cu^{1+} ions provide the $S = 0$ states required for spin transfer and the onset on magnetic frustration in the Cu^{2+} host lattice with $S_L = 1/2$. In addition, the existence of n-type superconduction significantly weakens the argument that high-T_c phenomena are based on hole transport through the oxygen sublattice, i.e., local peroxide O^{1-} formation. For the oxygen ligands to provide conduits for electrons, O^{3-} ions would have to be postulated – a situation even more unlikely than the peroxide case.

Another interesting feature of these compounds is the necessity to create oxygen defects as part of their preparation. In this case, the Cu^{1+} ions occupy square planar sites (Cu–O_4), with an occasional missing oxygen, As shown in Fig. 8.32 for $YBa_2Cu_3O_y$, the Cu(1) sites that are proposed to harbor Cu^{1+} ions are also square planar to linear, as y falls below 7. Since Cu^{1+} ions are usually not accepted in sites of higher coordination, e.g., octahedral O_6, because of their large radii $(\sim 0.96\,\text{Å})$, this result is entirely consistent with traditional metal-oxide chemistry.

A primitive layered n-type family $Sr^{2+}_{1-x}Nd^{3+}_x\left(Cu^{2+}_{1-x}Cu^{1+}_x\right)O_2$ was reported by Smith et al. [85]. In this case, the Cu ions reside in planar coordinations, with the *A* sites forming oxygen-free layers that are interleaved between Cu–O_4 planes. The n-type compositions are noteworthy because of the increased $T_c = 40$ K and the lower value of $x_{max} = 0.14$. In the context of the foregoing discussion, these results may be explained by an E_{hop} higher than that of the compound containing Ce^{4+}, but with a larger β parameter, as compared in Table 8.1. This interpretation remains in accord with the general conclusions that T_c through E_{hop} has a crystal-field dependence related to oxygen coordination, and that cation spatial ordering is essential for high T_c.

Mixed-valence combinations involving zero-spin transition ions were examined further in the context of HTS materials design [86]. Figure 8.21 suggests how various zero-spin polaron possibilities could be developed from low-spin states to participate in coherent spin tunneling. In certain lattice structures, the t_{2g} shell can be the source of superconductivity, as confirmed by the critical temperature of 4.7 K reported for low-spin Co^{3+} $\left(d^6, S = 0\right)$ and Co^{4+} $\left(d^5, S = 1/2\right)$ combinations in $Na_x^{1+}Co_x^{3+}Co_{1-x}^{4+}O_2$ that becomes n-type when hydrated, with likely composition $Na_{0.35}^{1+}Co_{0.35}^{3+}Co_{0.65}^{4+}O_2 \cdot 1.3H_2O$ [87].

Because these situations require strong cation–anion interactions, the possibility of utilizing cations with the longer reaching radial components of the $4d^n$ and $5d^n$ shells might be attractive candidates for future investigation. Larger U_p values together with greater η and b_p integrals could result in increased T_c values through favorable E_{hop}/W ratios in (8.39). Low-spin combinations of the strongly covalent Ru^{2+} $\left(4d^6, S = 0\right)$ and Ru^{3+} $\left(4d^5, S = 1/2\right)$ ions in highly reduced stoichiometry, for example, would satisfy the $U_{ex} = 0$ condition that allows $b_p^2 \gg U_p^2$, and charge ordering could then parameter optimize the polaron dispersal for spin transport. In the octahedral sites of a spinel lattice, a $T_c \approx 12$ K derived from t_{2g} shell d_{xy}–$d_{xy}\sigma$ delocalization exchange $\left(3d^1 \leftrightarrow 3d^0\right)$ was observed from n-type $Ti^{3+} \rightarrow Ti^{4+} + e^-$ transfers [88].

In summation, the mixed-valence manganites and cuprates are metallic if the polaron trap energy from antiferromagnetic exchange is eliminated. $La_{1-x}Ca_xMnO_3$ is metallic for $x < 0.5$ because of the occurrence of ferromagnetism that results from vibronic Jahn–Teller effects and delocalization exchange between partially filled e_g orbital states. Above $x = 0.5$, the crystallographic structures are influenced by a complexity of factors that include J–T distortions that in turn control the splitting of the e_g states. Magnetic order can then be established in a variety of antiferromagnetic configurations, often tenuous and subject to insulator–metal transitions when high magnetic fields reverse the sign of the resultant exchange field to restore the ferromagnetism and create and produce a significant magnetoresistance effect. An internal spontaneous magnetic moment acting as a magnetic source, however, would oppose the flux exclusion (Meissner effect) that is a fundamental property of superconductivity. p-Type $La_{2-x}^{3+}Sr_x^{2+}Cu^{2+(3+)}O_4$ is metallic because zero-spin Cu^{3+} "hole" carriers (or Cu^{1+} carriers in n-type $La_{2-x}^{3+}Ce_x^{4+}Cu^{2+(1+)}O_4$) can transport as large polarons in the charge-ordered B sublattice. These conditions remove the possibility of magnetic exchange trapping and frustrate any antiferromagnetic order above a concentration threshold of only a few percent. In both materials systems, large polarons can become itinerant, in one case contributing to a condition from which a magnetoresistance anomaly may occur at the Curie temperature, and in the other, making possible the coherent transfer among $S = 1/2$ and $S = 0$ charge-ordered and spin-frustrated ions that is necessary for a superconducting state. Other compounds where the absence of a magnetic exchange trap allows polaronic conductivity to condense into superconductivity are the mixed-valence Ti, V, and Bi oxides [89].

8.5 Supercurrents and Magnetic Fields

From this model, two-fluid functions can be developed to analyze penetration depths, critical magnetic fields, critical current densities, and other properties as a function of temperature [23]. The end result in each case is determined by the supercarrier density n_s. One central result of this approach to the superconducting state that arises from the Boltzmann-type partitioning of the carriers, however, is that the supercarrier population will decrease accordingly with temperature until the density of supercarriers falls below the threshold value of n_t and the superconducting state is extinguished. As a consequence, high supercurrent densities would be expected only at the lower temperatures regardless of the critical temperature value.

Although the basic requirement for zero resistance has been defined as $n_s \geq n_t$, this condition alone is not sufficient for superconduction. Superconductivity is a thermodynamic state of energy lower than the normal state, and condensation to the superconducting state occurs when this stabilization energy is transferred to a supercurrent flow. In the following sections, the role of magnetic field in limiting this current is described, and the factors that determine critical magnetic field, critical current density, and related phenomena are examined [9].

8.5.1 Supercurrent Formation

It is important to remember that a superconductor is not a "perfect" conductor. Perfect conduction implies simply zero electrical resistivity – the unrealizable case of universally unimpeded charge transport. For a material to be a perfect conductor, there can be neither scattering among carriers nor energy loss in the form of thermal dissipation through electron–phonon interaction. In reality, electrons in a solid can never be completely uncoupled from the lattice, and where so-called "free" electrons are involved, the state occupation limitations imposed by the Pauli exclusion principle create a repulsive action that also limits current flow. Superconduction, however, requires spatial rigidity of the supercurrent in order to fulfill the $\boldsymbol{E} = 0$, $\boldsymbol{B} = 0$ conditions. As pointed out in the discussion of (8.26), supercurrent rigidity ($\nabla \cdot \boldsymbol{i}_s = 0$) requires both $\nabla \cdot \boldsymbol{A} = 0$ (the London gauge) and $\nabla n_s = 0$, where the latter is also a sufficient condition that implies real-space ordering of supercurrent carriers

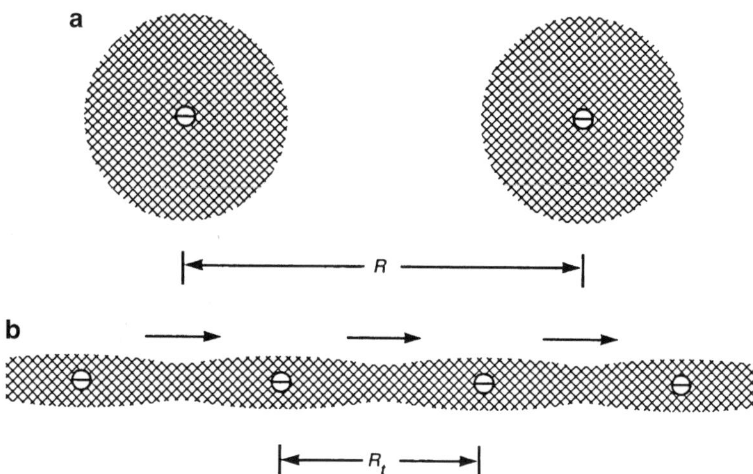

Fig. 8.36 Two-dimensional model of polaron condensation to superconducting state: (**a**) below percolation threshold and (**b**) after ferroelectric alignment with ordered supercurrent flow at $R = R_t$ [9]

that may also involve some form of dynamic spin ordering. The spatial ordering of charges ($\nabla n_s \to 0$) suggests a MO scheme as a source of rigid conduits for supercurrent flow.

If mixed valence can provide conduction among bonding electrons delocalized within the directed lobes of their covalent bonds, electrostatic ordering of these carriers must exist; lattice periodicity alone would suggest a regularity for any carriers participating in the bonding. The charge balance requirements for optimizing the Madelung energy also dictate such a spatial distribution of valence states. Since part of the trapping or stabilization energy of the carrier is due to the electrostatic attraction to its polaron source, i.e., the other half of its electric dipole, the carrier ordering is dictated by the dispersal of the fixed polaron sources, as depicted in Fig. 8.36a. For vanishingly small concentrations, the carriers are isolated, with radially symmetric cell profiles in an x–y plane (in reality, only the fourfold symmetry of a d_{x2-y2} orbital in the CuO_4 square-planar case). As the spatial density of polaron "cells" increases, the shapes of the overlapping regions converge into a chain to assume the minimum energy orientation of aligned electric dipoles. At a threshold density of fixed polaron sources, condensation of ordered chains would occur spontaneously in a manner depicted by Fig. 8.36b.

In the preceding sections, the necessary condition that $\rho = 0$ was discussed in terms of a series of parallel, independent chains that could be represented by a one-dimensional model, since only one completed chain would be required for zero resistance. For phenomena related to the buildup of supercurrent and its associated magnetic effects, however, orbital transfer can no longer be treated as an isolated linear chain, but rather as series of interconnected chains distributed across a macroscopic area.

The completion of a single path is contingent on a threshold supercarrier density n_t, occurring at $T = T_c$. Since a fraction of the electrons is in thermal activation at all times, the total n_t cannot be expected to complete every possible superconduction linkage simultaneously. Even if the chemical ordering were ideal ($P = 1$), the incidence of electron–phonon encounters is still random. An analog to this effect would be the electrical breakdown of a gaseous medium, which begins with a single irregular striation that moves about as dictated by random molecular collisions. As the ionizing collisions increase with density, multiple striations percolate and the gas is eventually transformed into a plasma continuum with a large fraction of the ionized gas participating in the current.

At this point, we may reasonably assume that the current increases continuously from zero, with only the excess of n_s over n_t initially contributing to the supercurrent. An effective carrier density at $T = T_c$ is therefore defined as $n_s^e(T_c) = n_s(T_c) - n_t(T_c) = 0$. At $T = 0$, however, there are no hopping electrons and we must assume that all of n_t is contributing to the supercurrent, i.e., $n_s^e(0) = n_s(0)$. To satisfy these boundary conditions in the most direct manner, the above relation may be generalized to all temperatures, according to

$$n_s^e(T) = n_s(T) - n_t(T), \tag{8.40}$$

where $n_t(T) = n_t(T_c)\left[1 - n_s^e(T)/n_s^e(0)\right]$ Thus, the fractional contribution of the threshold carrier density $n_t(T_c)$ to the supercurrent increases in direct proportion to the buildup of the effective carrier density $n_s^e(T)$ as $T \rightarrow 0$. The logic of (8.40) is displayed in Fig. 8.37, where the $n_s^e(T)$ population declines because of the

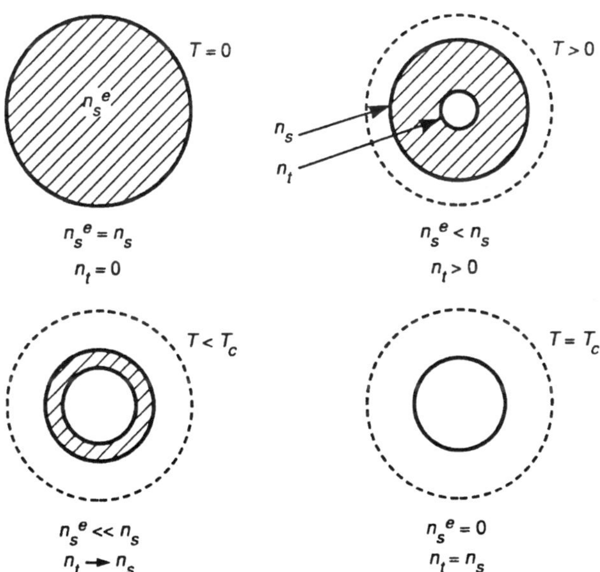

Fig. 8.37 Pictorial representation of the simultaneous decrease of $n_s^e(T)$ and the growth of $n_t(T)$ as $T \rightarrow T_c$, where n_s and n_t converge to establish the threshold density for the onset of superconductivity [9]

proportionate decrease of the $n_s(T)$ and increase of the $n_t(T)$. After rearrangement to isolate $n_s^e(T)$, (8.40) becomes

$$n_s^e(T) = n_s^e(0)\left[\frac{n_s(T) - n_t\,(T_c)}{n_s(0) - n_t\,(T_c)}\right]. \tag{8.41}$$

Based on this relation for the effective carrier density that contributes directly to the supercurrent, the various London current and magnetic-field-related parameters can be examined.

8.5.2 Condensation Energy

According to the London thermodynamic arguments, the condensation energy of the superconduction transition is the reduction in the Gibbs free energy per unit volume $\Delta G = G_n - G_s$. According to the London thermodynamic arguments, the condensation energy of the superconduction transition is the reduction in the Gibbs free energy per unit spontaneous magnetism that was originally proposed by London in defining his macroscopic molecule concept [90]. Unlike spontaneous magnetism, however, there is no fixed population of dipoles and no strong exchange field that may lend itself to a Brillouin function variation with temperature. Instead, the fractional population of superconducting polarons is exponentially dependent on $(kT)^{-1}$, and since the polarons are transported by covalent transfer that is not directly phonon-coupled to the lattice, the conduction mechanism is temperature independent. This situation permits the electrostatic potential energy of the dipole array to be released adiabatically as kinetic energy of carriers in an ordered state (i.e., dynamic ferroelectricity), without the randomness of the thermalization that produces the temperature change associated with the magnetocaloric or electrocaloric effects.

Since condensation occurs as a supercurrent, a two-dimensional model is sufficient to describe the physical situation. For a planar array of dipoles with moment $\boldsymbol{m}_d$, ΔG can be expressed as[15]

---

[15] This relation may be derived from the standard theory for the interaction energy density of a dipole array (with $K = 1$)

$$E_d = \sum_{j>k}\left[\frac{\boldsymbol{m}_j \cdot \boldsymbol{m}_k}{r_{jk}^3} - \frac{(\boldsymbol{r}_{jk} \cdot \boldsymbol{m}_j)(\boldsymbol{r}_{jk} \cdot \boldsymbol{m}_k)}{r_{jk}^5}\right],$$

where $\boldsymbol{r}_{jk}$ is the distance between neighboring dipoles j and k. For the case of an x–y square-planar array, we may consider the energy minimum to occur with the dipoles aligned with the x-axis. E_d in the x direction is $-2m_d^2/r_d^3$, in the y direction, $+m_d^2/r_d^3$, and in the z direction, 0. The net E_d would then resemble (8.42). In general, lattice structure, as well as polaron dispersal, must be examined to determine the best approximation for individual cases.

$$\Delta G \sim \left(\langle \boldsymbol{m}_{\mathrm{d}} \boldsymbol{m}_{\mathrm{d}} \rangle / K R_{\mathrm{d}}^{3}\right) n_{\mathrm{s}}^{\mathrm{e}} \approx \left(m_{\mathrm{d}}^{2} \langle \cos\theta \rangle / K R_{\mathrm{d}}^{3}\right) n_{\mathrm{s}}^{\mathrm{e}}. \tag{8.42}$$

Since $\langle \cos\theta \rangle = 1$ (perfect alignment), $R_{\mathrm{s}} = a \,/\, x_{\mathrm{s}}^{\mathrm{e}} = a \,/\, n_{\mathrm{s}}^{\mathrm{e}} V$, and $m_{\mathrm{d}} = eR_{\mathrm{d}}/2$ (at the midpoint between cell centers), (8.42) may be approximated by

$$\Delta G \approx (1/4)\left(e^{2} V / a K\right)\left(n_{\mathrm{s}}^{\mathrm{e}}\right)^{2} \approx (1/4)\left(e^{2} / V a K\right)\left(x_{\mathrm{s}}^{\mathrm{e}}\right)^{2}, \tag{8.43}$$

where the minimum value of $R_{\mathrm{d}} = R_{\mathrm{t}} = 2a$ as shown in Fig. 8.36b. For $a = 4$ Å, $K = 16$, and recalling that $V \approx 1.7a^{3}$ for $La_{2-x}Sr_{x}CuO_{4}$, it may be estimated that $\Delta G = 8.32 \times 10^{8} \left(x_{\mathrm{s}}^{\mathrm{e}}\right)^{2}$ ergs cm^{-3}. (For this computation and those that follow, $e = 4.8 \times 10^{-10}$ esu.). Since $x_{\mathrm{s}}^{\mathrm{e}}$ is a function of T, $\Delta G(T)$ may be evaluated at $T = 0$. With $x_{\mathrm{s}}^{\mathrm{e}}(0) = 0.068$ for $La_{1.8}Sr_{0.2}CuO_{4}$ (see Fig. 8.27), $\Delta G(0) = 3.77 \times 10^{6}$ ergs cm^{-3}; for $YBa_{2}Cu_{3}O_{7}$, where $x_{\mathrm{s}}^{\mathrm{e}}(0) = 0.09$ and $V \approx 1.5a^{3}$ [considering only the Cu(2) sites], and $\Delta G(0) = 7.54 \times 10^{6}$ ergs cm^{-3}.

The formation of the spatially ordered carriers ($\nabla n_{\mathrm{s}} = 0$) in a molecular chain results from the electrostatic dictates of the Madelung energy. For the creation of the supercurrent, however, the dynamic order of the carriers results from electrostatic repulsion between the mobile halves of the dipoles, possibly enhanced by the Pauli principle repulsive action. As the carriers transfer between cells, maintaining the required one carrier per cell ordering, there is no direct competition for quantum states and the Pauli repulsion serves as a propellant to charge transport, rather than as a cause of carrier scattering. As a consequence, supercurrent rigidity follows naturally from the constraints of the directed bonding orbitals which act as inflexible conduits for the passage of electrons.

8.5.3 London Penetration Depth

From the relation between current density and magnetic field stated in (8.28), the London penetration depth is now defined in terms of $n_{\mathrm{s}}^{\mathrm{e}}$, according to

$$\lambda_{\mathrm{L}} = \left(mc^{2} / 4\pi e^{2} n_{\mathrm{s}}^{\mathrm{e}}\right)^{1/2} \approx \left(mc^{2} V / 4\pi e^{2} x_{\mathrm{s}}^{\mathrm{e}}\right)^{1/2}, \tag{8.44}$$

which provides $\lambda_{\mathrm{L}}(0) = 5.32 \times 10^{5} V^{1/2} \left(x_{\mathrm{s}}^{\mathrm{e}}(0)\right)^{-1/2}$ cm, if the true electron mass is used. For the $La_{1.8}Sr_{0.2}CuO_{4}$ perovskite in Fig. 8.27, $V = 1.7a^{3}$, $a = 4$ Å, and $x_{\mathrm{s}}^{\mathrm{e}}(0) = 0.068$, which yields $\lambda_{\mathrm{L}}(0) \approx 2,130$ Å, in good agreement with the 2,000 Å value determined from experiment [91,92]; for $YBa_{2}Cu_{3}O_{7}$, where $V \approx 1.5a^{3}$ and $x_{\mathrm{s}}^{\mathrm{e}}(0) = 0.09$, $\lambda_{\mathrm{L}}(0) \approx 1,740$ Å, in accord with the 1,670 Å value derived from microwave stripline resonance measurements [93]. The familiar ratio then reduces to a new two-fluid function

$$[\lambda_{\mathrm{L}}(0)/\lambda_{\mathrm{L}}(T)]^{2} = n_{\mathrm{s}}^{\mathrm{e}}(T)/n_{\mathrm{s}}^{\mathrm{e}}(0). \tag{8.45}$$

From (8.38), (8.39), and (8.41), the following relation can be derived [72, 94]

$$\frac{n_s^e(T)}{n_s^e(0)} = \frac{\left[1 - \exp\left(-E_{hop}/kT\right)\right] - \left[1 - \exp\left(-E_{hop}/kT_c\right)\right]}{\exp\left(-E_{hop}/kT_c\right)}, \tag{8.46}$$

which leads to

$$[\lambda_L(0)/\lambda_L(t)]^2 = n_s^e(t)/n_s^e(0) = 1 - \exp(W - W/t), \tag{8.47}$$

where $t = T/T_c$.

In Fig. 8.38, generic curves of $n_s^e(t) / n_s^e(0)$ from (8.47) are plotted for a range of W values [including $W = 1.76$, i.e., the universal ratio between kT_c and the BCS energy gap $\Delta(0)$], and compared with curves of the empirical two-fluid $\left(1 - t^4\right)$ and BCS models [95]. At this point, it is instructive to compare the basic critical-temperature relations for the BCS and GET models:

$$kT_c = \Delta(0)/1.76\,(\text{BCS}), \tag{8.48a}$$

and

$$kT_c = E_{hop}/W\ (\text{CET}). \tag{8.48b}$$

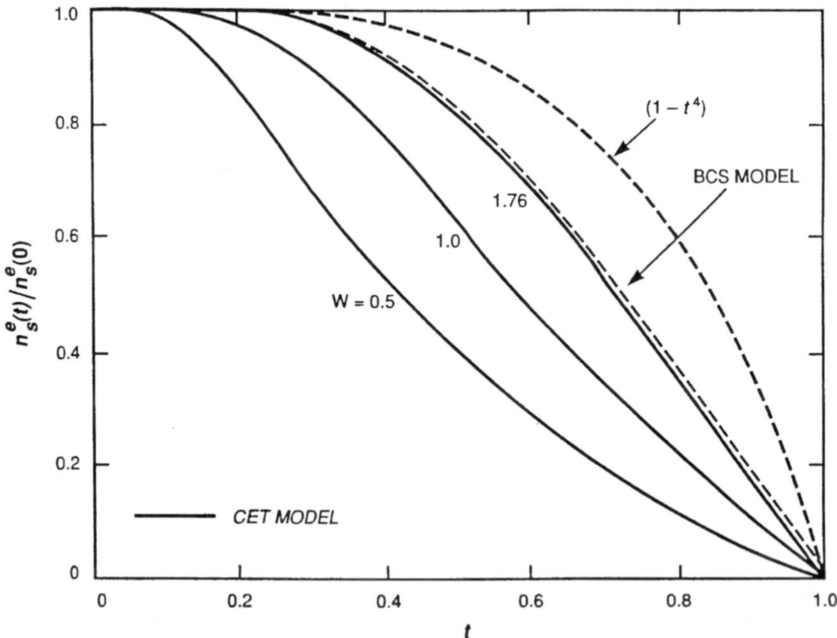

Fig. 8.38 Comparison of $n_s^e(T)/n_s^e(0)$ vs. t for $W = 0.5$, 1.0, and 1.76 (BCS value), the actual BCS function, and the empirical $\left(1 - t^4\right)$ two-fluid function. Figure reprinted from [94] permission. © 1993 by the IEEE

With the $La_{1.8}Sr_{0.2}CuO_4$ and $YBa_2Cu_3O_7$ systems examined in Sect. 8.4, working values of W determined from fits to data are 0.75 and 0.5, respectively. For the BCS model, the denominator is the constant, 1.76, calculated from π/γ, where γ is Euler's constant (= 0.577). All of the material-related information is contained in the gap parameter $\Delta(T)$ computed at $T = 0$ that appears in the numerator. This temperature-dependent gap energy, which determines the ratio of supercarriers to quasiparticles (normal electrons), is a maximum at $T = 0$ and falls to zero in a Brillouin-type curve as $T \rightarrow T_c$. In the CET treatment, a "gap" equivalent would be the fixed numerator E_{hop}, but the denominator is also a material-related variable that is strongly dependent on the polaron radius γ_p. Although their meanings differ somewhat, $\Delta(0)$ and E_{hop} both represent energy separations (the former a condensation gap for electrons paired in k-space and the latter a polaron trap barrier), the important differences in the two relations of (8.48) also lie in the W parameter vs. the fixed denominator 1.76 of the BCS model.

Although the regime in which the above analysis was focused is the "clean" limit $\xi_0 \ll \lambda_L$ of the high-T_c superconductors, for completeness it is appropriate to mention the relations for effective penetration depth λ and coherence length ξ that are used for the other classes of superconductors. For type-I pure limit with $\xi_0 \gg \lambda_L$ (formally expressed as $\xi^3 \gg \xi_0\lambda_L^2$), $\lambda = 0.65\lambda_L\,(\xi_0/\lambda_L)^{1/3}$, and for the type-II (London or dirty) limit where $\xi \rightarrow \ell$ (formally expressed as $\xi^3 \ll \xi_0\lambda_L^2$, $\lambda = \lambda_L\,(\xi_0/\xi)^{1/2} \approx \lambda_L\,(\xi_0/\ell)^{1/2}$.

8.5.4 *Critical Magnetic Field*

If a bulk specimen is placed in a magnetic field H, the field will be expelled from the interior during a superconduction transition (the Meissner effect). Consequently, the superconduction Gibbs free energy G_s will increase by the amount of the energy density $H^2/8\pi$ of the expelled flux [96]. As a function of H, G_s becomes

$$G_s(T, H) = G_s(T, 0) + H^2/8\pi. \tag{8.49}$$

For a nonmagnetic specimen in the normal state, however, $G_n(T, H) = G_n(T, 0)$, and since the condition for the return to the normal state is $G_n - G_s = 0$, the critical field may be defined according to Fig. 8.39 as

$$\nabla G(T, 0) = G_n(T, 0) - G_s(T, 0) = H_c(T)^2/8\pi \tag{8.50}$$

and

$$H_c(T) = [8\pi \Delta G(T, 0)]^{1/2}\,, \tag{8.51}$$

where $H_c(T)$ is the value of H that (with i_s) decays exponentially from the surface according to $\exp(-x/\lambda_L)$ as determined by the solutions of (8.27). After substitution from (8.43),

$$H_c(T) = \left[2\pi e^2 V/aK\right]^{1/2} n_s^e(T) = \left[2\pi e^2/VaK\right]^{1/2} x_s^e(T). \tag{8.52}$$

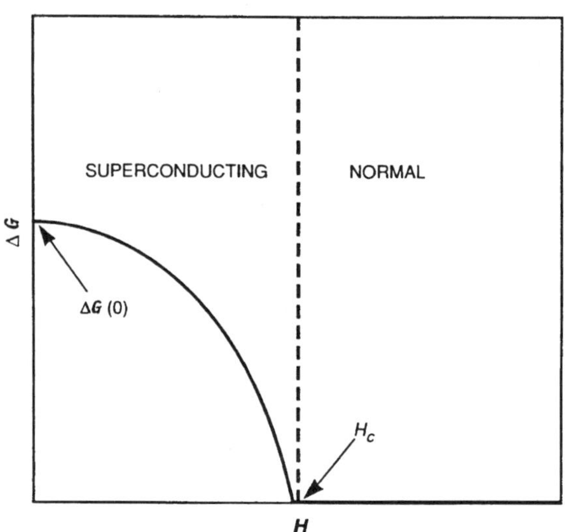

Fig. 8.39 Change in Gibbs free energy as a function of H, indicating a decreasing energy available for conversion to the kinetic energy of supercurrent as $H \rightarrow H_c$ [9]

To determine H_c, the specimen cross-sectional area A should exceed the penetration depth in order to avoid the necessity of a correction $(1 - A_{eff}/A)^{-1/2}$ to the effective volume of flux expulsion. If $A_{eff}/A \rightarrow 1$, as in the case of a fine wire or thin film, the flux penetrates much of the material and the effective H_c is substantially greater than the true value.

From the estimates of $\Delta G(0{,}0)$ beneath (8.43), substitution into (8.51) leads to $H_c(0) \approx 10\,\text{kOe}$ for bulk $La_{1.8}Sr_{0.2}CuO_4$, and $H_c(0) \approx 14\,\text{kOe}$ for $YBa_2Cu_3O_7$. An experimentally based value of $10 \pm 2\,\text{kOe}$ for bulk polycrystalline $YBa_2Cu_3O_7$ was derived from conventional theory [67].

A universal relation follows directly from (8.52):

$$H_c(T)/H_c(0) = [\lambda_L(0)/\lambda_L(T)]^2 = n_s^e(T)/n_s^e(0) \tag{8.53}$$

and

$$H_c(t)/H_c(0) = 1 - \exp(W - W/t) \tag{8.54}$$

to produce a temperature dependence the same as the two-fluid function of (8.45). The curves plotted in Fig. 8.40 for $W = 1.76$ and 2 are therefore from the same family as those of Fig. 8.38, but in this case it is more appropriate to compare them with the standard thermodynamic relation $H_c(t)/H_c(0) = \left(1 - t^2\right)$,[16] and also to examine the slopes at $T = T_c$. From the derivative of (8.54), it may be shown that

[16] Note that the universal relations for $H_c(t)/H_c(0)$ and $[\lambda_L(0)/\lambda_L(t)]^2$ are identical according to (8.45) and (8.54), but differ in their thermodynamic counterparts $(1 - t^2)$ and $(1 - t^4)$, respectively.

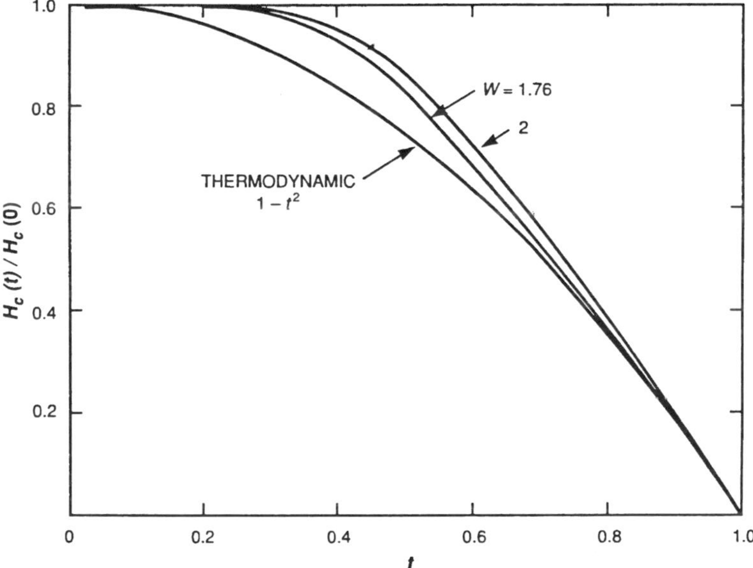

Fig. 8.40 Comparison of $H_c(t)/H_c(0)$ vs. t for $W = 1.76$ (BCS value) and the thermodynamic $(1 - t^2)$ two-fluid function [9]

$$[\partial H_c(t)/\partial t]_{t=1} = -H_c(0)W \tag{8.55}$$

or

$$\partial H_c(T)/\partial T = -W\,[H_c(0)/T_c] \text{ at } T = T_c. \tag{8.56}$$

To match the slopes of the CET curve to the thermodynamic relation, W must equals 2 slightly larger than the 1.76 BCS ratio. As indicated in Figs. 8.38 and 8.40, the curves fit well over the upper half of the t range. Based on these results, one concludes that lower values of W substantially alter the shape of these curves, including the appearance of a tail and an inflection point at $t = W/2$, i.e., concavity begins to appear for $W < 2$, and have an overall deleterious effect on the magnitudes of H_c.

8.5.5 Critical Current Density

From the hypothesis of dynamic ferroelectricity, critical current may be defined by equating carrier kinetic energy to condensation energy ΔG. From (8.50) and (8.51), the carrier kinetic energy [97] can be expressed as

$$\Delta G = (1/2)\,n_s^e m v_s^2 = H_c^2/8\pi = (1/4)\left(e^2 V/aK\right)\left(n_s^e\right)^2, \tag{8.57}$$

and

$$v_s = \left[(1/2)\,e^2V/maK\right]^{1/2}\left(n_s^e\right)^{1/2} = \left[(1/2)\,e^2/maK\right]^{1/2}\left(x_s^e\right)^{1/2}. \qquad (8.58)$$

For $La_{1.8}Sr_{0.2}CuO_4$ and $YBa_2Cu_3O_7$, with $a = 4\,\text{Å}$ and $K = 16$, $v_s = 7.3 \times 10^6$ and $8.4 \times 10^6\ \text{cm s}^{-1}$, respectively. Note that the relation $v_s \sim K^{-1/2}$ is consistent with the expected trends in local accelerating fields within materials of high or low dielectric constant, i.e., metals or insulators.

Since the general supercurrent density $i_s = n_s^e e v_s$, a critical current density $i_s\,(= i_c)$ can be defined from (8.58) as

$$i_c = \left[(1/2)\,e^4V/maK\right]^{1/2}\left(n_s^e\right)^{3/2} = \left[(1/2)\,e^4/mV^2aK\right]^{1/2}\left(x_s^e\right)^{3/2}, \qquad (8.59)$$

For the $La_{1.8}Sr_{0.2}CuO_4$ and $YBa_2Cu_3O_7$ superconductors of Fig. 8.27 with $K = 16$, $a = 4\,\text{Å}$, and $V = 1.7a^3$ and $1.5a^3$, $i_c(0) = 3.67 \times 10^8\ \text{A cm}^{-2}$ and $6.33 \times 10^8\ \text{A cm}^{-2}$ as the upper theoretical limits corresponding to $x_s(0) = 0.068$ and 0.09, respectively.[17]

From (8.47), (8.59) may be converted to a universal curve,

$$i_c\,(t)\,/\,i_c(0) = [1 - \exp\,(W - W/t)]^{3/2}. \qquad (8.60)$$

This relation is appropriate for film specimens where penetration depths exceed the thickness. For bulk specimens or films with greater thickness, (8.60) can be modified according to [98]

$$i_c\,(t)\,/\,i_c(0) = [1 - \exp\,(W - W/t)]^{3/2}\,A_{\text{eff}}\,(t)\,/\,A_{\text{eff}}(0), \qquad (8.61)$$

which simplifies to

$$i_c\,(t)\,/\,i_c(0) \approx [1 - \exp\,(W - W/t)]^{3/2}\,\lambda_L\,(t)\,/\,\lambda_L(0) \approx [1 - \exp\,(W - W/t)] \qquad (8.62)$$

In Fig. 8.41, the data of Lessure et al. [99] for a bulk $YBa_2Cu_3O_7$ specimen is plotted and compared favorably with (8.62) for $W = 1$.

In an external magnetic field H, part of the condensation energy density $H^2/8\pi$ is required to expell H from the interior of the material as shown in Fig. 8.39, and a more general expression for $\Delta G\,(H)$ may be constructed from (8.50) and (8.51):

$$\Delta G\,(H) = G_n\,(H) - G_s\,(H) = \left(H_c^2 - H^2\right)/8\pi. \qquad (8.63)$$

[17] From (8.28) it will be recalled that $i_S = \left(c/4\pi\lambda_L^2\right)A$. If $A_c \approx H_c\lambda_L$ (for specimens of dimensions greater than λ_L), then (8.59) may be derived directly from the London theory, with $i_c = (c/4\pi)\,(H_c\lambda_L) \approx \left[(1/2)\,e^4/mV^2aK\right]^{1/2}\left(x_s^e\right)^{3/2}$, after $\lambda_L = \left(mc^2/4\pi e^2 n_s^e\right)^{1/2}$ from (8.26) and $H_c = \left[2\pi e^2V/aK\right]^{1/2} n_s^e$ from (8.52) are substituted.

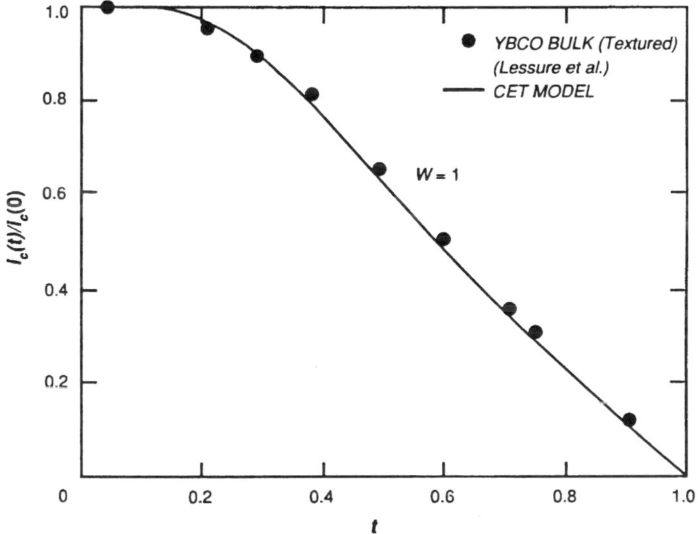

Fig. 8.41 CET universal plot of $i_c(t)/i_c(0)$ for the bulk case with $W = 1$ [9], compared with data of Lessure et al. [99]

From (8.58) and (8.63), (8.57) for the kinetic energy density may be generalized to

$$\Delta G(H) = (1/2)\, n_s^e m v_s^2 = \left(H_c^2 - H^2\right)/8\pi \tag{8.64}$$

If it is assumed that n_s^e is unaffected by H,[18] the reduction in ΔG occurs as a decrease in carrier velocity and

$$v_s = \left[\left(H_c^2 - H^2\right)/4\pi m n_s^e\right]^{1/2}, \tag{8.65}$$

from which a universal relation for critical current density as a function of magnetic field may be deduced as

$$i_c(H)/i_c(0) \approx \left[1 - (H/H_c)^2\right]^{1/2} \tag{8.66}$$

Thus, it is seen that the *supercurrent is limited by magnetic field which offsets the condensation energy, and by temperature which controls the density of available carriers.* Note that (8.66) describes the situation in the absence of a fluxoid lattice, i.e., type-I superconductors, where H_c is the thermodynamic critical field, and

[18] In type-II superconductors, there are situations where n_s^e may be considered to be a function of H, such that $\lambda_L(H) \to \infty$ as $n_s^e(H) = |\psi_S(H)|^2 \to 0$ [100].

should not be expected to predict accurately the behavior of type-II superconductors, in which a variety of magnetic-field-related effects may influence $i_c\,(t, H)$ [101].

Since $T_c = E_{hop}/kW$, the superconduction critical temperature may be increased in two ways (1) by raising E_{hop}, at the risk of introducing a magnetic exchange energy barrier that could upset the $b \geq E_{hop}$ requirement and (2) by reducing W, either through improved polaron dispersal ($P \rightarrow 1$) or smaller polaron radii ($n_t \rightarrow 0$), but at the expense of lowering in a relative sense H_c and i_c as $T \rightarrow T_c$.

8.5.6 Coherence Length

A natural segue from the discussion of critical magnetic fields is the subject of coherence and its role in type-II superconductors. As discussed in Sect. 8.4.1, the notion of coherence was introduced by Pippard [54], who proposed that the nonlocal nature of the superelectron may be characterized in terms of the uncertainty principle. In this approach as applied to free-electron systems, the superconducting electrons are drawn from the population with energies within kT_c of the Fermi level. In order to obtain a relation for the individual carrier momentum p_s in the superconducting state, Pippard reasoned that their momentum range could be estimated by dividing the condensation energy, assumed to be equivalent to kT_c, by the Fermi velocity v_F, i.e., $\Delta p_s \sim kT_c/v_F$) [102]. As a consequence, the position uncertainty (coherence length) becomes $\Delta x\,(= \xi_0) \geq (h/2\pi)\,(v_F/kT_c)$.

Another definition of coherence length was provided by Ginsburg and Landau (GL) from the solution of a Schrodinger-type equation with a nonlinear term [55]. The resulting aggregate eigenfunction for this differential equation contained an exponential decay [see (8.31)], $\psi_s\,(x) \sim \psi_s(0)\exp\left(-x/\xi\right)$, where $|\psi_s|^2 = n_s$ is the expectation value of a spatially varying superconduction ensemble wavefunction of coherence length ξ that reduces to the Pippard result for $T \ll T_c$. Thus, conceptual compatibility between Pippard and GL can be established if ξ is the average distance a carrier travels before losing coherence with the ordered state. In reality, the carrier rarely reaches this limit, but the attendant velocity range defines the degree of spatial order. In its essentials, the Pippard definition describes the coherence of a carrier chain composed of wavepackets with a spatial profile that may be assigned a de Broglie wavelength defined by

$$\lambda_{deB} = h/p_s. \tag{8.67}$$

This concept is compatible with the basic CET model of a chain of localized wavefunctions that link to form a single molecular-orbital function. Applying the uncertainty relation for space packets, we obtain

$$\Delta p_s \Delta x \geq h/2\pi, \tag{8.68}$$

or in this present context

$$p_s \xi_0 \approx h/2\pi, \tag{8.69}$$

where Δp_s is replaced by p_s.[19]

In the CET model, the condensation energy is not immediately determined by kT_c, so the use of the above Pippard relation for Δp_s must be modified. The role of the energy trap is different from the conventional band theory approach in that it determines directly the population of *available* supercarriers, but only indirectly their energy. As discussed in Sect. 8.5.2, ΔG is determined by $n_s^e(T)$ and is more closely associated with H_c than T_c. Because the CET model has conveniently produced a distinct relation for the carrier velocity, however, the question of estimating Δp_s by indirect means is unnecessary, and we may therefore define a coherence length similar to that of Pippard directly from the relation for superconductor carrier velocity of (8.58),

$$\xi_0 \approx (h/2\pi)/p_s = (h/2\pi)/mv_s \approx \left[\left(h^2/2\pi^2 me^2\right)(aK/V)\right]^{1/2} \left(n_s^e\right)^{-1/2} \tag{8.70}$$

If the perovskite parameters from previous estimates, $K = 16$, $V = 1.7a^3$, and $a = 4\,\text{Å}$, are applied to (8.70) (expressed in terms of x_s^e),

$$\xi_0(0) \approx 8 \times 10^{-8} x_s^e(0)^{-1/2} \text{ cm} \tag{8.71}$$

For $x_s^e(0) = 0.068$ ($La_{1.8}Sr_{0.2}CuO_4$) and 0.09 ($YBa_2Cu_3O_7$), $\xi_0(0) \approx 32$Å and 27Å, respectively, in general agreement with experiment [103]. For type-I metal superconductors, $\xi_0(0) \sim 10^4$ Å, which is also consistent with (8.71) through the direct dependence of ξ_0 on the dielectric constant $\left(K^{1/2}\right)$ that becomes very large in highly polarizable materials with loosely bound electrons.[20]

Since a discussion of the type-II superconductors will follow, it is appropriate to establish the analytical relationship between ξ_0 and λ_L from (8.70) and (8.44):

$$\xi_0 = \left[\left(2h^2/\pi m^2 c^2\right)(aK/V)\right]^{1/2} \lambda_L, = \left(2 \times 10^{-10}\right)(aK/V)^{1/2} \lambda_L \text{ cm}, \tag{8.72}$$

where lengths are expressed in cm. For the two perovskite systems

$$\kappa = \lambda_L/\xi_0 \approx 5 \times 10^9 (aK/V)^{-1/2}, \tag{8.73}$$

---

[19] An alternative derivation of (8.69) may be obtained from a standard quantum mechanics operator through the relation $-j\,(h/2\pi)\nabla\psi_s = p_s\psi_s$, where p_s is a good quantum number for a stationary state. Since the ensemble wavefunction $\psi_s(x) \sim \psi_s(0)\exp(-x/\xi_0)$ from (8.31), $\nabla\psi_s = -(1/\xi_0)\,\psi_s$ and we again arrive at $p_s\xi_0 \approx h/2\pi$. This result suggests that the ensemble concept of GL and the individual wavepacket idea of Pippard are equivalent as far as coherence is concerned.

[20] The dielectric constant in the context of conducting materials is viewed here as a parameter that represents the polarizability in a quasi-insulating state, with at least some of the "free" electrons condensed back on their parent ions for covalent transfer. In the metallic state, of course, macroscopic polarization effects can only be inferred, since any measurements are precluded by the presence of free electrons.

which gives values of 67 for $La_{1.8}Sr_{0.2}CuO_4$ and 63 for $YBa_2Cu_3O_7$, both in good agreement with the value of 62 derived from experiment [67]. These results serve to confirm the expectation that the high-T_c perovskites behave as natural type-II superconductors, with $\lambda_L/\xi_0 \gg 1$. (Metals with the large K values are "Pippard" superconductors, with $\lambda_L/\xi_0 \ll 1$.) Although κ is temperature independent, in accord with traditional theory [104], both λ_L and ξ_0 derive individual temperature dependences through the supercarrier density; as $T \to T_c$, λ_L and $\xi_0 \to \infty$ through their common dependence on $\left(n_s^e\right)^{-1/2}$ from (8.44). Although Pippard's uncertainty principle arguments were used in these derivations of coherence length, the CET formalism gives the above temperature dependence of ξ_0 that is more characteristic of the GL definition, which also produces a temperature-insensitive $\kappa = \lambda(T)/\xi(T)$ [105]. If Pippard's use of the carrier mean-free-path (here replacing the symbol ℓ from Sect. 8.4.1 with d, which could represent the thickness of a film or the diameter of a fine wire) is adopted to define a reduced effective coherence length ξ, we obtain

$$1/\xi = 1/\xi_0 + 1/d, \tag{8.74}$$

for which $\xi \to d$ where $\xi_0 \gg d$. If d represents the radius of a fluxoid domain in this model, i.e., $2d$ is the spacing between nonsuperconducting regions, a decline in superconduction efficiency can set in as d becomes increasingly smaller. Microscopic-scale d values can lower n_s by raising the polaron dispersal parameter β, and thereby increasing W to cause a lower T_c. Another effect of short mean-free-paths would occur in the size of the large-polaron radius r_p [106]. Since r_p determines n_0 (and hence n_t), the nominal threshold density for superconduction, any decrease in its value would also increase W.

At this point, we can compare the physical meanings and relationships of the large-polaron radius and the coherence length. The cell radius of an isolated polaron r_p is a normal-state parameter that determines the minimum density for which a polaron chain may condense to the superconduction state, and is directly dependent on the transfer integral b_p. The coherence length ξ_0, on the other hand, emerges from the uncertainty principle as applied to the wave mechanical nature of the superconducting state, i.e., ξ_0 is larger for smaller momentum values. Since it varies as $\left(n_s^e\right)^{-1/2}$ according to (8.70), it is dependent on the β dispersal parameter and would be generally smaller for oxides where $x_s < 0.1$.

8.5.7 *Type-II Superconductors*

As introduced in Sect. 8.4.1, the three general categories of superconductors are distinguished by the relations between ξ_0 and λ_L. In reality, only type-I materials ("pure" superconductors with $\xi_0 \gg \lambda_L$) feature a single critical magnetic field, in which the flux expulsion takes place completely once the thermodynamic H_c threshold is reached (in a bulk specimen). In materials of practical importance, however, flux expulsion occurs over a range of fields, with the limits defined by

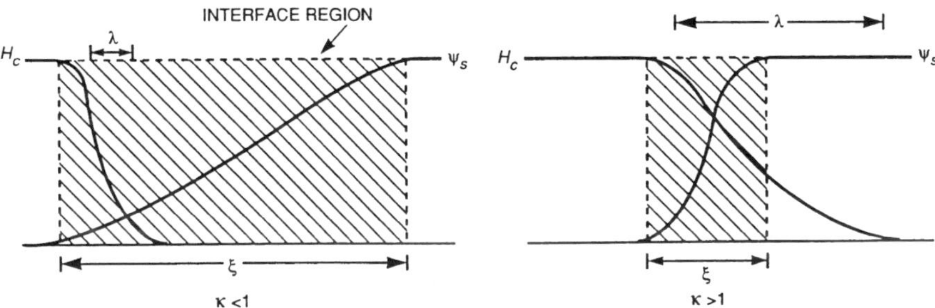

Fig. 8.42 Variation of H and ψ_s at the fluxoid/superconductor interface for $\kappa \ll 1$ (type I) and $\kappa \gg 1$ (type II) [9]

$H_{c1} \leq H_c \leq H_{c2}$. As indicated by the spatial variation in H_c and ψ_s at a normal/superconductor interface sketched in Fig. 8.42, type-II superconductors with $\kappa \gg 1$ have interface regions set by ξ narrow enough and λ_L large enough to permit the bulk volume of the material to divide into islands of "fluxoid" domains that harbor magnetic flux in their "normal" cores at a field less than H_c. With the presence of fluxoid domains, the thermodynamic critical field H_c can no longer be measured directly. Since the invasion of flux reduces the volume of material from which flux is expelled, a larger field H_{c2} will be required to offset the condensation energy defined in (8.50). As analyzed in detail by Abrikosov [107] and discussed by Kittel [108] and Tinkham [109] the critical magnetic fields for "clean" high$-T_c$ perovskite superconductors with $\xi \approx \xi_0$ and $\lambda \approx \lambda_L$ may be approximated by

$$H_{c1}/H_c \approx [(\surd 2)\, k]^{-1} \ln k \tag{8.75}$$

and

$$H_{c2}/H_c \approx (\surd 2)\, \kappa. \tag{8.76}$$

Since κ is temperature independent according to (8.75) and (8.76), H_{c1} and H_{c2} should both track with H_c as functions of temperature. As shown in Fig. 8.43, a universal curve with $W = 1$ is fitted to the H_{c2} data [110] from bulk polycrystalline (Dy,Eu)$Ba_2Cu_3O_y$.

Another comparison with experiment may be carried out with the values of critical fields estimated from measurements of $\partial H_c(T)/\partial T$ at $T = T_c$. For $W = 1$, the slope from (8.56) is simply $-H_c(0)/T_c$, and the results of Finnemore et al. [111], $\partial H_c(T)/\partial T = -165\,\mathrm{Oe\,K^{-1}}$ for $YBa_2Cu_3O_7$ conform closely to the calculated estimate of $-155\,\mathrm{Oe\,K^{-1}}$, with $H_c(0) = 14\,\mathrm{kOe}$ [computed beneath (8.52)] and $T_c =$ 91 K. If (8.75) and (8.76) are used with the above approximations, and with $\kappa = 63$ from (8.73), the ratio $[\partial H_{c2}(T)/\partial T] \div [\partial H_{c1}(T)/\partial T] \approx 2\kappa^2/\ln \kappa = 1,920$, which matches closely the result of the Cava et al. [67]: $\left(13\,\mathrm{kOe\,K^{-1}}\right) \div \left(7\,\mathrm{Oe\,K^{-1}}\right) =$ $1,860$. It also follows from (8.75) and (8.76) that for $H_c = 14\,\mathrm{kOe}$, $H_{c1} \approx 660\,\mathrm{Oe}$ and $H_{c2} \approx 1,250\,\mathrm{kOe}$.

The question of the "dirty" superconductor may now be examined in the above context. Traditionally, a type-II superconductor is a metal with a ξ reduced to values

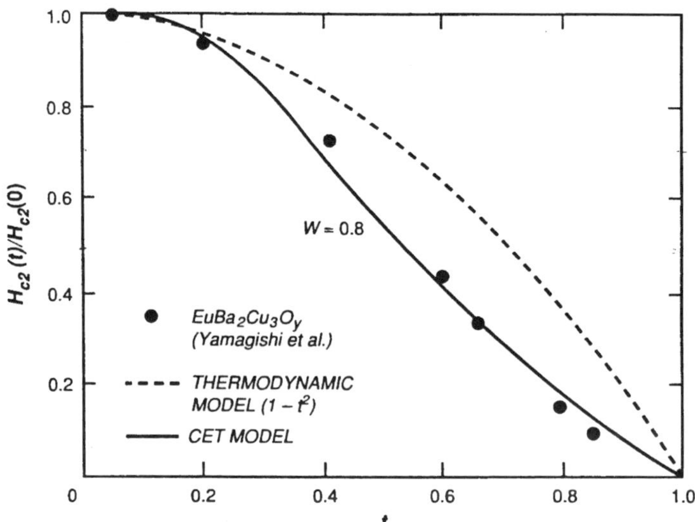

Fig. 8.43 CET-calculated curve of $H_{c2}(t)/H_{c2}(0)$ vs. t compared with data of Yamagishi et al. [110] for $W = 0.8$ [9]

below λ_L by the presence of inhomogeneities – "normal" regions that act as nucleation centers for the fluxoid lattice, in a manner similar to the formation of reverse magnetic domains about pores and grain boundaries in a ferromagnet. According to (8.75), the situation for which $\lambda_L/\xi \approx \lambda_L/d \gg 1$ occurs in metals where $d \ll \xi_0$, thus limiting the coherence length to a mean-free path. With the discovery of the high-temperature oxide superconductors with typical $\xi_0 \sim 30$ Å, the $\lambda_L/\xi \gg 1$ condition appears to exist naturally even in macroscopically homogeneous specimens.

The effect of local or dispersed inhomogeneities in the form of impurities or lattice defects has somewhat different implications for a metal and for an insulator. In metals, with the greatest charge screening effects, $\xi_0 \gg \lambda_L$, and the effect of inhomogeneities is significant in reducing ξ to set up the condition for establishing the fluxoid lattice, which can occur where $d \ll \lambda_L$. In effect, the impurities in a metal are *required* for the creation of a type-II superconductor through the nucleation and pinning of fluxoid regions at the impurity centers. In superconducting oxides with finite (measurable) dielectric constants, $\xi_0 \ll \lambda_L$, and the fluxoid lattice forms *in the absence of*impurities. Without pinning centers, they may be described as natural type-II superconductors.

Since the fluxoids are mobile, the inhomogeneities will only affect the fluxoid lattice if $d \leq \xi_0$ at which point the impurity regions would then serve as pinning centers for the fluxoids. The use of induced inhomogeneities as a practical design strategy for increased magnetic field and supercurrent capabilities of high-T_c materials may therefore be viewed as a stabilization of the fluxoid lattice due to the "pinning" of domains about the inhomogeneities. Without these centers, the fluxoid structure is likely to be fluid, and with the merging and collapsing of domains

Table 8.2 Superconductor parameter values and their dependence on carrier density

Parameter	n_s^e Dependence	Theory prediction	Experiment
$\Delta G(0)$	$(n_s^e)^2$	8×10^6	10^6–10^7 erg cm^{-3}
$v_s(0)$	$(n_s^e)^{1/2}$	8.4×10^6	– (cm s^{-1})
$i_c(0)$ Film	$(n_s^e)^{3/2}$	6×10^8	4×10^7 A cm^{-2}
Bulk	(n_s^e)		
$\lambda_L(0)$	$(n_s^e)^{-1/2}$	1,740	1,500–2,000 Å
$\xi_0(0)$	$(n_s^e)^{-1/2}$	27	20–35 Å
$\kappa\ (= \lambda_L/\xi_0)$	–	63	43–100
$H_c(0)$	n_s^e	14	10–20 kOe
$T_c(0)$	$n_s^e(0)$	–	–(K)
$\partial H_c/\partial T\ (T = T_c)$	$-H_c(0)/T_c$ $(W = 1)$	–155	-165 Oe K^{-1}
$(\partial H_{c2}/\partial T) + (\partial H_{c1}/\partial T)$	$2\kappa^2/\ln \kappa$	1,920	1,860
$R_S\ (T = 77\,\text{K})$	$n_n/(n_s^e)^{3/2}$	2×10^{-5}	$10^{-5}\Omega$

increasing as $T \to T_c$. Since $\lambda_L(T)$ increases more rapidly with T for small values of W, the melting of the fluxoid lattice at temperatures well below the critical temperature may occur even in materials with $d > \lambda_L(0)$.

To summarize the results of the above analyses, it has been shown that the magnitudes of λ_L, H_c, i_c, and ξ_0 are all determined by the effective density of supercarriers $n_s^e(T)$. The various relationships are organized in Table 8.2, together with the parameter values determined in the foregoing analyses and their projected limits where applicable.

Appendix 8A Magnetic Levitation

Magnetic levitation is the most visually dramatic manifestation of superconductivity, occurring when the inducement of a diamagnetic supercurrent in a magnetic field attends the expulsion of flux from the interior of the specimen. An important distinction should now be recalled: a material qualifies as a superconductor where $n_s \geq n_t$ (requiring that $T \leq T_c$); however, it does not become superconducting until the carrier chain condenses to form a Meissner supercurrent for $H \leq H_c$. Thus, the supercarrier density is controlled by temperature, but the supercurrent by magnetic field. Analogous to the concept of an image charge representing the effects of a metal plane beneath a real electric charge of opposite sign, the induced supercurrent in a specimen may be represented by a magnetic dipole, as sketched in Fig. 8.44. Since the current produces a diamagnetic moment, the dipole moment is a mirror image of the external magnet, and the force between them is repulsive.

As pointed out earlier, the spatial ordering constraint of the covalent bond furnishes the eigenstate rigidity required for the $\nabla\psi_s = 0$ condition. In this context,

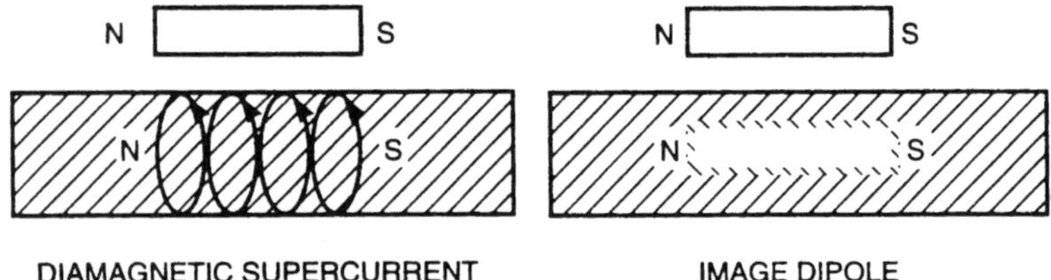

Fig. 8.44 Diagrammatic representation of the Meissner flux expulsion/levitation effect from induced diamagnetism in superconductor [9]

the celebrated levitation property of the Meissner effect can be appreciated in the context of a magnetomotive force imparted to a current-carrying coil in an inhomogeneous magnetic field. In a superconductor, however, the current loop will be established on the specimen according to the disposition of the external field relative to the specimen. Similar to the coil, a superconducting object will assume an appropriate equilibrium position and attitude in an external field; unlike the coil, however, there are no restoring forces acting on the specimen when it is disturbed from its "equilibrium" position. Since the current path will then adopt a different chain of covalent orbital lobes, a new equilibrium state may be established within the same specimen orientation by altering either the position of the specimen or the magnetic field conditions.

Without external mechanical assistance to alter the carrier condensation energy (or extent of polaron alignment) $\Delta G\left(= H_c^2/8\pi\right)$, the spatial relation between the specimen and magnet is fixed, except for rotation about the magnetic field axis. Rotation about an axis perpendicular to the magnetic field, or any translational adjustment, requires an energy input to establish a new equilibrium current condition (thus precluding any restoring-force effects). The mechanical aspects of the Meissner effect lend further credence to the conclusion that superelectrons are part of the binding forces of the lattice, and do not exist independently as unbound electrons in the superconducting state.

References

1. E.C. Stoner, *Proc. Leeds Philos. Soc.* **2**, 391 (1933)
2. K. Yosida, *Theory of Magnetism*, (Springer, New York, 1996)
3. A.W. Overhauser, *Phys. Rev.* **128**, 1437 (1962)
4. H. Fröhlich, *Adv. Phys.* **3**, 325 (1954)
5. N.F. Mott, *Metal–Insulator Transitions*, (Taylor & Francis, New York, 1990)
6. T. Holstein, *Ann. Phys.* **8**, 325 (1959)
7. J.B. Goodenough, *Prog. Solid State Chem.* **5**, 145 (1972)
8. G.F. Dionne, *Magnetic Interactions and Spin Transport*, A. Chtchelkanova, S. Wolf, and Y. Idzerda, eds., (Springer, New York, 2003), Chapter 1
9. G.F. Dionne, *Covalent Electron Transfer Theory of Superconductivity*, (MIT Lincoln Laboratory Techn. Rept. **885**, 1992), NTIS No. ADA2539757

10. M. Wolfsberg and L. Helmholz, *J. Chem Phys.* **20**, 837 (1952); also C.J. Ballhausen and H.B. Gray, *Molecular Electronic Structures, An Introduction*, (Benjamin/Cummings, Reading, MA, 1980), p. 161
11. L.D. Landau and E.M. Lifshitz, *Quantum Mechanics: Non-Relativistic Theory*, (Addison-Wesley, Reading, MA, 1958), p. 139
12. L. Friedman and T. Holstein, *Ann Phys.* **21**, 494 (1963)
13. J.B. Goodenough, *Magnetism and the Chemical Bond*, (Wiley, New York, 1963), Table XII
14. K. Siratori and S. Iida, *J. Phys. Soc. Jpn* **15**, 210 (1960)
15. S.M. Watts, S. Wirth, S. von Molnar, A. Barry, and J.M. Coey, *Phys. Rev. B*, **61**, 9621 (2000)
16. J.B. Goodenough, *Prog. Solid State Chem.* **5**, 359 (1972)
17. K.-I. Kobayashi, T. Kimura, H. Sawada, K. Terakura, and Y. Tokura, *Nature* **395**, 677 (1998)
18. Y. Moritomo, Sh. Xu, A Machida, T. Akimoto, E. Nishibori, M. Takata, and M. Sakata, *Phys. Rev. B* **61**, R7827 (2000)
19. R.R. Heikes and W.D. Johnston, *J. Chem. Phys.* **26**, 582 (1957)
20. R.J.D. Tilley, *Defect Crystal Chemistry and Its Applications*, (Blackie, Glasgow, 1987), Chapter 6
21. G.F. Dionne, *Transition-Metal Oxide Superconductivity*, (MIT Lincoln Laboratory Techn. Rept. **802**, 1988), NTIS No. ADA1970698
22. G.F. Dionne, *J. Appl. Phys.* **67**, 4561 (1990)
23. G.F. Dionne, *Covalent Electron Transfer Theory of Superconductivity*, (MIT Lincoln Laboratory Techn. Rept. **885**, 1992), NTIS No. ADA2539757, Section 5
24. S. Jin, T.H Tiefel, M. McCormack, R.A. Fastnacht, R. Ramesh, and L.H. Chen, *Science* **264**, 413 (1994)
25. J.B. Goodenough, *Phys. Rev.* **100**, 564 (1955)
26. G.H. Jonker and J.H. Van Santen, *Physica* **XVI**, 337 (1950)
27. C. Zener, *Phys. Rev.* **82**, 403 (1951)
28. P.-G. de Gennes, *Phys. Rev.* **118**, 141 (1960)
29. J.B. Goodenough, A. Wold, N. Menyuk, and R.J. Arnott, *Phys. Rev.* **124**, 373 (1961)
30. J.B. Goodenough, *Phys. Rev.* **117**, 1442 (1960)
31. J.B. Goodenough, *Magnetism and the Chemical Bond*, (Wiley, New York, 1963), Chapter III; also J.B. Goodenough and A.L. Loeb, *Phys. Rev.* **8**, 391 (1955)
32. J.B. Goodenough, *Rep. Prog. Phys.* **67**, 1977ff (2004)
33. M. Fäth, S. Freisem, A.A. Menovsky, Y. Tomioka, J. Aarts, and J.A. Mydosh, *Science* **285** (1999)
34. G.F. Dionne, *Anomalous Magnetoresistance in the Lanthanide Manganites and its Relation to High-T_c Superconductivity*, (MIT Lincoln Laboratory Techn. Rept. **1029**, 1996), DTIC No. ADA309080
35. G.F. Dionne, *J. Appl. Phys.* **79**, 5172 (1996)
36. Y-Q. Li, J. Zhang, S. Pombrik, S. DiMascio, W. Stevens, Y.F. Yan, and N.P. Ong, *J. Mater. Res.* **10**, 2166 (1995)
37. J.B. Goodenough and J.M Longo, *Landolt-Bornstein Tabellen*', *New Series III/4a*, K. Hellwege, ed., (Springer, Berlin, 1970), p. 231
38. J.B. Goodenough, *Aust. J. Phys.* **52**, 155 (1999)
39. M.I. Klinger and A.A. Samokhvalov, *Phys. Stat. Solidi (b)* **79**, 9 (1977)
40. G.F. Dionne, *Magnetic Interactions and Spin Transport*, A. Chtchelkanova, S. Wolf, and Y. Idzerda, eds., (Springer, New York, 2003), Chapter 1, Section 7
41. G.F. Dionne, *J. Appl. Phys.* **99**, 08M913 (2006)
42. G.F. Dionne, *J. Appl. Phys.* **101**, 09C509 (2007)
43. G.F. Dionne and R.G. West, *Appl. Phys. Lett.* **48**, 1488 (1986)
44. G.F. Dionne, *J. Appl. Phys.* **105**, 07A525 (2009)
45. G.F. Dionne, *J. Appl. Phys.* **50**, 4263 (1979)
46. G.F. Dionne and H-S. Kim, *J. Appl. Phys.* **103**, 07B333 (2008)
47. Y.K. Yoo, Q. Xue, H-C. Lee, S. Cheng, X-D. Xiang, G.F. Dionne, S. Xu, J. He, Y.S. Chu, S.D. Preite, S.E. Lofland, and I. Takeuchi, *Appl. Phys. Lett.* 86, 042506 (2005)

48. H-S. Kim, L. Bi, G.F. Dionne, and C.A. Ross, *Appl. Phys. Lett.* **93**, 1488 (2008)
49. J. Bardeen, L.N. Cooper, and J.R. Schrieffer, *Phys. Rev.* **106**, 162 (1957); **108**, 1195, (1957)
50. J.G. Bednorz and K.A. Muller, *Z. Phys.* B **64**, 189 (1986); also J.G. Bednorz, K.A. Muller, and M. Takashige, *Science* **236**, 73 (1987)
51. M.K. Wu, J.R. Ashburn, C.J. Torng, P.H. Hor, R.L. Meng, L. Gao, Z.J. Huang, Y.Q. Wang, and C.W. Chu, *Phys. Rev. Lett.* **58**, 908 (1987)
52. F. London and H. London, *Physica* **2**, 341 (1935)
53. F. London, *Superfluids*, vol. 1, (Wiley, New York, 1950), p. 143
54. A.B. Pippard, *Proc. R. Soc. (London)* **A216**, 547 (1953)
55. V.L. Ginsburg and L.D. Landau, *Zh. Eksperim. i Teor. Fiz.* **20**, 1064 (1950)
56. J.M. Blatt, *Theory of Superconductivity*, (Academic, New York, 1964), p. 129
57. J.B. Goodenough, G. Demazeau, M. Pouchard, and P. Hagenmüller, *Solid State Chem.* **8**, 325 (1973)
58. G.F. Dionne, *J. Appl. Phys.* **69**, 5194 (1991)
59. J.I. Budnick, B. Chamberland, D.P. Yang, Ch. Neidermayer, A. Golnik, E. Recknagel, M. Rossmanith, and A Weidinger, *Europhys. Lett.* **5**, 651 (1988)
60. J.M. Tranquada, *J. Appl. Phys.* **64**, 6071 (1988)
61. J.B. Torrance, Y. Tokura, A.I. Nazzal, A. Bezinge, T.C Huang, and S.S.P. Parkin, *Phys. Rev. Lett.* **61**, 1127 (1988)
62. A.M. George, I.K. Gopalakrishnan, and M.D. Karkhanavla, *Mater. Res. Bull.* **9**, 721 (1974)
63. M.A. Subramanian, C.C. Torardi, J.C. Calabrese, J. Gopalakrishnan, K.J. Morrissey, T.R. Askew, R.B. Flippen, U. Chowdhry, and A.W. Sleight, *Science* **239**, 1015 (1988)
64. M. Tinkham, *Introduction to Superconductivity*, (Robert E. Krieger, Malabar, FL, 1985)
65. J.M. Tarascon, L.H. Greene, W.R. McKinnon, G.W. Hull, and T.H. Geballe, *Science* **235**, 1373 (1987)
66. G.F. Dionne, *J. Appl. Phys.* **69**, 4883 (1991)
67. R.J. Cava, B. Batlogg, R.B. van Dover, D.W. Murphy, S. Sunshine, T. Siegrist, J.P. Remeika, E.A. Reitman, S. Zahurak, and G.P. Espinosa, *Phys. Rev. Lett.* **58**, 1676 (1987)
68. A.C. Westerheim, L.S. Yu-Jahnes, and A.C. Anderson, *IEEE Trans. Magn.* **27**, 1001 (1991)
69. G. Shirane, Y. Endoh, R.J. Birgeneau, M.A. Kastner, Y. Hidaka, M. Oda, M. Suzuki, and T. Murakami, *Phys. Rev. Lett.* **59**, 1613 (1987)
70. D.C. Johnston, A.J. Jacobson, J.M. Newsam, J.T. Lewandowski, D.P. Goshorn, D. Xie, and W.B. Yelon, *Chemistry of High-Temperature Superconductors*, D.L. Nelson, M.S. Whittingham, and T.F. George, eds., (American Chemical Society, Washington, DC, 1987), p. 136
71. B. Batlogg, R.J. Cava, C.H. Chen, G. Kourouklis, W. Weber, A. Jayaraman, A.E. White, K.T. Short, E.A. Rietman, L.W. Rupp, D. Werder, and S.M. Zahurak, *Novel Superconductivity*, S.A. Wolf and V.Z. Kresin, eds., (Plenum, New York, 1987), p. 653
72. G.F. Dionne, *Temperature Dependence of Large Polaron Superconductivity*, (MIT Lincoln Laboratory Techn. Rept. **1024**, 1995), NTIS No. ADA2972875
73. W. Low, *Paramagnetic Resonance in Solids*, Solid State Physics, Suppl. 2, F. Seitz and D. Turnbull, eds., (Academic, New York, 1960), p. 101
74. K.H. Johnson, D.P. Clougherty, and M.E. McHenry, *Mod. Phys. Lett.* **3**, 867 (1989)
75. K.H. Johnson, D.P. Clougherty, and M.E. McHenry, *Mod. Phys. Lett.* **3**, 1367 (1989)
76. A.W. Sleight, *Science* **242**, 1519 (1988)
77. R.J. Cava, B. Batlogg, R.B. van Dover, J.J. Krajewski, J.V. Waszczak, R.M. Fleming, W.F. Peck Jr, L.W. Rupp, Jr., P. Marsh, A.C.W.P. James, and L.F. Schneemeyer, *Nature* **345**, 602 (1990)
78. J.M. Tarascon, L.H. Greene, B.G. Bagley, W.R. McKinnon, P. Barboux, and G.W. Hull, *Novel Superconductivity*, S.A. Wolf and V.Z. Kresin, eds., (Plenum, New York, 1987), p. 705
79. L.E. Orgel, *Introduction to Transition-Metal Chemistry: Ligand Field Theory*, 1st Ed., (Wiley, New York, 1960), p. 66
80. I.D. Brown, J. *Solid State Chem.* **82**, 122 (1989). A linear model of polaron density predicts $x = 0.25y - 1.5$ for $YBa_2Cu_3O_y$, yielding $x = 0.1$ for $y = 6.4$. Analysis of data by the *bond valence sum* method of Brown suggests a slightly lower value of x

81. Y. Fujiwara, S. Hirata, M. Nishikubo, T. Kobayashi, H. Nakayama, and H. Fujita, *IEEE Trans. Magn.* **27**, 1166 (1991)
82. C.C. Torardi, M.A. Subramanian. J.C. Calabrese, J. Goplalakhrishnan, K.J. Morrissey, T.R. Askew, R.B. Flippen, U. Chowdhry, and A.W. Sleight, *Science* **240**, 631 (1988)
83. Y. Tokura, H. Takagi, and S. Uchida, *Nature* **337**, 345 (1989)
84. G.F. Dionne, *Transition-Metal Oxide Superconductivity*, (MIT Lincoln Laboratory Techn. Rept. **802**, 1988), NTIS No. ADA1970698, Table 1
85. M.G. Smith, A. Manthiram, J. Zhou, J.B. Goodenough, and J.T. Markert, *Nature* **333**, 836 (1988)
86. G.F. Dionne, *A Strategy for Higher Temperature Superconductivity in the Layered Cuprates*, (MIT Lincoln Laboratory Techn. Rept. **1021**, 1995), NTIS No. ADA2968857
87. K. Takada, H. Sakurai, E. Takayama-Muromachi, F. Izumi, R.A. Dilanian, and T. Sasaki, *Nature* **422**, 53 (2003)
88. D.C. Johnston, H. Prakash, W.H. Zachariasen, and R. Viswanathan, *Mater. Res. Bull.* **8**, 777, (1973)
89. G.F. Dionne, *Covalent Electron Transfer Theory of Superconductivity*, (MIT Lincoln Laboratory Techn. Rept. **885**, 1992), NTIS No. ADA2539757, Section 7
90. F. London, *Superfluids*, vol. 1, (Wiley, New York, 1950), p. 150
91. G. Aeppeli, R.J. Cava, E.J. Ansaldo, J.H. Brewer, S.R. Kreitzman G.M. Luke, D.R. Noakes, and R.F. Kiefl, *Phys. Rev.* **B35**, 7129 (1987)
92. W.J. Kossler, J.R. Kempton, X.H. Yu, H.E. Schone, Y.J. Uemura, A.R. Moodenbaugh, M. Suenaga, and C.E. Stronach, *Phys. Rev.* **B35**, 7133 (1987)
93. D.E. Oates and A.C. Anderson, *IEEE Trans. Magn.* **27**, 867 (1991)
94. G.F. Dionne, *IEEE Trans. Appl. Supercond.* **3**, 1465 (1993)
95. M. Tinkham, *Introduction to Superconductivity*, (Robert E. Krieger, Malabar, FL, 1985), p. 81
96. C. Kittel, *Introduction to Solid State Physics*, 3rd Ed., (Wiley, New York, 1966), p. 352
97. M. Tinkham, *Introduction to Superconductivity*, (Robert E. Krieger, Malabar, FL, 1985), p. 118
98. G.F. Dionne, *Covalent Electron Transfer Theory of Superconductivity*, (MIT Lincoln Laboratory Techn. Rept. **885**, 1992), NTIS No. ADA2539757, Section 6.6
99. H.S. Lessure, S. Simizu, P.J. Kung, B.A. Baumert, S.G. Sankar, and M.E. McHenry, *IEEE Trans. Magn.* **27**, 942 (1991)
100. J. MannHart, P Chaudhari, D. Dimos, C.C. Tsuei, and T.R. McGuire, *Phys. Rev. Letts.* **61**, 2476 (1988)
101. M. Tinkham, *Introduction to Superconductivity*, (Robert E. Krieger, Malabar, FL, 1985), p. 124
102. M. Tinkham, *Introduction to Superconductivity*, (Robert E. Krieger, Malabar, FL, 1985), p. 7
103. T.K. Worthington, W.J. Gallagher, and T.R. Dinger, *Phys. Rev. Lett.* **59**, 1160 (1987)
104. M. Tinkham, *Introduction to Superconductivity*, (Robert E. Krieger, Malabar, FL, 1985), p. 11
105. M. Tinkham, *Introduction to Superconductivity*, (Robert E. Krieger, Malabar, FL, 1985), p. 113
106. G.F. Dionne, *Covalent Electron Transfer Theory of Superconductivity*, (MIT Lincoln Laboratory Techn. Rept. **885**, 1992), NTIS No. ADA2539757, Section 5.1.3
107. A.A. Abrikosov, *Soviet Phys.-JETP* **5**, 1174 (1957)
108. C. Kittel, *Introduction to Solid State Physics*, 3rd Ed., (Wiley, New York, 1966), p. 364
109. M. Tinkham, *Phys. Rev.* **129**, 2413 (1963)
110. A. Yamagishi, H. Fuke, K Sugiyama, M. Date, Y. Tajima, and M. Hikata, *Phys. B* **155**, 174 (1989)
111. D.K. Finnemore, M.M. Fang, J.R. Clem, R.W. McCallum, J.E. Ostenson, L. Ji, and P. Klavins, *Novel Superconductivity*, S.A. Wolf and V.Z. Kresin, eds., (Plenum, New York, 1987), p. 627
112. C.W. Chu, L. Gao, F. Chen, Z.J. Huang, R.L. Meng, and Y.T. Xue, *Nature* **365**, 323 (1993)

Index

A

Activation energy
- in double exchange, 119, 121, 124, 389, 410, 412
- hopping electron, 125, 389, 400, 405, 409, 421
- from polaron trap, 396

Anderson, E.E., 161, 312
Anderson, P.W., 116, 118, 127
Angular momentum, L, S, J, 9–11, 25, 37, 42, 50, 52, 55, 63, 64, 71, 88, 187, 188, 222, 224, 289, 356, 358, 362, 371
Anion-cation, 24, 107, 113, 122, 127, 137, 193, 438
Anisotropy
- crystal-field, 201, 202, 204, 205, 207–213, 216, 220–222, 224–227, 237, 240, 241, 244
- magnetocrystalline, 7, 37, 52, 133, 174, 176, 201, 204, 226, 228–250, 254, 274, 294, 309, 406–407, 410
- shape, 201, 249, 251, 252, 258, 259, 265, 293, 295, 298, 309

Approach to saturation, 234, 256–257, 299
Aufbau occupancy diagrams
- high and low-spin states, 75–79, 89, 108, 217
- J-T and S-O stabilized, 84–86, 217, 219, 221
- octahedral site, 97, 98, 109, 123, 129, 217
- tetrahedral site, 97, 129

B

Ballhausen, C.J., 64, 69, 91, 113, 243, 263, 264
Bardeen, J., 418
BCS ensemble wavefunction, 418, 419, 444, 445, 447
$Bi^{3+} - Fe^{3+}$ M–O hybrid excited state, 371–375
Bismuth Faraday anomaly, 376
Bi-YIG anomalous Faraday effect, spectral model, 365–367
Bleaney, B., 215, 225
Bloch–Bloembergen, 7, 19, 28, 252, 295, 297, 334, 356, 385, 416
Boltzmann statistics, 11, 237
Bozorth, M., 5, 229, 256
Brillouin theory
- curve shapes, 17, 18, 445
- function, 13, 15, 19, 130, 131, 133, 155, 156, 165, 196–197, 202, 215, 238, 265, 289, 302, 411, 442

Button, K.J., 23, 24, 137, 192, 329, 334, 351

C

Canting
- iron sublattice, 185, 190, 193, 195, 197, 221, 237–239, 246, 249, 256, 265, 302, 303, 305, 345, 365, 369, 377, 379, 386, 398, 401, 421, 433, 437, 438
- rare-earth garnet effect, 184–189

Charge carriers, 142, 386, 396, 417, 422
Charge transfer probability, 386–389, 392, 393, 397–399, 403, 405, 409, 411, 413, 418, 420, 422, 426
Chemical bonding
- covalent, 23, 42, 88–101, 103, 107, 109, 113, 121, 122, 124, 129, 134, 168, 174, 222, 228, 344, 388, 392, 402, 419, 424, 431, 440, 455
- ionic, 40, 49, 73, 90, 92, 99, 103, 112, 387
- metallic, 88, 107, 408

Chikazumi, S., 229, 236

Co^{2+} exchange isolation
in LiTiferrite, 176, 307
in YIG, 176, 179
Coherence length, 416–418, 424, 445, 450–452, 454
Colossal magnetoresistance (CMR), 386, 401–403, 405, 407, 409
Concentration threshold $x_s > x_t$, 428
Condensation, 86, 171, 411, 419, 420, 439, 440, 442–443, 447–451, 453, 456
Conductivity
normal *vs* super, 386
Cooper, L.N., 418
Correlation exchange, 121
Covalent electron transfer, 94, 122
Covalent electron transfer theory (CET), 413, 437, 444, 445, 447, 449–452, 454
Critical current density, 447–450
Critical magnetic field, 439, 445–447, 450, 452, 453
Critical rf field, 311, 337
Cross transfers, 365
Crystal, 37, 40, 43–45, 48–73, 75, 76, 80, 82, 83, 86–92, 97–99, 101, 107–109, 113–115, 118, 121, 129, 130, 132, 134–136, 138, 141–143, 147, 152, 169, 175, 177, 182, 183, 185–192, 201, 202, 204, 205, 207–213, 216, 220–222, 224–228, 231–234, 236, 237, 240, 241, 244, 254–256, 258, 261–265, 267, 268, 273, 274, 284, 286, 287, 289–292, 295, 304, 306, 307, 319, 325, 360, 372–374, 385, 386, 388, 396, 397, 402, 404, 412, 420, 431, 432, 438
Crystal sites, 360
Crystal systems, 44–45

D
Degenerate orbital states, 49, 80
Delocalization, 24, 94, 114, 120–124, 126, 379, 380, 398, 402, 413, 430, 438
Demagnetizing factor, 5, 308
Demagnetizing field, 5, 6, 234, 257–259, 273, 292, 312, 313, 315, 317, 318, 345, 411
D_{4h} *vs.* C_{3v} distortions, 420
Dillon, F., 317, 360, 381, 382
Domain wall resonance, 273, 319, 338

E
Effective linewidth in polycrystals, 299
Elastic constants, 81, 232, 233
Electric-dipole transition, 357, 358, 360, 361
Electron correlation energy, 123
Energy level diagrams, 69, 97, 226, 261, 289, 348, 357, 375
Exchange
double, 119, 121, 124, 389, 410, 412
energy, 19, 24, 113, 116, 118, 122, 143, 161, 162, 168, 186, 196–197, 216, 235, 237, 243, 252, 265, 302, 304, 320, 339, 346, 347, 350, 365, 388, 389, 394, 405, 419–420
integral, 21, 25, 92, 113, 119, 226, 375, 390, 394, 399, 424, 436
resonance, 29, 264, 344–351, 355
semicovalent, 122, 123
super, 23, 24, 107, 109, 111, 113, 115–117, 119–122, 126, 127, 129–131, 133, 134, 136, 139, 140, 142–144, 146, 152, 161–168, 173, 183, 193, 201, 222, 366, 371, 372, 375, 376, 378, 379, 385, 397–399, 401, 402, 410, 430
trap, 124, 421, 438, 439
Exchange resonance, 344–351
Extinction coefficient k, 355, 361, 381

F
Faraday, 42, 70, 193, 305, 318, 330, 332–333, 343, 353–355, 366, 367, 369, 374–376, 379–381, 413, 414
Faraday rotation, 42, 70, 193, 305, 318, 330, 332–333, 343, 353–355, 366, 374–376, 379–381
Ferrites
garnets, 155, 156, 158–161, 165, 167–170, 175–185, 189, 190, 192–194, 197, 202, 220, 238–240, 245, 246, 249, 250, 294, 298–300, 302, 307, 343, 360–366, 372, 373, 379, 380
hexagonals, 48, 153, 169, 190–193, 240, 260, 295, 322, 323
spinels, 46, 77, 153, 167–169, 175, 176, 178, 179, 216, 224, 240, 244, 249, 257, 305, 307
Ferromagnetic limit, in CMR manganites, 401, 402

G
Geller, S., 159–161, 166, 175, 177, 178, 183–185
Giant molecule concept, London's $\nabla n_s = 0$ condition, 419

Gilbert, 297, 334, 335
Gilleo, A., 158, 160, 161, 175–178
Ginzburg–Landau theory, 416, 417, 419, 450
Goodenough, J.B., 119, 124–129, 141, 144, 243, 386, 397, 399, 402, 403, 409
Goodenough–Kanamori (G–K) rules, 125–129
Griffith, J.S., 202, 222, 291
Group theory, 57, 64–68, 70, 71, 79, 204

H
High-magnetization limits, 195–196
High-temperature superconductivity (HTS), 1, 386, 407, 409, 413, 423, 426, 438
Hole carrier, 386, 429
Holstein polaron theory, 394–396
Hopping electrons, 170, 176, 179, 386, 394, 395, 397
Hund's rule, 24, 40, 42, 49–51, 53, 55, 59, 68, 72, 75, 89, 91, 92, 98, 107, 109, 119–123, 129, 130, 143, 211, 213, 291, 385, 389, 397, 398, 402, 403, 420
Hysteresis loop parameters, 251

I
Initial permeability, 192, 251–253, 319
Integral
 exchange, 21, 25, 119, 390, 394, 424, 436
 overlap, 22, 55, 88, 90–92, 95, 96, 104, 110–114, 119, 126, 135, 138, 147, 305, 321, 354, 371, 374, 375, 379, 394, 410, 412, 426
 transfer, 92, 95, 107, 109, 115, 119, 124, 173, 365, 371, 429, 452
Intersublattice exchange, 163, 345
Intersublattice optical transitions, 376–380
Intrasublattice exchange, 159, 163
Ionization potential, 75, 92, 99, 379, 387
Ising approximation, 24–25, 118, 347
Itinerance, 24

J
Jahn-Teller effect
 static *vs.* vibronic, 80, 81, 403, 404, 412
 vs. S-O stabilization, 84–86, 219–221
Jonker, G.H., 145, 192, 402, 403, 409
J-T *vs.* S-O stabilization in exchange field, 219, 265

K
Kerr, 343, 361–363, 366, 367, 381, 382
Kittel, C., 33, 99, 121, 157, 233, 323, 453

L
Landau–Lifshitz, 297, 334
Landé *g*-factor, 173, 204
Langevin theory, 12–15, 18, 202, 289
Laporte's rule, 358, 376, 380
Large-polaron radius, 424, 428, 452
Larmor precession, 26–27, 327, 330, 356
Lattice energy, 73, 90, 339, 373, 407
Lax, B., 23, 24, 137, 192, 329, 334, 351
Linear combination of atomic orbitals (LCAO) model, 89, 90, 94
Line broadening, homogeneous *vs.* inhomogeneous, 33, 297, 298
Local *vs.* collective properties
 electronic, 1, 49, 89, 288
 magnetic, 1, 3, 37, 39, 40, 42, 43, 47, 49, 50, 55, 64, 70, 71, 73, 89, 96, 101, 107, 113, 139, 151, 156, 168, 180, 191, 201, 202, 210, 288, 380, 414
 magnetoelastic, 43, 70, 80, 82, 84, 87, 142, 174, 176, 179, 201–269
London equations, 413–416, 424
London, F., 20
Lorentzian and Gaussian shapes, 33, 297, 313, 360, 363, 380
Low-field magnetic loss, 322
Low-spin states, 55, 73, 75–79, 89, 108, 305, 420, 423, 438
Low, W., 64, 211, 225, 226

M
Magnetic dilution
 in ferrites, 157, 161, 168, 171, 306, 412
 in low concentrations, 306
 with magnetoelastic percolation, 307, 410–412
Magnetic exchange trap, 438, 439
Magnetic sublattices, 121, 131, 151, 156, 303
Magnetism
 Antiferro, 4, 8, 15–16, 19, 21, 22, 109, 113, 116, 117, 119, 123, 126, 129–140, 143–145, 151–154, 398–403
 Dia, 8–11, 15, 26, 422, 428, 456
 Ferri, 4, 8, 34, 129, 131, 151–197, 273, 346

Magnetism (*cont.*)
  Ferro, 4, 8, 15–16, 21, 22, 24, 102, 108, 109, 116, 119–126, 129, 131, 140, 144, 146, 149, 151, 152, 385, 399, 401–404, 413, 420, 438
  para, 8–11, 15, 16, 108, 137, 202–212, 219, 220, 275, 289, 292, 348, 380, 399, 412
Magnetoelastic condensation
  in octahedral sites, 77, 80, 84, 87
  in tetrahedral sites, 77, 82, 175, 176, 213
Magnetoelastic constant, 232
Magnetoelastic percolation, 307, 425, 428, 440
Magneto-optical transitions
  charge transfers, 378
  diamagnetic, 358, 359
  paramagnetic, 358, 359
Magnetoresistance, 19, 25, 144, 386, 401–402, 406, 408, 438, 439
Magnetostatic waves, 307, 317
Magnetostriction Local-site distortion model, 246
Magnons, 307, 309, 315, 379. *See* Spin waves
Meissner effect, 414, 438, 445, 456
Metal-insulator transition, 401, 402, 408, 438
Metallic oxides
  perovskites, 397, 399, 401
  simple, 397, 400
Microwave signal damping
  Bloch–Bloembergen, 295, 297, 334, 356
  Gilbert, 297, 334
  Landau–Lifshitz, 334
Molecular
  bonding, 75, 115, 344, 351
  field, 19–24, 131–134, 153–156, 164–165, 193–196
  orbitals, 89, 114, 115
Molecular-field coefficient, 25, 130, 131, 164–166, 168, 170, 171, 174, 180, 183, 184, 196, 197, 265, 302, 345, 406, 407
Molecular-orbital model
  bonding/antibonding orbitals, 102, 103, 379
  for H_2^+ ion, 97, 124, 171, 365, 371, 398
  normalization issues, 95, 103, 262
  one-electron approximation, 76, 213
Morrish, A.H., 133, 154, 229

N
Na^+Cl^-ionic molecule, 99, 104
n-type superconductivity, 436, 437

O
Occupation probability, 12, 347
O_h to D_{3d} symmetry, 47, 67
O_h to D_{4h} symmetry, 64, 66, 143, 431
Operator equivalents, 63, 64, 71, 211, 212, 226, 227
Orbit, 9, 26, 37, 50, 79–82, 215, 219, 222, 226, 361, 371
Orbital shell, 10, 38, 50, 104
Orgel, E., 68, 69, 265–267
Oscillator strength, 357

P
Pauli principle indistinguishability, 19–20, 80, 114, 345
Penetration depth, 414, 443–445, 448
Percolation threshold, 425, 428, 440
Periodic table, 1, 38–40, 99, 101, 151
Permittivity tensor, 343, 352–353, 366, 376
Perovskites, 1, 46, 48, 77, 129, 131, 135, 136, 140–142, 146, 175, 226, 387, 399, 401, 407, 420, 430, 433, 437, 443, 451, 453
Perturbation hierarchy, 54–55, 210
Phonons, 81, 274, 280–287, 289, 300, 301, 304, 307, 309, 355, 439, 441, 442
Pippard theory, 416, 417, 450–452
Point-charge model, 49, 89, 97, 211, 227, 263, 431
Polarization
  circular, 27, 30, 31, 299, 312, 317, 327–333, 335, 336, 344, 346, 353, 355, 356, 358, 361, 362, 381
  linear, 27, 30, 31, 329, 330, 332–333, 343, 353
  rotation, 193, 371
Polarons
  range, 389, 395, 411, 422
  small *vs.* large, 388–392
Propagation constant, 329, 352, 353, 355
Pryce, M.H.L., 61, 68, 209, 219

R
Rare-earth impurities in YIG
  loss from Mn^{3+} ions, 174
  low-temperature loss, 174
Rare-earth ions ($4f^n$), 43, 62, 71–73, 121, 177, 180–185, 187, 188, 197, 202, 203, 227, 228, 241, 249, 302, 303, 306, 380, 422
Refractive index n, 355, 361, 381

Relaxation
direct process, 284
longitudinal, 28, 34, 274–278, 287, 296, 319
Orbach process, 282, 284–286, 291, 303
Raman process, 281–285, 287, 291, 303, 309
spin–lattice, 28–30, 32, 34, 78, 87, 88, 173, 215, 219, 241, 244, 246, 274, 275, 278–287, 291, 295, 299–301, 305, 318, 319, 326, 385
spin–spin, 274, 291
transverse, 34
Relaxation time
temperature dependence, 274, 278, 282, 300
Remanence ratio, 233, 234, 254–257
Remanent magnetization, 254
Resonance
antiferromagnetic (AFMR), 288, 351
domain wall, 273, 318–320, 338
exchange, 29, 264, 344–351, 355
ferromagnetic (FMR), 31, 231, 240, 267, 281, 292–295, 300, 318, 343, 348, 370
line broadening, 30, 33, 297–300
line shape, 33, 328, 356, 385
paramagnetic (EPR), 27, 33, 80, 99, 204, 207, 209, 215, 224, 261, 288–292, 295, 332, 343, 431
rf peak-power threshold, 311, 315
RKKY exchange mechanism, 121
Rutile lattice, 135, 137, 138

S
Schloemann, E., 298, 299
Schrieffer, J.R., 418
Selection rules
electric-dipole, 13, 288, 292, 343, 344, 354–360, 376, 377, 379, 380
magnetic-dipole, 288, 343, 348, 352, 355, 357, 376, 379, 380
Semicovalent, 121–123, 126, 413
Single ion anisotropy, 205, 228, 236, 240, 242
Slater, J.C., 69, 70, 73, 124, 127, 222
Smit, J., 169, 191, 192, 229, 333
Snoek's law, 324–327
Spherical harmonics, 52, 53, 56, 61, 62, 64, 70
Spin dynamics, 80, 216, 274, 397, 412, 419, 440
Spin-echo in EPR
spin decoherence time (τ_2), 291, 296
spin relaxation time (τ_1), 290, 291
Spin-flip, 30, 122, 123, 189, 287, 357
Spin Hamiltonian
D and *E* parameters, 210, 221, 290
for d^1 ion, 207
for d^3 ion, 221
for d^5 ion, 224
Spin orbit coupling
and S-O stabilization, 83, 215
and spin-lattice relaxation, 78, 219, 241, 244, 286, 306
Spin-order frustration, 161
Spin waves, 19, 29, 307–311, 313–315, 317, 385
Stevens, K.W.H., 63, 211, 212, 216, 225, 226, 228
Subsidiary absorption at high power, 315–317
Sugano, S., 78, 79, 222, 266, 267, 373
Suhl, H., 312, 314–317, 337
Superconductor condensation energy, 439, 440, 442, 448–451, 453, 456
Superexchange
correlation, 121, 410
delocalization, 121, 124, 126, 379, 380, 398, 402, 413, 430, 438
virtual *vs.* real, 116, 379
Susceptibility tensor, 297, 327, 328
Symmetry
cation sites, 46, 47
lattice, 25, 40, 45, 47, 80, 87, 137, 138, 175, 188, 190, 204, 214, 215, 228, 231, 232, 235, 251, 290, 291, 319, 322, 377, 385, 396, 404, 440

T
Tanabe, Y., 78, 79, 222, 266, 267, 373
Temperature
critical, 417, 421, 422, 426–430, 432, 434, 436–439, 441, 444, 445, 450–455
Curie, 15–18, 22, 25, 128, 129, 134, 144, 145, 147–149, 154, 155, 158, 161, 170, 171, 173, 174, 178, 183, 192, 193, 215, 236, 252, 296, 305, 320, 347, 350, 375, 376, 380, 389, 394, 397, 401–403, 405–408, 410, 412, 439
Debye, 283, 303, 394, 395, 407, 410, 412, 426
Néel, 22, 128, 131, 132, 135–141, 146, 148, 149, 193, 215, 397, 421
Tensor
combined [ε]·[μ], 352–355
permeability, 3, 328, 333–336, 352

Tensor (*cont.*)
permittivity, 352, 366, 376
susceptibility, 297, 327, 328
The $\Delta S = 0$ transition rule, 376–380
Thermomagnetism, 156, 171, 173, 177, 178, 180, 182, 184, 196, 406, 410, 412
Thermomagnetism of dilute magnetic oxides, 410, 412
The $T_{\rm N,}$ $T_{\rm c} = 0$ condition, 421, 422, 428
Ti^{3+} in alums
Cs alum, 207
Rb alum, 206, 207, 209, 210, 215, 261, 285, 287, 289, 290
spin-lattice relaxation, 285–287
Tinkham, M., 351, 453
Transition-metal ions ($3d^n$), 22, 24, 39–43, 53–55, 59, 62, 63, 73, 76, 78, 84, 87, 89, 109, 126, 127, 139, 141, 167, 174, 182, 183, 187, 193, 202, 204, 209–211, 222, 228, 241, 242, 274, 289, 291, 305, 306, 350, 386, 396, 400, 413, 422, 423
Transition probability, 378
Transport
polaronic, 94, 99, 125, 409, 413, 432
spin, 99, 115, 116, 121–125, 143, 302, 385–456
Trigonal bipyramid site, in M-type Ba ferrite, 191, 225, 240
Two-fluid model, 426
Type-II superconductors, 418, 450–455

V

Valence-bond model Heitler–London H_2 molecule, 20, 99, 103–105, 107
Van Santen, J.H., 145, 192, 402, 403, 409
Van Vleck, J.H., 55, 63, 188, 202, 203, 234, 265, 286, 287

W

Wijn, H.P.J., 169, 191, 192, 229
Wittekoek, S., 361, 367, 368
Wolf, W.P., 187, 188, 236
Wolfsberg–Helmholtz approximation, 109

Z

Zeeman effect
and Boltzmann statistics, 289, 348
and Kramers doublets, 205, 206, 208, 225, 262, 284, 289, 348, 349
Zero-spin polarons, 419–423
and carrier statistics, 418–419

Printed by Printforce, the Netherlands